TEXT BOOK OF MOLLUSCA

By

Dr. Amita Saxena

Associate Professor

College of Fisheries

G.B. Pant University of Agriculture & Technology

Pantnagar (Uttranchal)

D P H

DISCOVERY PUBLISHING HOUSE

NEW DELHI-110002

First Published-2005

ISBN 81-8356-017-2

Published by

DISCOVERY PUBLISHING HOUSE
4831/24, Ansari Road, Prahlad Street,
Darya Ganj, New Delhi-110002 (India)
Phone: 23279245 • Fax: 91-11-23253475
E-mail:dphtemp@indiatimes.com

Printed at:
Amit Enterprises

PREFACE

Phylum Mollusca has great potential as for as their taxonomy, biology, ecology and applied approaches are concerned. There are variety of animals present in the world terrestrial and aquatic, both fresh water and marine in habitat. The detailed classification and characteristics of three classes Gastropoda, Bivalvia and Cephalopoda are taken into consideration. The digestive, respiratory, excretory, osmo-regulation Nervous, sensory organs and reproductive system are described elaboratively. The understanding of these systems along with their physiology has tremendous significant for the students, teachers scientists, farmers who are interested to go for farming for molluscs for their livelihood or as a hobby. It is also useful for the students and researchers interested to work on Phylum Mollusca.

Dr. Amita Saxena is highly thankful to her sisters Honable Vice Chancellar Deans students, teachers, colleagues, friends, for constant encouragements and needful help.

The publisher deserves a sincere thanks for cooperation and timely publication without whom it is not possible to do such kind of work.

It is all blessing of God who inspired me to do such piece of work in the field of science and welfare of human.

Contents

1

MOLLUSCA—CLASSIFICATION

GENERAL CHARACTERS OF MOLLUSCA

Definition (Latin word Mollis-soft)

Definition of Molluscs

Coelomate, bilateral often secondarily altered to an asymmetrical condition, without (or rarely with) evidences of metamerism, with a soft body usually protected by a calcareous shell of one or more pieces, secreted by a fold of body wall, the mantle, with ventral body wall altered into a muscular organ of locomotion, the foot, with pharynx commonly provided with a toothed band, the radula and with gills, heart, open, circulatory system and metanephridia.

1. Molluscs are essentially aquatic, mostly marine, few freshwater and some in terrestrial forms.
2. Body is soft, bilaterally symmetrical and consists of head, foot mantle and visceral mass.
3. The shell, when present usually univalve or bivalve, constituting an exoskeleton internal in some.
4. Body is commonly protected by an exoskeleton calcareous shell of one or more pieces, secreted by the mantle.
5. Body is triploblastic, coelomate and unsegmented (except in Monoplacophora).
6. System grade of body organization.
7. Mantle or pallium is a fold of body wall that leaves between itself and the main body mass, the mantle cavity.
8. Head is distinct bearing the mouth and provided with eyes, tentacles and other sense organs except the Pelecypoda and Scaphopoda.
9. Ventral body wall is modified into a muscular flap or plough like surface the foot, which is variously modified, for creeping, burrowing and swimming.

10. Visceral mass contains the vital organs of the body in a compact form taking the form of a dorsal hump or dome.
11. The body cavity is haemocoel, the true coelom generally limited to the pericardial cavity and the lumen of•the gonads and nephridia.
12. Digestive tract is simple with an anterior mouth and posterior anus but in gastropods, scaphopods and cephalopods the intestine becomes U-shaped bringing the anus to an anterior position.
13. Pharynx contains a rasping organ the radula except in pelecypoda.
14. Circulating system is open except in cephalopods which shows some tendency towards a closed system.
15. Respiratory system consists of numerous gills or ctenidia usually provided with ospharadium at the base. Lung is developed in terrestrial forms.
16. Excretory system consists of a pair of metanephridia which are true coelomoducts and communicate from pericardial cavity to the exterior by nephridiophore.
17. Nervous system consists of paired, cerebaral, pleural, pedal and visceral ganglia joined by longitudinal and transverse connectives and nerves,.
18. Sense organs include eyes statocysts and receptors for touch, smell and taste.
19. Dioecious or monoecious one or two gonads with gonoducts, opening into renal ducts or to exterior.
20. Fertilization is external or internal.
21. Development is either direct or with metamorphosis through the trochophore stage called veliger larva. (Glochidium larva is common us Bivalvia).

CLASSIFICATION OF MOLLUSCA

The classification scheme for the larger groups of mollusca appears to have reached a seasonable degree of stability but the arrangement into smaller units in the usual state of flux.

The scheme have adopted is a the conservation side, following generally Thiele (1929-1935), Wenz (1938-1960) and Morton (1958).

PHYLUM-MOLLUSCA

Classes

Aplacophora | Polyplacophora | Monoplacophora | Gastrophora

Scaphopoda | Polecypoda | Cephalopoda

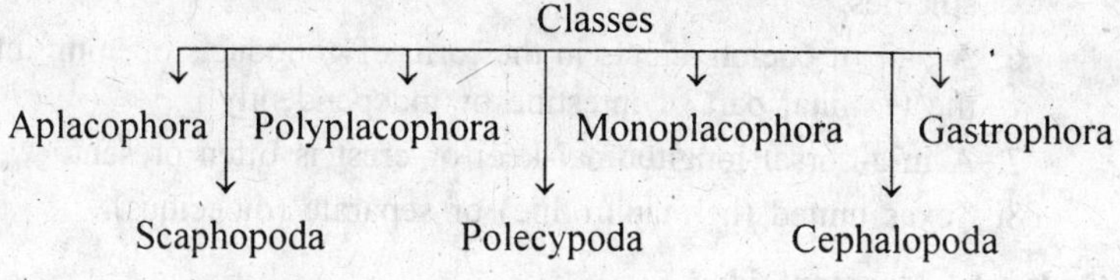

Class 1. Aplacophora—Solenogasters

Definition : Untorted, bilaterally symmetrical, vermiform, molluscs, without head mantle, foot, sheath nephridia, with a cuticle beset with calcareous spicules with a straight digestive tract, generally provided with calcareous spicules with a radula and with a pair of coelomoducts serving as gonoducts.

Characters

1. Body worm-like, bilaterally symmetrical and cylindrical.
2. Head, mantle, foot, shell and nephridia are absent.
3. Mouth and anus are terminal or subterminal at opposite end.
4. Digestive tract straight generally provided with a radula.

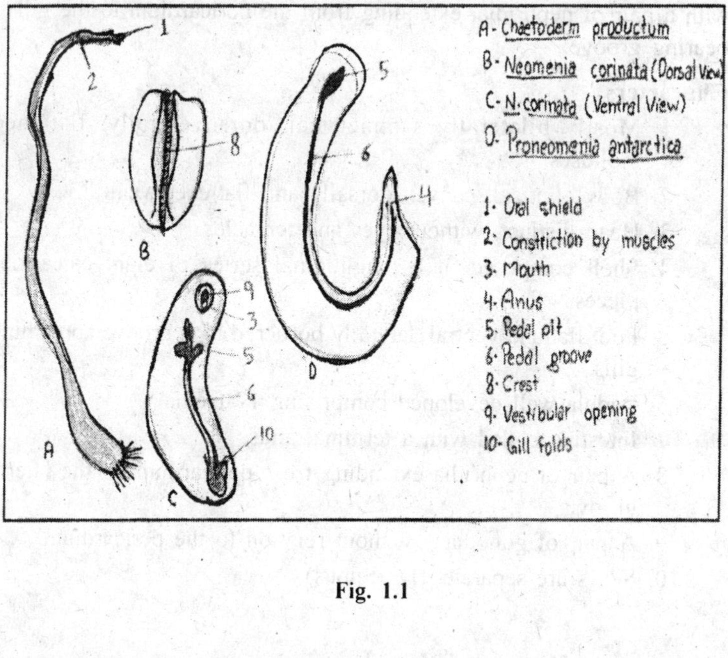

Fig. 1.1

5. Body covered with cuticle beset with numerous calcarious spicules.
6. A pair of coelomoducts in the form of gonoducts opening into the terminal part of intestine or independently.
7. A mid-dorsal longitudinal keel or crest is often present.
8. Sexes united (hermaphrodite) or separate (dioecious).

Order 1—Neomenioidea

With a mid-ventral longitudinal groove.

e.g., Neomenia, Proneomenia, Lepidomenia

Order 2—Chaetodermatoidea

With a mid-ventral groove,

e.g., Chatoderma, Prochaetoderma.

I. CLASS 2—POLYPLACOPHORA:/CHITONS

Definition : Untroted bilaterally symmetrical, dorsoventrally flattened molluscs , with a distinct head (devoid of eyes and tentacles), with a shell composed of a longitudinal series of 8 pieces, with a flat foot laterally bordered by a groove containing gills, with a well developed radula of 17 teeth in the transverse dissection with a coiled intestine and terminal anus with a pair of gonoducts unrelated to the pericardium, and with a pair of nephridia, extending from the pericardium to the gill—bearing groove.

Characters

1. Mostly bilaterally symmetrical, dorsoventrally, flattened molluscs.
2. Body eliptical, convex dorsally and flattened ventrally.
3. Head distinct, without eyes and tentacles.
4. Shell composed of a longitudinal series of eight calcarious pieces.
5. Foot flat and ventral, laterally bordered by a groove containing gills.
6. Radula well developed comprising 17 teeth.
7. Intestine coiled with a terminal anus.
8. A pair of nephridia extending from pericardium to the lateral groove.
9. A pair of gonoducts without relation to the pericardium.
10. Sexes are separate. (Dioecious).

Order 1—Lepidopleurida

Values of the shell without insertion plates, or if present without insertion teeth. Ctenidia a few and posterior.

e.g., Lepidepleurus, Hanleya.

Order 2—Chitonina

Valves of the shell with insertion plates and teeth. Gills are along the whole length of mantle groove.

e.g., Tonicella, Chiton, Cryptochiton, Choneplex.

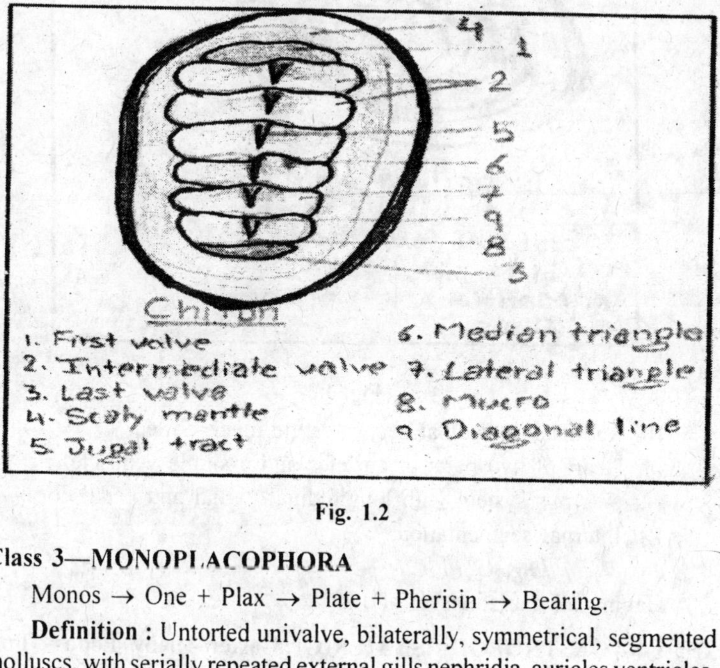

Fig. 1.2

Class 3—MONOPLACOPHORA

Monos → One + Plax → Plate + Pherisin → Bearing.

Definition : Untorted univalve, bilaterally, symmetrical, segmented molluscs, with serially repeated external gills nephridia, auricles ventricles, foot retractor muscles and nerve branches and with median terminal anus.

ORDER

Characters

1. Trybiliodea: Body bilaterally symmetrical and segmented.
2. Shell comprises single piece or value.
3. Head without eyes and tentacles.
4. Foot flat and ventral.

5. Mantle encircles the body as a circular fold of the body wall.
6. Gills external and serially arranged.
7. Sexes separate. (Dioecious).
8. Five pairs of gills in pallial groove.
9. Six pairs of nephridia, two of which are gonoducts.

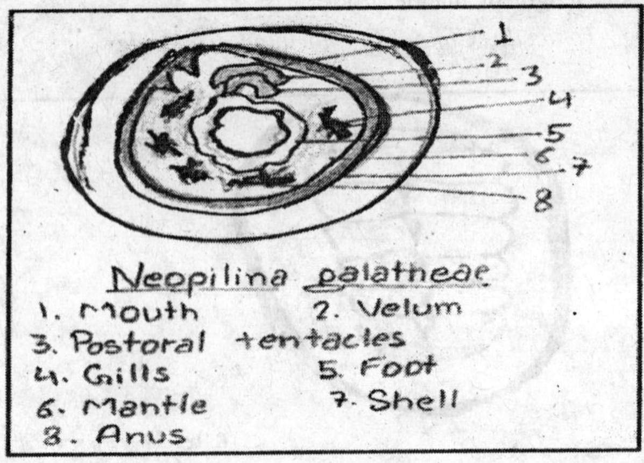

Fig. 1.3

10. Radula in a radular sac, intestine much coiled.
11. Heart of two pairs of auricles and a single ventricle.
12. Nervous system with longitudinal pallial and pedal cords.
13. Internal segmentation.

e.g., Neopiling, Galatheae.

Having all characters of above .

Class 4—GASTROPODA:SNAILS (Gr. Gaster→belly+Podos→Foot)

Definition : Torted/detorted unsegmented, asymmetrical molluscs typically provided with a univalve, spirally coiled shell with a more or less delimited head bearing tentacles and eyes, with a well developed foot whose ventral surface forms a flat, creeping sole with mantle, lining the last whorl of the shell, with radula, anteriorly displaced anus and spirally coiled visceral mass, with paired or unpaired heart auricle, ospharidium, gill and nephridium and with a single reproductive system.

Characters

1. Gastropods are marine, freshwater, terrsestial and few parasitic on echinoderms.

2. Body unsegmented, asymmetrical typically with a univalve, spirally coiled shell.
3. Head distinct bearing tentacles, eyes and mouth.
4. Foot is ventral, broad flat and muscular forming the creeping sole and often bearing dorsally a hard piece, the operculum on its posterior end.
5. Visceral mass spirally coiled exhibiting torsion.
6. Mantle is a collar-like fold of body wall, lining the body whorl leaving a space, the mantle cavity between itself and the body.
7. Buccal cavity contains an odontophore with a bearing rows of chitinous teeth.
8. Digestive system comprises a muscular pharynx, oesophagus, stomach, long coiled intestine and anteriorly placed anus.
9. Respiration by gills (ctenidia) in most forms through the wall of the mantle cavity in some forms and placed anus.
10. Circulatory system is open and the heart is enclosed in a pericardium.
11. Excretory organs comprise metanephridia which are paired in primitive forms and reduced to a single nephridia in most forms.
12. Nervous system comprise distinct and pleural beside buccal, pedal, parietal and visceral ganglia.
13. Sexes are separate dioecious in most forms, while in some forms united (hermaphrodite).
14. Development includes trochophore and veliger larval stages.

Class—GASTROPODA

Sub-class

I. Prosobranchia	II. Opisthobranchia	III. Pulmonata
1. Archaeogastropoda	1. Onchidiacea	1. Basommatophora
2. Mesogastropoda	2. Cephalaspidea	2. Stylommatophora
3. Stenoglossa /Neogastropoda	3. Anaspidea/Aplysiacea	
	4. Pteropoda	
	5. Acochlidiacea	
	6. Philinoglossacea	
	7. Saccoglossa	
	8. Notaspidea	
	9. Nudibranchia	
	10. Rhodopacea	
	11. Pyramidellacea	
	12. Parasita.	

Sub-Class I—Prosobranchia

Definition : The prosobranch (fox-gilled) snails also called streptoneura, are torted gastropods in which the mantle and contained pallial complex are anteriorly located, in front of the visceral mass.

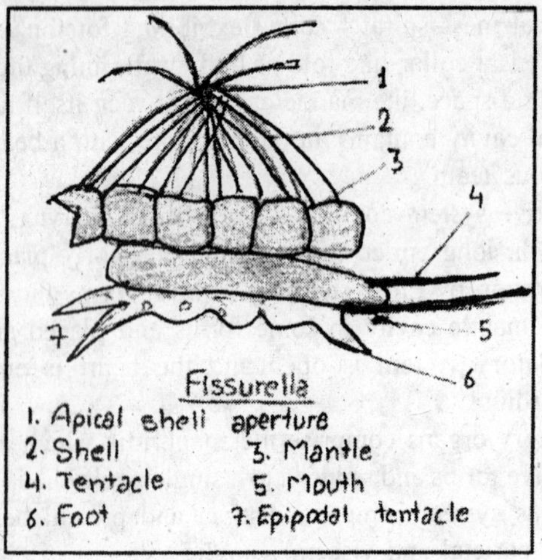

Fig. 1.4

Characters

1. Mostly marine, few fresh water or terrestrial forms.
2. Owing to torsion, of the visceral mass, the visceral nerve commissures are twisted into a figure of 8'.
3. Mantle cavity opens anteriorly in front of the visceral mass.
4. Shell is generally conical and spirally coiled with an operculum.
5. Head distinct with snout bearing a pair of tentacles and a pair of eyes.
6. Foot is muscular, forms the ventral part of the body.
7. Ctenidia or gills, if present, are situated in front of the heart.
8. Sexes are separate (Dioecious).

Order 1—ARCHEOGASTROPODA

1. Prosobranchs without, siphon penis and prostrate glands.
2. Operculum is also absent in many forms with few exceptions.
3. One or two bipectinate internal gills.
4. Heart mostly present with two auricles.

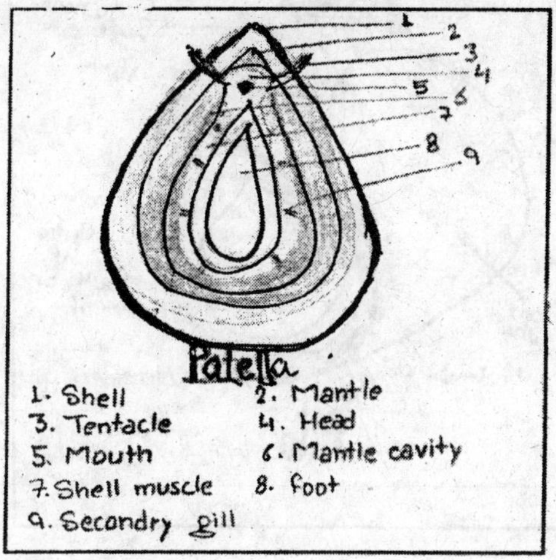

Fig. 1.5

5. Two osphradia usually present.
6. Two nephridia present.
7. Nervous system not concentrated usually with pedal cords.
8. Sex cells discharged directly into the sea by way of the right nephridium.

e.g., (Ear shell) Haliotis, Keyhole limpet, Fissurella, Acmara, Patella, Trochus, Nerita, Astraea, Turbo (Cat's eye).

Order 2—Mesogastropoda

1. Prosobranchs usually with siphon, penis and a non-calcified operculum.
2. Radula tactin glossate type having 7 teeth in each row.
3. One monopectinate gill present.
4. Heart present with one auricle.
5. Single osphradium.
6. Single nephridium.
7. Nervous system is concentrated without pedal cords.

e.g., Viviparus, Ampullarius, Pila, Volvata, Trumcatella, Littorina, Hydrobia, Jonthina, Cyprea.

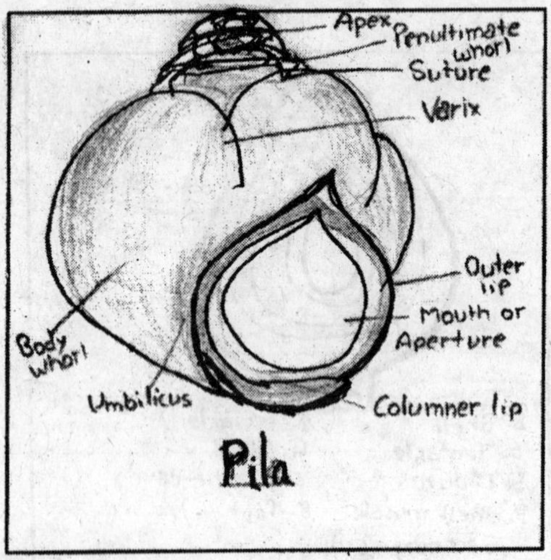

Fig. 1.6

Order 3—Stenoglossa/Neogastropoda

1. Shell is with more/less elongated siphonal canal.
2. Radula consists of rows with two or three teeth in each row.

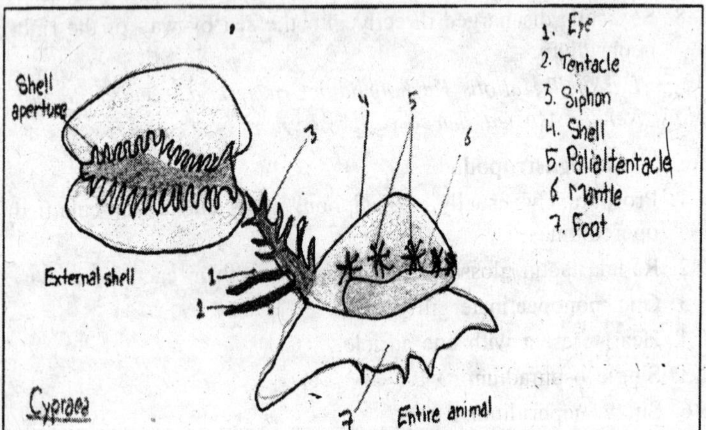

Fig. 1.7

e.g., Murex, Magilus, Buccium, Melogena, Conus, Terebra.

Sub-class II—Opisthobranchia

Definition : Detorted rarely torted marine gastropods, with exposed conical shell/shell concealed in the mantle/commonly wanting, the mantle cavity retreated along the right side or usually without operculum with one gill often respiratory substitutes in the form of dorsal outgrowths often with parapodia with one auricle always posterior to the ventricle and with one hermaphroditic reproductive system.

1. Exclusively marine gastropod.
2. Shell often reduced/wanting when present often covered with mantle/ pedal folds.
3. Operculum usually absent
4. Single gill/often replaced by secondary branchiae in the form of dorsal outgrowths.
5. Heart with one auricle posterior to the ventricle.
6. One to detorsion the mantle cavity rated to the right side of radula often lost.
7. Nervous system concentrated due to detorsion.
8. Hermaphrodite *i.e.,* sexes united.

Order 1—ONCHIDIACEA

1. Appearance Slug-like, naked or without shell opisthobranchs.
2. Mantle projects widely beyond the foot.
3. Head bears a pair of retractile tentacles each tipped with an eye.
4. Pulmonary sac, anus and female gonopore located at the posterior end.
5. Male gonopore placed anteriorly.

e.g., Onchidium, Onchidella.

Order 2—CEPHALASPIDEA

1. Shell is generally present but may be paired or wholly enclosed by mantle.
2. Parapodial lopes present absent.
3. Head with tentacular shield.

e.g., Acteon, Hydatina, Bulla.

Order 3—ANASPIDEA/APLYSIACEA

1. Found mostly in tropical and subtropical waters.
2. Shell small, more or less covered by mantle.

3. Parapodia likes well developed.
4. Anterior end bears a pair of tentacles, a pair of rhinophores and a pair of eyes.
5. Sperm duct open, running the body length to the penis, located anteriorly.

e.g., Aplysia, Akera.

Order 4—PTEROPODA

1. Pelagic snails with/without shell.
2. Swim by a pair of lateral expansions.
3. Protandrous hermaphrodite with an open sperm groove.

e.g., Spiratella, Cavolina, Peraclis, Clione.

Order 5—ACOCHLIDIACEA

1. Minute with shell/naked snail.
2. Gills parapodia and visceral sac projecting behind the fool.
3. Sexes united/separate in few.

e.g., Acochlidium.

Order 6—PHILINOGLOSSACEA

1. Minute naked snails.
2. Head appendages absent.
3. Gill absent.
4. Visceral mass separated form the foot only by a shallow groove.

e.g., Philinoglossa.

Order 7—SACCOGLOSSA

1. Animals with/without shell..
2. Pharynx suctorial.
3. Sperm duct is closed.
4. Parapodia and cerata present.

e.g., Oxynoe.

Order 8—NOTASPIDEA

1. Shell present/absent.
2. Parapodia absent.
3. Gills bipectinate and osphradium present on right side.
4. Mantle present but devoid of mantle cavity.

e.g., Tylodina, Pleurobranchus

Order 9—NUDIBRANCHIA

1. Shell absent/naked.
2. Mantle/mantle cavity absent.
3. Internal gills or ctenidium and osphradium absent.
4. Respiration by secondary branchiae usually arranged in a circle around the anus.

e.g., Doris, Tritonia, Armina, Eolis

Order 10—RHODOPACEA

1. Vermiform snail.
2. Without external appendages.
3. Nephridia protonephridial type.

e.g., Rhodope.

Order 11—PYRAMIDELLACEA.

1 Shell spirally twisted.
2 Long invaginable proboscis.
3 Operculum present.
4 Gill and radula absent.

e.g., Turbonilla, Odostomia.

Order 12—PARASITA

1. Endoparasitic gastropods found in the interior of hobthurians.
2. Extremely degenerated snails.

e.g., Entoconcha, Thyonicola.

Sub-class III—PULMONATA

Definition : Detorted, mainly freshwater and terrestrial gastropods, with typical spiral shell or redivided shell partially/wholly encircled by mantle overgrowth, or without a shell, without an operculum, without a gill, with mantle cavity altered into a pulmonary sac by fusion with the neck region, leading to a contractile opening the pheneumostome, with one. auricle anterior to the ventricle and with one hermaphroditic reproductive system.

Characters

1. Mostly freshwater/terrestrial, a few marine members, operculum is absent.
2. Shell typically spiral /reduced/ absent it present partly/completely concealed by mantle.

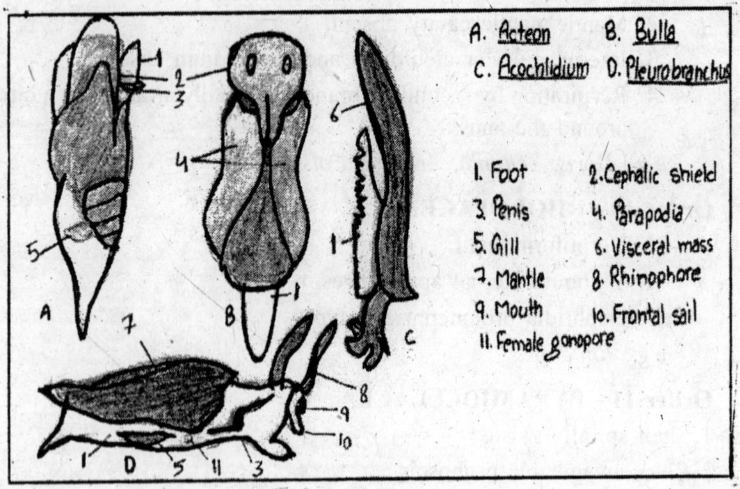

Fig. 1.8

3. Mantle cavity transformed into a pulmonary sac with a narrow pose on the right side. Gills absent.
4. Heart with one auricle anterior to the ventricle.
5. Nervous system secondary symmetrical owing to the shortening of connectives and concentration of ganglia into a circum-oesophageal ganglionic complex.
6. Hermaphrodite.

Order 1—Basommatophora

1. Freshwater or brackish water and a marine in habitat.
2. Shell delicate with a conical spiral large aperture.
3. One pair of non-inveginable tentacles with the eyes at their bases.
4. Male and female gonopores generally separate.

e.g., Siphonaria, Lymnae, Planorbis (has haemoglobin in blood).

Family Amphibolidae—Brackish water, Amphibola

Family Lyminaeidae—Fresh water with conical dextral shell.

Systellommatophora—Lymanae, Synatelous, 3-3cm, contains haemoglobin Planorbis.

Veronecellidae—Elongated slugs, with flattened shells, no frame, shell or Mantlecavity *Siphonaria* japonica.

Order 2—Stylommatophora

1. Terrestial pulmonates.
2. Shell with a conical spire, internal or absent.
3. Two pairs of invaginable/renactile tentacles with the eyes at the tips of the posterior pair.
4. Male and female gonoposes usually united.

e.g., Limax, Helix, Partula, Retinella, Aoion

Class 5—SCAPHOPODA

Definition : Torted, symmetrical molluscs, without eyes, tentales or gills, enclosed in a tusklike shell, open at both ends, mantle tabular through fusion of its edges, foot elongated, mouth surrounded by lobular out growths, base of snout bears a pair of clusters of protrusible filaments, heart rudimentary dioecious, exclusively marine.

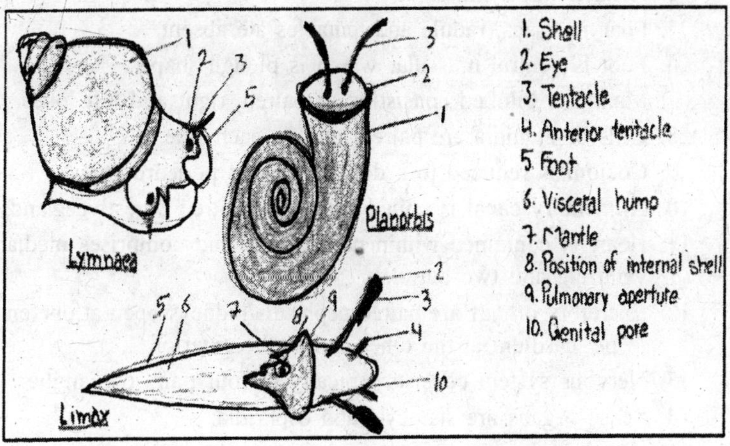

Fig. 1.9

Characters

1. Exclusively marine.
2. Body bilaterally symmetrical, elongated, and endorsed in a tusklike shell open at both ends.
3. Eyes, tentacles and gills are absent.
4. Mantle tubular completely endorsing the body.

5. Mantle surrounded by lobular processes out growths.
6. Foot is reduced used for digging.
7. Heart rudimentary
8. Sexes separate.

e.g., Dentalium.

Class 6—PELECYPODA

Definition : untorted body bilaterally symmetrical molluscs, endorsed in a shell of two lateral valves hinged together mid-dorsally, with correspondingly bilobed mantle, without head, pharynx, jaws, radulal tentacles, with two auricles, and paired gills, nephridia and gonads, dioecious/hermaphoditic, marine or freshwater.

Characters

1. Aquatic mostly marine some freshwater forms.
2. Body is bilaterally symmetrical and laterally compressed.
3. Shell consists of tow lateral valves, gingled together mid-dorsally.
4. Head is not distinct.
5. Pharynx, sacs, radula and tentacles ate absent.
6. Foot is ventral muscular which is plough shape.
7. Mantle is bilobed consisting of paired, right and left lobes.
8. Gills or ctenidia are paired, one on each side.
9. Coelom is reduced to a dorsally placed pericardium.
10. Alimentary canal is coiled with large paired digestive glands.
11. Heart is contained within pericardium and comprises median ventricle and two auricles.
12. Excretory organs are paired nephridia/kidneys open at one end in pericardium at the other end to the exterior.
13. Nervous system consists typically of four pairs of ganglia.
14. Sense organs are statocyst and osphridia.
15. Sexes ate separate/united.
16. Development is accumpained by metamorphosis which usually includes a trochphore larvae.

Order 1—PROTOBRANCHIATA

1. Single pair of plume-like ctenidia, each consisting of two rows of flattened gill filament.
2. Foot is not compose but go to flattened ventral surface/sole for creeping.

3. Two adductor muscles present.

e.g., Nucula Solenomya.

Order 2—FILIBRANCHIATA

1. Single pair of plate- like gills formed of distinct V-shaped filaments.
2. Two adductor muscles present, anterior may be reduced/absent.

e.g., Mytilus, Arca.

Order 3 PSEUDOLAMELLIBRANCHIATA

1 Gills are plaited so as to form vertical folds.

2 Single large posterior adductor muscle present.

3 Foot rudimentary/feebly developed.

e.g., Pecten, Ostraea, Melagrina.

Order 4—EULAMELLIBRANCHIATA

1. Gills are firm and basket-like.
2. Two equal-sized adductor muscles present.
3. Foot large, Byssus small/absent.

e.g., Anodonta, Unio, Cardium, Venus, Mya, Teredo.

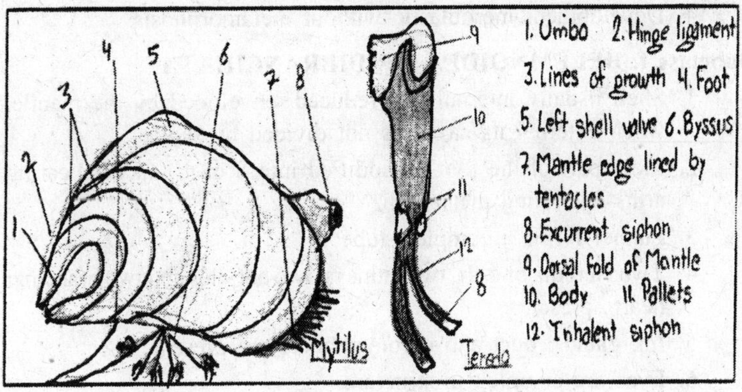

Fig. 1.10

Order 5—SEPTIBRANCHIATA

1. Gills reduced to a horizontal muscular partition dividing the mantle cavity.
2. Two adductor muscles present.

3. Foot long and slender and byssus rudimentary/absent.
   e.g., Poromya, Cuspidaria.

Class 7—CEPHALOPODA

1. Exclusively marine.
2. Body bilaterally symmetrical with head and trunk.
3. Shell spiral, chambered or usually with or without reduced shell embedded in the mantle.
4. Head bears large eyes and mouth.
5. Trunk consists of symmetrical and uncoiled visceral mass.
6. Mantle encloses posteriorly and ventrally a large mantle cavity.
7. Foot altered into a series of sucker bearing arms or tentacles encircling the mouth.
8. Mouth bears jaws and radula.
9. Row or four pairs of bipectiate gills.
10. Circulatory system closed, heart with two or four auricles.
11. Excretory system comprises two or four pairs of nephridia.
12. Nervous system is highly developed and the principal ganglia are concentrated around the oesophagus.
13. Sexes are separate.
14. Development meroblastic without metamorphosis.

Subclass I. BELEMNOIDEA OR DIBRANCHIATA

1. Shell usually internal and reduced, enveloped by the mantle, when external its cavity is not divided by septa.
2. Main part of the foot is modified into 8 or 10 sucker bearing arms encircling the mouth.
3. Funnel forms a complete tube.
4. Two ctenidia or gills, two kidneys, two auricles and two branchial hearts present.
5. Ink glands, duct and chromatphores are present.
6. Eyes are complex in structure.

II. ORDER 1—DECAPODA

1. Body is generally enlongated often with lateral line.
2. Arms are 10 of which 8 short and 2 long. Two longer arms or tentacles are retractile bearing suckers at their distal ends.
3. Eight smaller arms bear stalked suckers provided with hormy arms.

4. Shell is internal and well developed.
5. Nidamental glands are usually present.
6. Heart enclosed in the well developed coelom.

e.g., 1. Sepia, 2. Loligo, 3. Spirula.

III. ORDER 2—OCTOPODA

1. Body usually globose and devoid of lateral fins.
2. Eight arms with sessile suckers and devoid of homy rims.
3. Shell is absent except in female Argonauta.
4. Nidamental glands absent.
5. Heart does not lie in the reduced coelom.

Examples: 1. Octopus, 2. Argonauta.

I. Subclass II—NAUTILOIDEA OR TETRABRANCHIATA

1. Shell is external, spiral and chambered.
2. Main part of the foot encircling the mouth, divided into lobes bearing numerous tentacles.
3. Funnel does not form a complete tube.
4. Four ctenidia or gills, four kidneys and four auricles present.
5. Ink gland and chromatophores absent.
6. Eyes are simple.

Example: Nautilus.

Subclass III—AMMONOIDEA

1. Shell is external spiral and chambered.
2. Main part of the encircline the mouth, divided into lobes bearing numberous tentacles.
3. Funnel dose not form a complete tube.
4. Four ctenidia or gills, four kindly and four auricles present.
5. Ink gland and chromatophores absent.
6. Eyes are simple.

Example: Ammonites.

CLASSIFCATION

Class—GASTROPODA

Order—Pelellogastropoda

Superfamily Lottioidea

Lotiidae

Acmaeidae

Lepetidae

Superfamily Nacelloidea

Nacellidae

Superfamily Patelloidea

Patellidae

Order—Cocculiniformia

Superfamily Lepatelloidea

Addisoniidae

Order—Neritoida

Superfamily Neritoidea

Neritidae

Neritopsidae

Phenacolepadidae

Order—Neomphalida

Superfamily Neomphaloidea

Neomphalidae

Order—Vetigastropoda

Superfamily Pleurotomarioidea

Pleurotomariidae

Scissurellidae

Haliotidae

Superfamily Fissurelloidea

Fissurellidae

Superfamily Trochoidea

Trochidae

Trochinae

Calliostomatinae

Stomatellinae

Turbinidae

Turbininae

Angariinae

Liotienae

Phasianellinae

Skeneidae

Superfamily Seguenzioidea

Seguenziidae

Order—Caenogastropoda

Suborder—Neotaenioglossa

Infraorder—Discopoda

Superfamily Luxonematoidea

Abyssochrysidae

Superfamily Cerithioidea

Campanilidae

Diastomatidae

Dialidae

Litiopidae

Obtortionidae

Scaliolidae

Cerithiidae

Cerithideidae

Potamididae

Batillariidae

Fossaridae

Planaxidae

Modulidae

Tunitellidae

Siliquariidae

Superfamily Vermetoidea

Vermetidae

Superfamily Littorinoidea

Littorinidae

Skeneopsidae

Superfamily Cingulopsoidea

Cingulopsidae

Eatoniellidae

Superfamily Rissooidea

Anabathridae

Barleeidae

Emblandidae

Caecidae

Elachisinidae

Epigridae

Falsicingulidae
Iravadiidae
Rissoidae
Asemineidae
Truncatellidae
Tomidae
Vitrinellidae

Superfamily Stromboidea

Aporrhaidae
Strombidae
Struthiolariidae
Seraphidae

Superfamily Xenophoroidea

Xenophoridae

Superfamily Calyptraeoidea

Calyptraeidae
Capulidae

Superfamily Vanikaeoidea

Vanikaridae
Hipponicidae

Superfamily Cypraeoidea

Cypraeidae
Ovulidae
Ovulinae
Eocypraeinae

Superfamily Velotinoidea

Triviidae
Triviinae
Eratoinae
Velutinidae

Superfamily Naticoidea

Naticidae

Superfamily Tonnoidea

CASSIDAE

Bursidae
Personidae

Ranellidae
Ficidae
Tonnidae
Laubierinidae

Infraorder—Heteropoda

Superfamily Carinarioidea

Atlantidae
Carinariidae

Infraorder—Ptenoglossa

Superfamily Triphoroidea

Triphoridae
Cerethiopsidae
Triforidae

Superfamily Epitonioidea

Epitoniidae
Aclididae
Xanthinidae

Superfamily Eulimoidea

Eulimidae

Suborder—Neogastropoda

Superfamily Muricoidea

Muricidae
Muricinae
Thaidinae
Coralliophilinae
Turbinellidae
Turbinellinae
Columbariinae
Vasinae
Buccinidae
Fasciolariidae
Melongenidae
Nassariidae
Columbellidae
Harpidae
Harpinae

Moruminae
Marginellidae
Cystiscidae
Pleioptygmatidae
Mitridae
Costellariidae
Olividae
Olivinae
Ancillinae
Volutidae
Volutomitridae

Superfamily Cancellarioidea

Cancellariidae

Superfamily Conoidea

Conidae
Terebridae
Turridae
Turrinae
Borsoniinae
Cochlespirinae
Crassispirinae
Daphnellinae
Drilliinae
Mangeliinae

Order—Heterostropha

Superfamily Valvatoidea

Orbitestellidae

Superfamily Rissoelloidea

Rissoellidae

Superfamily Omalogyroidea

Omalogyridae

Superfamily Architectonicoidea

Architectonecidae
Mathildidae

Superfamily Pyramidelloidea

Pyramidellidae

Amathinidae

Order—Opisthobranchia

Suborder—Cephalaspidea

Superfamily Ringiculoidea

Ringiculidae

Superfamily Acteonoidea

Acteonidae

Bullinidae

Hydatinidae

Superfamily Philinoidea

Philinidae

Cylichnidae

Superfamily Retusoidea

Retusidae

Superfamily Diaphanoidea

Diaphanidae

Superfamily Bulloidea

Bullidae

Superfamily Haminoeoidea

Haminoeidae

Superfamily Cylindrobulloidea

Cylindrobullidae

Suborder Anaspidea

Superfamily Aplysoidea

Aplysiidea

Akeridae

Suborder—Sacoglossa

Superfamily Julioidea

Juliidae

Superfamily Oxynooidea

Oxynoidae

Volvatellidae

Suborder—Notaspidea

Superfamily Umbraculoidea

Umbraculidae

Tyiodinidae

Suborder—Thecosomata

Infraorder Euthecosomata

Superfamily Limacinoidea

Cavoliniidae

Limacinidae.

Class—BIVALVIA

Infraorder—Pseudothecosomata

Subclass—Paleotaxodonta

Order—Nuculoida

Superfamily Nuculoidea

Nuculidae

Superfamily Nuculanoidea

Nuculanidae

Malletiidae

Tindariidae

Subclass—Cryptodonta

Order—Solemyoida

Superfamily Solemyoidea

Solemyidae

Subclass—Ptèriomorphia

Superorder—Isofilibranchia

Order—Mytiloida

Superfamily Mytiloidea

Mytilidae

Superorder—Prionodonta

Order—Arcoida

Superfamily Arcoidea

Arcidae

Cucullaeidae

Noetiidae

Parallelodontidae

Superfamily Limopsoidea

Limopsidae

Glycymerididae

Philobryidae

Superorder—Eupteriomorphia

Order—Pterioida

Superfamily Pterioidea

Pteriidae

Malleidae

Isognomonidae

Superfamily Pinnoidea

Pinnidae

Order—Limoida

Superfamily Limoidea

Limidae

Order—Ostreoida

Suborder—Ostreina

Superfamily Ostreoidea

Ostreidae

Gryphaeidae

Superfamily Dimyoidea

Dimyidae

Superfamily Plicatuloidea

Plicatulidae

Suborder—Pectinina

Superfamily Anomioidea

Anomiidae

Placunidae

Superfamily Pectinoidea

Pectinidae Class—**CEPHALOPODA**

Subclass—Nautiloidae

Order—Nautilida

Nautilidae

Order—Octopoda

Argonautidae

Order—Sepiida

Sepiidae

Spirulidae

Propeamussiidae

Spondylidae Class—SCAPHOPODA

Order—Dentaliida

Dentaliidae
Gadilinidae
Laevidentaliidae
Omniglyptidae

Order—Gadilida

Gadilidae
Entalinidae
Pulsellidae
Siphonodentalidae

Subclass—Paleoheterodonta

Order—Trigonioida

Superfamily Trigonioidea

Trigoniidae

Subclass—Heterodonta

Order—Veneroida

Superfamily Lucinoidea

Lucinidae
Fimbriidae
Thyasairidae
Ungulinidae

Superfamily Cyamioidea

Cyamiidae
Gaimardiidae

Superfamily Galeommatoidea

Galeommatidae
Lasaeidae
Montacutidae

Superfamily Cardetoidea

Carditidae

Superfamily Chamoidea

Chamidae

Superfamily Crassatelloidea

Crassatellidae
Astartidae

Superfamily Cardioidea
- Cardiidae
- Hemidonacidae
- Tridacnidae

Superfamily Mactroidea
- Mactridae
- Mesodesmatidae
- Cardilidae
- Anatinellidae

Superfamily Solenoidea
- Solenidae
- Cultellidae

Superfamily Tellinoidea
- Tellinidae
- Donacidae
- Psammobiidae
- Semilidae

Superfamily Arcticoidea
- Arcticidae
- Trapeziidae

Superfamily Glossoidea
- Glossidae
- Vesicomyidae

Superfamily Corbiculoidea
- Corbiculidae

Superfamily Veneroidea
- Veneridae
- Petricolidae
- Cooperellidae
- Glauconomidae

Order—Myoida

Superfamily Myoidea
- Myidae
- Corbulidae

Superfamily Gastrochaenoidea
- Gastrochaenidae

Superfamily Hiatelloidea

Hiatellidae

Superfamily Pholadoidea

Pholadidae

Teredinidae

Subclass—Anomalodesmata

Order—Pholadomyolda

Superfamily Pholadomyoidea

Pholadomyidae

Superfamily Pandoroidea

Pandbridae

Lyonsiidae

Cleidothaeridae

Myochamidae

Superfamily Poromyoidea

Poromyidae

Cuspidariidae

Verticordiidae

Superfamily Thracioidea

Latemulidae

Periplomatidae

Thraciidae

Superfamily Clavagelloidea

Clavagellidae

Superfamily Cymbuolioidea

Peraclidae

Order—Pulmonata

Suborder—Archaepulmonata

Superfamily Ellobioidea

Ellobiidae

Suborder—Basommatophora

Superfamily Amphiboloidea

Amphibolidae

Superfamily Siphonarioidea

Siphonariidae

Trimusculidae

Gastropoda Gross Structure

The shell is an elongated structure consisting of tubular whorls, coiled around a central axis called the columella. The various whorls lie in different planes so that the shell is called conspiral is contrast to the planospiral shell found in the protozoan, $6^{1/2}$ whorls in the shell of *Pilaglobosa*. The smallest and the oldest whorl lies at the apex of the shell, which represents the first shell or the protoconch laid down by the larva. The whole series of whorls, exclusive of the body whorl, is known as the spire. Internally, all the whorls communicate freely with another, whorl.

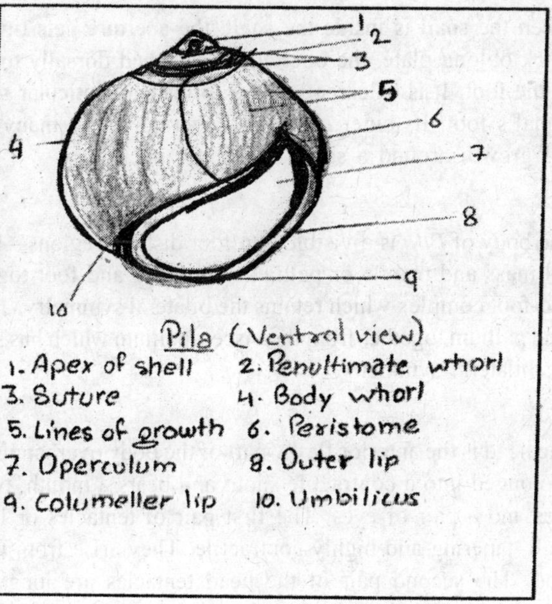

Fig. 1.11

The body whorl opens to the exterior by a wide opening, the aperature or mouth, which is situated on the ventral surface of the shell. The smooth and continuous margin of the aperature is called the peristome. Its outer margin is called the outer lip, while the inner one next to the columella is the inner or columellar lip. The columella is shallow, twisted and rod-like structure, which opens to the exterior narrow aperature, the umblicus, situated near the columellar lip at the end opposite the apex of the shell.

When the shell of *Pila* is held with the apex upward and the aperature facing the observer, it displays a clockwise coiling from the apex to the aperture so that the latter comes to lie to the right-hand side. Such a shell is called dextral, clockwise or right-handled and is of normal occurrence in *Pila* and a big majority of gastropods.

Microscopic Structure

Chemically a molluscan shell is composed of:

(i) Conshiolin. An albuminoid hormy substance which serves as the matrix, and

(ii) calcium carbonate, which occurs as crystals of calcite or aragonite.

Operculum

When the snail is inside the shell, the aperture gets tightly closed by a thick oblong plate, the operculum, attached dorsally to the hinder part of the foot. It is calcareous plate formed by cuticular secretion of the animal's foot. Its outer chitinous covering shows many concentric rings of growth around a small subcentral nuclecus.

BODY

The body of *Pila* is divisible into four distinct regions—head, foot, visceral mass and mantle or pallium. The head and foot together from the head-foot complex which retains the bilateral symmetry. The visceral mass and pallium together from the visceropallium which has secondarily lost the bilateral symmetry.

1. Head

The head is the anterior fleshy part of the body overhanging the foot. It is prolonged into a contractile snout and bears a mouth, two pairs of tentacles and a pair of eyes. The first pair of tentacles or labial palps are small, tapering and highly contractile. They arise from the sides of the head. The second pair of the head tentacles are long fleshy and cylindrical and arise from behind the snout from the dorsal surface of the head. They are also contractile and can be withdrawn by invagination. From near the base of each tentacle projects a small, stumpy eye-stalk or ommatophore bearing a small but prominent eye at its tip.

2. Foot

The large, strong, muscular, ventral part of the body forms the foot when fully expanded, it is roughly triangular in shape with the apex directed backwards. The foot is flat smooth and grey ventral on surface or the sole creeping. The foot bears the operculum dorsal on its posterior end, called the operculerous lobe. When the foot is withdrawn the operculum completly fits into the mouth of the shell and closes it.

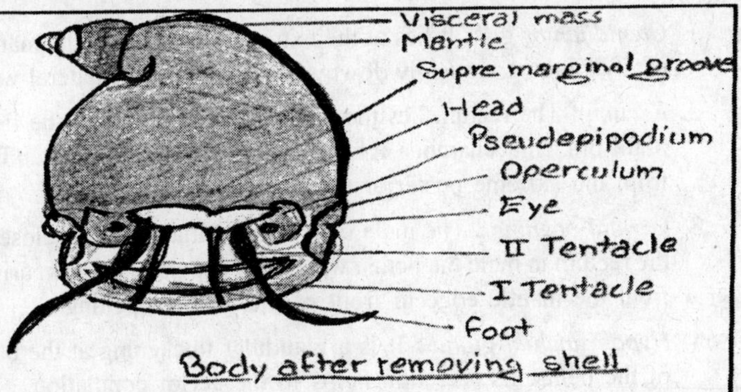

Fig. 1.12

The foot the locomotory organ of snail. It is highly contractile with the muscle fiber arranged crosswise and lengthwise. Inside the foot are present numerous pedal gland-cells which secrete a slim rail for locomotion.

3. Visceral Mass

It constitutes a short figure on the dorsal side. Containing all the visceral organ. It is soft and grey to dark brown in colour. It is spirally coiled like the shell in which it lies, occupying the body whorl as well as the spire. Coiling results from faster growth on the side of the visceral mass away from the columella.

4. Mantle

The skin of the visceral mass forms a thin and delicate covering, called the mantle or pallium, which is a characteristic molluscan organ. Anteriorly, the mantle becomes thickened and pigmented and serves as protective clock, however, the head and appendages when the animal is retracted. The mantle secretes and lines the shell. The periostracan the ostracum the shell and secreted by the row and shell glands respectively, while the hypostracum is secreted by the general epithelial covering of the mantle.

Organs of Branchial Chamber

1. *Ctenidium or gill* : It lies at the extreme right side of the mantle cavity hanging vertically downwards from its dorsolateral wall.
2. *Rectum* : The rectum lies the left of the.ctenidium on the floor of the branchial chamber. It is a raised tube-like organ, extending from the extreme posterior ent of the mantle cavity
3. *Genital opening* : The male or female gential duct lies close to the rectum in male the penis, which is a copulatory organ, arises from the mantle edge in front of the gential opening.
4. *Hypobranchial gland* : It is a glandular thickening at the base of the penis. Its secretion helps in the act of copulation.
5. *Renal opening* : The anterior chamber of the renal organ projects into the branchial chamber as a small area of reddish colour.

Organs of Pulmonary Chamber

1. *Pulmonary sac* : A bag like-large organ hangs downwards from the roof of the mantle cavity, occupying a larger area of the pulmonary chamber. This is the pulmonary sac or lung. It communicates with the pulmonary chamber through an elongated opening helps in aerial respiration.
2. *Nuchallobe arising from the mantle*, adjacent to the left nuchal lobe and situated on the left side of the pulmonary chamber is a gill-like osphradium. It is bipectinate feather-like and consists of 22 to 28 fleshy and some what triangular leaflets. It helps in testing the physical and chemical qualities of the entering water and aids in the selection of food.

LOCOMOTION

Pila moves with the creeping start activity of the muscular sole of its foot. Waves of muscular contractions from behind forward or vice versa, provide the principal power of movement. Pedal mucous glands leave behind a slime.

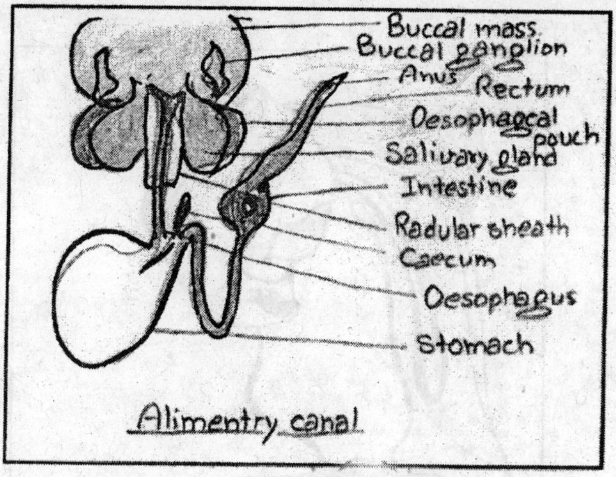

Fig. 1.13

DIGESTIVE SYSTEM

I. Alimentary Canal

The alimentary canal of pila is a coiled tube extending from the mouth and terminating at the anus. Its anterior part is especially modified. The entire canal may be divided into three regions—the foregut comprising buccal cavity and oesophagus, the midgut including stomach and intestine, and the hindgut consisting of rectum.

1. Buccal Cavity

The chamber into which the mouths lead is the buccal cavity. It is lined by cuticle and surround by a large, thick-walled highly muscular and pear-shaped structure, the buccal mass. Its wall is provided with several sets of muscles for its movement and the movement of radula.

(a) Buccal musculature out of several sets of muscles, the protractors are well developed. They include:

   (i) a medium dorsal, three pairs of anterior dorso-laterals and two pairs of posterior dorso-laterals on the dorsal surface, and

   (ii) three anterior muscles and a pair of long and strong latero-ventral forward muscles on the ventral surface.

(b) *Vestibule and jaws* : The buccal cavity is regionated into an anterior tubular part, called vestibule, and a posterior part.

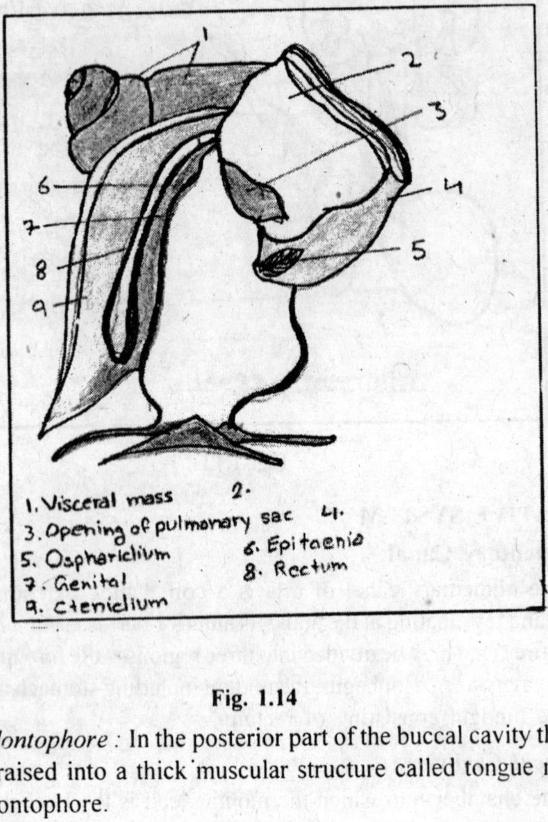

Fig. 1.14

(c) *Odontophore* : In the posterior part of the buccal cavity the floor is raised into a thick muscular structure called tongue mass or odontophore.

(i) The structure is supported by two sets of cartilages lying below the epithelium at the top of the odontophore, and

(ii) A-pair or s-shaped lateral cartilages, with thick ventral edges ant thin dorsal edges, lying in the sides.

(d) *Radula* : The buccal cavity contains a brownish, chlinious, curved, ribbon-like structure, called the radula or lingual ribbon. Its anterior end, bearing a pair of wing-like flaps. Its posterior end is lodged in band-like and below the buccal mass. Below the radula lies a delicate and elastic, sub-radular membrane. The dorsal surface of the radula bears teeth arranged in numerous transverse rows. Each row contains seven teeth, one central (rachidian), and one lateral and two marginals on its either side, giving the formula 2, 1, 1, 1, 2. The radula is moved forward and backward on the odontophore for rasping food particles.

2. Oesophageous

The oesophagus is a narrow and long tube emerging dorsally from the buccal mass.

3. Stomach

The stomach lies on the left side of the visceral mass, below the pericardium. Its cavity is U-shaped which is regionated into aboard posterior cardiac chamber and a narrow anterior pyloric chamber. The lining of the stomach is folded. A short rounded and blind pouch, the caceum, arises from the lower outer wall of the pyloric chamber. At the junction of two chambers of the stomach opens a duct from the digestive gland.

4. Intestine

The pyloric stomach is followed by a long and coiled intestine. It means backward into the visceral mass.

5. Rectum

It comprises a thick-walled tube which extends into the branchial chamber of the mantle cavity between the ctenidium and gential duct,

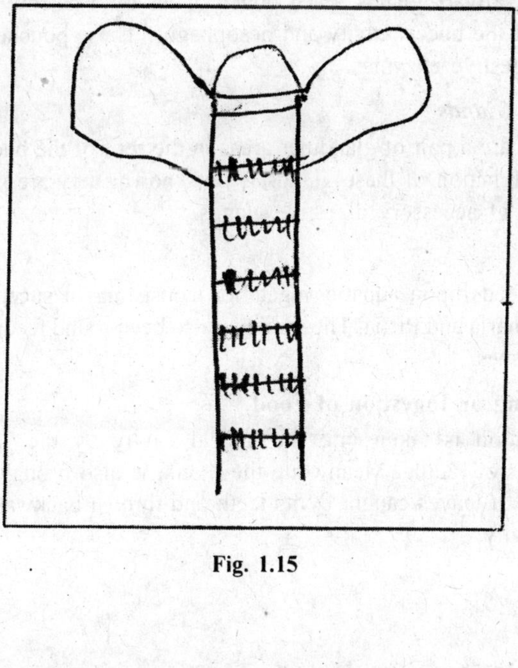

Fig. 1.15

trail during creeping. So the snail does not move directly on the surface, but upon the slime path. Retraction occurs by the action of muscles and the absence of blood pressure turgor.

II. Digestive Glands

1. Salivary Glands

These are two in number and lie on either side of the posterior part of the buccal mass. Each gland looks like a branching white mass. A duct from each gland enters the muscles of the buccal mass and then opens into the buccal cavity. The salivary secretion contains mucin-like substance and a carbohydrate enzyme.

2. Digestive Gland

A somewhat triangular plate of cone with a convex outer and more or less flattened inner surface. This structure is a digestive gland (often referred to as the liver or hepatopancreas). It has two main lobes, the smaller in contact with the stomach and the larger extending to the apex of the spiral. Two separate ducts arise from the two lobes which unite together to form a common duct before opening into the stomach.

3. Oesophageal Pouches

A pair of simple rounded, cream-coloured oesophageal pouche lies below the salivary glands. Each pouch opens by a narrow duct at the junction of the buccal cavity and oesophagus. These pouches probably secrete digestive enzymes.

4. Buccal Glands

These are a pair of glandular areas in the roof of the buccal cavity, the exact function of these glands is not known; they are probably of the nature of accessory digestive glands.

III. Food

Pila feeds upon aquatic vegetation consisting of succulent plants like Vallisnaria and Pistia. The sanil has also been found feeding on dead animal tissues.

IV. Feeding or Ingestion of Food

The food is taken into the buccal cavity by the "chain-saw" movements of radula. Meanwhile the radula is also brought forwards, the pieces of leaves caught by its teeth and thrown backwards into the buccal cavity.

RESPIRATORY SYSTEM

Pila shows both aquatic as well as aerial modes of respiration. To carry on with theses two modes of respiration efficiently and successfully, the animal possesses a gill or ctenidium for a aquatic respiration and pulmonary sac or lung for aerial respiration.

(I) Respiratory Organs

1. Ctenidium

It is a monopectinate gill (comb-like) and is situated on the right side of the branchial chamber hanging from its dorsolateral wall. It consists of along axis, the ctenidial axis. The ctenidial axis is traversed by an afferent blood vessel and an efferent blood vessel. The axis bears a long series of flat triangular leaflets known as lamellae. Each lamella is attached by abroad base to the ctenidial axis. The two free sides of the lamella are unequal; the shorter side receives blood from the afferent vessels and is thus called efferent side as form it the blood flows into the efferent vessels. In the middle the lamellae are larger and decreases in size towards the two ends. The ctenidium is innervated by nerves from the left pleural and supra-intestinal ganglia.

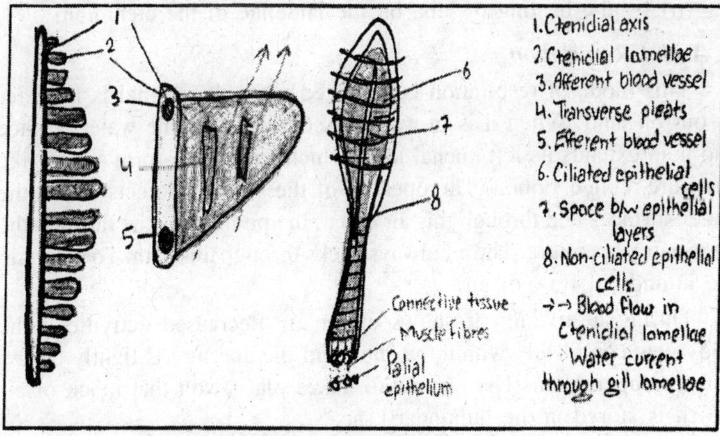

Fig. 1.16

2. Pulmonary Sac

It is a bag-like structure and developed from the mantle wall. It hangs into the pulmonary chamber of the mantle cavity into which it opens rough a large, oblique pulomanry aperature. These walls are muscular and highly vascular, consisting of blood spaces.

3. Nuchal Lobes

These are highly contractile structures of the mantle; situated one on either side of the head.

II. Mechanism of Respiration

1. Aquatic Respiration

The aquatic respiration takes places when the snail is at the bottom or is floating or lying suspended in midwater or attached to plants and weeds in water. The two nuchal lobes channel-like structures. The water current first flows beneath the osphradium to the posterior part of the pulmonary chamber. After washing the entire length of the ctenidium, it flows out through the right nuchal lobe. The flow of water current is maintained by:

(i) the alternate lowering and raising of the floor of the mantle cavity, and

(ii) by the beating of cilia on the lamellae of the ctenidium.

2. Aerial Respiration

This mode of respiration is observed when the animal is in water or out on land. When it is in water, it comes nearer the water surface and then extends its left nuchal lobe, which rolls up to form a tube like structure, called siphon. The opening of the siphon projects above the water surfaces and through this air enters the pulmonary sac through the pulmonary aperature. The air always backs through this path. To maintain the in-and-out flow of air.

During aestivation it shows extremely decreased activities. The body gets completely within the shell and the aperture is tightly closed with the operculum. The respiration takes place with that much of air which is stored in the pulmonary sac.

BLOOD-VASCULAR SYSTEM

The circulatory system of *Pila* has attained a great complexity due to its double mode of respiration. Involving the gill as well as lung. It consists of :

(i) pericardium,
(ii) heart,
(iii) arteries,
(iv) sinuses, and
(v) veins.

1. Pericardium

The pericardium is a thin walled, roughly avoid chamber lying dorsally on the left side of the body whorl behind the mantle cavity. It extends anteriorly up to the stomach and the digestive gland. The pericardial cavity represents the true coelom.

2. Heart

The heart, lying enclosed within the pericardium, consists of two chambers, an auricle and ventricle.

(a) Auricle

It is a thin-walled highly contractile and roughly triangular sac, situated in the dorsal part of the pericardium. The dorsal apex of the auricle receives blood. Ventrally, the auricle opens into the ventricle through an auriculoventricular opening, provided with two semilunar valves, so arranged as to permit the flow of blood into the ventricle but not vice versa.

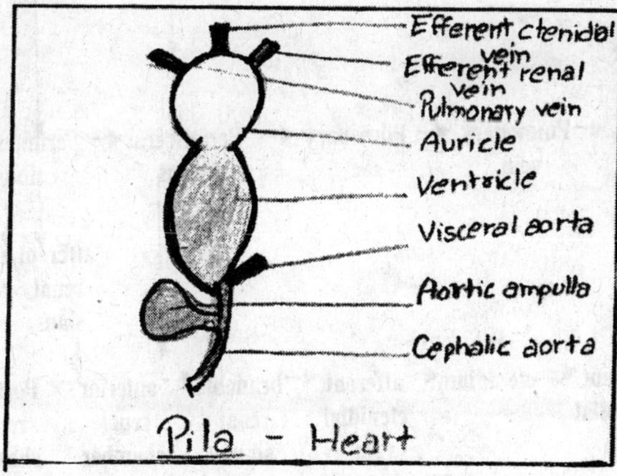

Fig. 1.17

(b) Ventricle

It is an ovoid sac lying below the auricle. It has thick, spongy and muscular walls and a reduced cavity due to a coarse meshwork of muscular strands, from the lower end of the ventricle arises a large artery, the aortic trunk.

3. Arteries

The,aortic trunk divides immediately into two branches, an anterior cephalic aorta and a posterior visceral aorta, each breaking up into numerous arteries.

(a) Cephalic Aorta

The base of the cephalic aorta bears a bulb-like, thick-walled and highly contractile structure, the aortic ampulla. The rhythmic contractions of the ampulla help in the circulation of blood.

(b) Visceral Aorta

The visceral aorta runs posteriorly into the visceral mass and supplies the visceral organs through a series of numerous branches.

4. Sinuses and Veins

Blood supplied by arteies and their branches to various parts of the body collects in small spaces called lacunae. Some of the lacunae, like

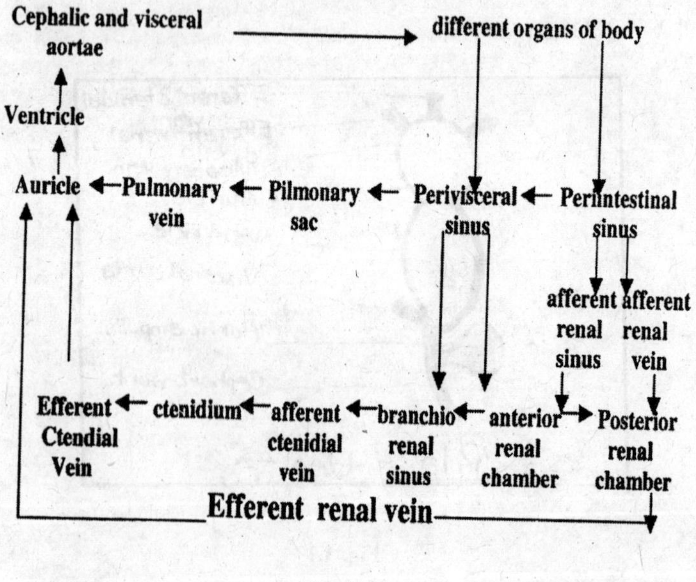

Fig. 1.18

capillaries, connect arteies with veins while others join to form larger spaces called sinuses. The sinuses are without definite walls. They collectively constitute the haemocoel. Blood from sinuses is drained by veins.

Blood

The blood serves to transport nutrients, oxygen and waste product form one part of the body to the other. It contains some colourless stellate amoeboid cells and a respiratory pigment called haemocyanin. The amoeboid cells are phagocytic in nature.

EXCRETORY SYSTEM

Renal Organ

In *Pila globose* the excretory organ is a large kidney or renal organ or organ of Bojanus the gonoduct. It communicates with the exterior on one hand and with the pericardial cavity representing coelom on other

1. Anterior Renal Chamber

It is more or less an oval organ, in colour and lies anterior to the pericardium. It opens into the branchial chamber of the mantle cavity through a slit-like opening near the epitenia. The internal cavity of the anterior chamber is very much reduced due to the presence of many triangular leaf like a processes or lamellae,

2. Posterior Renal Chamber

It is a broad brownish to grey and hook-shaped chamber. Situated behind the anterior renal chamber, between the rectum on the right and the pericardium, large internal cavity encloses a part of the genital duct and a few coils of the intestine.

NERVOUS SYSTEM

The nervous system of pila consists pained ganglia, commissures and connectives uniting them, and nerves running from these central organs to all parts of the body.

1. Ganglia

The paired ganglia which are aggregation of nerve cells are as follows:

(a) Cerebral Ganglia

A pair of roughly ganglia, situated anteriorly on the dorso-lateral sides of the buccal mass.

(b) Buccal Ganglia

A pair of small, triangular, lying dorso-later on either side at the junction of the buccal mass and oesophagus, partly embedded in the muscles.

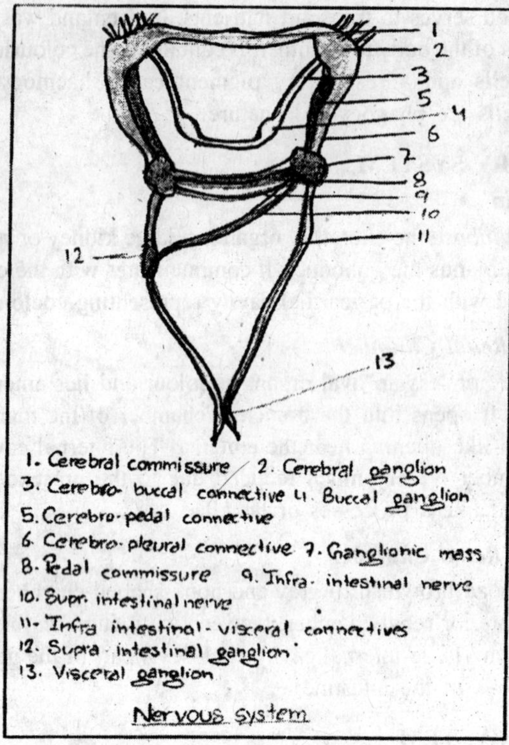

Fig. 1.19

(c) Pleuro-Pedal Ganglia

A pair of large, somewhat triangular ganglionic masses present one on either ventro-lateral side of the buccal mass. Each one is formed by the fusion of an outer pleural and an inner pedal ganglion, separated by a faint noth.

(d) Supra-intestinal Ganglia

An unpaired fusiform ganglion, lying in a sinus behind the left pleuro-pedal ganglionic mass.

(e) Visceral Ganglia

A single ganglionic at the lower end of the visceral mass.

2. Commissures

The commissures are those nerves which establish connection between two similar ganglia. In *Pila* these constitute.

(a) Cerebral Commissure

A thick band of nerve connecting two cerebral ganglia and lying dorsally to the buccal mass.

(b) Buccal Commissure

A fine nerve which connects the two buccal ganglia and runs transversely on the ventral side of the oesophagus.

(c) Pedal Commissures

Two thick nervous bands that lie one above the other underneath the buccal mass and connect the two pedal ganglia together.

3. Connectives

The connectives are those nerves which connect two different ganglia.

(a) Two Cerebro-buccal Connectives

These connect, on either side, the cerebral ganglion and buccal ganglion together.

(b) Two Cerebro-pleural Connectives

These connect on either side, the cerebral ganglion and buccal ganglion together. Two cerebro-pleural connectives. These connect, on either side the cerebral and outer pleural ganglion of the pleuro-pedal ganglionic mass.

(c) Two Cerebro-pedal Connectives

These connect, on either side the cerebral and inner pedal ganglion of the pleuro-pedal ganglionic mass.

(d) Pleuro-infra—Intestinal Connective

Also called infra-intestinal nerve, it is a nerve connective between the pleural ganglion of the left pleuro-pedal mass and the infra-intestinal ganglion which is fused with the right pleuro-pedal mass.

(e) Infra-intestinal—Visceral Connective

Present below the intestine, it is a long nerve that connects the visceral, ganglion with the infra-intestinal parts of right pleuro-pedal-infra- intestinal ganglionic mass.

(f) Supra-intestinal—Pleural Connective

Running above the intestine, it is a slender nerve that connects the visceral ganglion with the supra-intestinal ganglion.

(g) Supra-intestinal—Pleural Connective

Also called the supra-intestinal nerve, it connects the supra - intestinal ganglion with the pleural part of the right pleuro-pedal infra-intestinal ganglionic mass.

(h) Zygoneury

It is a nerve connection between the pleural part of the left pleuropedal ganglionic mass and supra-intestinal ganglionic mass.

SENSE ORGANS

A snail is diffusely sensitive due to the sensory cells which are distributed all over the head, foot and various other parts of the body.

(I) Osphradium

The single osphradium is situated on the left side of the animal suspected from the roof of the mantle cavity close to the entrance through the left nuchallobe. It is small, elongated, oval structure. It is bipectinate consisting of 22-28 thick, fleshy, and roughly triangular leaf lets, arranged in two rows along a slightly raised median or central axis. The leaflets are largest in the middle of the osphradium.

In a transverse section, the osphradium consists of an outermost covering of a single layered epithelium, internally lined by a thin basement membrane, the interior filled up with nerves, connective tissues and blood spaces. The epithelial cells are elongated possessing basal nuclei and they are of three types:

(i) sensory,
(ii) ciliated, and
(iii) glandular.

Probably serves as an olfactory organ. Meaning to smell. It serves to test the chemical nature of the inspiratory water current.

(II) Eyes

The snail's head carries a pair of short fleshy and stump-like stalks or ommatophores. Each ommatophore bears a small, black and somewhat circular eye, slightly below its tip on the outer side. In a section, an eye looks a pyriform cup-like invagination of epidermis, called the optic vesicle. The optic nerve enters the vesicle. The cavity of the vesicle is filled up by an oval, structureless hyaline body, the lens. Eyes of *Pila* are probably not true organs of sight. The sense of sight is greatly limited in range and the snail does not seem to distinguish objects.

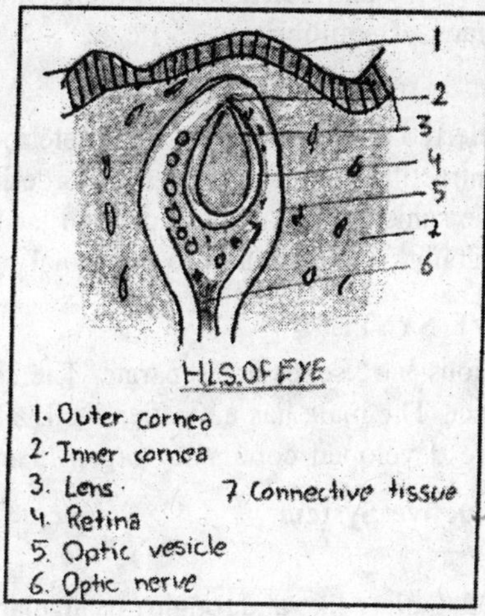

Fig. 1.20

(III) Statocysts

The statocysts are a pair of small, pyriform and cream-coloured sacs, lying one on either side attached to the pedal ganglion of that side by a band of connective tissue. Each statocyst is a hollow capsules surrounded by an outer thick, though leathery covering of connective

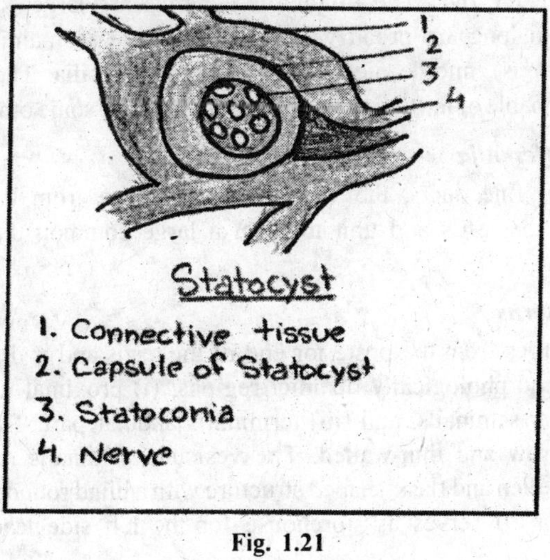

Fig. 1.21

tissue. The cavity of the capsule is filled with a fluid and available number of minute, oval and calcareous particles, the statoconia. The statocysts are organs of equilibrium.

(IV) Tentacles

As already referred to, the snout of *Pila* is anteriorly prolonged into a pair of short, contractile an conical processes bordering the mouth. The tentacles are of the same colour as the snout and tactile in function,. A sense of taste is doubtfully attributed to the labial palps.

REPRODUCTIVE SYSTEM

Pila is dioecious, *i.e.*, sexes are separate. The sexual dimorphism is slight but distinct. The male has a smaller shell, a less swollen body whorl, and a more developed copulatory organ than the female.

(I) Male Reproductive System

1. Testis

It is a single, flat, plate-like and roughly triangular whitish structure, occupying the upper part of the first two or three whorls of the shell. It lies in close association with the brownish of greenish digestive gland. A thin cutaneous membrane separates the testis from the shell.

The testis of *Pila* produces two kinds of sperms, eupyrene and oligopyrene. The eupyrene sperms are small, thread-like, about 25m, long and 1.2m, broad, and with distinict head (containing a twisted nucleus), middle piece and tail with a single cilium. They are motile and can fertilize the ova. The oligopyifee sperms are large, spindle-like, about 32.5m long 3m broad with a distinct head (containing a broad curved nucleus), middle piece, and tail of 4 or 5 cilia. They are non-motile, incapable of fertilizing ova and probably have some other function.

2. Vasa Efferentia

Several fine ducts, the vasa efferentia, arise from the different regions of the testis and unit to form a large common duct, the vas deferences.

3. Vas Deferns

It emerges from the posterior end of the testis and is differentiated into three morphologically distinict regions: (i) proximal tubular part, (ii) vesicular seminails, and (iii) terminal glandular part. The proximal part is narrow and thin-walled. The vesicular seminalis is somewhat curved, woollen and flask-shaped structure with a blind rounded posterior prolongation. It serves as storehouse for the left side leads into the terminal, thick-walled, glandular part of the vas deferens, which runs forward into the left side of the rectum. It finally opens by the male genital aperture.

4. Copoulatory Organ

The copoulatory organ or penis arises separately from the mantle edge in front of the anus. It is a long. About, slightly curved and flagellar structure with a swollen base and a tapering free end. The penis remains enclosed in penis sheath formed by a thick, glandular flap of a yellow colour attached to the mantle along its side.

5. Hypobranchial Gland

At the base of the penis sheath is an oval glandular thickening with pleated surface, the hypobranchial gland. It consists of tall cells containing small basal nuclei.

(II) Female Reproductive System

1. Ovary

The ovary occupies the same position as the testis in male. It is much branched and coloured light-orange in the young, becoming darker in mature individuals.

2. Oviduct

The narrow and transparent oviduct arises from about the middle of the ovary. It runs anteriorly just below the skin along the inner margin of the digestive gland. Near the renal organ it turns downwards and then upwards to open into the receptaculum.

3. Receptaculum Seminalis

The small, bean-shaped receptaculum seminanlis (or seminal receptacle) lies enclosed within the cavity of the posterior renal chamber, closely attached ventrally to a thin-walled pouch from the walls of the uterus, pouch of the receptaculum.

4. Uterus

The large yellow and pear-shaped uterus lies in the body whorl below the intestine and to the right of the renal chambers.

5. Vagina

The vagina is a white cream-coloured, band-like tube running forward just beneath the skin.

6. Copulatory Apparatus

The rudimentary penis is not enclosed within a well-developed penis-sheath, but lies beneath a glandular fold of the mantle.

7. Hypobranchial Gland

The hypobranchial gland of the female is poorly developed with a rudimentary glandular thicking.

(III) Copulation

Pila breeds during the rainy season, when the long period goes aestivation comes to an end. Copulation occurs under water or on the moist ground of the banks and may continue for three of four hours. During copulation, the male and female snails approach each other in such a manner that the right nuchal lobe of one comes to lie opposite that of the other. The tip of the penis enters the female gential aperature and the sperms from the vesicular seminalis of the male are transferred to the receptaculum seminis of the female.

It is suggested that the veliger larva gains by torsion, as after torsion it can pull the velum into the shell and fall to the bottom if pursued by enemies. There are serious objections to this idea. As molluscan cilia are generally under nervous control the velum need not be retracted to stop swimming movements. Furthermore, torsion does not occur in other classes of molluscs but their larvae have been able to survive. Torsion brings the mantle cavity to he front of he body. As a result, water currents resulting from forward movement tend to bring water into the mantle cavity, while they work against water entering when the mantle cavity is behind. This would be especially important in fresh-water gastropods with the habit of moving against the current in a stream. But there are serious drawbacks to this idea also. Most gastropods more so slowly that currents set up by their own movements are not a very important factor. Besides, many snails have a narrow, tubular siphon for drawing water into the mantle cavity. In these snails, the opening of the mantle cavity is smaller, and advantage coming from its anterior position is lost.

The most obvious disadvantage of torsion is that the anus is brought directly over the head and threatens to foul the respiratory organs and the head sense organs. It is not surprising to find the first adaptational trend evoked by torsion involved attempts to get the swept away from the front door. Torsion and the consequent anterior position of the mantle cavity gave gastropods a new look. They are no longer Protogastropoda, but Prosobranchia. The first prosobranchs are fossils from the upper Cambrian. For the next 200,000,000 years or so, until about the end of the Paleozoic, most prosobranchs belonged to the order Archeogastropods. The anus is bent upward, fitting into a special notch in the shell. As the shell grows, the anal notch becomes a slit band (selenizone). The most ancient prosobranchs were often flat spirals with conspicuous slit bands.

In modem species with arrangements of this kind, respiratory currents enter the mantle cavity from in front, flow up between the gill filaments and depart dorsally, past the anus carrying away the feacal material. The site of the anus varies considerably in these gastropods. The anus may change position as the animal grows. Haliotis has a series of openings in the selenizone. Not all of the archeogastropods have a perforate shell, however. Many modern species have a solid shell. The gill, auricle, and nephridium on the larval left (adult right) side are reduced and disappear. Why this adaptive tendency should have appeared is not clear, but when completed, a new routing of water through the mantle cavity becomes possible. Water enters on the left, flows over the persisting gill first, and then passes, the anus and nephridiopore on its way out of the mantle cavity, sweeping the wastes, away neatly enough. Reduction and loss of the organs on the larval left side occur among some of the archeogastropods. Many gastropods have further reduced the gill surface. The remaining left ctenidium becomes one-sided, is built into the mantle wall, and eventually may be reduced to a vascular area in the mantle wall. The vascularized wall of the mantle cavity then serves as the respiratory surface, acting as a water-or air-breathing lung.

Some gastropods undergo detorsion. This reverse movement occurs during development, and while it does not wholly eliminate the results of torsion, it greatly reduces them. Gastropods that have followed this line have no shells or reduced shells, and some have wholly lost the mantle cavity. In this case, the body surface is the site of respiratory exchange, and may bear simple or complex secondary gills. A strong tendency to untwist the nervous system is evident. It grows naturally out of detorsion, but is also achieved by the centralization of the nervous system and the shortening of the visceral nerves.

2

BIVALVIA

The bivalvia evolved in the process of colonization of deposits of sand or sandy mud in the pre-combination epoch, these animals feeding directly upon organic matter in these unstables, Adaptations to suit the new mode of life included enclosures of the entire body by paired, lateral folds of the mantle which secreted lateral shell valves articulating mid-dorsally; other adaptations were the loss of head, the buccal mass, the radula, and all cephalic sense organs. Sensory functions were taken over by the margin of the mantle. These early bivalves were protobranchiate in from and collected particulate food directly from the substance by ciliary collecting organs which may have been closely comparable with the pulp proboscides of modern comparable with the pulp proboscides of modern protobranchs. From this ancestry three main lineages distinguished by their feeding mechanism have survived to the present day; these are:

The protobranchia,

The septibranchia,

The polysyringia.

The protobronchia respected a continuation to the day at the ancestral protobrachiate stock. With the exception of the solenomyidae, the feeding mechanisms of which are not yet known, the remainder of the protobrachia feed primarily by means of the ciliated palp proboscides upon organic matter in the substraturm. The ctenidia principally subserve respiration and are only slightly concerned in the collection of food policies.

The septibrachia evolved directly from a protoboanchiate ancestry by adoption of the ctnedia to from a muscular pumping septum, the original function of which was probably to argument the respiratory water stream, by the action of which the carcases of small crustaceans etc., could be sucked into the mantle caving and then consumed. There are strong grounds for believing that this new scavenging or carnivorous mode of life must have arisen directly from a protobranchiate ancestor and not from auciliary feeding eulamellibranch as has hitherto been supposed. The remainder of the Bivalvia, with filibranch, pseudo-lamellibranch, or eularmellibranch ctenidia, constitute the polysyringia (no other name is available for these bivalves-although the term "lamellibranchia is suitably descriptive it is undesirable since it has long been used as the name of the whole class). The polysyringia arose directly from a protobranchiate ancestry by development of the ciliary food-collecting mechanism of the ctenidia and the association between the ctenidia and the labial palps, thereby initiating the collection of foot particles suspended in the supernatant water and liberating the polysyringia from dependence or sedimentary deposits. This development initiate an extensive phase of adaptive radiation in the exploitation of the new source of food material, various lineages become adopted to life on rocky source on coral reefs in sedimentary deposits, or in fresh water etc.

FEEDING METHOD

The direct stage in the adaptive radiation of the Bivalvia, which concerns the evolution of two feeding methods, is set out in diagrammatic form.

Protobranchia	Septibranchia	Polystringia
(relation the ancient habit of feeding principally on sedimentary deposits	(have acquired a new feeding habit trapping larger objects drifting in water near bottom)	(have specialized progressively in ctenidial ciliary sorting mechanism)
↑	↑	↑

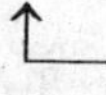

PROTOBRANCHIATE ANCESTOR

(fed upon bottom deposits by means of labial palps and palp proboscides)

The acquisition of new feeding mechanisms and exploitation of new sources of food, had a profound effect on the evolution of class. Bivalvia

EXTERNAL FEATURES OF BIVALVIA

The following account is of the anatomy of the west African *Aspathoria brumpti*, is member of the family Mytilidae, the account is closely applicable also to the European *Anodonta cygnea* which belongs to the related family Unionidae.

Any differences noted are attributable to the differences between the two genera, while the very extensive similarities are due to the close affinity between the Mulelidae and Mytilidae and the Unionidae, both of which are assigned to the order Unicnacea.

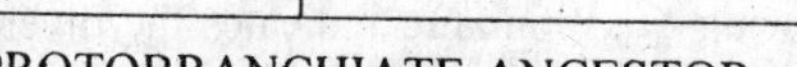

In a similar way. But with a greater degree of adaptation, the account could be applied to the investigation of bivalves in other orders and families for in many respects the account concerns the general features of the majority of bivalves.

External Features of *Aspatharia brumpti*

The body is protected by left and right shell valves which are approximately oval in outline, and are attached to each other at a hinge on the dorsal margin, the shell valves are similar in shape and size *i.e.,* they are equivalve. Growth of the shell valves is by accretion at their margins and occur regularly the size and shape of the shell valves at earlier stages in the life of the animal begin indicated by growth lines which were caused by some interruption in the growth of the shell at that time. The greater part of the shell is covered with a layer of periostracum may have been worm away, the underlying prismatic layer of calcareous material may also have been eroded, exposing the innermost nacreous layer of the shell. This area of erosion occurs around the unbo, a small backwardly pointing beak situated near the dorsal edge of the shell. Draw the shell, represent the antero-posterior axis of the animal by drawing a horizontal line indicating the greatest length of the shell. Drop a perpendicular to this line from the umbo, and note that it divides the shell into a smaller anterior part and a large posterior part. The shell valves are therefore inequilaterally.

Diversity of Form In the Bivalvia

The following questionnaire indicates the diversity of form which may be encountered in the class Bivalvia, and this diversity will be most fully appreciated by examining in turn a number of bivalve genera such as *Mytilus, Cardium, Scrobuicularia, Mya, Ostrea* etc., and completing the questionnaire separately for each genus studied.

Shell valve : Equivalve/inequivalve
Equilateral/inequileteral

Hine line : Straight/arcuate

Axis of movement of valves : longitudinal/vertical/both longitudinal and vertical.

Outline of valves : Orbicular/oval/triangular/emarginated anteriority.

Sculpture: Radial ridges/concentric ridges/decussations/spins rows of fine teeth.

Ligament : External/internal if external, prosodetic/opisthodetic.

Mantle

Fusion : Mantle isthmus only/also below exhalant opening/also short fusion below inhalant siphon/fusion below inhalant siphon extends for forwards.

Nature of mantle fusion: Inner mantle fold only/ inner and middle mantle folds/inner middle and outer folds all involved.

Cruciform muscle : Present/absent.

Sensory elements : Pallial tentacles present/absent pallial eyes present/ absent.

Siphonal orifices : Not raised upon siphons two short separate siphons two long siphons united at base only siphons united throughout.

Inhalant siphonal orifice : With in curved straining tentacles/without straining tentacles.

Surface of siphons where present : Periostracal sheath present/ absent.

Muscles

Adductor muscles : Is omyarian/anisomyarian/monomyarian if isomyarian, muscles contracting Simultaneously/muscles contracting alternately.

Retractor muscles : Anterior retractor muscles present/absent anterior protractor muscle present/absent Posterior retractor muscle present/absent byssal retractor muscle present/absent

Orbicular muscle : Pallial sinus absent/shallow/very deep.

Digestion in Bivalves

Although at first sight the Bivalvia may seen to be a fairly homogenous class, it has-recently been established that the protobranchia comprise an independent lineage to which might even be accorded the status of a sub-class, studies on the diverticula of a variety of bivalves have revealed striking differences between the Nuclidal on the one hand, and the Anisimyaria and Eulamellibranchia, on the other. Again, highly specialised conditions are found in the septibranchia which alone among the Bivalvia have embarked upon a carnivorous mode of life lesser specialization in feeding and digestion occur in the Teredinidae, or ship-worms and in the Tridacnidae.

Digestion in *Nucula hanleyi* (Bivalve)

Nucula hanleyi, seen from the right side after removal of the right shell valve and the right mantle lobe. In addition to the organs in the mantle cavity, the course of the organs in the mantle cavity, the course of the alimentary canal is also shown in fig. The stomach (s) consists of two portions, the lower conical part containing a viscous mass, not a well-defined crystalline style. The stomach wall is muscular, and is relatively free from attachments to the surrounding tissue; the upper globular portion is extensively protected internally by a chitinous secretion, and muscular contraction of the stomach wall serve to mix the ingested material with the digestive enzymes and to triturate the food material.

Ctenidia

Flat/delicate/pelicate Homorhabdia/ heterorhabdia

Structure and ciliation of ctenidial surfaces : At kins type A. G. labial palps: Large with ridged ciliary sorting surface/greatly reduced

Foot : Wedge shaped/tongue shaped/cylindrical with terminal sucker/ rudimentary.

Byssus apparatus : Present, active in adult present, inactive in adult absent.

Alimentary canal : Stylesac: Separate from mid- gut/conjoined with mid gut

Crystalline style : Present firm present, soft and reduced absent.

Stomach : Internal structure type.

Pericardium : Pericardial glands: On auricles one pericardial walls one auricular and pericardial walls

Posterior aorta : .Aortic bulb present/absent.

Excretory organs : Reno-pericardial apertures median fenestra between left and right distal limbs present/absent.

Excretory apertures : Separate from genital aperture/united with genital aperture.

Gonads : Male/female/hermaphrodite

Conditions of gametes : Spermatozoa active/inactive/not found Ova abundant ripe/otherwise. Modification of demibranchs to from brood pouches: Inner demibranch modified/not modified. Outer demibranch modified/ not modified.

Nervous system : Cerebral ganglia, pedal ganglia visceral ganglia connectives and nerves present

SCHEME OF CLASSIFICATION OF THE BIVALVIA

Only a few orders, families and genera listed, these having been chosen to illustrate the main lines of diversification of the class. Only the principal references in the text are cited.

Class—BIVALVIA

Sub Class	Order	Family	Genus
Protobranchia	Nuculacea	Nuculidae	*Nucula*
		Nuculanidae	*Nuculana*
		Solenomyidae	*Solenomya*
Septibranchia	Poromyacea	Cuspidariidae	*Cuspidaria.*
Polysyringia	Myticacea	Mytilidae	*Mytilus*
(flitter feeding		Pinnidae	*Lithophaga*
bivalves)			*Pinna*
	Pectinacea	Pectinidae	*Pecten*
		Limidae	*Lima*
	Anomiacea	Anomiidae	*Anomia*
	Ostreacea	Ostreidae	*Ostrea*
	Unionacea	Unionidae	*Anodonta*
	Mutelidae		*Aspatharia*
	Sphaeria	Shalriidae	*Sphaerium*
	Dreissenacea	Dreissnidae	*Pisidium*
	Erycinacea	Montacutidae	*Dreissena*
	Cardiacea	Cardiidae	Montacuta
		Tridacnidae	*Cardicam*
			Tridacna
	Venerracea	Veneridae	*Hippopus*
			Venus
			Paphia
	Mactracea	Petricolidae	*Petricola*
		Mactridae	*Mactra*
	Tellinacea	Donacidae	*Lutraria*
			Domax
		Semelidae	*Egeria*
		Tellinidae	*Scrobicularia*
			Tellina
	Solenacea	Solenidae	*Macoma*
			Solen
	Saxicavacea	Saxicavidae	*Ensis*
	Myacea	Myidae	*Hiatella*
		Aloididae	*Mya*
			Aloidis
	Gastrochaenacea	Gastrochaenidae	*Rocellaria*
	Adesmacea	Pholadidae	*Pholas*
			Barnea
			Martesia
		Xylophaginidae	*Xylophaga*
		Teredinidae	*Teredo*
	Pandoracea	Thraciidae	*Bankia*
	clavagellacea	Clavgellidae	*Thracia*
			Brechites

Class—BIVALVIA
Sub-Class—Protobranchia

Order Nuculacea Solenomyacea	**Families Nuculidae, Malletidae**
	Sub-Class Lamellibranchia
ORDER	Toxodonta
Super family	Arcaidae, Glycymeridae.
ORDER—ANISOMYRIA	
Mytilacea	mytilidae
Pteriacea	pteridae, Vulsellidae, Pinnidae
Pectinacea	Pectinidae. limidae
Anoiniacea	Anomiidae
Ostreacea	Ostreidae
ORDER—SCHIZIDONTA	
Trigoniacea	Tringoniidae
Unionacea	Unionidae, mutelidae, Aetheriidae.
ORDER—HETERODONTA	
Astartacea	Astartidae
Carditacea	Carditidae
Sphaeridacea	Sphaexiidae, Corbiculidae
Isocardiacea	Isocardiidae
Cprinacea	Cyprinidae
Cyamiacea	Cyamiidae
Gaimardiacea	Gaimcerdiidae
Dreissenacea	Dreissendidae
Lucinacea	Lucinidae
Erycinacea	Erycinidae, Galeommatidae
Chamacea	Chamidae
Cardiacea	Cardiidae, Tridacnidae
Mactracea	Mactridae, Maphidesmatidae
Tellinacea	Donacidae, Asaphidae, Semelidate, Tellinidae.
ORDER	**ADAPEDONIA**
Solenacea	Solenidae, Glavcomyidae.
Myacea	Aliididae, Myidae
Saxicavacea	Saxicavidae
Adesmacea	Pholadidae, Xylophaginidae, Teredinedae.

THE BIVALVIA

Though the bivalve are wonderfully disease in form and habit their basic pattern is unmistakable. No bivalve has a head buccal mass or radula. In nearly all of them the mantle enclose the whole body and is itself covered by a two piece shell the role of food-catching has passed from the head to the gill, and the main sense organs have moved to the head to edge of the mantle. In higher forms the except for the siphon and the foot gape and the animal burrows deeply into the substrate great emphasis is placed on mucus and cilia in the life of bivalves, and the pallial organs and the stomach are especially elaborate. The chief muscular organs besides the foot are the all important shell adductors of the first and oldest group of bivalves, the protobranchia, only a few living genera remain, all-as we have seen rather specialized the *Nuculidae* and Malletiidae for example feed on surface deposits by means of peculia palp proboscides.

Malletia and Yoldia have highly enlarged pumping ctenidia . The Solenomyidae with their long tubular mainly poriostracal shell, have developed a power of darting and swimming with the piston like foot. But in spite of these peculiarities, Which the most primitive ancestors of lamellibranches could not have possessed, palaeonotology and comparative morphology agree in pointing to the protobranch bivalves as the forerunners of all the rest. The Nuculidae already existed in the Silurian and related shells are found in the Cambrian.

In living forms the ctenidia arrangement of the mantle cavity, and structure of the fill and reproductive organs, have an indelible primitive stamp. In all their characters-whether primitive or specialised-protobranchs stand clearly apart from other bivalves; on many grounds they deserve to be separated into their own sub- class.

Above the Protobranchia we run into difficulties in classification. With such a constant basic structure we shall expect to find much parallel evolution in lamellibranchs and there are three different types of characters of which we must take account. First there are 'adaptive' character which are immediate modifications for a particular mode of life and thus dangerous to use in classification. Secondly, there are 'prograssive' characters, those which usually show a definite trend of advance often running parallel through several unrelated groups: and thirdly, 'conservative' characters-those that persist unchanged or stable over long periods. It is the last that will be most useful in showing real affinities.

In the last hall century there have been two main bases of classification of lamellibranches, employin either gills or the hinge characters, neither by themselves fully satisfactory. Pelseneers's system (1889)208 attached particular importance to the condition of the gills, whether 'protobranch', 'filibranch', 'pseudolamellibranch', eulamellibranch, or 'septibranch'. French conchologist Douville, who sketched a provisional classification without proposing any formal names. He was the first to emphases that in such a variously evolving class, the plasticity of each line may lead frequently to convergent cross resemblance's. His arrangement was almost overlooked until brought to notice again in 1982 by Davies 1986 in a stimulating essay on bivalve classification.

Douville's classification of lamellibranchs struck on ecological note, beginning with three parallel stream, normal, sessile and deep burrowing. Each line formed in own, like a many stranded rope, the strands in general advancing together but fraying out repeatedly into adaptive radiations. Douville was aware of pitfalls of convergent cross-resemblance, but underestimated them; with the insistent adaptive variation and mosaic evolution a many the lamellibranchs the broad lineages become deeply and completely divided.

First among what used to be Pelseneer's Filibranchia and pseudolamillibranhia, we can recognise a recusring and set its characters. Attachment by the byssus or secondarily by shell cementation is the leading habit: only a few forms are free moving. The shell are rather thin and often highly pigmented.

To lay stress on these characters Pelsenees made the basis of his time order. Pseudolamellibranchia, distributing same families to the filibranchi another to the Eulamellibranchia. He failed to realise that the condition of the gill is a progressive character, passing through successive stages of evolution in several parallel lines. It leads to a 'horizontal' classification rather than a natural or 'vertical' grouping.

British conchologists, such in past replaced pelseneer's system drawing on various classification that use the structure of the hinge, such as that of Cossmam (1914) and Dall (1895 and 1916). This has the advantage of keeping the palaeontologist in harmony with the Zoologist. It's great weakness is in the neglect of soft parts: For example, the obviously distinct subclass of protobranchs is placed in one group with same or all of the filibranchs.

In recent times Atkins has suggested regrouping based on the ciliation of the gills, and Purchon has put forward a classification with mew ordinal manes on the basis of internal stomach structure.

Most single-based systems break under stain at same points: what is needed is as many reliable souses as possible, one such assignment was attempted by the nacreous as well as the prismatic layer is highly developed.

The remaining bivalves form a large assemblage with many parallel trends but all agreeing in increasing commitment to burrow, with consequent fusion of mantle margins affection also siphon and ligament. Mores generalized members have solid white porcellanous shells and usually burrow shallowly as with the plumps *Cardium, Venus* or *Ostrea*. Further adaptations to deep burrowing involve continuation of the same trends, extension of fused siphons lightening and streamlining of the shell and sometimes reduction of the foot. This has produced modern civilization of bivalves with long fused siphons, a closed mantle cavity, and a gaping shell with poorly developed hinge, Such forms include the immobile *Panopea* and *Mya* and the *paddocks* (Pholadidae) which burrow into firm substrates.

CLASSIFICATION

(I) Surface Attached Lamellibranchs

Orders taxodonta and Anisomyaria. These, after the prosobranchs, include fashionable in the middle and later paleozoic and early Mesozoic, when most of the living families were established. The majority are filibranchs but *Ostrea* and *Lima* develop Pseudolamillibranch gills. They vary greatly in appearance but all agree in the widely open mantle and lock of siphons, the importance at some stage at least, and generally throughout life, of the byssus (though the Oysters lose it after settlement), and the tendency to asymmetry between the anterior and posterior parts of the body. In an old but still useful study, Jackson (1890) 152 united the 'Auidulidae and their allies' with the Arcidae and the constituent families have been loosely grouped over since. L.R. Cox in a recent Palaeontological classification recognizes four orders (Eutaxodonta, *i.e.*, taxodonta; Isotilibranchia)

ORDER—ANOMALODESMATA

Pandoracea	Lyonsiidae, Pandoridae, myochamidae, Chamosteridae, throciidae, Latemulidae, Verticordiidae.
Clavagellacea	Clavagellidae.

ORDER—SEPTIBRANCHIA

Poromyacea	Poromyidae, Cuspidariidae.

(II) Shallow-Burrowing Lamellibranchs Orders Heterodonta and Shizodonta

The bivalves of the 'eulamellibranch' series have avoided the extreme specialization due to attached life. Yet they are not to be thought of as primitive and have on the whole flourished later than the attached forms. In later mesozonic tertiary, Recent times. A majority of species lie freely at or near the surface, or burrow actively in sand or mud. An instant trend in the tendency to closer of the mantle and development of siphons Through in such a large group any general statement will oversimplify, their earlis habit with already in the codes burrowing we have not with already in the cockles and the venus shells. Two such early groups with the mantle widely open and the serbions short or wholly lacking are the astartacea, well represented in Britain and the carditacea. Here the primitive genus is cockle like Venericardia, while cardita is attached by a byssus and has before mytilised or herronyarian with the posterior and much expended.

(III) Deep Burrowing: Immobile Lamellibranchs

Order—Addpecddonta

The strongest evolutionary force in this order has been to modify the shell for deep penetration, after with permanent sacrifice of mobility. The mantle is always extensively fused , the hinge is weak with its teeth degenerate or unseparated. The values generally gape freely at either end. The immobile habit is frequently obvious from the shell, which tends to be thin and fragile with nothing of the streamlining or solidify of the Heterodonta.

In the superfamil, Myacea, the deep burrowing clams *Panopea, Mya* and *Platyodon* have a very small foot, together with long fused leathery-cased siphons and a closed mantle cavity *Panopea* is generally held to be primitive deep burrower. Mya with its assymentric hinge, to have re-adopted a burrowing habit after a recumbent period upon the surface; *Platyodon* burrows in stiff mud by rocking the shell valves sideways to abrade the wall of the burrow; *P. cancelatus*, a Californian mysid, continues throughout life to bore efficiently in hard clay. Mya with vestigial foot, becomes finally stationary. The myacea show a certain adaptive radiation. *Sphenia*, like sexicave (in the nearby Sexicavacea), has short siphons and nestlesin shallow crevices. It is also becoming mytillised or hereromyarian by abyssal attachment.

The main habit of the myacea; that of deep permanent burrowing, has been passed on to the super family *adesmacea*. This consist of the pholadiadae or pidetochs (boring in soft clay) and 2 families of wood—borers, the less modified xylophaginidae and the very specialised shipworms, Teredinidae. The ligament is much reduced in the adesmacea and the shell valves rock upon their hinge pits. In the transverse plane.

The foot is used as an attachment disc whilst rotary boring movements are made evolution cuminates in the extraordinary modification of the genus *Teredo.* The naked siphon dominates the whole morphology of the animal the visceral mass is small and anterior, the shell being used only as on abrading tool.

A very different habit of burrowing, with rapid vertical mobility in shifting substrata, is found in the solenacea, or razir shells. Here the shell is long right and razor shaped. With the ambones and hinge at the anterior end, and the posterior territory of the shell enormously elongated. By contrast the siphons themselves are short. The foot is a long compressed plug that can be thrust downwards from the gape at the anterior end of the shell.

(IV) The Last Lamellibranchs

Order—Anomalodesmata and Septibranchia

These bivalves are apparently the latest evolved of the lamillibranch series. The relationships and the mode of life of the anomalodesmata are still imperfectly understood and the group may be found to be polyphytetic: its members agree, however, in being all hermophrodite and in having the outer demibranch reduced and upturned dorsally from the frequent weakness or loss of the hinge teeth and the fusion of the mantle margins, they may be suspected of having part—abandoned a former deep burrowing habit. As in allodia among the myacea—the two shell valves may become unequal. Some species may lie recumbent on one side. In the pandoridae and myochamide the valves show the assymetrical pleurobranch condition and myodora may burrow shallow by inserting the shell horizontally or obliquely into the sand. The fragile valves of the burrowing family laternulidae (*Cochloclesma, Latemuta* and *Periploma*) are also unequal, the right being more convex. British cochlodesma praetenue embeds in send to a depth of 3 inches and lies horizontally on one or other side. Its siphons are long and separate, making mucous tubes as in throcidiae; the inhalant one alone reaches the surface, the exhalent opening into a blind horizontal gallery. The lyonsiidae shows a progression towards sessile life: lyonsice lives freely at the surface, entodes may attaches by its byssus crevices and the American mytilimerinestles deeply in ascidian tests, becoming spherical and hetromyarian. The reciidae have returned to deep burrowing and line their siphonal tubes with mucous secretion. The Australian chamosteridae have-like the chamidae become family cemented to the substrate by the deep right valve, the left forming a lid.

A stage offshoot is seen in the family clacagellidae, the coatering pot shells, surely the least recognisable of all bivalves. As in Teredo the conjointed siphons are much the largest part of the cenimal and secrete a strong calcareous tube. *Clavagella*, which may reach 3 feet in length and 2 inches in diameters is embedded but cannot actively burrow. The valves are functionless and excessively small, the left one in *Clavegella* and *Brechiles* both being fused to one side of the tube.

For the mytilides; collcanchida for the oysters; Spteroconhida for the rest)

All four are usefully embraced in the subclass pteriomiophia. In all but the earliest of this series the primitive symmetry is greatly modified. The most ancient superfamily is the *Arcacea (noah's ak shells and their relatives),* which may be placed alone in the order taxodanta. These have taxodant dentition, A long row of uniform teeth as in *nucula*. The two adductor muscles are equal and the anterior and posterior halves of the body approximately equally developed as in the British *Glycymeris*, which has the round shape of a venus shell *Glycymeris and Iomopsis* burrow shallowly but *Arca* rest at the surface with the dorsal side uppermost, attached by a byssles springing form the whole edge of the foot in the remaining families (order Anisomyaria, if we many these include that mytilidae),

The sessile habit has caused profound changes with the mussels (mytillidae) the byssus and foot have moved to the anterior mussels to a small size (the heteromyarian condition) In *Modiolus and Mytilus* the animal is still attached upright to the substrate, and especially in mytilus-the anterior end is small and pointed the posterior end broad and rounded, with a large posterior adductor. The mytillidae show considerable adaptive radiation. The genus Modiolaria contain small nut shaped bivalves, embedded in a nest of their own byssus threads or in ascidian test.

Two genera have become narrow and elongate, burrowing in the rocks. *Botula* is attached by byssus threads within its burrow in soft of the ligament. Lithophega, the date mussel, bores by rotating the shell in calcareous rocks: the fused inner lobes of the mantle are glandular and appear to secretory. In more advanced Anisomyaria, the shell usually lies upon its right side, becoming flattened, with the two valve no longer quite alike. The pearly oysters pteriidae (Super family pteriacea), have a short byssues emerging by a deep notch in the right (lower) valve. The anterior adductor muscle has been lost, the foot is very small and functionless, and the whole anterior half of the animal is now of minor dimensions. The single posterior adductor muscle lies at the center of the valves in the monomyarian condition. The mantle cavity head widely open and water enters around 2/3rd or more of the rounded margins of the shell. The pearly oyster itself (*Pinctada*) has almost circular valves, but with a long straight hinge. Later genera, such as the wing shells (*Pteria* or *Avicula*) show a strong tendency to elongate along the hinge line. In the vulsellidae-bye the same emphasis of the hinge at exhalant point is carried posteriorly on a long salient, until in the *hammeroysten* (Malleus) the shell finally becomes T-shaped or hammer-like.

Like the pearl oysters, the Pectinacea or scallops lie upon the right side. Both valves may be concave and similar or the left one may be flat. Here however, the valves have again become symmetrical about hinge and beak. In the viscera distortion resulting from anterior fixation by the byssus remains but the shell mantle have been emancipated from the rest of the body, and unlike the pteriacea have a new superficial symmetry. An early representative such as *Chlamys varians* still shows a slight asymmetry with a byssus as in *Pteria,* passing through a notch in the right or lower valve, and the squared 'slugs' at either side of the hinge unequally developed. The exhalant current is expelled at -these points, immediately to the sides of the narrow hinge, incoming water can thus enter right round the rest of the circumference and flow straight to the hinge side.

ECONOMIC IMPORTANCE

Some mollusca are indirectly harmful to man but most of them are beneficial. The harmful molluscs are slugs and shipworms. Slugs are injurious in gardens and cultivation. They not only eat the leaves but also destroyed plants by cutting up their roots and stems. *Teredo*, the shipworm burrows into wooden structures immersed in sea, an caused serious damage to piers and ships. But molluscs are a great source of human food in various parts of world, Millions of mounds of clams, oysters, scallops and mussel are eaten in China, Japan, Malaya, Europe and America oysters being regard as a elicacy. Shells of fresh water mussels are used in the pearl button industry in all parts of the world, they are made from nacrerous layers of shell no other material stands laundering as these buttons. Shells of oysters are mixed with tar for matching roads in America and time from of their egg shell, lime is also used in building. In many parts of the world molluscan shell are used for making ornaments and jewellery in some parts shells of cyprea (cowrie) are used a money and as ornaments. Many fresh water clams and marine oysters produce parts, but the most vaiuable natural pearls are produced by pearl oysters. *Pinctada, margaritifera* and *Pictada mertensi* which inhabit the warmer parts of Indian and pacific oceans along the coasts of China, India, Ceylon and Japan. A pearl is made when a small foreign object, such as a particle of sand or parasite, lodges between the shell and the mantle, in this manner a pearl is formed. But pearls are also produced by most pelecypods including fresh water bivalves clams. In Japan pearl culture is practiced by artificially introducing a small solid or liquid irritant below the mantle of the oyster, the resultant one year old pearl is then transplanted to another oyster a pearl of good size is obtained in three years after transplantation in this manner a pearl is formed. But designed pearls are also produced by most pelecypods including fresh water clams Bhunedmias in India even Pantnagar Uttranchal by author. In Japan pearl culture is practiced by artificially introducing a small solid or liquid irritant below the mantle of the oyster, the resultant one year old pearl is then transplanted to another oyster a pearl of good size is obtained in three years after transplantation.

REFERENCES

1. Hyman Henrietia Libbie [1967], The Invertebrates Mollusca-I [vol. VI], McGraw Hill Book Company.
2. Jordan L.E. [1967], Invertebrate Zoology, S. Chand and Co.
3. Morton E.J. [1967], Mollusc, Hutchinson and Co.
4. Gardines Mary. S [1968], Biology of Invertebrates. Hindustan Publishers.
5. Rakshpal. R [1970],. Invertebrates.
6 Bigelo. A.S. [1968]. classification and phylogeny. Pergamon press.
7. Zurich G.A. [1970]. Encyclopedia [Animal Life]. Reinhold Co.
8. Chandrasekaran S.V. [1989], Fishery Bulletin, Seatle, Washington.
9. Kotapal R.L. [1994-94]. Text Book of Mollusca, Rastogi Publication [Meerut].
10. Snodgrass E.R. [1965]. A Text Book of arthropod Anatomy, Hafner Publishing Company.
11. Prasad S.N. [1955], A Text Book of Invertebitic Zoology, Allahabad.

3

CEPHALOPODA

INTRODUCTION

There is relatively small number of living cephalopod species 650 are currently recognized. Nearly 1000 have been described but the number has been subsequently reduced. This may be compared with approximately 10500 fossil forms and probably over 100000 living species of molluscs in total. Cephalopods are a class of Mollusca with the exception of those belonging to genus *Nautilus*, all of the living species are included in the subclass Coleoidea.

Although, cephalopods are soft-bodied invertebrates, their large size, extreme mobility and the complexity and versatility of the behaviour of those most commonly encountered, combine to give a characteristic impression of activity and vigour. They range in size from tiny sepioid *Idiosepius*, which reaches an adult length of 15 mm, to the giant squid *Architeuthis*. They can certainly attain an overall length of 16 m and weight over 500 kg. Cephalopods are present in all oceans, particularly over regions of the continental shelf. They are benthic and others are pelagic. Cephalopods occur in surface waters but many species live in the bathy-pelagic zone at depths ranging down to several thousand meters.

Originating in the Cambrian period, very early in the fossil record, the early evolution of cephalopods was as occupants of heavy coiled shells broadly classified into the Nautiloidea and Ammonioidea. Of the many thousands of species contained in these two subclasses, only four accepted species are living today in the single genus *Nautilus*. Although contemporary cephalopods in the sense that they are extinct, they have apparently changed little since the nautiloids were flourishing and thus provide a fascinating insight into the probable life of the fossil forms.

Class—CEPHALOPODA

All marine molluscs; bilateral symmetry; well-developed coelom; primitively a chambered, external shell, reduced in modern forms. Radula enclosed within chitinous mandibles; ring of prehensile appendages around the mouth; 1 or 2 pairs of gills; water circulation in mantle cavity expelled through ventral funnel tube for jet propulsion. Centralized nervous system; highly organized sense organs; complex behaviour. Sexes separate; fertilization internal; sperm contained in complex spermatophores; eggs large and yolky; direct development.

Subclass—Nautiloidea

Cephalopods with heavy, external chambered shells which may be straight or coiled. Appeared in the Cambrian period and became very numerous in species and individual throughout warm seas. All now extinct except for one family, the Nautilidae, with restricted distribution in the Indo-pacific Oceans. Typically for *ectocochleate* cephalopods the body occupies the terminal chamber of the shell which also acts as a buoyancy organ. Retraction of body into shell displaces water from mantle cavity for jet propulsion. Two pairs of gills and nephridia; head with numerous unsuckered appendages. *Nautilus* (fam. Nautilidae)

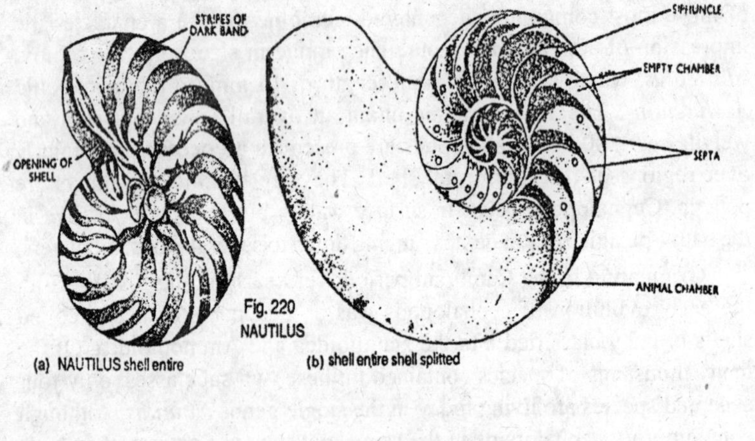

Fig. 3.1: *Nautilus*

changes in buoyancy and the presence of both liquid and gas in the camerae, thus suggesting an active control comparable to that found in *Sepia.* By inference, these observations muse raise the possibility of similar mechanisms being present in extinct shelled cephalopods. But even without active buoyancy control, the camerae must have provided an effective flotation uevice, and jet locomotion was probably soon achieved. [In the Ordovician *Orthonybyoceras* we can see in the rocks the traces made by an orthoconic nautiloid propelled by funnel jets and can find evidence of its proficiency at steering by side deflections of the funnel (Flower, 1955).]

Nautiloids and Ammonoids

The first nauriloids had short cap-like external shells with a slight endogastric coil but other forms appeared early in the history of the group, some with straight shells, and some coiled with the gradient of incremental growth greatest posteriorly, producing an exogastric shell as in Nautilus. Bilateral symmetry is disturbed by skewed coiling only in a few later forms, and the mantle cavity is never displaced by torsion. The coiling of the camerate gas-filled shell has evolved on lines very duTerenr from the heavy viscera-filled gastropod coil.

A wide range of shell form is found among Palaeozoic nautiloids, and it is presumed that each represents a definite relationship between the buoyant properties of the shell and the postural requirements of the animal. In a tightly coiled shell like that of *Nautilus* (Fig. 3.1), the relationship between the center of gravity and the center of buoyancy determines thac the body chamber lies almost horizontally below the gas-filled chambers, but in some of the more loosely coiled cyrtocones the aperture probably faced obliquely downward. Several groups of Palaeozoic nautiloids independently developed long straight shells, and it appears that in some of these the apical end was counterbalanced by siphuncular or cameral deposits which enabled those animals to adopt a horizontal posture. Perhaps the most remarkable shell form is that of *Ascoceras*, in which the orthoconic juvenile stage was shed from the inflated flask-like adult shell. Within this shell the septa curved obliquely forward and so the gas-filled camerae lay in a saddle-like bulge above the horizontal body chamber.

Many different groups of nautiloids and ammonoids have abandoned the coiled shape and fast swimming existence to return to a neobenthonic habit at the bottom, or to take on a life of passive floating. In many of the short, vertically floating brevicones of Ordovician nautiloids

(*Mandaloceras, Trimeroceras*) the apertural margins became so inflected as to allow only restricted openings for funnel, eyes, tentacles, and mouth, a facies from which one might suspect a microphagous plankton-feeding mode of life. Some genera of Mesozoic ammonoids, such as *Macroscaphites* and *Hamites*, abandoned the coiling of the vounger shell to give a long, straight, or ultimately recurved terminal chamber. Such forms may have swum slowly or remained hung suspended with the mouth raised just above the bottom in a highly advantageous posture for leisured foraging, but planktonic feeding seems more likely for the recurved forms in which the aperture faced upward. Certain ammonoids (such as Turrilites) and nautiloids (Trochoceras) developed a skewed spiral with a coiled turbinate shell from which inust be interred a reversion to a crawling benthonic habit.

With the pallial cavity still enclosed within the terminal chamber of an external shell, none of the extinct ammonoids and nautiloids could have developed the efficient contractile mechanism of the pallial musculature seen in the modern naked cephalopods, where the water jet is expelled powerfully through the tubular funnel. If we may judge by modern *Nautilus*, jet expulsion must have been effected by the contraction of the funnel musculature itself, which here forms not a closed tube but an approximation of two fleshy lateral lobes in the central mid-line (Fig. 3.1 Fu).

Modern Cephalopoda

The living Cephalopods, with the exception of *Nautilus*, belong to the subclass Colcoidea (Fig. 3.2). In all of them the shell (P) is reduced and invested by the fleshy integument, being altogether lost in the living Octopoda.. The strong pallial musculature is now set free for the contractians that provide the jet mechanism, beautifully coordinated, especially in the fast-swimming squids, with a system of giant neurones that secures both speed and synchronization of the passage of the nerve impulses in pailial contraction. Here, with the loss of the chambered shell, speed of movement can overcome gravity and the dominating evolutionary trends have been concerned less with buoyancy than with greater emcienced of pallial contraction, fast swimming, and the development of enhanced sensory and cerebral powers. A squid such as *Loligo* displays the greatest emancipation from benthic life and from the constraint of the external shell. Every line of the body is beautifully subservient to speed and maneuverability. *Loligo* is circular in cross section, pointed at the end remote from the head like a torpedo, where there are also a pair of triangular stabilizing fins that can be shut down in fast motion. The shell is fully internal, lying just beneath what has become the upper surface in swimming (the former anterior aspect). It is a light and chitmized "pen" (P). which points the body like a rigid arrow, and becomes expanded laterally into a flexible vane giving some support to the pallial musculature. No chambered portion survives or is even indicated iin development. The cephalic end of the body, which

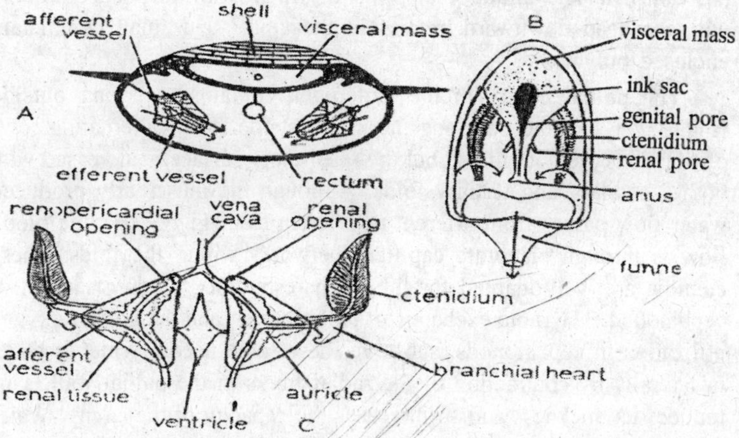

A. Sepia, cross section of mantle cavity

B. Sepia, mantle cavity in ventral view

C Heart, Ctenidia and renal organs of a coleoid cephalopod

Fig. 3.2: Diagrams showing essential features of medars acladid Cephalopoda.

in slower swimming with the funnel (FU) turned back may be directed foremost is also streamlined with the eight short arms completing the smooth, wedge-shaped contour. The two longer, tencaculate arms are kept ensheathed, being extruded with a lightning movement, mediated by giant neurones, to take active prey. The tubular funnel is freely movable, the animal moving in the opposite direction to that in which water is expelled.

The modern sepioid shell or "cuttlebonc" (Fig. 3.2C) is much more primitively complete than the reduced pen of the squids (Fig. 3.2E). It displays the persistent outer or uppermost side of the chambered shell of extinct forms, now broad and flat with the septa very numerous and closely crowded together. Its smooth inner surface represents tlie siphuncle so widely opened out that the lower wall of the chambered shell has vanished. The unique small *Spirula*, which swims or floats obliquely with the head downward, retains at the apical and small comnletely enclosed ophiocone.

The pallial cavity in the Coleoidea contains two long outside (Fig. 3.2, CT) attached along the afferent side. Their alternating rows of filaments are not ciliated but the respiratory surface is increased with both secondary and tertiary folds. Although the muscularly produced water flow passes from afferent to efferent side (Fig. 3.2B), the blood flow is through elaborate capillary networks within the folds. These ctenidia are well adapted for the, high respiratory needs of an active cephalopod. The rapid exchange of pallial water, and the greater oxygen utilization in cephalopods, subserve these same needs. Water is taken in laterally by contraction of the radial fibers in the pallial wall. This reduces its thickness and so increases the capacity of the cavity. Water is expelled through the funnel in the mid-ventral line by contraction of the circular muscles of the mantle. The sides of the funnel are produced into a "collar".

Subclass Ammonoidea

Chambered external shells, usually coiled and with complex septa and sutures separating the chambers. Very numerous form the Devonian to the Cretaceous periods but now all extinct.

Subclass—Coleoidea

Modern forms, Devonian period to present. Shell internal and reduced. Muscular fins and muscular mantle forming a sac enclosing the viscera. one pair of gills and kidneys; large mantle cavity; ink sac typically present; skin containing pigment organs, chromatophores, variably expanded by neuro-muscular control.

Order—BELEMNOIDEA

Internal shell, straight and with a solid posterior portion; commonly fossilized; all extinct.

Order—SEPIOIDEA

Calcareous chambered shell present internally and functioning as buoyancy organ in some; shell greatly reduced to a purely organic pen in others. 8 suckered arms, 2 long tentacles with suckered club; suckers pedunculate with horny rims. Includes cuttlefish and sepioids.

Sepia (fam. Sepiidae)

Sepiola (fam. Sepiolidae)

Euprymna (fam. Sepiolidae)

Sepietta (fam. Sepiolidae)

Order—Sepioldea

e.g., Cuttleflsh, Sepiolids.

The common Cuttleflsh, *Sepia officinalis* is one of the best known Cephalopods of old world, The species is abundant In the eastern Atlantic and in the mediterranean Sea. *Sepia officinalis* occurs in coastal waters and on continental shelf at depth not greater than 150 m. The status of the subspecies *S. officinalis hierredda* of the West African coast is uncertain. Further to the South, around Africa, South and East Asia and Australia more than one hundred species of *Sepia* are known. No SEPIA occur around North and South America (Voss 1974).

Sepiola robusta-bottle tailed squid.

Euprurona scolopes-Berry

Behaviour is noctural and cryptic, most observations were based on animals reared or maintained in lab.

Order—Teuthoidea

e.g., Squid.

The market squid *Lolego opalescens* is small squid that attains a maximum reported mantle length of 230 mm. It is distributed along the west coast of North America from the middle of Barrage California northward to South-eastern Alaska. This species has been fished commercially since at least 1863 in the vicinity of Monetary, California and today it is widely believed to be an under-utilized fisheries resources in north-eastern pacific.

Order—TEUTHOIDEA

Shell reduced to chitinous 'pen' (gladius) lying dorsally. Elongate body usually finned. Eight suckered arms, two long tentacles with suckered club; suckers pedunculate and with homy rims, some with hooks. Squids.

Suborder—MYOPSIDA

Loligo (fam. Loliginidae)

Loligo Pealei Commosgrud

Suborder—OEGOPSIDA, a large assemblage with many families.

Gonatus (fam. Gonatidae)

Fabricii—Most abundant squid of artic and sub artic water of North Atlantic

Illex (fam. Ommastrephidae)

Dosidicus (fam. Ommastrephidae)

Gigae Peru—Chilean quant squid measure to 115-120cm in mantle bag it is endemic.

Teuthowenia (fam. Cranchiidae)

Order—OCTOPODA

Internal shell drastically reduced and split into two lateral rods. Eigh arms only with non-pedunculate suckers. Globular body with or without fins. Octopuses.

Suborder—CIRRATA, deepwater forms with fins.

Suborder—INCIRRATA, large group with several pelagic families, without cirri and fins. The common octopods dealt with in all belong to the only benthic family.

Octopus (fam. Octopodidae)

Eledone (fam. Octopodidae)

Bathypolypus (fam. Octopodidae)

Order—VAMPYROMORPHA

Eight long arms united by a swimming web, two small tendril-like arms in dorso-lateral position.

Octopus briareus

The Atlantic reef octopus. Octopus briareus is a common shallow-water species inhabiting coral reefs, rocks and seagrass beds in the tropical West Atlantic from the east coast of Florida, throughout the southern Gulf of Mexico, the bahama Islands and the Caribbean Sea to the northern coast of South America. *Octopus briareus* is benthic

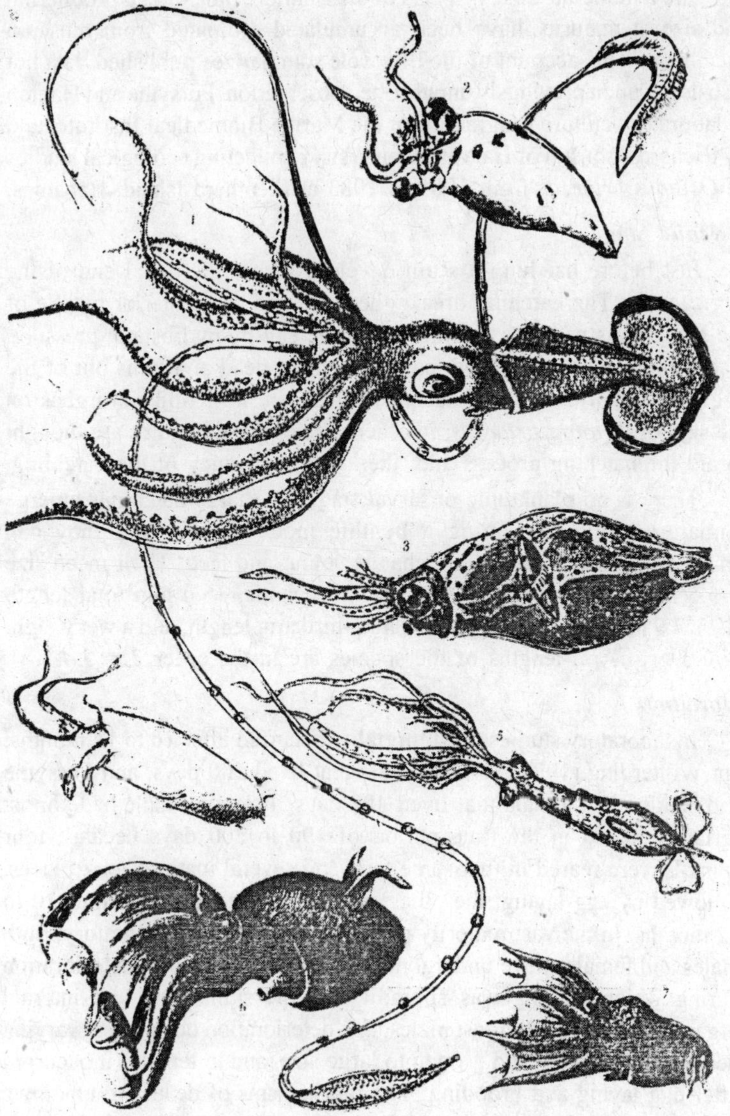

A few Cephalopods.

Fig. 3.3

throughout its life cycle, it produces very large eggs and hatching and grows to a moderate size. Aspects of distribution, time of egg production and size at maturity have been accumulated estimated from museum specimens. This account of the life cycle summarizes published data but also draws upon previously unpublished observation. Forsythe and Hanlon in laboratory culture programme at the Marine Biomedical Institute, and by Richard Aronson of Harvard University conducting ecological studies of *Octopus briareus* from 1981 to 1983 at Eleuthera Island, Bahamas.

Juvenile Stage

Just before hatching a seam develops across the distal end of the egg capsule. The hatching breaks the seam and the posterior mantle of the hatching is partially ejected out of the egg by hydrostatic pressure. The octopus pushes first the mantle, then the head and arms out of the egg case usually within 15 seconds. There are no Kollikers organs on the skin of *Octopus briareus*; in other species these bundles are thought to aid the hatching process and, thereafter, buoyancy of the hatchling.

There is no planktonic or larval stage. The large hatchings emerge miniature adults, assuming a benthic mode of life, and they can immediately crawl, swim, ink, change colour and feed. Their mean size is 5.5 mm mantle length (ML), 5.0 mm mantle, 15.0 mm total length (TL), 7.9 mm first arm length, 9.0 mm third arm length, and a wet weight of 0.095 g. Arm lengths of the species are in the order *2 = 3.4.1*.

Mortality

In laboratory studies most mortality occurred after 10 to 12 months, but Wolterding (1971) had one male that lived 500 days, and Forsythe and Hanlon had a male that lived 492 days. The latter male had almost certainly mated in the time period of 190 to 200 days because four animals were reared in that same tank and several matings took places, followed by egg laying; the other animals in that tank died after 10 to 12 months. In the vast majority of natural deaths in the laboratory, both males and females have undergone a 2 to 4 week period of deterioration during which feeding was sporadic and the skin, arms and internal organs degenerated. In most males, this deterioration occurred at varying periods after mating and growth to large size, and in females it occurred after egg laying and brooding. The mechanisms of death are unknown but were probably linked to the hormone system regulating sexual maturation. In summary, natural death seems to be closely related to reproduction.

1. *Octopus briareus* has a distinctive foraging method in which it moves over considerable distances making 'speculative attacks' on small clumps.
2. *Octopus joubini*—Pygmy Octopus :

   It is small benthic Octopus and low distributed from Georgia on the east coast of North America. It is lytic behaviour and strongly nocturai habits.
3. *Octopus maya*—Eggs are largest known for any species.
4. *Octopus vulgaria*—Common Octopus. It Is undoubtedly the best known of more than 100 species actually described.
5. *Eldone cirrhosa*—A medium sized, benthic Octopod. It Is widely distributed over the shelf regions of the northeastern Atlantic and Mediferremean sea.

A stomodaem and the end of rectum from a protodaeum. Ectodennal fields corresponding to the ganglia gives rise to the nervous system. Development is so greatly modified that relationships to other molluscs cannot be determined by embryonic stages coral and likely food spots every few meters. These attacks are not visually directed at specified prey. The *octopus* swims or walks over the spot, spreads its arms and web radially in parachute fashion, and completely envelopes the site. It then uses the arms to root out small crabs and fish that are either grasped or caught in the arms interbranchial web as they flee.

Egg laying occurs from January through April. Egg development requires 50-80 days at ambient temperatures of 19-25°C, so that hatchlings emerge between late February and early June. The juveniles grow through the springs and summer when temperatures are high and food is radially available. Since males and females can mature is as little as $4^{1/2}$ months, *Octopus* reach maturity between July and October. As temperatures drop in the autumn, mating takes place from about October. As temperatures drop in the autumn, mating takes places from about October through April. Some females store the sperm (up to at least 100 days) and others lay eggs immediately. At the conclusion of egg brooding between January and April, females die within several weeks. During winter and spring, males reach old age and die also.

Nautilus macromphalus

Nautilus, the last extant genus of externally shelled cephalopods, is among the most poorly known of living cephalopods. Because of its deep and geographically isolated habitats, investigation into the life history of Nautilus has proven costly and difficult. Most information about the four accepted a living species comes from aquarium observations and studies.

Nautilus pompilius Linnaeus is the most widespread species, occurring at scattered islands and continental fringes across a wide region of the Indo-Pacific,. *Nautilus macromphalus* Sowerby, which differs from *N. Pompilius* in shell characteristics (but appears to be identical anatomically), occurs only around New Caledonia and the Loyalty Islands. By historical accident, however, far more studies have been conducted on *N. macromphalus* than on the far more widely ranging *N. pomilius*. Because of its unique external shell, most investigation of *Nautilus macromphalus* has centered on aspects of buoyancy regulation and maintenance.

Egg Stage

No *Nautilus eggs* have ever been observed in nature. Numerous eggs have been laid in aquarium settings, but it is unknown if they are of normal appearance, for none have ever turned out to be fertile. Eggs are laid singly, and attached to a hard substrate, generally in a concealed area.

Nautilus macromphalus eggs have been described in detail by Willey, and more recently by Mikami and Okutani. The eggs are always enclosed in a milky-white capsule, which is flexible when wet, but brittle and easily breakable when dried. The egg cases have a double wall, with the inner being oval in shape and the outer layer having a pelated appearance. The eggs of nautilus are notable because of their great size. The normal eggs were much larger than the others, with a mean length of 36.8 mm, diameter 24.8 mm, and weight of 3.7g.

The infertility of eggs laid in various aquaria, coupled with the lack of observation and collection of eggs in nature, continuous to stifle any sort of embryological understanding of this important group of cephalopods. However, later maintenance of *Nautilus macromphalus* at 16°C.

Juvenile Stage

The hatching of a nautilus has never been observed because of the very large egg size it has been postulated that the juvenile emerges at a very large size. Until hatching is observed. Information about this important question must be based on other evidence. The two most commonly used have been:

(i) changes in shell ornament, and

(ii) changes in the isotopic composition of the early shell.

A nautilus egg could hold a shell of about 25 mm across. At about this same diameter all nautilus shells show a pronounced growth discontinuity, termed the nepionic construction, which separates two very different patterns of shell micro-ornament. Internal to the construction, the outer shell in all species is ornamented with a canella shell sculpture of fine transverse growth lines and finer lira running perpendicular to the growth lines the colour of the shell is generally yellowish-brown in appearance Post-constriction shell showing the first presence of an ocular sinus and a much deeper hyponomic sinus that on the pre-constriction shell the shell colour also changes becoming darker Perhaps most importantly with interpretation of hatching size post-constriction shell always bears numerous shell-breakage scars. Generally these shell breaks are shallow concave indentations (although larger scars can be present) suggesting that small portion of the shell were broken off with later shell growth continuing from the point of the break. The shell breaks, which are common on all post-nepionic constriction stages, are absent or vary rare on the pre-construction shell. Since shell breaks could occur only after hatching, the hatching *Nautilus* would be about 23-25 mm in diameter, and have up to seven septa already completed, according to this interpretation.

Growth

At the time of writing there is no information available about natural growth rates of any species of *Mautilus*. However, there is now a great deal of information about growth rates of *Nautilus macromphalus* in aquarium captivity. Growth studies of *Nautilus macromphalus* have been conducted at the Aquarium of Noumea. New Caledonoa, and at Yomuriland Aquarium, followed by Hamada *et al.,* Kanie *et al.,* and ward *et al.,* The specimens maintained by Martin *et al.,* head the advantage of being freshly captured at the start of the observation, but suffered from fluctuating water temperatures that are now known to be far higher than those naturally encountered by the *Nautilus*. In the other studies, water temperatures were maintained between 15 and 20°C, which appear to be similar to water temperatures at the depths where *Nautilus macromphalus* can be found. All specimens in these studies were in the last phase of growth, having already between 23 and 30 septa. There is no information about growth rates in specimens smaller than this.

The growth experiments of Martin *et al.*, were conducted at temperatures between 22 and 29°C which are much higher than the normal temperature experienced by *Nautilus macromphalus*. The remarkable correlation between growth and cameral liquid removal has recently been demonstrated in *Nautilus macromphalus*. New chambers always begin to from when the liquid in the last formed chamber has been half emptied. By the time the new chamber is ready to begin emptying its own liquid, the liquid volume in the preceding chamber has been reduced to 30% or less of its initial volume. The chamber formation process takes between two and three months (between any two identical stages). This estimate agrees quite closely with the Martin *et al.*, (1978) estimates based on aperture shell growth and long-term weight increase.

Maturation

It appears that all chambered cephalopods reach a maximum size, at which time the organism is termed mature. In *Nautilus*, Maturity is marked by the following shell features: thickening of the aperture edge, a black band at the edge of the aperture, depending of the ocular sinuses, increased thickening of the last two septa, and septal approximation.

Maturation of the sexual organs occurs after the formation of the final septum, but prior to cessation of aperture shell growth that marks full maturity. In both males and females of *Nautilus macromphalus*. The reproductive organs (testes, spadix, and ovaries) show a marked increases in size prior to full shell growth. Measured as proportion of total soft-part volume, the sex organs occupy less than 1% of the animal's volume throughout growth. Immediately after formation of the last septum, the sex organs grow rapidly, until they occupy between 5 and 7% of the tissues volume. This rapid growth occurs over a two to there month period. The spadix in the males also rapidly enlarges at the time. The air weights of the immature testes are about 1g in the matures, they are between 20 and 35g in females, the ovaries enlarge from about 1 g to between 15 and 25 g the enlargement of the sex organs coincides with a widening of the body chamber to make room for the larger organ system. My observation shows males to have functional spermatophores immediately before the attainment of the full shell growth.

Reproduction

Copulation behaviour is apparently quite similar in *Nautilius pompilius* and *N. macromphalus*. Copulation can take as long as 24 hours. *Nautilus* produces spermatophores, which can remain viable for long periods of time. The spermatophores are produced by an organ called the accessory gland, and stored in the spermatophore sac, which communicates with the penis. The sperm themselves are morphologically distinct from other cephalopod sperm in a number of details. There is no information on how the ultimately transferred to the female during copulation. When next seen, attached to the ventral surfaces of the female's cephalic sheath, the spermatophore is once again without capsule. There is no information about the subsequent movement of the spermatophore or about the area of fertilization within the female.

Ecology

Most general textbook discussion about *Nautilus* make some references to diurnal ventricle migration. The following is typical: 'it is known the *Nautiuls* spends the hours of daynight on or near the seabeds at depths up to 600 m and comes to the surfaces at night.' And has had enormous influence on the thinking about the palaeobiology of extinct-chambered cephalopods.

Nautilus macromphalus: All impressions derive from the number of specimens caught in baited traps, or observed at night be drivers. Aquarium observations suggest that *N. macromphalus* has acute powers of chemoreception. Finding food material on the vast expanse of the fore-reef slope probably involves tracking faint scent traces in the water perhaps from long distances, the large number of *N. macromphalus* that traditionally have been trapped outside the barrier reefs of new Caledonia suggest that large populations may be present. Alternatively, these catches may represent individuals that have been attracted from large areas of the fore-reef slope to the rich scent of the baited traps. This latter conjecture is strengthened by observations about trapping methods traps. This left overnight rarely result in high yields; commercial trapping efforts in new caledinia rely on from four days to a week of bottom-time for good yields. Relatively small numbers of *Nautilus*, scattered across the force-reef slope in perpetual search for food resources that are attracted from perhaps very long distances to the traps. Such populations should be very vulnerable to uncontrolled fishing. Just such a situation appears to be occurring in New Caledonia.

Sepia officinalis

The common cuttlefish, *Sepia officinalis* Linnaeus, is one of the best-known cephalopods of the Old world. The species is abundant in the eastern Atlantic and in the Mediterranean Sea. It is easily identified by its slipper shape and long, narrow fins that from a undulatory margin along each side of the mantle. The largest specimens measure nearly half a meter without the long tentacles. The latter are retracted in special pouches and are ejected only two-size prey. In young individuals the species can be identified from the brown skin colour. Other-features that distinguish these animals from the small-sized, more reddish *Sepia orbignyana* and *Sepia elegans* include the shape of the chalky cuttlebone and the arrangement elegans in the tentacles club. From the North Sea and English channel southwards.

Sepia officinalis occurs in coastal waters and on the continentals shelf at depths not greater than 150 m. Thus status of the subspecies *S. officinalis* Hierredda Range. Further to the south, around Africa, South and East Asia, and Australia, more than one hundred species of Sepia are known. No Sepia occur around North and South America (Voss, 1974).

Egg Stage

Septa officinalis generally lays its eggs in shallow water. At depths rarely greater than 30 or 40 m. each egg is fixed to any oblong object, not more than one cm in diameter, by means of a ring-shaped basal structure of the envelop.

Juvenile Stage

The hatching of *Sepia officinalis* has a mantle length (ML) that may vary from 6 to 9 mm, depending primarily on the ovum size. The cuttlebone normally contains 8 or 9 competed 'chambers'. The total length of the animal is c. 1.7 times ML, the head-width c. 0.8 times ML. These body proportions are typically juvenile the head and arm crown being much larger relative to the mantle complex than in the adult where total length is c.1.4 times ML, head width c. 0.45 times ML, Nevertheless, the overall aspect of a Sepia hatching is strikingly similar to the adult. These young animals already produce different colour patterns that from the organizational basis of the more diversified adult patterns.

Reproduction

The most spectacular part is the elaborate courtship behaviour of male *Sepia* and the corresponding colour patterns. Both sexes show a typical Zebra pattern that is particularly conspicuous in the courting male. When approaching a potential partner for copulation, the male extends one ventral arm towards the courted individual. If the latter happens to be another male, it will signal so by stretching out the ventral arm facing the approaching male; if it is a female not willing to copulate, it will swim away, whereas a female willing to copulate will stay. At the start of copulation the male grabs the female willing to copulate will stay. At the start of copulation the male grabs the female from the side and manoeuvres her into a head-to-head position, holding her with dorsal and lateral arms intertwined. The male then reaches with his left ventral and lateral arm, the hectocotylus into his own mantle cavity to seize spermatophores extruded from the Needhams sac through the penis. The proximal part of the hectocotylus is modified in that normal-sized suckers are replaced by series of minute suckers scattered over a large area of transverse folds. The spematophores are transferred to a special pouch under the buccal mass of the female.

REFERENCES

1. R.D. Purchon, (1968), The biology of mollusca, Pergamon press, London.
2. Mary S. Gardiner, (1968), The biology of Invertebrates McGraw Hill, book company.
3. Jerome J. wolken, (1971), Biology of Intertidal animal Logos Press London.

4

DIGESTIVE SYSTEM

DIGESTION

Alimentary Systems

Food is processed to various degrees among different gastropods as it passes from the buccal cavity, along the oesophagus, into the stomach. In micro-algal grazers, food is lubricated with mucus secreted by the oesophageal epithelium and by the salivary glands, which may also secrete analyses as in *Littorina littorea.* Ciliated tracts carry the food-laden mucus to the stomach where carbohydrate-splitting enzymes, sometimes including cellulose, secreted by the digestive gland, partially break down the food. An area of ciliated ridges in the stomach wall, well developed in all microphagous gastropods, directs small particles to the openings of the digestive gland ducts. At least in some species, hydrostatic pressure generated by muscular contractions of the stomach forces the particles into these ducts. Muscular contraction of the ducts then moves the particles into the fine lamina of the digestive gland where they are phagocytosed and digested intracellularly. In other species, particles may be transported to the digestive gland by ciliary action. Larger, undigested particles are directed by the ciliated folds away from the digestive gland openings, towards the intestine. Waste material excreted by the digestive gland is transported by cilia outwards along the digestive gland ducts into the stomach, to join the rejected particles entering the intestine. Where the stomach merges with the intestine, rejected material and excreta are bound with mucus and rotated by ciliary action into a faecal rod. Passing along the intestine, the faecal rod is fragmented by muscular contraction of the intestinal wall and the pellets are expelled from the rectum. In deposit-feeding, filter-feeding and some grazing mesogastropods (*e.g.,* Rissoacea, Cypraeidae and Strombidae) a rotating crystalline style slowly liberates amylases and serves as a windless, pulling food strings into the stomach. The globular protein nature of the style indicates that no proteases are present in the stomach of these gastropods. Difficult to reconcile with this view, however, is the presence of a crystalline style in certain neogastropods such as *Ilyanassa obsoleta*. The style is secreted and broken down on a daily cycle, but although, *I. obsolete*, ingests large amounts of plant material and detritus that would be digested by amylases from the style, it also readily consumes flesh that requires proteases for digestion. Perhaps the style is present only when *I. obsoleta* is feeding on detritus.

A carnivorous diet is associated with more extensive extracellular digestion, especially the breakdown of proteins. Carnivorous prosobranchs have highly developed salivary and oesophageal glands (Fig. 4.1) whose secretions lubricate and begin to digest the food on its way to the stomach. The predominance of extra cellular digestion is associated with simplification of the stomach, which often lacks any vestige of a ciliary sorting area.

Opisthobranchs show the greatest alimentay diversity. Nudibranchs such as *Triionia* feeding on soft, animal tissues have relatively simple alimentary systems in which the foregut and hindgut are merely conduits for the food and faeces. Salivary glands secrete a lubricating, enzyme-containing fluid that passes with the food into a simple sac-like stomach, which also receives enzymes from the digestive gland. Digested fluid is absorbed across the stomach wall and fine particles are taken up by the digestive gland. *Gymnosomes* also have a simple alimentary system, able to digest the flesh of their *thecosome* prey with remarkable efficiency. Flesh extracted from the prey's shell receives salivary secretions in the buccal cavity and passes directly along a simple, highly dilatable oesophagus to the digestive sac. This is a spacious chamber formed by the digestive gland and stomach, the latter being reduced to a small ciliated area.

Other Opisthobranchs have a more elaborate foregut equipped for temporary storage, mastication or digestion of food prior to its entry into the stomach. This has enabled bullomorphs to exploit large or hard-shelled prey or to become macro-herbivores. Among bullomorphs, the Aglajidae are active predators of polychaetes and of other gastropods, which are swallowed whole. In the case of gastropod prey, the shell passes intact through the gut. The oesophagus is dilated into a crop region (Fig. 4.2C) where most of the digestion occurs. Muscular contraction of the stomach moves enzymes from the digestive gland to the crop. Conversely, muscular contractions of the crop force digested fluid into the small, simplified stomach. Subsequent muscular contraction of the stomach moves the fluid into the digestive gland for absorption, and ciliary tracts carry enzymes and excreta out of the digestive gland

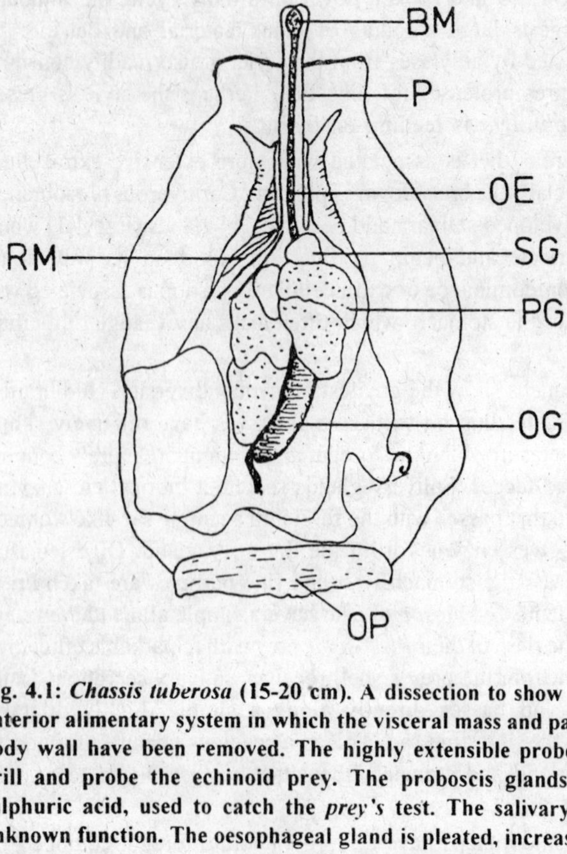

Fig. 4.1: *Chassis tuberosa* (15-20 cm). A dissection to show the plan of the anterior alimentary system in which the visceral mass and part of the dorsal body wall have been removed. The highly extensible proboscis is used to drill and probe the echinoid prey. The proboscis glands secrete 0.1 N sulphuric acid, used to catch the *prey's* test. The salivary glands are of unknown function. The oesophageal gland is pleated, increasing its internal surface area, and is involved with digestion. BM—Buccal mass, OE—oesophagus, OG—oesophageal gland, OP—operculum, P—proboscis, PG—proboscis gland, RM—proboscis retractor muscles. After Hughes and Hughes (1981)

into the stomach. The Philinidae are burrowing carnivorous bullomorphs that cruŝh bivalves with a muscular gizzard lined with calcareous plates (Fig. 4.2B). *Haminoea zeiandiae* is a herbivorous bullomorph that commonly feeds on *Enteromorpha* (Fig. 4.2A). The jaws grasp the algal filament and the radula snatches a short length and passes it to the crop region of the oesophagus, where it is packed ready for entry into the gizzard. Three spiny plates lining the gizzard regulate the influx of algal segments from the crop and grind the food into small fragments. These are conveyed by muscular and ciliary action of the posterior foregut to the stomach.

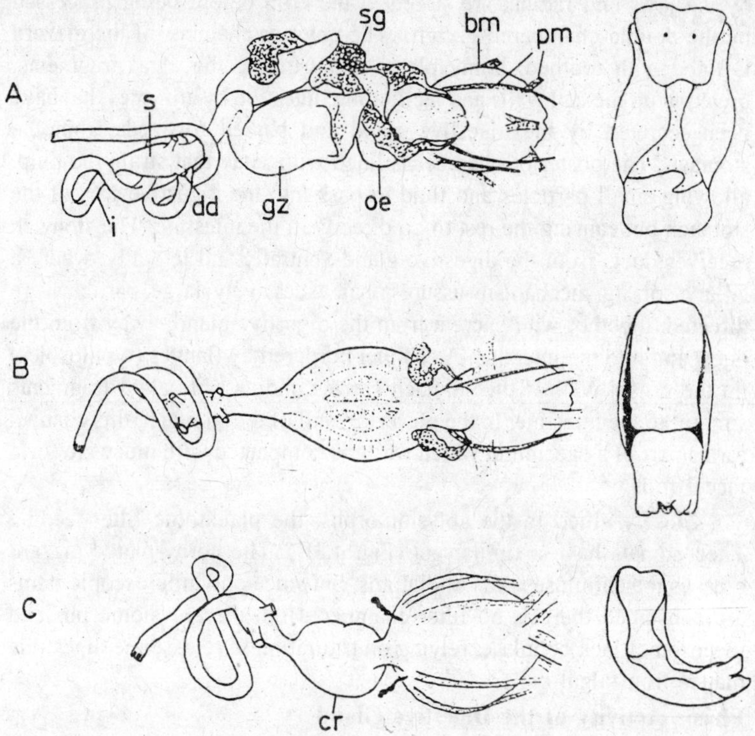

Fig. 4.2 : Bullomorph Alimentary Systems. ***A. Haminoea zelandiae*** **(animal 2 cm). Large salivary glands secrete amylolytic enzymes that begin to digest the algal carbohydrate during passage through the oesophagus. The muscular gizzard contains toothed, horny plates that grind up algal filaments into a chyme that is passed to the stomach for further digestion. After Rudman (1971).** ***B. Philine auriformis*** **(animal 8 cm). Bivalve prey are swallowed by the muscular, bulbous buccal mass, and their shell's are crushed by large calcareous plates lining the gizzard. The salivary glands are relatively small, as is typical of carnivorous gastropods. After Rudman (1972a).** ***C. Melanochlamys cylindrica*** **(animal 2 cm). Polychaete and nemertine prey are sucked in by the bulbous buccal mass and stored in the crop, where they are digested by enzymes passed forward from the stomach. The soft-bodied prey do not need crushing and there is no gizzard. After Rudman (1972b). bm—Buccal mass, cr—crop, dd—digestive gland duct, gz—gizzard, i—intestine, oe—oesophagus, s—stomach, sf—salivary gland, pm—buccal mass protractor muscles.**

Aplysiomorphs display the most elaborate modifications of the alimentary system for macro-herbivory (Fig. 4.3A). Pieces of alga, cut by the jaws and radula, are stored in the crop before being processed by the double-chambered gizzard. The anterior chamber of the gizzard is lined with toothed, homy plates that triturate the algal fragments, breaking up the cell walls and facilitating digestion by enzymes that have been secreted by the digestive gland and passed forwards from the stomach. The posterior chamber is lined with setae that strain the pulp, allowing small particles and fluid to pass into the digestive part of the stomach but causing the rest to go directly to the intestine. The stomach receives ducts from the digestive gland ventrally and laterally, where a ciliary sorting mechanism assures that excessively large particles are directed, together with excreta from the digestive gland, away from the ducts and into the intestine. A channel bordered by flanges (typhlosoles) .in the dorsal wall of the stomach carries undigestible algal fragments straight to the intestine. In the posterior wall of the stomach, this channel forms part of a caecum in which waste is compacted with mucus to form faecal rods.

Closely allied to the aplysiomorphs, the planktonic filter-feeding Thecosomata have a similar gut (Fig. 4.3B). The horny-plated gizzard crushes the cellulose walls of diatoms, but since only microscopic items are consumed there is no filter chamber. Both the aplysiomorphs and thecosomes lack cellulase, relying on trituration to release the digestible matter from algal cells.

Phasic Activity of the Digestive Gland

The digestive gland is usually far larger than the stomach. occupying much of the visceral mass, where it is recognisable as dark-brown tissue. Ciliated ducts connecting with the stomach branch into fine, blind-ending tubules around which are grouped digestive and secretory cells. Both are derived from columnar, basophilic cells near the ends of the digestive gland tubules in *Rissoa parva*. Digestive cells are tall, cylindrical, with a basal nucleus and an 'apical brush' of microvilli, whereas the secretory cells are conical with a wide base and a narrow tip reaching to the lumen (Fig. 4.4). Like the digestive cells, the secretory cells have a basal nucleus and apical microvilli, but they have denser cytoplasm with more extensive endoplasmic reticulum and Golgi systems, associated with protein synthesis.

Phasic activity of the digestive gland is probably widespread among molluscs and has been described in detail for *Littorina* spp. Three phases of activity are recognisable.

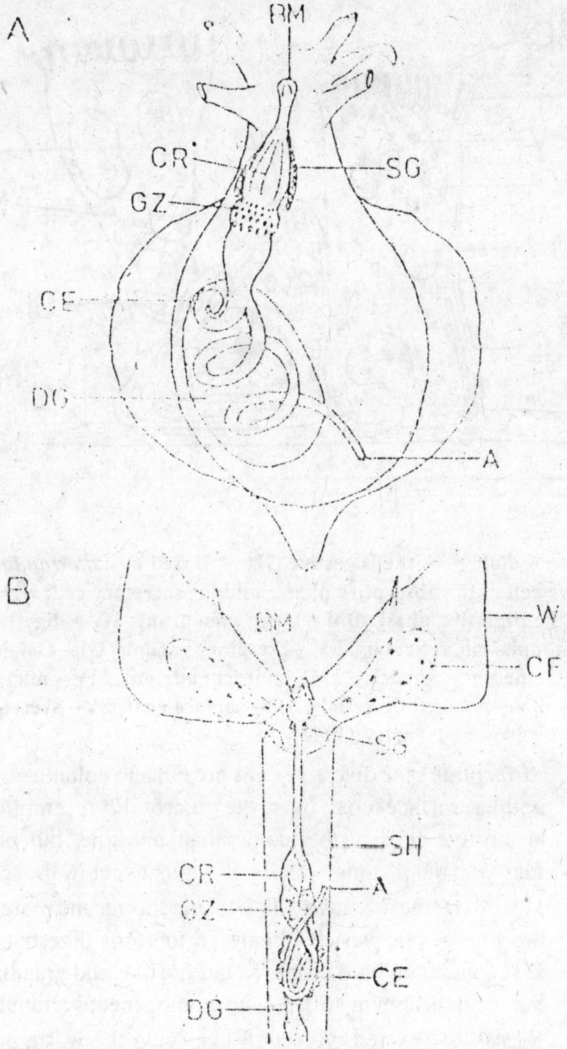

Fig. 4.3: *A. Aplysia* **(10 cm). Algal food accumulates in the crop and is passed to the gizzard for trituration. Fine particles enter the digestive gland and larger particles and digestive waste are compacted into faecal rods in the caecum. After Thompson (1976). B. Thecosome Pteropod** ***Creseis acicula*** **(6 mm across the 'Wings', Posterior Third not Shown). The pteropod swims by beating the wings, which are covered in mucus that traps diatoms and other small particles. Ciliary fields transport the mucus to the mouth. Ingested diatoms are crushed by the gizzard and the fine, semi-digested particles are passed to the digestive gland. Digestive waste is compacted into faecal rods in the caecum. After Yonge (1926). A—Anus, BM—buccal mass, CE—caecum, CF—ciliary field, CR—crop, DG—digestive gland, GZ—gizzard, SG—salivary gland, SH— shell, W—wing.**

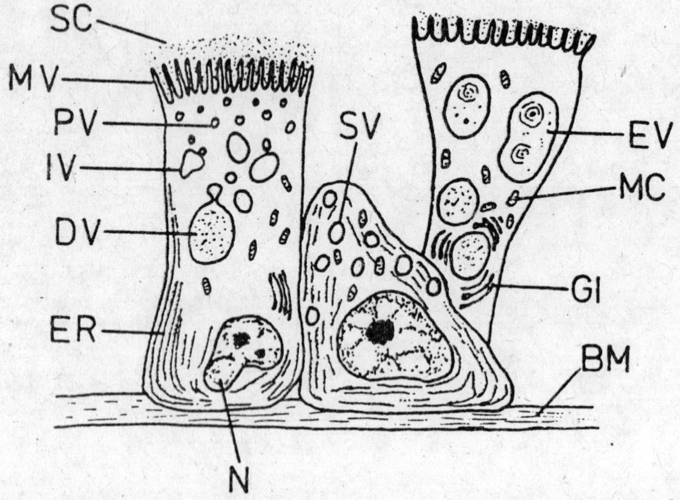

Fig. 4.4 : Cytology of the Digestive Gland, Based on *Littorina littorea*. Left, digestive cell in the absorptive phase: middle, secretory cell; right, digestive cell in the digestive phase. BM—Basal membrane, DV—digestive vacuole, ER—endoplasmic reticulum, EV—excretory vacuole, GI—Golgi apparatus, IV—intermediary vacuole, MC—mitochondrion, MV—microvilli, N—nucleus, PV—pinocytotic vesicle, SC—surface coat, SV—secretory vesicle. After Merdsoy and Parley (1973)

1. *Absorption* : the digestive cells are tall and columnar, sometimes with a surface coat over the microvilli resembling that of absorptive cells in the mammalian intestine, but of unknown function; pinocytotic vesicles accumulate below the apical brush.
2. *Digestion* : the digestive cells become shorter and more distended; the pinocytotic vesicles coalesce to form digestive vacuoles, first containing lipid droplets, and fibrillar and granular material but later becoming filled with homogeneous granules.
3. *Excretion* : excretory vacuoles carrying the waste products .of intracellular digestion burst into the lumen and the digestive cells regain their elongated shape.

The secretory cells also show phasic activity. During the absorptive to mid-digestive phase, numerous zymogen-like granules appear in the lower- and mid-apical regions and are released into the lumen. At this time extracellular digestion occurs in the stomach. In the excretory phase, the endoplasmic reticulum of the secretory cells becomes fragmentary and swollen, representing the end of protein synthesis and the reorganisation of organelles ready for the next feeding cycle.

Phasic activity of the digestive gland is correlated with the tidal cycle in *Littorina saxatilis* of high shore levels but not in *L. littorea* of mid- to low shore levels. The significance of tidal synchrony is unclear, since although it is absent in *L. littorea* it occurs in *Rissoa parva* from extremely low levels on the shore. The possible association of phasic activity of the digestive gland with foraging activity deserves further investigation.

ABSORPTION

Food is absorbed at a rate equal to the product of the ingestion rate and the absorption efficiency, the two factors often being interdependent. When passed more quickly through the gut, food is digested less thoroughly. But if food is plentiful, incomplete digestion may be unimportant because the extraction of nutrients is usually an asymptotic function of digestion time, governed by the 'law' of diminishing returns. This applies particularly to the more refractive foods, so herbivores and detritivores usually process food faster and in larger quantities than carnivores.

Animal tissue and bacteria are often more digestible than others foods. *Hydrobia* has absorption efficiencies of 75% of bacteria,

60% of diatoms

50% for BGA

8% for chrococcus

BGA protected from digestive enzymes but coat. The interesting variations of absorption efficiency may occur among closely related herbivores. In *Patella*, microalgae and red algal at different levels or shores grazing act found to be different.

Differential absorption of nutrient is of common in Nudibranchs, absorption takes place 58% of available carbohydrate and 93% of nitrogen which is correlated with uniformly highly absorption of different amino acids. It is correlated with uniformly high absorption of different amino acid.

In *Aplysia punclata* there is less restricted diet and less consistent pattern of absorption of amino acid from a single food species. Different kinds of food eaten by heterophagus spp. tend to be absorbed with different efficiency. The quality of abundance of food physiological state of consumer may also be important.

In snails when it enters for reproductive phase the absorption efficiency of *Planorbis alderi* feeding on *Tellina tenuis* increase from 40% to 50%.

Hatchings of *Planorbis* absorbs bacteria with efficiency of 94% decreasing to 75% throughout the later stages of development.

The absorption efficiency does not appear to vary systematically with body size in post-larval stages or with temperature.

Digestion

In *pila*, foregut, midgut and hidgut is +. In more true lips are absent and in buccal cavity raticula. Present which part forward the food also crishes it. Oesophegus is a long narrow tube followed bv stomach and from stomach narrow coiled tube started which is intestine and goes to rectum.

Which is thick wall tube. Digestive buccal glands + in the root of buccal cavity scaretic necessary digestive twices silivary glands. There are pair of white lying on the inner side of oesophagus buccal mass. Their secretion is partially mucus imparting digestive fluid containing enzyme acting on Mucous helps in lubrication in reduce in food to transport it.

Digestive Gland

The greater apart of coiled visceral mass is occupied by large, soft triangular gland of brownish and dirty green colour. There are some tubules are glandular called alveolus which contains types of cells.

(1) *Secretory Cells*—produce and ferment which breaks down the cellulose plant rare phenomenon amongst animal.

(2) *Reabsorption Cells*—produce proteiolytic enzyme and they take up and digest intracellularly the protoplasm of plant kills.

(3) *The Lime Containing Cells*—store the phosphatics of lime. In digestive gland of *Pila* is thus more than liver oesophageal pouches lie beneath the salivary gland one on each side of oesophagus. This secrete digestive enzymes and serve for a temporary storage. *The food consist of aquatic vegetation food is taken buccal cavity by the chains of movement of radula which are present in Pila. The action of sphincter and protector muscles of buccal mass two jaws move up to the mouth opening and cutup leaves caught by teeth and thrown backward to the buccal cavity.* Salivary glands pour their secretions to buccal cavity where it mixes with food. It contains enzyme of conversion of starch into sugar. *The stomach food is digested by secretion of digestive gland containing enzyme comparable to those of pancreas. Thus, extracellular digestion occurs*

inside the digestive glands. The reabsorptive cells of which digest the cellulose portion of food. And protein portion is digested by prototypic enzymes.

Absorption efficiency therefore depends on the digestion time and on the digestibility of the food. Animal tissue and bacteria are often more digestible than other foods. *Clione limacina* completely digests the tissues of *Spiratella retroversa,* absorbing over 90 per cent of the carbon and virtually 100 per cent of the nitrogen *Hydrobia ventrosa* has absorption efficiencies of 75 per cent for bacteria, 60-71 per cent for diatoms, 50 per cent for blue-green algae, and only 8 per cent for *Chrococcus,* a blue-green alga protected from digestive enzymes by a thick mucous coat. In general, however, there is complete overlap between the absorption efficiencies of carnivorous and herbivorous species (Table 3.1). Interesting variations of absorption efficiency may occur among closely related herbivores. South African species of *Patella* graze micro-algae and red algal turf at different levels on the shore. Branch (1981) classified these limpets into 'exploiters' living among an abundance of food, which is processed rapidly in large quantities but with less efficiency, *e.g., Patella oculus* and *P. granatina* (absorption efficiency 72-79 per cent) and 'conservers' with restricted supplies of food that is processed rapidly but more efficiently, *e.g., P. granularis* (absorption efficiency 78-81 per cent), *P. longicosta* (86 per cent) and *P. cochlear* (93 per cent). Whereas *P. longicosta* and P. *cochlear* graze a sustained yield from a 'garden' of red algae, *P. granularis* lives high on the shore where algal growth is restricted.

The deposit-feeder *Hydrobia ventrosa* digests bacteria and diatoms efficiently but in common with other detritivores, much of its ingested material consists of non-living organic particles. *Hydrobia ventrosa* absorbs boiled hay, representing detritus, with an efficiency of 34 per cent, and although bacteria are absorbed with efficiencies of at least 70 per cent, the absorption efficiency for a mixture of bacteria and boiled hay is only 56 per cent. It is noticed that fine estuarine sediments often consist largely of faecal material, so the deposit-feeders must be consuming organic matter that has already been exploited by various animals. The enigma was solved by measuring the carbon and nitrogen content of faecal pellets both freshly produced by *Hydrobia ulvae* and after incubation in seawater. Evidently *H. ulvae* absorbed microbial protein from the detritus, so new faeces were low in nitrogen. Incubated faeces were recolonised by microbes that fixed nitrogen while using the detrital carbon as a substrate, making the faeces a rich source of nitrogen for deposit-feeders including *H. ulvae* itself.

Subsequently, it has been found that several detritus-feeders can absorb a significant proportion of their carbon requirements from the non-living detritus, but microbes are the main source of protein. At the present state of knowledge, however, generalisations require caution: one species, *Hydrobia totteni*, has been shown capable of subsisting entirely on micro-algae, apparently deriving little nutrition from bacteria even when these are abundant.

Differential absorption of nutrients is of common, if not universal, occurrence. The nudibranch *Archidoris pseudoargus* feeds entirely on *Halichondria panicea* and related sponges, absorbing 58 per cent of the available carbohydrate and 93 per cent of the nitrogen. This highly efficient nitrogen absorption is correlated with a uniformly high absorption of different amino acids (Fig. 4.5A), reflecting the ability of *Archidoris* to meet its nutritional requirements from a single species of prey. It is similar to the high efficiency of nitrogen absorption by *Clione*, which also feeds on a single species of prey (see above).

Table 4.1: Absorption Efficiencies of Gastropods, Defined as the Proportion of Ingested Energy Absorbed across the Gut Wall.

Species	Food	Absorption efficiency (%)	Reference
Herbivores			
Patella vulgata	micro-algae	41	Wrightand Hartnoll (1981)
Fissurella barbadensis	micro-algae	34	Hughes (1971b)
Nerita peloronta	micro-algae	42	Hughes (1971a)
N. tessellata	micro-algae	40	Hughes (1971a)
N. versicolor	micro-algae	39	Hughes (1971a)
Tegula funebralis	micro-algae	70	Paine (1971)
Littorina irrorata	micro-algae	45	Odum and Smalley (1959)
L. planaxis	*Ulva lactuca*	86	North (in Grahame, 1973)
L. obtusata	fucoids	73	Wrightand Hartnoll (1981)
Carnivores			
Polinices alderi	*Tellina tenuis*	52	Ansell(1982)
Nucella lapillus	*MytHus edulis*	66	Bayne and Scullard (1978a)
Archidoris pseudoargus	*Halichondria panicea*	52	Carefoot (1976b)
Navanax inermis	*Haminoea virescens*	50-70	Paine (1965)

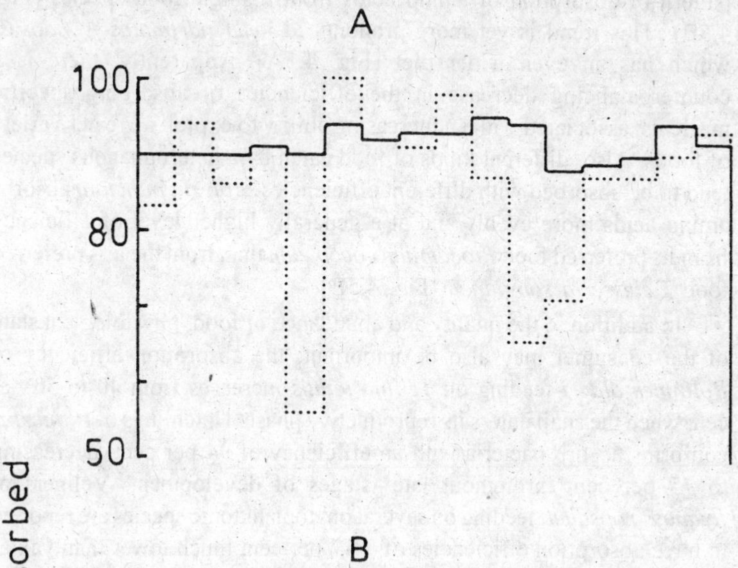

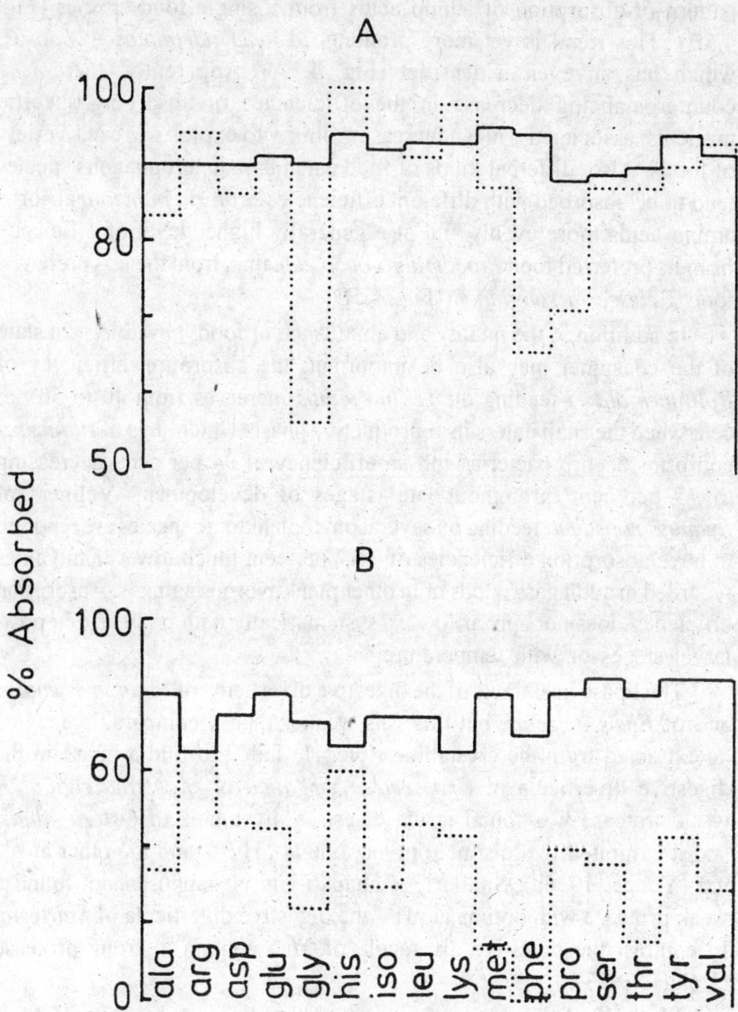

Fig. 4.5 : Efficiencies with which Nudibranchs Absorb Amino Acids from their Food. A. Solid histogram, *Archidoris pseudoargus* feeding on *Halichondria panicea.* A specialist feeding only on this and closely related prey; absorption efficiencies are high and consistent. Dotted histogram, *Dendronotus frondosus* feeding on *Tubularia larynx.* A less specialised feeder taking a variety of prey; absorption efficiencies are extremely variable. After Carefoot (1967b). B. Solid histogram, *Aplysia punctata* feeding on the preferred alga in its natural diet, *Plocamium coccineum.* Absorption efficiencies are high and relatively even. Dotted histogram, *A. punctata* feeding on a less preferred alga, *Delesseria sanguina;* absorption efficiencies are more variable. After Carefoot (1967a)

Aplysia punctata has a less restricted diet and a less consistent pattern of absorption of amino acids from a single food species (Fig. 4.5B). This trend is yet more pronounced in *Dendronotus frondosus*, which has an even wider diet (Fig. 4.5A). Apparently, there is a counterbalancing decrease in the efficiencies of absorbing specific nutrients, associated with an increasing ability to exploit a greater variety of foods. Also, different kinds of food eaten by a heterophagous species tend to be absorbed with different efficiencies. *Aplysia punctata* absorbs amino acids more evenly and at a generally higher level of efficiency from its preferred food *Plocamium coccineum* than from the less preferred food *Delesseria sanguinea* (Fig. 4.5B).

In addition to the quality and abundance of food, physiological state of the consumer may also be important: the absorption efficiency of *Polinices alderi* feeding on *Tellina tenuis* increases from 40 to 50 per cent when the snail enters its reproductive phase. Hatchlings of *Planorbis* conforms absorb bacteria with an efficiency of 94 per cent, decreasing to 75 per cent throughout later stages of development. Veligers of *Ilyanassa obsoleta* feeding on several phytoplanktonic species are reported to have absorption efficiencies of 9-17 per cent much lower than values recorded in adult gastropods or in other planktivorous animals. Absorption efficiency does not appear to vary systematically with body size in post-larval stages or with temperature.

The lipase in extracts of the digestive diverticula of *Vinus mercenaria* and of *Ensis directus*, but this was weaker than a comparable enzyme he extracted from the crystalline style. He failed to find a lipase in the digestive diverticula of *Crtissostrea virginica* or of *Atrina rigida*. A weak protease was found in the diaestive divenicula of *Ostrea edulis*. and it exhibited two distinct optima, one at pH 3.7 and the other at pH 9.0 (Yonge. 1926a). Similarly, Ganapati and Nagabhushanam found a weak protease with optima at pH 4 the digestive diverticula of *Martesia*. In contrast the digestive divenicula of *Mya* contain a strong protease (Yonge. 1923).

Reid (1964) made a comparison of the digestive systems of *Lima* and of *Mya* and foimd ihem to be significantly different. In *Lima* which appears to fit the traditional theory, digestion of polysaccliarides occurs largely in the stomach, and the digestion of fats and proteins occurs intracellularly in the digestive diverticula. In *Mya* on the other hand, the emphasis seemed to be on extracellular digestion. This comparative study by Reid draws attention to the fact that digestive systems are evolving independently in the various phylogenies of bivalves, and that as yet we have reliable data on too few examples to permit generalisation.

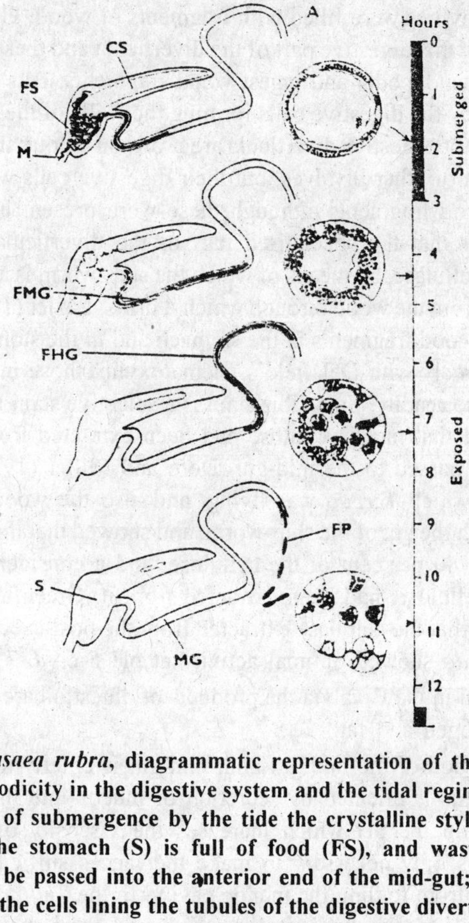

Fig. 4.6 : *Lasaea rubra*, diagrammatic representation of the relationship between periodicity in the digestive system and the tidal regime. During the short period of submergence by the tide the crystalline style (CS) is well developed, the stomach (S) is full of food (FS), and waste material is beginning to be passed into the anterior end of the mid-gut; absorbtion is occurring in the cells lining the tubules of the digestive diverticula. Shortly after the animal is exposed by the falling tide the style begins to diminish in size; the cells of the digestive diverticula are fully loaded. Finally, the gut becomes empty and the digestive diverticula are discharging.

The occurrence of a cellulose has been demonstrated in *Teredo* in a number of different ways, the earlier literature having been reviewed by Potts. who found that one part of the digestive diverticula of *Teredo* is highly modified for the purpose of digesting small fragments of wood.

The left posterior lobe of the digestive diverticula differed from the remainder in possessing larger lobules with wider *lumina* and with thinner walls composed of flattened cells. This part of the diverticula, which was white in appearance, was described by Potts as the digestive part, and its cavities were filled with fragments of wood. Phagocytes lie in the lumen of this digestive part of the diverticula and these phagocytes can put out pseudopodia and ingest wood fragments. Bits of wood are also ingested by the digestive cells forming the walls of the tubules. The other lobes of the digestive diverticula are brown in colour; they resemble the diverticula of other bivalves, but their digestive cells were not seen to contain wood fragments although these were present in the lumen. It seems clear that the specialised part of the diverticula contain an intracellular cellulase, by means of which the ship-worm is able to derive nourishment from the wood through which it drills. Lazier (1924) showed that whereas wood fragments in the stomach and in the stomach caecum would stain freely with Delafield's haematoxylin. those in the mid-gut did not stain so readily. Since ihis stain is a selective stain for cellulose, this suggested that much cellulose had been extracted from !he wood before it was passed to the mid-gut. Dore and Miller (1923) analysed the wood in which *Teredo* was living, and also the wood which was passed through the gut of the ship-worm, and showed that in the digestive process some 80 per cent of the cellulose, and a considerable amount of the hemicellulose, had been extracted from it. Greenfield and Lane (1953) found that the cellulase extracted from the post-caecal part of the body of *Teredo* showed optimal activity at pH 5-6—6-7.

In general in the Bivalvia the products of digestion are stored in the form of glycogen and fat.

The function of the mid-gut and hind-gut is chiefly to consolidate the faeces. This is effected by secretion of mucus into the lumen and by a change in the pH which increases the viscosity of the mucus. Typically it is only necessary to make the faeces sufficiently firm to prevent them from fouling the mantle cavity. In the *Tellinacea* the mid-gut is notably longer in those species which take great quantities of mud into the mantle cavity. In such bivalves as *Scrobicularia, Macoma, Abra*, etc., the inhalant siphon collects food material from the surface of the substratum, which is usually muddy. Here it is important that the faeces should retain their form for a long period after they have been extruded to ensure that the same material is not collected again by the inhalant siphon in the near future. It is for this reason, doubtless, that the mid-gut is especially long in such forms.

The feeding mechanisms of bivalves do not include qualitative selection of the food material, and much may be swallowed at the mouth which is not digestible. For this reason one cannot judge the basic food requirements of a bivalve from an examination of gut contents. Naked flagellates will be very quickly digested and will leave no trace, whereas indigestible organisms may remain easily recognisable. Nelson (1933) found enormous numbers of a small nematode, *Chromadora*, in the stomach of *O. edulis*. and thought that these were being digested by some undetermined enzyme which could penetrate the cuticle of the nematode. It is far more probable that the *Chromadora* were collected and swallowed inadvertently, were dying quickly in the acid stomach contents, and were undergoing auto-digestion. Conversely, the sorting mechanisms of the stomach are not 100 per cent perfect, and food organisms may chance to pass quickly through the stomach and be discovered undigested, and perhaps unharmed. in the mid-gut or in the faeces. The occurrence of such an organism in the mid-gut or faeces does not prove that it is not acceptable food to the bivalve in question.

The Tridacnidae probably feed to a certain extent in the normal way upon material collected by the ctenidia. swallowed at the mouth and digested by the extracellular amylase of the crystalline style, followed by intra-cellular digestion in wandering phagocytes and in the epithelium of the digestive diverticula. The crystalline style of a 3-ft long *Tridacna derasa* was 34 cm long and 0.5 cm in maximum width, which suggests that this organ is not without importance. The digestive diverticula, however, are greatly reduced in numbers although they are normal in histological structure. The main source of food in the Tridacnidae is derived from the symbiotic association with zoo-xanthellae which has had so marked an effect upon the morphology of the family. The zoo-xanthellae only occur within phagocytes which lie in great numbers in blood spaces in the mantle. The zoo-xanthellae are "farmed" in the superficial tissues of the inner lobe of the mantle margin, being particularly abundant around the lens-like hyaline organs which may transmit sunlight deeper into the mantle tissues. The zoo-xanthellae contain a small assimilation product and a small pyrenoid, but have a relatively large accumulation of starch; the cell wall apparently lacks cellulose, so there is no hindrance to the ultimate digestion of the plant cell by the phagocyte.

The zoo-xanthellae in the mantle tissues are always found to he undigested. whereas vast numbers of zoo-xanthellae in various stages of digestion lie in phagocytes around the gut and in the spaces among the digestive diverticula. At some stage in the history of these symbiotic plant cells they are evidently transported bv the phagocytes to the vicinity of the gut. to be digested there by the phagocytes. Transport of symbionts from the mantle to the region of the gut must be rapid, for phagocytes containing plant cells were very rarely found in samples of blood taken from the heart. This additional source of food and method of digestion is doubtless responsible for the great size that can be reached in this family; *T. derasa* is the largest bivalve ever to have occurred in the history of the world, and this great size could not have been gained solely by the orthodox ctenidial feeding mechanisms of bivalves. Symbiosis also occurs in the Horse-shoe Clam, *Hippopus*, but the condition is less advanced here than in *Tridacna. Hippopus* possesses no hyaline organs, it has fewer zoo-xanthellae. fewer phagocytes in the visceral mass, and the digestive diverticula are correspondingly better developed than in *Tridacna*.

Digestion

Site of digestion : In primitive gastropods microfeeding is common and ingested material is subjected to gastric cavity to an initial Extracellular digestion which is completed intracellularly following phagecytosis by the cells of digestive diverticula and in some case by ameobocytes.

The relative importance of initial extracellular digestion varies considerably in *Patella*. The only extracellular enzyme appear to be an amylase and possibly cellulase. The juice contains proteases and lipases in addition to carbohydrases.

In remaining gastropods the relative importance of extracellular and intracellular methods of digestion is closely relative to the mode of feeding. In those herbivores mesogastropod which feed more or less continuously generally microfeeders the phagocytic activity of cells of digestive diverticula is of prime importance mesophageal glands are reduced or last and the gut possess a true crystalline style which provides for continuous release of *amylolitic* and *lipolitic* enzymes, generally, the only extracellular enzyme active in gastric cavity, but in those herbivores in mesogastropods which feed discontinuously and particularly in carnivores genera which will in general attain their food at unpredictable intervals, extracellular digestion appears to predominate in *Natica oesophageal* glands secrete a *proteinase* while the digestive diverticula secrete *amylolytic, lipolytic, proteolytic* enzymes. Similarly in carnivorous *Neogastropoda* a wide range of enzymes are active in lumen of gut. In most cases, the salivary glands, and glands of Liver secrete *proteolytic enzymes* but no *amylolytic* and lipolytic enzymes the gastric juice contains carbohydrases, lipases as well as proteases presumable derived from digestive diverticula. Carbohydrases occur in digestive juices salivary glands, digestive diverticula of most gastropods and also in crystalline style were present. These are also reported from so called stomach plates *i.e.,* plates of oesophageal gizzard. The common carbohydrases, such as amylase and maltase but many gastropods particularly herbivores possess a wide array of carbohydrases of the 30 or more enzymes associated with digestive tract of *Helix*, more than 20 carbohydrases which are reported to include a and P amylase, pH 6.2—6.8 and 4.5 respectively.

Cellulases, Chitinase and a variety of glycosidases are also reported. The tegula ones its abundance in littoral zone in part of this array of enzymes which enables it to digest a wide variety of algae. The crop juice, and extracts of digestive diverticula of slug, *Arion, Arter* have similar range of enzymes. The general a glucosidic bond present in α glucosidases while β linked polysaccharides and glucosides may be acted upon by a 1,4-β polysaccharidases and β glucosidase respectively.

The Chitinase activity of alimentary canal of *Helix* is also reported. The presence of Chitinolytic flora in digestive juice is found. Further, noted death and increase in concentration of bacteria which occur during periods of rest and a starvation, was paralled by an increase in Chitinolytic activity. This is solely of bacterial origin. The comparison of Chitinolytic activity of gastric juice, intestinal juice of digestive diverticular extract showed no significant difference between animals. It may thus considered that the contribution of a Chitinolytic bacterial flora to the production of intestinal Chitinase is negligible and digestive diverticula during actual site of Chitinase secretion. The β glucosonidase and sulphatases the physiological significance is not yet established but to some extent they are related with macroherbivores. These enzymes assist inbreakdown of sulphated polysaccharide which occurs in many sea weed and are widely distributed in animal tissue.

Alginic acid, a high polymer guluronic and mannuronic acids is the structural polysaccharides of many from algae and constitutes an important part of food of marinephytophagous gastropods. Alginases have being detected in digestive juices and in the diverticula of number of gastropods and as for Chitinases would seen to be permanent constituents of enzymic digestive equipment of gastropod. Alginases are known to occur not only in phytophagous marine spp. but also carnivorous marine spp. *e.g., Purpura*, *apillus* and *Nassa reticulata*. Phytophagous freshwater *Paludina, Lymnaea*. There appear to be correlated with dietary habits and amount of alginase present in the digestive tract. The proteolylic enzymes represented in carnivorous extracellularly in *Murex*, protease, aminopeptidase, dipeptidase occurs carboxypeptidase in last two glands. The proteinase appears to be trypsin pH 7.8 to 8.2. But unlike mammalian enzyme it is apparently present in the glands in an activated form also essentially similar distribution of proteases extracellular demonstrated in mesogas-tropods. The lipases, and esterases are also reported in all 3 species like *Murex, Planorbis, Helix*, enzymes is activated by $CaCl_2$ and is inhibited by Na^- and citrate. Its relativity on four substrate is as follows: Tributyrin (morethan) methyl > butyrate > Tween 60 > olive oil.

There are two individuals one kept on a carbohydrate rich boiled potato or protein rich horse meat diet showed no significant difference in cathespin esterase activity, but paired to produce more amylase than the starch later repeated the experiment this time the state of computing the enzyme activity on basis unit of dry weight gland powder. It was expressed per unit weight of lining substance as indicated by protein—Nitrogen content. In these experiment the meat fed snails produce less proteases cathespin than the starch fed animals but than later produced more carbohydrases. The effects of diet on utilization of food *i.e.*, amount of food absorbed was also investigated the starch fed animals showed a definite decline in utilization despite an increase in amylase content of digestive diverticula. This may be due to some dietary deficiency which causes a decline in absorption.

Absorption of digestive food takes place chiefly in digestive glands and intestine.

The undigested food from rectum passes through the thus into the branchial cavity and finally to outside along with outgoing current of water through the right syphon.

GASTROPODA

The Gastropoda are remarkable for the diversity of their feeding mechanisms, and for the wide variety of foods on which they subsist. The main classificatory units of the Gastropoda cannot be distinguished from one another on the basis of feeding habits and food preferences. The Prosobranchia in particular display almost every conceivable food preference, while the Opisthobranchia are nearly as diversified in their requirements. It seems clear that the Gastropoda underwent an explosive radiation at a very early stage in their evolutionary history, this explosion largely concerning food selection and methods of feeding. Both the Opisthobranchia and the Pulmonata must have originated in very ancient

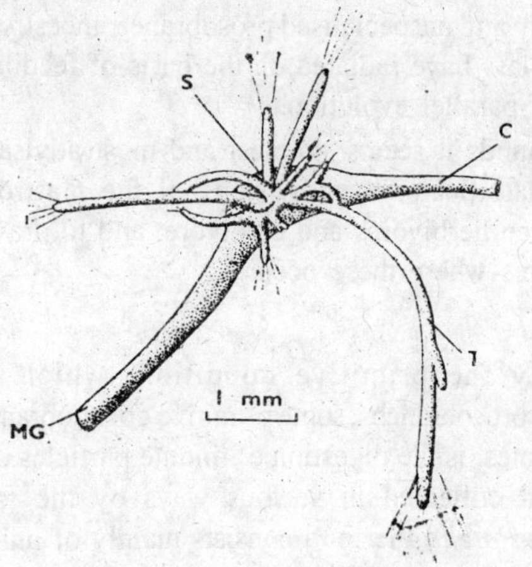

Fig. 4.7 : *Solemya parkinsoni*, enlarged view of the anterior part of the alimentary canal, seen from the right side. The stomach (S) is only a very slight dilation as compared with the oeoshagus (O) and the mid-gut (MG), while there are very few and small tubules to the digestive diverticula (T).

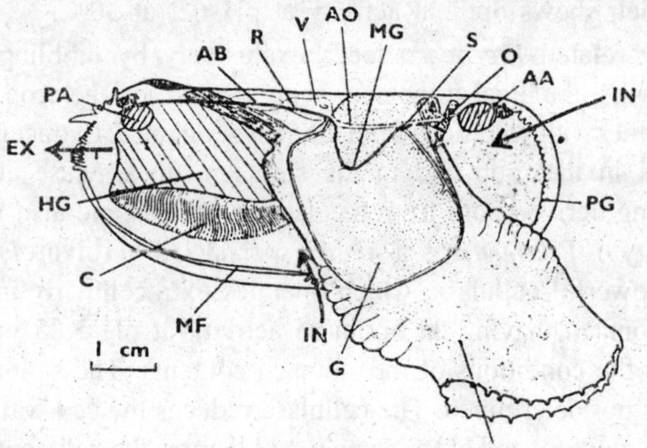

Fig. 4.8 : *Solemya parkinsoni*, seen from the right side after removal of the right shell valve and right mantle lobe, and after further dissection to display the course of the alimentary canal. The animal lives deeply buried in intertidal sandy beaches, near low-water mark, and the enormous size of the foot (F) is doubtless associated with burrowing habit. The ctenidium (C) is suspended from the ventral border of a very large hypobranchial gland (HG). The methods of feeding and digestion in Solemya are not understood. The whole alimentary canal is greatly reduced; *e.g.*. the oesophagus (O), the stomach (S), the mid-gut (MG), and the rectum (R).

times from an early and unspecialised prosobranch ancestry, and all three divisions of the class have radiated on the basis of feeding habits with some measure of parallel evolution.

On these grounds it seems simplest and most advisable to ignore here the main classificatory subdivisions of the Gastropoda, but to distinguish between herbivores and carnivore, and to draw attention to specialized features where these occur.

Herbivores

Undoubtedly the primitive condition, which is found in archaeogastropod prosobranchs, some primitive opisthobranchs, and some primitive pulmonates, is the digestion of minute particles of plant origin which have been collected in various ways by the radula. In the archaeogastropod *Patella* the food consists mainly of unicellular algae and diatoms, with only rare fragments of a larger seaweed. As this material is drawn into the buccal cavity by the radula this is lubricated by a mixture of secretions from the four salivary glands, which do not contain any enzymes. The mixture passes backwards in a centrally sited food-groove which is flanked by a series of lateral-pouches of the fore-gut. These oesophageal pouches pour on to the food a secretion containing An amylase which shows optimal activity at pH 6.2 at 30°C.

The closely related *Pterocera* feeds exclusively by nibbling the finest algal growths, the algal fragments being passed via the crop into the stomach together with secretions from the salivary glands. Extracellular digestion occurs in the crop and in the stomach, an amylase and a glycogenase being derived both from the salivary glands and also from the crystalline style. *Pterocera* is a highly specialised herbivore since it possesses a powerful cellulase, which operates extracellularly in the lumen of the stomach, having an optimum activity at pH 5.85 which approximates to the conditions of the stomach contents. The source of this enzyme was not determined. The cellulase reduces the cell walls of plant material to glucose, and at the same time liberates the cell contents for digestion by other enzymes. A weak extracellular lipase which operates in the stomach is probably derived from the digestive diverticula.

Among the Basommatophora the aquatic *Lymnaea stagnalis* cuts suitable sized pieces of plant material and swallows these without much trituration. Sand in the gut is essential to digestion, serving for trituration of the gut contents; sand is most abundant in the gizzard, but also occurs in the crop and in the recurrent passage of the pylorous. If deprived of sand, *L. stagnalis* loses its appetite and eventually dies. Secretions from

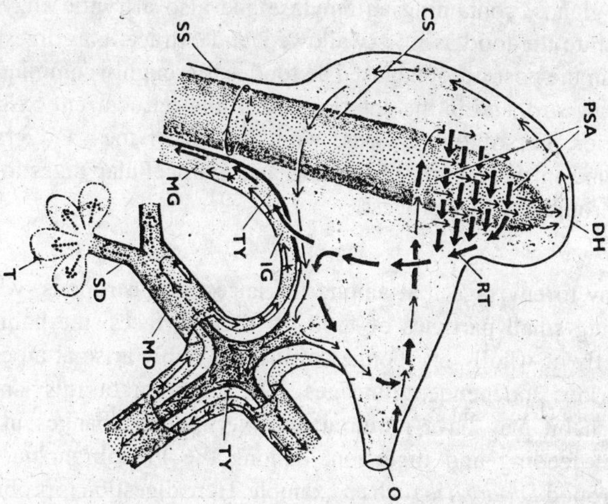

Fig. 4.9 : Diagrammatic representation of the probable circulation of particles in the stomach and digestive diverticula of a bivalve with a gastropemptan type of stomach (ref. Fig. 81, type 5). Heavy arrows represent the course followed by coarse particles; fine arrows represent the course taken by fine particles; tailed arrows represent material that is being rejected; dotted arrows represent movement due to absorption of material by the digestive cells. For interpretation of lettering, see pp. 263-4. (Originally published in Owen, 1955, *Quart. J. micr. Sci.* 96, 533, fig. 13.)

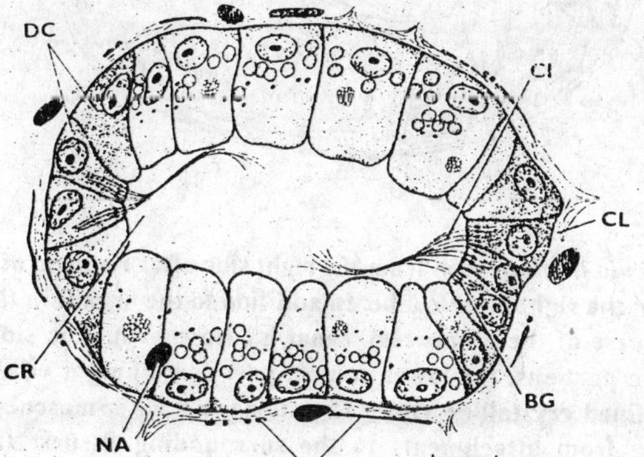

Fig. 4.10 : Transverse section through a tubule of the digestive diverticula of *Venerupis pullastra,* showing the disposition of two tracts of darkly stained ciliated cells in the tubule. (Originally published in Owen, 1955, *Quart. J. micr. Sci.* 96, 533, fig. 13.)

the salivary glands, containing an amylase and also a tryptic enzyme, are poured on to the food as it is swallowed, and extracellular digestion commences in the post-oesophagus. The food is thoroughly comminuted in the crop, gizzard, and in the anterior part of the retrocurrent passage of the anterior part of the pylorus, with the aid of the sand grains normally found in these parts of the gut, and extracellular digestion is continued throughout this process.

Carnivores

It is easy to envisage that among the ancestral gastropods, which fed by rasping small particles of food off rock surfaces, the habit of feeding chiefly or wholly on sponges would inevitably arise at an early stage in various independent lineages. The adoption of this simple carnivorous habit may have required only very slight changes in the processes of feeding and digestion. Among the Prosobranchia, the archaeogastropod *Diodora* is such an example. Here digestion is probably

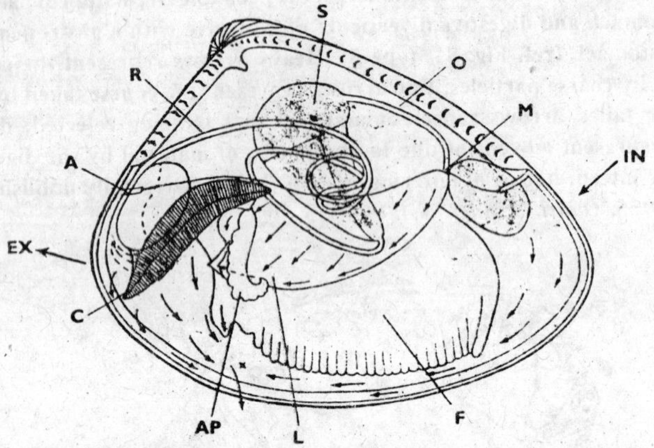

Fig. 4.11 : *Nucula hanleyi*, seen from the right side after removal of the right shell valve and the right mantle lobe. In addition to the organs in the mantle cavity, the course of the alimentary canal is also shown. The stomach (S) consists of two portions, the lower conical part containing a viscous mass, not a well-defined crystalline style. The stomach ".all is muscular, and is relatively free from attachments to the surrounding tissues; the upper globular portion is extensively protected internally by a chitinous secretion, and muscular contractions "of the stomach wall serve to mix the ingested material with the digestive enzymes and to triturate the food material. (Originally published in Yonge).

largely extracelluar by enzymes derived from the oesophageal pouches and perhaps also from the digestive diverticula. Digestion of the cells of the sponge materials must be a comparatively simple matter, but the elimination of masses of sharp sponge spicules without damage to the epithelium of the gut is a major consideration. These spicules are compounded into a faecal rods which superficially resembles a crystalline style, and this extends backgrounds through the mid-gut. Such a faecal rod may be distinguished from a crystalline style by its mode of formation, the direction of its movement through the gut and by the lack of absorbed extracellular enzymes.

As a good example of a predacious carnivore, we may consider *Philine aperta*, which feeds on diatoms, foraminifera, small molluscs, and other organisms living in sub-littoral sands. The prey is swallowed whole, together with a certain amount of sand. The posterior part of the oesophagus is modified to form an anterior crop, a gizzard, and a posterior crop, the exit from the latter being guarded by a sphincter muscle. The walls of the gizzard are highly muscular and are lined internally by three calcareous plates, each of which has a boss on its inner surface. A sheet of muscle surrounding the gizzard operates these plates by means of which food is crushed. Small molluscs, such as the snail *Hydrobia*, may remain in the gizzard for many hours while the shell is being smashed.

In general, in Bivalvia the products of digestion are stored in form of glycogen and fat in primitive gatropods microfeeding is common and engested material is subjected to gastric cavity for extracellular digestion *e.g.*, *Patella*, the completion of it occurred intracellular following phagocytosis by the cells of digestive diverticula and in amoebocytes. The only extracellular enzymes appear to be amylase and possibly cellulose. The gastric juice contains protease, lipase in addition to carbohydrates in remaining gastropods it is closely related to mode and type of feeding in herbivores mesogastropods microfeeder the phagocytic activity of cells of digestive diverticula is of prime importance. Their oesophegeal glands are reduced or lost and gut passes a true crystalline style which provide a continuous release of amylotic and lipotic enzymes.

SPECIFIC DYNAMIC ACTION

Upon ingesting a meal, the O_2 consumption of a molluscs increases markedly without any increase in activity .The events producing this increase in O_2 consumption are called specific dynamic action (SDA). One might suppose at first that the increased energy consumption relates to the synthesis of the enzymes, mucus and other digestive liquids needed for digesting a meal,, but this is not the case. The enzymes are already present intracellularly prior to eating and are only released during digestion. The size of SDA relates to the amount of protein in a meal and to the proportion of that meal which is used for energy than for growth. Protein that is to be used for energy must first have amino-group (NH_2) removed and excreted as (NH_3), both processes requiring an input of energy. Thus, the pulse of increased O_2 consumption is followed by a pulse of increased ammonia excretion.

Table 4.2 : Carbohydrates hydrolyzed by gut extracts from the Snail *Tegula funebralis*[a]

Substrate	Where found	Type of linkage	Foregut[b] (pharynx + salivary gland)	Midgut[b] (midintestinal region + digestive diverticula)
Starch	Widespread	1, 4-and 1, 6α-Glucose	+	+
Glycogen	Widespread in animals. yeast	1, 4-and 1, 6α-Glucose	+	+
Laminarin	Brown algae	1, 3β-Glucose	+	+
Cellulose	Widespread in plants	1, 4β-Glucose	+ – ?	+
Alginate	Brown algae	1, 4β-Mannuronic acid	–	+
Fucoidin	Brown algae	1, 2α-Fructose	–	+
Inulin	Red plants (composites)	1, 2 ? Fructose	–	–
Agar	Red algae	1, 3-β-Galactose	–	?
Carrageenin	Red algae (*e.g.*, *Gigartina*, *Chondrus*	1, 3-α-Galactose Kappa fraction only	–	–
Iridophycin	Red algae (*Iridophycus*)	1, 3-α-Galactose Kappa and Lambda fraction	–	+
Maltose	Widespread	α-Glucoside	+	+
Sucrose	Widespread	α-Glucoside (fructoside)	+ – ?	+
Cellobiose	Breakdown of cellulose	β-Glucoside	+	?
Salicin	Willow	β-Glucoside	+	?
Lactose	Widespread	α-Galactoside	–	?
Malibiose	Widespread	α-Galactoside	–	?

[a]Data from Galli and Giese (1959).

[b]A plus sign indicates positive evidence; a minus sign. lack of evidence.

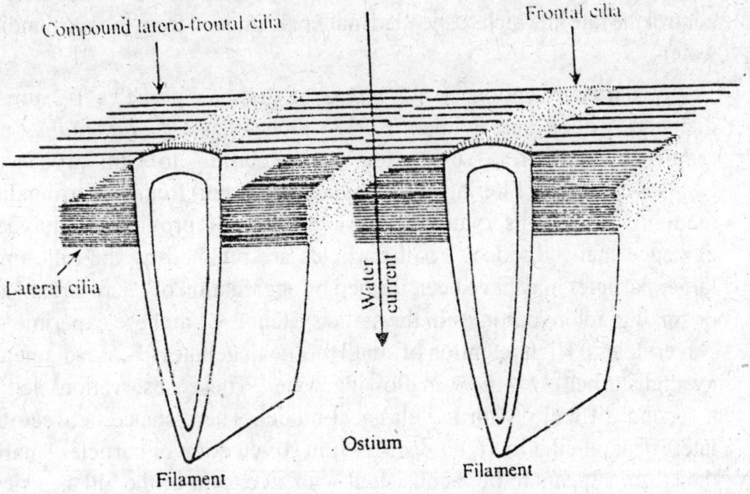

Fig. 4.12 : A schematic representation of two filaments on the gill of *Mytilus edulis*. The cilia are at rest. (Redrawn from Dral, '1967.)

The digestive gland envelops the stomach and part of the intestine; the intestine passes through the heart and pericardial cavity, over the posterior adductor muscle and opens into the suprabranchial chamber, adjacent to the exhalant siphon.

Function of Ctenidia in Feeding

The mechanism by which the lamellibranch gill removes suspended particles from the water pumped through the mantle cavity aroused considerable interest when it became apparent, in the early nineteenth century, that the ctenidia play a role in feeding as well as in gas exchange suggested that the gill acts as a sieve, the latero-frontal cilia forming the meshwork. This concept provided the basis of the 'classical' view of filtration by the bivalve gill, a view which has only recently been modified. Observations of filtering behaviour and measurements of filtration rates revealed two possible anomalies in the classical theory. Firstly, considerable variations were observed in filtration rates recorded from mussels of the same species under similar conditions. In some cases, fluctuations in filtration rates were observed in short-term experiments with individual animals. Although, some of this variation was attributable to differences in experimental method, and in some instances to poor technique, it became evident that bivalves are able to control the rate at which suspended matter is removed from the surrounding water

A started retention by *M. edulis* of graphite particles 1-2 μm in diameter. Zobell and Feltham (1938) found that *M. califomianus* can remove bacteria from suspension, and according to Duff (1967) *M. edulis* is capable of filtering viruses (diameter 25 nm) from the surrounding medium. These observations, however, do not provide unequivocal evidence that individual small particles are retained by the gill, since larger particles may have been formed by aggregation of micro-organisms or small graphite particles in the static system used in these experiments. Nevertheless, High retention of small motile flagellates (1-2μm diameter) by undisturbed *M. edulis* in flowing water. These observations led to a second difficulty with the classical model. The distance between the latero-frontal cilia in *M. edulis* is 2-3μm so retention of particles smaller than 2μm appears to be inconsistent with a concept of the gill as a sieve is that particles may be removed by adhesion to the latero-frontal cilia rather than by retention in a meshwork has observed that alumina particles (5μm diameter) often adhere to gill cilia in *M. edulis*. Although, this mechanism appears to play a role in feeding, it is unlikely to account for the high retention efficiencies recorded for small particles.

An alternative mechanism for the retention of particles by the lamellibranch gill was proposed by MacGinitie (1911), who maintained that a mucus sheath was secreted over the gill during feeding. Such a mechanism would account for removal of very fine particles from suspension, and from this point of view the hypothesis was not unattractive. Attempt to reconcile MacGinitie's observations with the conventional 'sieve' concept by envisaging the mucus sheet in operation only under conditions of poor food supply was made whereas the latero-frontal meshwork would function at higher particle concentrations. Nevertheless. It is difficult to visualise latero-frontal ciliary function in the presence of a sheet of mucus, so that the two proposed mechanisms appear to be mutually exclusive. Subsequent detailed analyses of the function of latero-frontal cilia in filtration have shown how small particles are efficiently retained without the formation of a mucus net, and stereoscan electron microscope studies have failed to demonstrate any mucus sheets on the gill different from zero. *Mytilus edulis* and by Winter (1966) for *Modulis*

Table 4.3 : A comparison of assimilation, efficiencies by *Mytilus edulis* calculated by Thompson and Bayne (1972) and Foster-Smith (1975b).

Concentration of suspension (cells ml^{-1})	Cells ingested per day ($\times 10^6$)	Cells assimilated per day ($\times 10^6$)	Assimilation efficiency Thompson and Bayne	Foster-Smith
1000	31	27.8	0.89	0.83
2500	78	62.4	0.80	0.78
10000	312	187.2	0.60	0.45
15000	468	187.2	0.40	0.40
20000	624	124.8	0.20	—
25000	780	0	0	—

Direct Absorption

Marine invertebrates can absorb from solution organic compounds such as amino acids and sugars. *Mytilus edulis* removed dissolved, labelled amino acids and glucose, most of which were absorbed onto the gills before transfer to the mantle and digestive gland. Acid phosphatase activity in the gill epithelium of *M. edulis* has been demonstrated by Pastee's (1968, 1969), who claims that the enzyme is located in the goblet cells of the epithelium and is exuded with the mucus into the pallial cavity. In other words, Pasteels considers that the mucus secreted onto the gill has a digestive as well as a mechanical function. The absorptive mechanism is chemically selective, but small particles are removed non-selectively by pinocytosis. Amoebocytes located in the gill tissue also show acid phosphatase activity, and are responsible for translocating material to other sites.

Although, there is little doubt regarding the mussel's ability to absorb dissolved organic compounds, it is difficult to assess the nutritional significance of such material. Available information suggests that the concentration of dissolved organic matter in seawater is normally very low, so that its contribution to the energy input in invertebrates may only be a fraction of the total.

Feeding and Digestion

Gill structure in some bivalves, especially the Ostreidae, has been examined but the gross morphology of the gill in the Mytilidae is not particularly well described. Paradoxically, there are several excellent accounts of the fine structure of the gill filaments in *Mytilus edulis*. including details of their ciliation and innervation.

The gills consist of four pairs of demibranchs which separate the pallial cavity into inhalant and exhalant (or suprabranchial) chambers throughout its length. Each demibranch comprises two lamellae, one attached to the gill axis (descending limb), the other lying free (ascending limb). The two lamellae forming each demibranch are held together by connective tissue junctions (interlamellar junctions) which link the ascending and descending limbs. Individual lamellae are formed by rows of ciliated filaments. Lateral cilia are responsible for moving water through the ostia (the spaces between pairs of adjacent filaments), latero-frontal cilia remove particles from the inflowing water and frontal cilia transport the particles to acceptance or rejection tracts at the margins of the lamella. In the Filibranchia (which includes the Mytilidae), individual filaments are discrete, although adjacent ones are loosely joined by ciliated discs.

Water enters the pallial cavity through the inhalant siphon (which is continuous along the entire length of the ventral surface) before passing through the gill ostia into the suprabranchial chamber, and is expelled through the exhalant siphon, which is narrower than the inhalant and lies at the posterior edge of the mantle, immediately dorsal to the inhalant margin. Both inhalant and exhalant siphons possess a velum which can regulate the current flows. The inhalant velum serves another important function, since it can deflect the incoming current to prevent it striking the gill directly.

Food particles are bound into mucus strings on the gill lamellae and are carried to the labial palps via ciliated grooves on the lamellar margins. The palps regulate the amount of food which enters the mouth, and direct surplus material on to the rejection tracts of the mantle surface. From the palps, food is directed into the mouth, passes through the oesophagus and enters the stomach, which is a complex structure containing a system of ciliary tracts which sort the particulate material and direct it either towards the digestive tubule duct openings or towards the intestine. Intracellular and extracellular digestion occurs in the digestive gland, which also stores nutrient reserves and regulates their transfer to other tissues.

The extended phase of elevated oxygen consumption would include any 'specific dynamic action'.

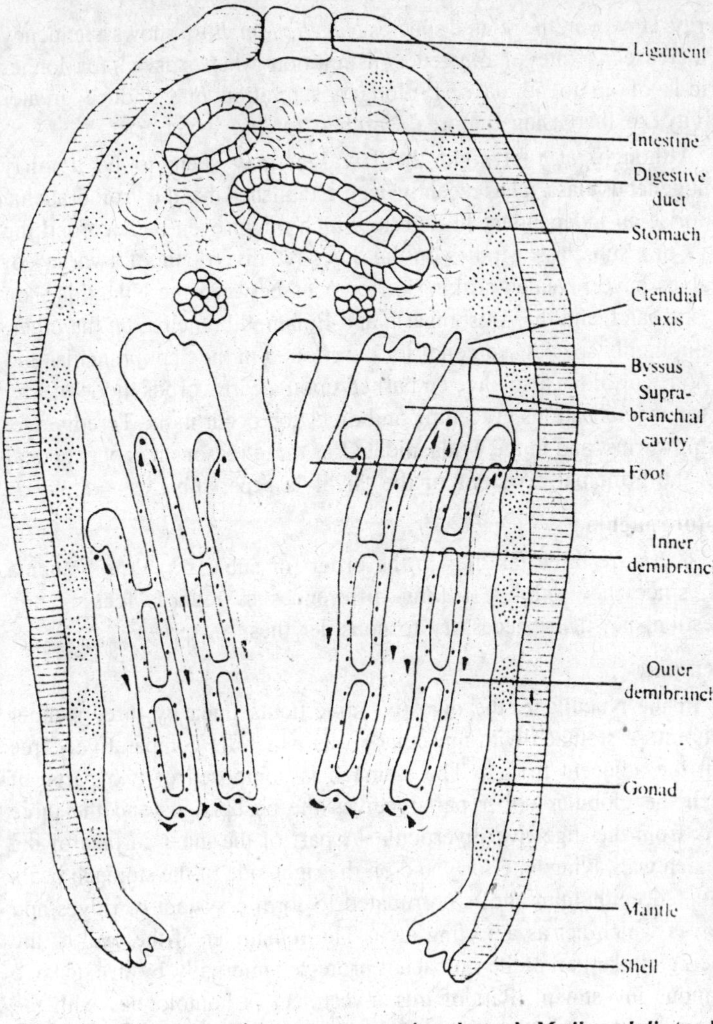

Fig.4.13 : Diagrammatic transverse section through *Mytilus edulis* to show the form of the gills and the direction of the main ciliary currents, represented as arrows.

Following the increased levels associated with a meal, the oxygen consumption rate of *Nucella lapillus* continues to decline until the next meal, at least for up to 18 days. Periods of this length without food are commonly endured by *N. lapillus* when wave action or low temperature prevent foraging, the declining metabolic rate being a means of saving energy. However, the related species *Nucella lamellosa* shows a tendency to increase its rate of oxygen consumption when starved for longer periods of up to 53 days. Prolonged starvation may induce greater activity, so increasing the area searched for prey.

Although, at first sight the Bivalvia may seem to be a fairly homogeneous class, it has recently been established that the Protobranchia comprise an independent lineage to which might even be accorded the status of a sub-class. Studies on the digestive diverticula of a variety of bivalves have revealed striking differences between the Nuculidae, on the one hand, and the Anisomyaria and Eulamellibranchia, on the other. Again, highly specialised conditions are found in the Septibranchia, and the Verticordiidae also have embarked upon a carnivorous mode of life. Lesser specialisations in feeding and digestion occur in the Teredinidae, or shipworms, and in the Tridacnidae. For these reasons it is best to deal with the principal divisions of the Bivalvia separately.

Protobranchia

There are three families in the order (or sub-class) Protobranchia and, since they exhibit striking differences as regards feeding and digestion, it will be necessary to consider these in turn.

Nuculidae

In the Nuculidae, the digestive diverticula open by three slender ducts into a stomach with muscular walls-which lie comparatively free from the adjacent tissues. The stomach is composed of two parts, of which the globular upper part receives the oesophagus and the three ducts from the digestive diverticula. A part of the inner surface of the stomach wall, lying anteriorly and on the right side of the stomach bears a ciliated epithelium and is corrugated to form a system of ridges and grooves which act as a sorting area. The remainder of the wall of the upper, globular part of the stomach is protected internally by an extensive chitinous investment. Part of this investment is homologous with the gastric shield which occur throughout the Bivalvia, while the remainder is a special development in the Protobranchia. The lower part of the stomach is a conical structure which is lined by a ciliated columnar epithelium and which corresponds with the combined style-sac and anterior mid-gut of many of the filter-feeding bivalves. There is no firm crystalline style in the Nuculidae, but instead there is a viscid mass which comprises secretions from the style-sac and from the digestive diverticula, mixed with food material, organic debris, and sand grains, etc., which have been garnered from the substratum by the palp proboscides. The contents of the stomach are thoroughly mixed by muscular movements of the stomach wall and by rotation imparted by the ciliated epithelium of the style-sac.

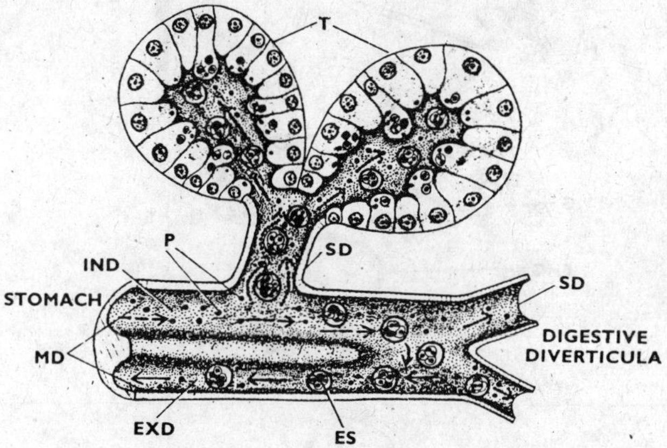

Fig. 4.14 : The probable circulation of fluid and particles within the digestive diverticula of filter-feeding bivalves. Unbroken arrows represent outwardly directed ciliary currents; broken arrows represent an inwardly directed compensation current; dotted arrows represent movement due to absorption of material by the digestive cells.

A sphincter muscle at the distal end of the style-sac prevents the untimely escape of material into the mid gut, except for small quantities which are allowed to pass out via a ciliated groove which lies on one side of the style-sac.

The cilia around the orifices of the ducts the digestive diverticula are very active and they serve to expel fluid from the ducts into the stomach and to prevent any particles in the stomach from entering the ducts. Digestion is exclusively extracelluar, and absorption is solely by the epithelium of the stomach and anterior mid-gut. There are two types of cell in the digestive diverticula: one of these is dark-staining, and bears a single flagellum and these cells may perhaps be secretary: the other type of cell is taller and goes through a phase during which the distal ends of the cells are nipped off and passed down the duct to the lumen of the stomach. These fragmentation spherules may be partly excretory, and may also the source of extracellular enzymes in the lumen of the stomach. Food reserves are stored in the form of protein spheres, fat droplets, and glycogen in the cells lining the alimentary canal.

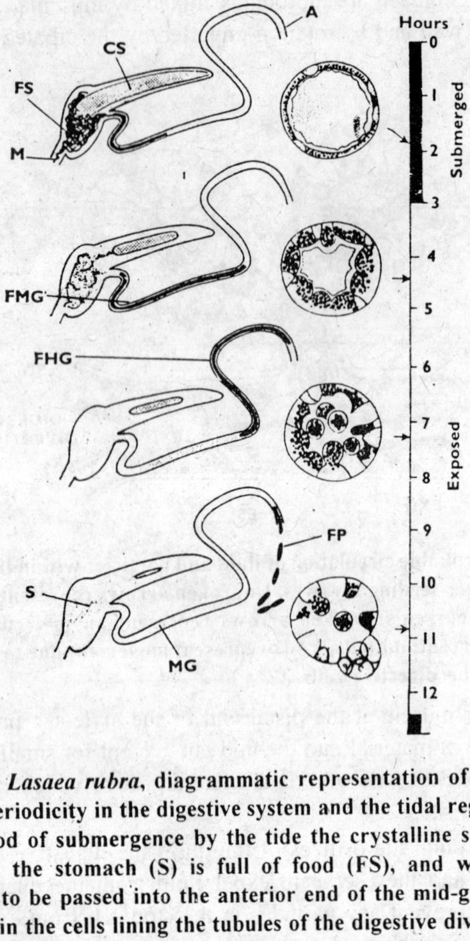

Fig. 4.15 : *Lasaea rubra*, diagrammatic representation of the relationship between periodicity in the digestive system and the tidal regime. During the short period of submergence by the tide the crystalline style (CS) is well developed, the stomach (S) is full of food (FS), and waste material is beginning to be passed into the anterior end of the mid-gut; absorption is occurring in the cells lining the tubules of the digestive diverticula. Shortly after the animal is exposed by the falling tide the style begins to diminish in size; the cells of the digestive diverticula are fully loaded. Finally, the gut becomes empty and the digestive diverticula are discharging. For interpretation of other lettering, see pp. 263-4. (Originally published in Morton, 1956, *J. mar. biol. Ass. U.K.* 35, 583, fig. 8.).

Nuculanidae

It is clear that the conditions in the Nuculanidae are partly different from those in the Nuculidae. The relatively large particles be swallowed by a combination of ciliary and muscular activity, and that the oesophagus is somewhat enlarged for this purpose. In general the internal architecture of the stomach is comparable to that in the Nuculidae, but one of the three masses of digestive diverticula differs from the other two and resembles the digestive diverticula of the septibranch *Cuspidaria*.

Solenomyidae

The digestive mechanisms of Solenomya remain a mystery. The alimentary canal of *S. togata* as being greatly reduced and simplified; the stomach is very small and has no ciliary sorting areas, no gastric shield, and although there is a style-sac there is apparently no crystalline style. *S. parkinsoni* the gut was similarly extra ordinarily small for so substantial an animal; the stomach was represented merely by a slight dilation associated with the point of entry of the two ducts from the greatly reduced digestive diverticula, while the mid-gut was a "slender greenish thread", less than half the diameter of the anterior aorta. The digestive diverticula consisted of a small number of distinct tubules on to the distal ends of which were inserted a number of rather stout muscle fibres. Owen found the stomach of *S. parkinsoni* to possess the normal internal features including a dorsa hood, and a gastric shield, in spite of its minute size. The style-sac, which tapered backgrounds towards the mid-gut, exhibited the normal ciliation.

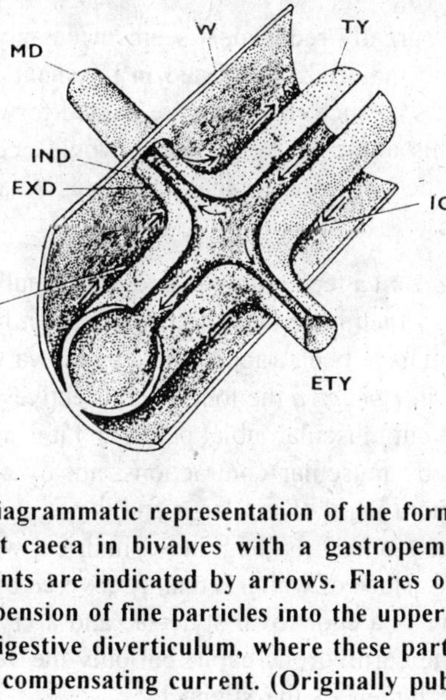

Fig. 4.16 : Diagrammatic representation of the form of the interior of the left and right caeca in bivalves with a gastropemptan type of stomach. Ciliary currents are indicated by arrows. Flares of the major typhlosole deliver a suspension of fine particles into the upper, unciliated part of the duct to the digestive diverticulum, where these particles are accepted into the inhalant, compensating current. (Originally published in Owen, 1955, *Quart. J. micr. Sci. 96*).

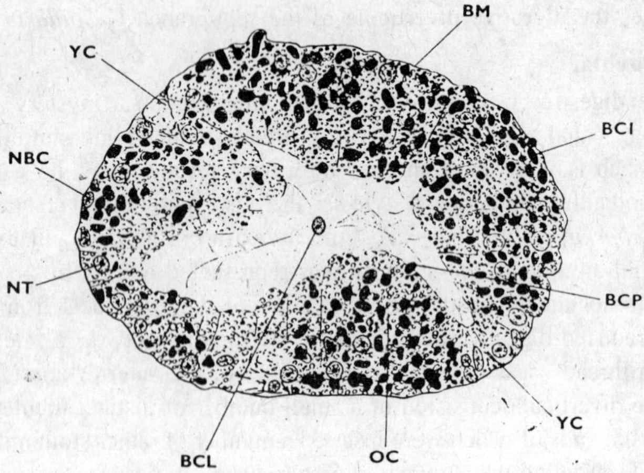

Fig. 4.17 : *Cuspidaria cuspidata*. Transverse section through a single tubule of the digestive diverticula. The specimen had been fixed after feeding for 10 hours on fish blood, and red blood corpuscles (BCL) can be seen in the lumen of the tubule, in process of being ingested by the digestive cells (BCP) and deep within the cytoplasm of the digestive cells (BCl).

Since the gut is so greatly reduced, it seems necessary to consider the possibility that food material is digested in the mantle cavity, and that the soluble products of such digestion are passed forward in liquid form to the mouth. This alone, it would seem, would account for the extreme reduction of the alimentary canal and its associated glands.

Septibranchia

Cuspidaria and *Poromya* feed upon the bodies of small crustaceans, such as copepods, etc., which have been drawn into the infra-septal part of the mantle cavity and have been trapped there by the valve at the base of the inhalant siphon. In *Poromya* the food mass is actively pushed into the mouth by the small but muscular labial palps, and it is then conveyed along the oesophagus by muscular contractions, not by ciliary action. The mouth in both genera is relatively large and is not obscured by the bases of the palps. The stomach is long and cylindrical with muscular, folded walls, the inner surface of which is mainly covered by a protective chitinous sheath. There is a short oval style-sac and a crystalline style is present inspite of the carnivorous habit, but only the very tip of the style projects into the lumen of the stomach.

The digestive diverticula tend to unite with one another to form a compact structure, and the lumina of the diverticula are strikingly wide. They comprise only one kind of cell in which no traces of secretary activity. The digestive diverticula of *Cuspidaria* open into the stomach by only two ducts which were thought to be rather wide, but I myself considered them to be no wider than in most other bivalves (Purchon, 1956).

Cephalopoda

Although, all the Cephalopoda are carnivorous, the class exhibits considerable variety in mode of life and in feeding habits. The familiar *Octopus* is a benthic hunter associated with rocky shores, which lurks in some hollow or crevice in the rocks and emerges from this den to attack crabs which venture near. The *octopus* crawls over the rocks by means of the suckers on its arms, and attacks small moving objects such as crabs, throwing itself upon the crab and enveloping it with the membrane which joins the bases of the arms. The crab is bitten and killed, and is carried back to the den to be consumed, the discarded shells of the prey accumulating on a midden in the vicinity of the den.

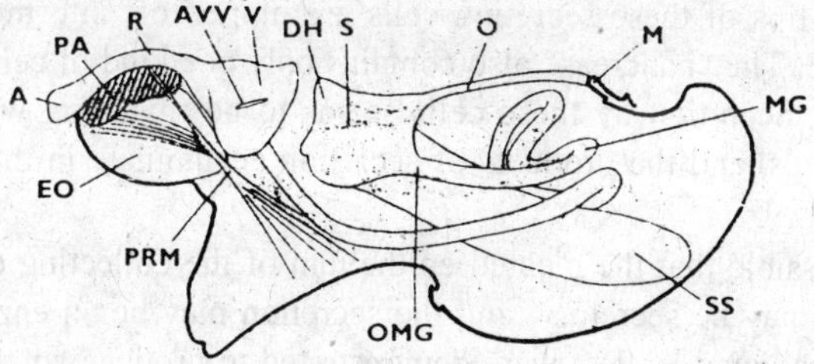

Fig. 4.18 : The course of the alimentary canal in the rock-boring eulamellibranch *Pholas daclylus*, the outline of the shell, mantle, ctenidia, etc., having been omitted from the figure. The relatively enormous style-sac (SS) has become completely separate from the mid-gut (MG) due to fusion of the major and minor intestinal typhlosoles, followed by separation of the two tubes so formed. This advanced condition has been achieved independently in many unrelated lineages and is without any phylogenetic significance.

These various examples differ from one another in occupying different habitates, in having different modes of life, and by capturing different kinds of prey by different techniques. They also evidently differ in their degrees of appetite, for *Octopus* will attack and kill another crab when it has only recently fed, or when it still has a crab held under its interbranchial membrane, while *Sepia* may not feed for some considerable time after having had a meal. The duration of digestion is also very variable: digestion of a meal takes 18 hours in *Octopus*, 12 hours in Sepia, and only about 6 hours in *Loligo* (Bidder, 1950). It follows that there will be other differences, perhaps very important ones, in the digestive processes of the different subdivisions of the Cephalopoda, and it is urgent that generalisations should not be made at the present time on the digestive processes the Cephalopoda as a whole from observations on Individual genera.

The digestive gland consists of two distinct parts, the "liver" which forms a compact mass on each side of the oesophagus, and the "pancreas" which is a diffuse gland investing the walls of the ducts which drain the "liver", and discharging into these ducts. The "liver" is pervaded by a system of blood capillaries, and the organ serves partly as a source of digestive enzymes and partly as a storage organ for food reserves. The epithelium of the tubules contains only one type of cell with distinct immature and mature phases. There is histological evidence that these cells produce two different types of secretion, which mingle in the hepatic ducts. As has been reported for various other molluscan types, the swollen tips of these secreting cells are nipped off into the lumen of the tubule. The "pancreas" also contains only one kind of cell, which is secretory. Secretion by these cells seems to be rhythmic, with long resting periods, and the products of secretion accumulate in the caecal sac.

It is possible that the pleated epithelium of the collecting organ in the caecum may be secretory, and the secretion may be an enzyme. In *Sepia* these ciliated leaflets were demonstrated to produce an activator.

Table 4.4 : Characteristics of Cephalopod[a] Proteases

Origin	Animal and reference	Substrate	Effect	pH optimum	Activation
Digestive Tract					
Crop extract	*Octopus*[a]	Gelatin	–		
	(Sawano,1935)	Chloroacetyl-*l*-tyrosine	+		
		Leucyldiglycine	+		
		Glycylglycine	+		
Stomach extract	*Sepia*[a]	Casein	–[b]		

(Contd...)

Origin	Animal and reference	Substrate	Effect	pH optimum	Activation
	(Ro nijn,1935)	Peptone	–[b]		
		Gelatin	–[b]		
	Ommastrephes[a] (Takahashi,1960)	Casein	+	8-8.5	
Caecum extract	*Octopus*	Gelatin	–		
	(Sawano,1935)	Chloroacetyl-*l*-tyrosine	+		
		Leucyldiglycine	+		
		Glycylglycine	+		
	Sepia	Casein	–[b]		
	(Romijn,1935)	Peptone	–[b]		
		Gelatin	–[b]		
Intestinal extract	*Sepia*	Casein	–[b]		
	(Ro nijn,1935	Peptone	–[b]		
		Gelatin	–[b]		
Intestinal + caecal extract	Ommastrephes (Takahashi,1960)	Casein	+	8-8.5	
Fluids From the Digestive Tract					
Crop fluid	*Octopus*	Gelatin	+	[8.0][c]	
	(Sawano,1935)	Chloroacetyl-*l*-tyrosine	+	[8.2][c]	
		Leueyldiglycine	–		
		Glycylglycine	+		
Caecal fluid	*Octopus*	Gelatin	+		
	(Sawao,1935)	Chloroacetyl-*l*-tyrosine	+		
		Lencyldiglycine	+	[8.2][c]	
		Glycylgycine	+		
	Sepia	Casein	+	6.8	
	(Romijin,1935)	Peptone	+	6.2	
		Gelatin	+	5.6	
		Chloroacetyl-*l*-tyrosine	+		
		Leueyldiglycine	+		
		Glycylgycine	+		

(Contd...)

Origin	Animal and r	Substrate	Effect	pH optimum	Activation
Hepatopancreas					
Liver extract	*Octopus*	Gelatin	+	7.8	
	(Sawano,1935)	Gelatin	+	4.7	Accelerated by KCN
		Chloroacetyl-*l*-tyrosine	+	8.1	
		Leueyldiglycine	+	7.7	
		Glycylgycine	+	8.1	
	Sepia (Romijin,1935)	Casein	+	6.0	Not activated by H_2S, glutathion, or extracts of *Sepia* digestive tract, or enterokinase
		Peptone	+	6.1	
		Gelatin	+	5.5	
		Chloroacetyl-*l*-tyrosine	+	5.5	
		Leueyldiglycine	+	8.3	
		Glyeylglycine	+	8.3	

[a] Unless otherwise stated the species nietioned in Tables II-!V are: *Octopus vulgaris* Lam., *Sepia officinalis* Linn., *Loligo forbesi* Steenstrup, *Ommastrephes sloani pacificus* (Steenstrup).

[b] No substrate mentioned, only absence of proteinase.

[c] pH optimum not given; pH presumed that of experiment.

5

EXCRETORY SYSTEM AND OSMOREGULATION

Molluscs are triploblastic coelomate, unsegmented (except in monoplacophora) and bilaterally symmetrical. Body divisible into head, mantle, foot and visceral-mass. Body divisible into head, mantle foot and visceral mass. Shell when present, usually univalve or bivalve, constituting an exoskeleton, internally in some coelom reduced and represented mainly by pericardial cavity, gonadal cavity and kidney. Digestive System complete with a digestive gland or liver (hepatopancreas) a rasping organ, the radula, usually present circulatory system mainly of closed type, but some emptying into sinuses; heart with one or two auricles and one ventricle; blood with amoebocytes and haemocyanin. Respiration direct or by gills or lungs or both. Excretion by paired metanephridia (Kidneys). Nervous system of paired ganglia, connectives nerves. Ganglia usually from a circumpharyngeal ring. Sense organ include eyes, statocysts and receptors for touch, smell and taste. Fertilization external or internal: development direct or through free larval forms .

The excretion in the molluscs, the opportunity has been taken to discuss some obvious lacunac in our knowledge of molluscan excretion which could be filled by using existing techniques. Although, Picken (1937) estimated the colloid osmotic pressure of the blood of two species of freshwater molluscs, there are few measurements of the colloid osmotic pressure in the literature. It is desirable now that osmometers are available, that measurement of the colloid osmotic pressure of a wide variety of molluscs should be made. Although the evidence suggests that urine is first formed by ultrafiltration. The exact sites of the ultrafiltration membranes are not known in any molluscs. It is suggested that the suspected sites of ultrafiltration should be examined with the electron microscope. The high molecular weight of molluscs blood proteins suggest that the filtration membrane may be permeable to small protein molecules. By the use of a variety of compounds of various molecular weights the permeability properties of the filtration membrane could be defined with the exception of Bouillon's brief work (1960) in *Helix* the fine structure of molluscans excretory organ has not been examined with the electron microscope. The *Helix* Kidney shows some interesting similarites with the verterbrate and crustaceans excretory organs but also some striking differences.

In 1962 the late professor Munro Fox asked to review excretion in Molluscs since this review has been published (Potts 1967), the opportunity will be taken here to draw attention to certain gaps in our knowledge, beginning with a discussion of some aspects of Kidney structure and function.

Although some larval molluscs possess protonephridia, adult molluscs always process Kidneys derived from coelomoducts. The evidence available suggests that urine is produced in the molluscs by ultrafiltration and is later modified by resorption and secretion. Ultrafiltration can be recognized by various criteria. Among these are:

1. The hydrostatic pressure of the blood should exceed colloidal osmotic pressure of the blood at the site of filtration.
2. At the site of filtration there should be a membrane permeable to water and to small solute molecules but impermeable to proteins. If these two criteria are fulfilled a projection free ultrafiltration should be produced in Donnan equilibrium with the blood.

EXCRETION

By definition, excretion means the separation of waste matter from tissues and body fluids and its elimination from the organism. The waste to be eliminated is the result of cellular metabolism. It is, primarily, the end products of catabolic process but may also include unusable or undesirable by products of anabolism as well as of catabolism. The nature of end products is related to the nature of the material that is metabolically degraded or assembled carbon dioxide and water as end products of oxidative metabolism are among them, as well as nitrogenous compounds resulting from the deamination of amino acids prior to oxidation of the carbohydrate residues. Chief among these nitrogenous compounds are ammonia, urea, and uric acid, and according to which of them is proportionately greatest in the mixture of excreted products, animals are conveniently classified as ammonotelic, urecotelic, or uricotelic. Included among excretory, products, are organic acids, as partial products or by-products of oxidative metabolism, and the products of nucleic acid degradation, for these complex compounds are broken down as well as synthesized by cells. While the nitrogen from their pyrimidine bases is usually released as ammonia or incorporated into uric, the purines often are unchanged and, if not further syntheses, excreted as such. All these compounds must be transported in solution or in suspension from the areas in which they are produced to the areas in which they are produced to the areas from which they are, therefore, closely related to the availability of water. In most animals the excretory products are flushed out of the body in a volume of water whose loss may be advantageous in some circumstances, deleterious in others. Along with these product and the water in which they are of value and whose loss, if it occurs, must be compensated for in one way or another. Excretion is, therefore, closely bound up with nutrition, and even more with the preservation of stability in the internal medium at multicellular animals and with their ionic and osmoregulation.

Any exposed surface of an animal's body may serve as an excretory area, as well as one through which water and its solutes can enter or leave the body. The entire integument of animal without exoskeletons probably takes care of the final elimination of much of their nitrogenous waste as well as of that of carbondioxide. Gills and the alimentary canal may be excretory as well as respiratory in their function but in many invertebrates there are tubular system, diffuse or compact, to which the function of excretion has been attributed. These system include contraction vacuoles; nephridia; the urocoeles that are the "kidney" of molluscs; the internal and maxillary glands of crustaceans and the coxal glands of arachnids; and the malpighian tubules of insects and myriapods. It has been generally assumed, partly because of their tubular nature and liquid content ,that these are all functionally analogous to the vertebrate nephron, but in only a limited number of cases has experimental evidence fully justified assumption.

Factors in the Operation of Excretory Systems

Separation of material from body or tissue fluids necessitates its passage across cell membranes. Water salts, and other solutes of appropriate physico chemical constitution also pass across them, so that the fluid in the excretory tubules may contain .initially at least, much that is of value to the organism as well as waste to be removed. This passage of material may be effected by means of diffusion, active transport secretion, or filtration, provided that there is a head of pressure adequate for this. The tendency to assume that invertebrate mechanisms operate in the same way as vertebrate ones has led to the implication that the urine of invertebrates is produced in the same way as it is in the vertebrates kidney, by filtration, secretion and selective reabsorption. Yet in only a limited number of invertebrate species are conditions knows to be possible for ultrafiltration to take place between body fluid and tubular system.

OSMOREGULATION

Ionic regulation is an important part of osmoregulation, although not all animals with ionic regulatory abilities have osmoregulatory ones as well. Osmosis derived from the Greek word osmos, meaning 'impulse" is a special case of diffusion, or diffusion under special circumstances .As defined in Webster's New International Dictionary, it is "The flow or diffusion that take place through a Semi-permeable membrane (as of a living cell) typically separating a solvent (as water) and a solution or dilute solution and a concentrated solution and thus bringing about conditions for equalizing the concentration of the components on the two sides of the membrane because of the unequal passage in the two directions until equilibrium is reached." A semipermeable membrane is strictly speaking ,one through which only water can pass and which act as a barrier to all other molecules. If then, an aqueous solution of any kind is separated by such a membrane from pure water, the rate of diffusion of water molecules from the compartment containing the pure water will be greater than the rate of their diffusion from the compartment containing the solution. Thus, there well be essentially a flow of water across the membrane in to the solution. A hydrostatic pressure that just prevents this flow defines the osmotic pressure of solution. This is a function of its concentration, which is usually expressed in terms of the quantity of solute per unit weight of water. The most useful and generally accepted way of expressing the quantity of solute is in terms of gram molecules or moles (m). This gives the number of particles of any given solute in a solution for according to Avogadro's number, each mole of a substrane contains 6.025×10^{23} particle. For convenience, the concentration of ions solution containing electrolytes are also expressed as moles, mole in these conditions being accepted as synonymous with gram ion.

Osmoconformers

Marine gastropod and bivalve molluscs are also osmoconformers and show some volume regulation, principally also through leakage of salts to the external medium. Soft bodied gastropods such bodied gastropods such as *Aplysia*, the sea hare, and glochidium, apulmonate, increase in weight when exposed to diluted Sea water, but never to the theoretical limit that would be possible nor does their reduced weight in concentrated water reach the theoretical minimum. *Aplysia* attains equilibrium with 75 per cent seawater in 2 to 3 hr, but after this and its initial gain, loses weight until it may becomes lighter than it was before transference to the dilute medium. That this decrease in weight is due to loss of salts have been shown by analyses of the chloride content of the hemolymph and the external medium at the beginning and the end of the experiment.

Osmoregulatory by Water Elimination Through Urocoeles

Among molluscs and arthropods with urocoels, there are species that are marine, estuarine brackish, and fresh water as well as terrestrial in habitat. These phyla, therefore, offer an excellent opportunity for comparative studies of ionic and osmoregulation and of the adaptability to different conditions of an excretory organ constructed from a coelomic space from the species that have been investigated , it seems evident that the common mechanism of urine formation is ultrafiltration from the vascular system into the pericardial cavity and modification of the filtrate after its passage into the urocoele.

All cephalopod molluscs are marine and have no problem of water elimination or water conservation. Tests have shown that the hemolymph, pericardial fluid, and urine of *Sepia*, *Eleclone* and *octopus* , representative genera of the class, are isosmotic with each other and with seawater, although differing from it and from each other in respect to nearly all the ions. Production of urine ceases when mesonopericardial duct is ligated, providing cogent evidence for the origin of urine from the pericardial fluid.

The source of the pressure that would make ultrafiltration possible is problematical ,but the fact that in *Octopous dofleini* the rate of formation of the pericardial fluid rises from a base level of 5 to 10 ml/hr to values as high as 48.6 ml/hr 1 hr after the injection of 100ml of seawater or of hemolymph shows a definite relationship between blood volume and urine production. Although, pressure in them would be adequate, filtration through the thick, muscular walls of the systemic or branchial hearts is unlikely, nor does it seem possible from the renal appendages for the hydrostatic pressure in them is probably too low. The most acceptable current hypothesis is that it occurs in the branchial heart appendages, the homologues of the pericardial glands in bivalves. Although, it has not been possible to measure directly the pressure the appendages, it can be measured in the branchial hearts. Such measurements in *octopus dofleini* have shown that the rate of filtration is related to the hydrostatic pressure within the heart and that the production of pericardial fluid stops when that pressure corresponds to one of about 4ml of water. This is about equivalent to the colloidal osmotic pressure of the blood, attributable to its content of protein and other molecules of colloidal dimensions. It is possible also to insert catheters into the renal organs and take samples, either intermittently or continuously, of the fluid in them. Analyses of the pericardial fluid and urine of *Eledone* have provided evidence for resorption of K^+, Ca^{++}, Mg^{++}, and Cl^- and for secretion of unocoele. Water and ions are taken in through the gills and other permeable areas, including the gut. There is no evidence for resorption of water in the urocoele, and the fact that insulin injected into the bloodstream is recovered in the same concentration in the urine may be taken as evidence that no water is lost from the initial filtrate. Ammonia is the principal nitrogenous product excreted, and it is eliminated in comparatively large amounts.

Excretion

Like many coelomic surfaces, the mallurcans. Pericardium take on excretory function. The type of excretory organe are developed first, the pericardial epithelium may itself is thickened to form deep brown-colour pericardial gland. In lower gastropodes and in many lamellibranchs there usually lie over the article wall in higher gastropods the gland are on the pericardial side-wall and in some lamellibranchs (*e.g.*, Unionidae) they may form extensions leading from the pericardium into the tissues of the mantle, known as kebers organs. In cephalopods they constritute the glandular appendages of the accessory branchial heart.

In coelomoducts which apen by renopericardial apertures from the pericardium are uselly glandular, forming a renal organ or plied by the rencel portal system.

Mallusca kidneys have various form. In chitons they are parire symmetrical tubes, each pralonged forword from its pericardial apening near the renal pare and provided with numerous small dendritic outgrowth ramifyling along its whole length. In gartropoda the kidney is usually a thick walled sac much expirided and massively folded within. In higher gartrapoda only the left part-torrional kidney remains. In Heteropoda and some Gymnosomata the kidney is thin-walled and transparent while in the dorid nudibranches it is greatly subdivided into rlender branches ramifying among the vircera. In the prosobranchia the kidney open into the hinder part of the mantal cavity. In higher apirthabranchia directly to the exterior on the right side; In the pulmonata its duct is narrouely drawn out alongride the rectum to open with the anus out side the mental cavity. The kidneys of bivalves are paried tube each ent on itself as a ⊃; the lower (or proximal limb is grandular and sperms from the pericardium while the upper (or distal) limbs is a thin walled bladder appening into the mental cavity. In primitive protobranchia the whole tube is excretery. In the cephalopoda the kidneys are infalted sacs with smooth out side wall their glandular tissue had shifted to form the spongy mass of renal appandages surrounding the afferent branchial veins which pursue their course through the middle of either kidney.

The blood in the heart has a high hydrostatic pressure from its through the wall of the auricles and ventricles, is continuously transferred a clear filtrate into the pericardium.

This fluid is isotonic with the blood but with the much less non-mineral with the recretions of the pericardial gland. It passes through the renopericardial spenings to the kidneys. Here the production of nitrogenous excretion are added to it by the actin of the lining cells. Inorganic substances may be taken back into the wood and the composition of the resulting urine is determined by the extent of renal recretion and of reabsorption of inons[209] in fresh water lamillibranchs such as Anodonta and in gastropods like lymnaea where-retention of inorganic ions is important most of the salt are resorbed in the kidney so that the urin becomes hypotonic to the blood and pericardial filtrate. And even in marine Molluscs some amount of ionic regulation exists resulting in the conservation of physiology varible substances thus the blood of marine gastropods Buccinum, Neptunea and Pleurobranchus calcium and potassium exceed the equilibrium values with sea-water. There is a slight elimination of magnessium while Na and Cl are in almost identical conc^{n} with that of sea water. In the active bivalves pectin and Ensis there is a similar picture but there is less accumulation of K in the more redentary Mga. In the very active cephalopods regulation extends to all the ions K^{+} and Ca^{++} being much in excess of there sea-water value Mg and Cl only slightly above and sodium (Na) * SO_4 rather below Molluscs appear to take up in organic ions from sea water by absoption through the gills and expored by surface rather than by the gut. At least with K, Mg, and Cl are usually with calcium, this can take place against a concentration gradient by performance of osmotic work.[229]

Through capable of ionic regualtion, marine Molluscs are in general permeable and in total osmotic equilibrium with the sea water[213] they have little power of osmotic regulation.

Fresh water gastropods and bivalves work with a very low ion concentration. Anodonta has indeed the lowest blood concentration recorded for any animal (0.08°C equivalent to 4-5% sea water). The problem of continuous osmotic inflow of water is solved in two ways. They ball themselves out continually by producing a copious urin about 25% of shell/body wieght daily or 65% of the extracellular fluid this uring is strongly hypo-osmotic with the recovery of salf by the renal organ in addition there is hyper-osmotic regulation especially with the uptake of ions. Such as Na^+ and Cl^- at the body surface dilute urine can reduce the asmoregulatory requirement by 80-90% but in Anodonta this still consumes 1.2% of the total metabolic energy.

Some fresh water aperculates such as viviparus furciatus (35g) and Bithynia tentaculata (150g). Where found by Needham to have relatively large amount of uric acid and this agrees with other evidence painting to a common evolutionary organ of some groups of terrestrial and fresh water prosobranchs *Hydrobia Jenkinsi e* has only recently—since about 1900-migrated to riverer from brakish water has no uric acid[189].

Most aquatic Molluscs excrete amonia either directly or in part converted to amines and urea. Ammonia is a paironous substance requiring solution in a large volume of water and the fresh water bivalve Anodonta appear partly ammonotelic, Bivalves in fact, being thoroughly aquatic seem rarely to form uric acid at all Needom found the nitrogenous excrete of Mga arenaria distributed as 21.5 parts ammonia 4.5 parts urea and 18.0 part amino acid and creatine.

The renal sac fluid of cephalopods yield a very high proportion of ammonia, sometime one third or 2/3 of the non-protein nitrogen, with smaller amounts of purines, amines urea and sometime uric acid there is high rate of conversion of ammonia to urea seen for example in mammals. In the water of a cephalopod aquarium there may be evern a higher production of amonia to other nitrogenous matter than in the urine; additional amonia is thus probably excreted by the gills Ar will as the pericardial gland and kidneys, the cephalopod digestive gland is an important excretory organ in 100gm of fresh 'liver' of octopus where detected 30-50 mg of amonia 6-25 mg of urea as well as 3-17 gm of uric acid as in the vertibrate liver excretory proudcts may be primarily formed in the digestive gland they are excreted from the blood by the pericardial glands, renal organs and gills.[24, 299]

In other Molluscs especially in gastropods where the kidney is an in some opisthobranchs, reduced the digested gland may discharge excretory products straight into the gut. Special excretory cells are developed, particularly in those herbivores whrere the digestive gland returns to the lumen chlorophylous pigment taken up by the blood. The gastric fluid of Aplysia is known to contain teeth urea and uric acid and the digestive gland was found to yield 13mg urea to 100 g fresh weight.

EXCRETION

Like many coelomic surfaces, the molluscan pericardium takes on excretory functions. Two types of excretory organ are developed. First, the pericardial epithelium may itself be thickened to form deep brown-colured pericardial glands. In lower gastropods and in many lamellibranchs these usually lie over the auricle wall. In higher gastropods the glands are on the pericardial side-well, and in some lamellibranches (*e.g.*, Unionidae) they may form extensions leading from the pericardium into the tissues of he mantle, known as Keber's organs. In cephalopods they constitute the glandular appendages of the accessory branchial hearts.

The coelomoducts which open by renopericardial apertures from the pericardium are usually glandular, forming a renal organ or kidney, extracting nitrogenous waste from the blood which is supplied by the renal portal system. Molluscan kidneys have various forms. In Chitons they are parted symmetrical tubes, each prolonged forward from its pericardial opening near the renal pore, and provided with numerous small dendritic outgrowths ramifying along its whole length. In Gastropoda the kidney is usually a thick-walled sac, much expanded and massively folded within. In higher Gastropoda only the left post-torsional kidney remains. In Heteropoda and some Gymnosomata the kidney is thin-walled and transparent, while in the *Ooris* nudibranchs it is greatly subdivided into slender branches ramifying among the viscera. In the Prosobranchia the kidney opens into the hinder part of the mantle cavity, in higher Opisthobranchia directly to the exterior on the right side; in the Pulmonata its duct is narrowly drawn out alongside the rectum to open with the anus outside the mantle cavity. The kidneys of bivalves are paried tubes, each bent on itself; the lower (or proximal) limb is glandular, and opens from the pericardium, while the upper (or distal) limb is a thin-walled bladder opening into the mantle cavity. In the primitive Protobranchia the whole tube is excretory. In the Cephalopoda the kidneys are inflated sacs with smooth outside walls. Their glandular tissue had shifted to form the spongy mass of renal appendages surrounding the afferent branchial veins which pursue their course through the middle of either kidney.

The blood in the heart has a high hydrostatic pressure. From it, through the walls of the auricles and ventricle, is continuously transfused a clear filtrate into the pericardium. This fluid is isotonic with the blood but with much less non-mineral matter. With the secretions of he pericardial glands, it passes through the renopericardial openings to the kidneys. Here the products of nitrogenous excretion are added to it by the action of the lining cells. Inorganic substances may be taken back into the blood, and the composition of the resulting urine is determined by the extent of renal secretion and of reabsorption of ions. In freshwater lamellibranchs such as *Anodonta* and in gastropoda like *Lymnaea*—where retention of inorganic ions is important—most of the salts are resorbed in the kidney so that the urine becomes hypotonic to the blood and pericardial filtrate. And even in marine molluscs some amount of ionic regulation exists, resulting in the conservation of physiologically valuable substances. Thus, in the blood of the marine gastropods *Buccinum, Neptunea* and *Pleurobranchus* calcium and potassium exceed the equilibrium values with sea water. There is a slight elimination of magnesium, while sodium and chlorine are in almost identical concentration with mat of sea water. In the active bivalves *Pecten* and *Ensis* there is a similar picture; but there is less accumulation of potassium in the more sedentary *Mya.* In the very active cephalopods regulation extends to all the ions, potassium and calcium being much in excess of their sea-water value, magnesium and chlorine only slightly above, and sodium and sulphate rather below. Molluscs appear to take up inorganic ions from sea water by absorption through the gill and exposed body surface rather that by the gut. At least with potassium, magnesium and chloride, and usually with calcium, this can take place against a concentration gradient by performance of osmotic work.

Though capable of ionic regulation, marine molluscs are in general permeable and in total osmotic equilibrium with the sea water. They have little power of osmotic regulation when placed in lowered salinities, though over long periods they may osmotically adjust their blood concentration to more dilute media. Many marine molluscs such as *Patella, Nucella, Cardium, Ostrea* and *Mytilus* have been reported to acclimatize to very low salinities. Most bivalves such as *Mytilus* and *Ostrea* prevent loss of salts over short periods in fresh-water by tightly closing the shell. The marine *Scrobicularia plana* can live in brackish estuaries; its internal osmotic pressure varies with that of the medium over a wide range; through after an equilibration period there is some suggestion of active control of blood concentration.

Freshwater gastropods and bivalves work with a vary low ion concentration. *Anodonta* has indeed the lowest blood concentration recorded for any animal (0.08°C, equivalent to 4-5% sea water). The problem of continuous osmotic inflow of water is solved in two ways. They bale themselves out continually, by producing a copious urine, about 25% of shell/body weight daily, or 65% of the extracellular fluid. This urine is strongly hypo-osmotic with the recovery of salts by the renal organ. In addition there is hyper-osmotic regulation, especially with the uptake of ions such as Na^+ and Cl^- at the body 80-90% but in *Anodonta* this still consumes 1.2 % of the total metabolic energy.

In land pulmonates water must be no longer baled out but conserved. Here the renopericardial aperture is very small and little if any fluid is sacrificed from the pericardium to the kidney. A form of nitrogenous excretion is evolved which requires little water, like birds.

In cephalopods the pericardial glands have become attached to the accessory branchial hearts. The chief sites of filtration are the glandular renal appendages laying on the afferent branchial vein. This vessel is peristaltic, and its contractions appear to forces small amounts of fluid through the walls of the small blood vessels which open in a labyrinth from this vein.

Uricotely occurs whenever snails migrate to land, irrespective of whether or not they are true pulmonates. In land operculates such as *Cuclostoma*, derived near littorinids, the uric acid content is very high (more than 1000 mg per g dry weight of kidney). Among shore littorinids there is an ascending series in uric acid content, the comparable figures being 1.5 g in *Littorina littorea*, 2.5 g in *L. obtusata* 5 g in *L. rudis* and 25 g in the supra-tidal *L. neritoides*. The figures for Helicidae are extremely high-as much as 600, 720 and 800 g in different individuals of *Helix pomatia*. In pulmonates that have returned to aquatic life uricotely is progressively lost. The series *Lymnaea stagnalis* (115 g), *Planorbis corneus* (41 g), *Ancylastrum fluviatile* (4 g), and *Lymnaea peregra* (0.2g) shows this though apparently not in strict order of aquatic readaptation.

Some freshwater operculates such as *Viviparus fasciatus* (35 g), and *Bithynia tentaculata* (150 g), were found by Needham to have relatively large amounts of uric acid; and this agrees with other evidence pointing to a common evolutionary origin of some groups of terres trial and freshwater prosobranchs. *Hydrobia jenkinsi*, which has only recently-since about 1900-migrated to rivers from brackish water has no uric acid.

Most aquatic molluscs excrete ammonia, either directly or in part converted to amines and urea. Ammonia is a poisonous substand requiring solution in a large volume of water, and the freshwater bivalve *Anodonta* appears partly ammonotelic. Bivalves, in fact, being thououghly aquatic, seem rarely to form uric acid at all. Needham found the nitrogenous excreta of *Mya arenaria* distributed as 21.5 parts ammonia, 4.5 parts urea and 18.0 parts amino-acids and creatine. The renal sac fluid of cephalopods yields a very high proportion of ammonia, sometimes one-third to two-thirds of the non-portion nitrogen, with smaller amounts ofpurines, amines, urea and sometimes uric acid. There is never the high rate of conversion of ammonia to urea seen, for example, in mammals. In the water of a cephalopod aquarium there may be even a higher proportion of ammonia to other nitrogenous matter than in the urine; additional ammonia is thus probably excreted by the gills. As well as the pericardial glands and kidneys, the cephalopod digestive gland is an important excretory orgąn. In 100 g of fresh 'liver' of *Octopus* 30-50 mg of ammonia, 6-25 mg of urea as well as 3-17 g of uric acid were detected.

In other molluscs, especially in gastropods where the kidney is, as in some opisthobranchs, reduced, the digestive gland may discharge excretory products straight into the gut. Special excretory cells are developed, particularly in those herbivores where the digestive gland returns to the lumen chlorophyllous pigments taken up by the blood. The gastric fluid of *Aplysia* is known to contain both urea and uric acid, and the digestive gland was found to yield 13 mg urea to 100 g fresh weight.

NITROGENOUS EXCRETION

The nitrogenous end product of protein metabolism in marine gastropods is mostly ammonia (Fig. 5.1), much of which diffuses across external epithelia. In so doing, ammonia raises the pH of mantle epithelium and probably helps the process of shell secretion by enhancing the conversion of soluble bicarbonate into insoluble carbonate ions. Urea is not usually produced, but uric acid occurs in the tissues in amounts proportional to the body mass and is perhaps used as a depot of nitrogen rather than as a dump for waste material.

Littorina littorea, however, excretes uric acid and some urea when exposed to the air at low tide, reverting to the excretion of ammonia when immersed by the rising tide. Excretion of nitrogen as uric acid or urea is metabolically more expensive than producing ammonia, but conserves water. The alternation between forms of nitrogenous excretion

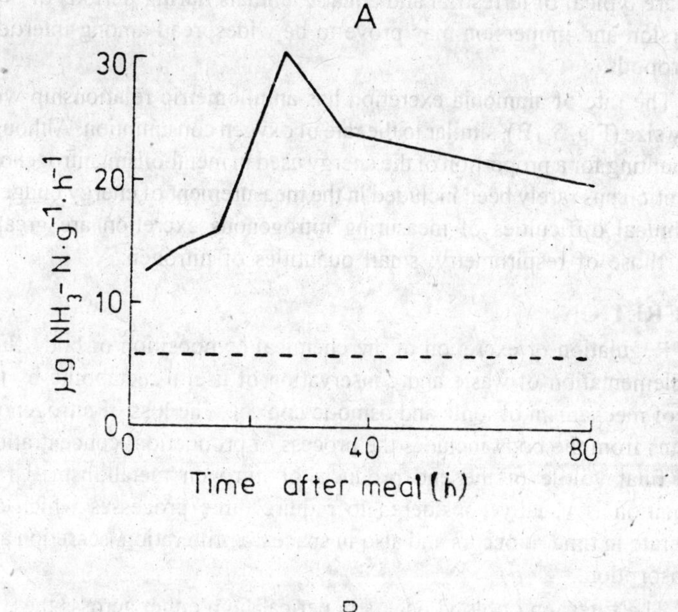

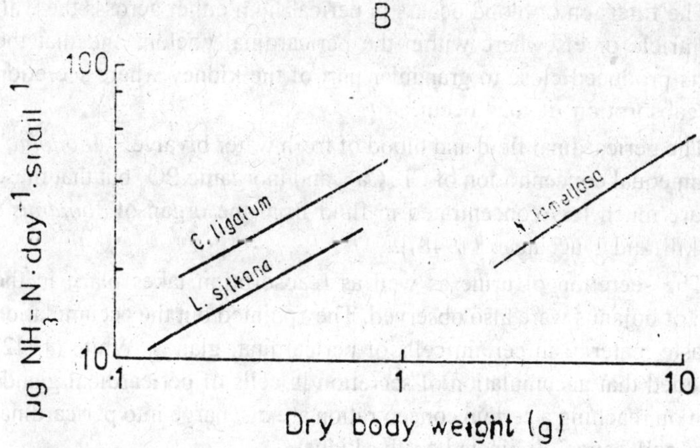

Fig. 5.1: (A) A Pulse of Ammonia Excretion by *Buccinum undatum* Occurs Immediately after Feeding. The rate of ammonia excretion per unit dry body weight then gradually declines towards the 'starvation' level (dashed one). After Crisp *et al.*, (1981). B. The Rate of (Ammonia Excretion is Allometrically Settled to Body Weight. Giving a Straight Line on a Double Logarithmic Plot. The slope of the regression line is 0.59 for *Calliostoma ligatum*, 0.57 for *Littorina sitkana* and 0.63 for *Nucella lamellosa*. After Duerr, (1968).

that are typical of terrestrial and aquatic animals during periods of tidal emersion and immersion may prove to be widespread among intertidal gastropods.

The rate of ammonia excretion has an allometric relationship with body size (Fig. 5.1B), similar to the rate of oxygen consumption. Although accounting for a proportion of the energy used in metabolism, nitrogenous excretion has rarely been included in the measurement of energy budgets. Technical difficulties of measuring nitrogenous excretion are greater than those of respirometry: small quantities of nitrogen.

EXCRETION

Regulation or excretion of the chemical composition of body fluid' by elementation of waste and conservation of useful metabolite by the fin. of mechanism of ionic and osmotic con:.:ol. The loss of nitrogenous wastes from the body includes the process of production, concentration, and final voitile of the end products of nitrogen metabolism. Urine formation is usually considered to require three processes which are separate in time. It occurs and also in space *i.e.,* filteration, secretion and reabsorption.

The filter ion of blood occurs in pericardium either acrosss the wall of ventricle or elsewhere within the pericardinal coelom and that the fluid is produced close to grandular part of the kidney where secretion and reabsorption of ions occur.

The pericardinal fluid and blood of fresh water bivalve. "*Anodonta*" contain equal concentration of Cl^-, Ca^{++} and inorganic PO_4 but that these ions are much less concentrated in fluid from the organ of "*bojanus*". [Eforkill and Duchateau (1948)].

The secretion of urine as well as reabsorption takes place in the organ of bojanus were also observed. They pointed out the accumulation of waste material in certain cells of pericardinal glands. White (1942) recoreded that accumulation of secretion in cells of pericardinal glands which on reaching a certain concentration are discharge into pericardinal cavity and hence eliminated to the kidney.

The condition based on a pieces of evidences:

1. The pericardinal fluid was isotonic with blood.
2. Pericardinal fluid contain less non-mineral substance than fluid as based on refractive index.
3. The blood had a high hydrostatic pressure in range of 3-8cm of water.

The identification of ventricle wall as a site of ultrafiltration of blood. It measured high systolic pressure in ventricles of number of bivalves.

The blood of *Modiolus spp.* forms to be significantly hyper-osmotic to the pericardinal fluid. In such situation a grandis of osmotic pressure would result between pericardinal fluid and the blood and in order for filtration occur across the ventricle wall the systolic Blood Pressure would have to exceed this osmotic pressure differences.

The ultrafiltratrion across the wall of the heart from the ventricle to the pericardial cavity could not occur. The rate of filteration of blood in "*Andonta*" by opening and pericardinal cavity, and collecting the fluid as it is formed by so abolishing the pericardial back pressure and here the measurement based on rate of filteration by insulin clearance. The filteration rate of 23ml/kg-1ml/kg/hr. for *Mytulus californiasis* for most invertibrate filter^{n} rate in kidney lie in the range of 1-0% of the body wt./hour. In "*Mytilus*", there is a complete complement of uricolytic enzymes "*Xanthin oxidase*" which is active in conversion of Xanthin to uric acid.

All the uricolytic enzymes are present in the digestive diverticula of bivalves. Nichelas. Though they could find no activity in extracts of kidney or mental tissue. It appears that the classical pathway of purine catabolism or purinolysis occur in muscles. It is suggested that all the enzymes or ornithin cycle and purinolysis are therefore, probably occur in at least in *mitules* but other possibilities for degradation of arginine also exist.

The physiological importance of these two cycle (Oknithin and Pyrolitic) in bivalves is not known but there are certain possibilities that during prolonged periods of wall closure capacity to dietoxify NH_3 by conversion of well would have adaptive value. The presence of urase may be functional in degrading to ammonia any excess used that accumulates. The alternative role of urase is free ammonia might react with hydrogenase formed by the action of carbonic anhydrate on bicarbonate so releasing ions for deposition in cell as $CaCO_3$.

NEUROSECRETION

Neurosecretory area exist in cerebral commissure of *Perna-perna* but the major neurosecration function in bivalves is centered on certain cells in the nerve ganglion. Neurosecrating cells in cerebral and visceral gangalion of *Mytilus spp.* The number of expected location of these less after from spp. to spp. The presence of neurosecreating cell in pedal ganglion is controversial. They are in *Mytilus edulis* or *Mytilus gallo provincia* but they are present in pedal ganglion of *Perna-Perna. Two types of neurosecrating cells* have been identified in *Mytilus. Some cells are pearshaped, uninuclear and upto 25μ in length when other are small and multiple nuclear.* The transport and fate of neuroceratory products is largely unknown. Movement of neurosecretory substances by axon intermediate cells and possibly by *glial*. Neurose cells in *hanells* has been associated with representative growth metabolism and responsed to environmental streak. Annual neuroceratory cycle in pear-shaped neurose. Cells of cerebral gangalion demonstrated the small multipular neurons neurosecratory cells of visceral gangalion showed continuous activity throughout the year. The annual neuroceratory cycle and gemetogene cycle appear to be closely co-related secretory material is accumulated in the cerebral ganglion during gametogenesis and that evaquated. From the cells when the gametes become fully matured removal ablation of cerebral ganglion during the resting phase of gemetogenetic cycle and at the beginning of activity gametogenesis delays maturation. Width many oocytes undergoing lysis before spawning ablation of cerebral gangalion the end of the gametogenesis before spowing hastens the maturation and gametes release. It is impossible to excite the mechanism of nervous and or hormonal in nature.

It the activities of neurohormones are in P. removal of internal inhibition such as—neurosecratory products of cerebral ganglion may allow the animal to be receptive to the external *stimulli* $\overline{c}$ then included the release of gametes. In *fresh water, mussll, Doreissna polymorphes* premature spawning may be and fin. of intensity of operative shock when abliation the ganglion. The + of a hormone active in inducing spawning linked with clear-cut spermetogenesis was also noted.

GROWTH AND METABOLISM

There is a close parallel between a seasonal cycle of neurosecretory cells activity and the rate of O_2 consumption as with a spauning and gametogenesis cycle whether there is a positive relationship. It is not conformed *Removal of cerebral ganglion has little/no effect on shell or body growth or on glycogen metabolism and storage.* However, abliation of cerebral *ganglion does result in disorders of lip metabolism, particularly, and red" in lipid accumulation.* The abliation of visceral *ganglion* has more or severe consequences. The mussels do not survive for more then 3-4ml environmental stress. The effect of salinity and temp. change on neurosecretory activity within the cerebral *ganglion* of

Mytilus, galloprovincialis is surprise in temp. 10°C maintained for one hr. or a sudden reaction in the salinity upto 20% resulted in an emptying of neurosecretory of cerebral ganglion. A falling temperature marease in salinity to 45%. were followed an accumulation of neurosecratory products. There was no variation in neurosecratory activity in visceral ganglion following similar experimental treatment. Ablation of cerebral ganglion paired the mussel. The mussels capacity for iso-osmotic intracellular regulation at resalinity resulting in an increase in ami-nitrogen content in the tissues. *Neurohormonal play a significant role in physiology of mussel* but the limited range of techniques available ablition of whole ganglion indirect of neurose activity with physiological events.

Neurohumoless (Control of cilliary activity) : The ciliated gill epithelium of lamellibranch provide a favourable system for investigating the mechanism and the control of ciliary motion in mussels.

Ciliary activities on gill of Mytilus is a feed by stimulation of branchial nerves. The fibrid from the branchial nerve lie adjacent gill filament and penetrate the fibrous basal lamilla under the epithelium. The electrical stimulation and the pattern of ciliary activation of the visceral ganglion and branchial nerves indicates that filament even small groups of adjacent small filament are independently innervated permitting for discrets control of ciliary activity on the different parts of the gills. The cilioexcitation is mediated by five hydroxyl/ptamine; *small quantity of 95 HT) in gill and the enzymes of 5HT synthesis and degradation 5HT has no effect on musculature of gill.* The gill cilid of bivalve molluscs may continue to be at for several days after excitation from the animal. Energy is a rich substant stored endogenously within the cell. The endogenous (glycogen) formation of ATP from glycogen may be linked to an anecerobic pathways or oxidative phophorylation 5HT has been shown to stimulate both anaerobic glycolysis and the rate of O_2 consumption by the gills. O_2 was required for prolonged ciliary beating. Bicarbonate and corbonic acid buffer system to prevent organic acids which might accumulate anaerobic end products from lowering pH.

The lateral cilia of *Mytilus* are capable of prolong activity without O_2 with add^n to micro molar can be blocked by inhibitors of glycolysis indicating that 5HT may exerts its effect through a rate limiting step in glycolysis 5HT may activate the enzyme phosphoglucomutase.

Acetyl choline is widely distributed inhibitory neurotransmeter. In vertebrate but it falls to stop cilliary beating when applied on the gills of *Mytilus edulis*. Catacholamines, dopamine are natural inhibitory neurotransmeter *specially, dopamine in the nerve ganglion Mytilus edulis* and other bivalves. The localized catacholamine in branchial nerve within gill epithelium showed that gill has capacity to metabolize dopamine. Its also reported that other catacholamine such as nor epinephrine or epinephrine has no significant on ciliary beating *5HT and dopamine are two neurohumous for cellular excitation and cellius inhibition respectively.*

Functional Organisation of Brain

The nervous system is flat in a general pattern as high degree of cephalization. The parts of brain contain three groups:

1. Suboesophageal mass
2. The higher motor sensors of supraoesophageal mass.
3. Optic lobes and correlation centres of supraoesophegeal mass.

1. *Suboesophageal Mass : It consist of both lower and intermedia motor centres anterior and posterior chromatophore lobes are lower.* Centres in that they issue nerve fibres which pass directly to the effector organ and low level. *Mechanical stimulation or thus lobes may cause very local responses by chromatophores. The reminder crest of part of supoesophageal mass does not fin. in the manner and is deemed to comprise intermediate motor centres.*

   At the ant end of supoesophageal mass the branchial lobe is penetrated drops of fibrics which paw from ant. Pedal lobe into the army Bg avoiding these nerve tracts and stimulating the substance and branchial lobe itself no resources were obtained from the arms or from the chromophores. *Stimulation of the ant pedal lobe result in the movement of arms and tentacles.* Some of the response being recongnisable as compounds of normal behaviour. *Stimulation of post pedal lobe resulted in restruction of the head and the movement of the funnel. The lateral pedal lobe admister the extensing and interensing muscles of eyes and stimulation of these lobes cause movement in both eyes or of ipsee. The ant chromatophore lobe lies in front of the ant pedal lobe and when it is stimulated this brings about expansion of chromatophore. On the front of the head and the arms. Movement of the skin creation of skin, skin papilli of the head and the arms. The pelvis visceral lob fin. both as a lower and as an intermediate motor centre. Superficial stimulation of the lobe at various points may cause and violent expiratory spawn. Retraction of the head movements of the callar, movements of the funnel or ejection of the ink.*

Stimulation of neuropila of the lobe may cause any or all of these fin. but without coordination. In posterior chromatophore lobe lie lateral dorsal lower to paliovisceral lobe. It function as motor centre threshold *stimulation of the lobe causes expansion of chromatophores of the ipsee lateral side. There is a commissure between left and right lobe and stimulation of one of the lobe at higher voltage result in expansion of chromatophore bilaterally.* Due to presembly to diffusion of nervous pulse across the commissure. Mechanical stimulation of the lobe show that they are responsible for local groups of chromatophores in comparison to mussel mantde.

In fin lobes lateral dorsal to pelia-visceral lobe these are lower motor centre But doesn't behave that way only unnature movements of the whole fin was noted. *There is no commisure between left and right fin lobe in the stimulation of one lobe never products bilateral responses of the fins.*

The characteristic undulating movement of the fin and diverse form of coordination of left and right fin must be controlled from the supraesophageal brain mass.

The nerve tract which pass indirect to the fin lobes via posterior pedal lobes. The magno cellular lobe contains the cell bodies of two first order giantfibres of Decapod. In octopoda, No giant fibre system. *In octopus, stimulation of post part of magnocellular lobe cause an inspiration but for expiration is caused by periovisceral lobes. These two lobes therefore together control the respiratory rhythems in octopods.*

2. *Higher motor centres of Supraoesophageal Mass: The fin. of breathing swimming, timing, attacking the pray and adoption of particular colour patients can't be administered by lower intermediate motor centre but are dependent upon coordinating actions provided by higher motor centers* $\overline{c}$ issue the necessarily complex interaction to the effecter organ via intermediate and lower.

These fin. are not depended in any way upon the correlation centre four animal has $\overline{c}$ has suffered the correlation centres by experimental reason can function normally in all respects so long as higher motor centres remain undamages. *Loss of correlation centres only upsets the capacity of animal learn and remember.* Many of the activities of body and are built up by coordination of activity of separate effector organs resp. swimming and feeding.

Sometimes median basal lobe is a higher motor centres $\bar{C}$ controls the suboesophageal centre concerned with the movement of the stimulation cause gentile respiratory movement or alternatively giant fibre system may be distributed. Thereby invoking rapid escape reaction and perhapes also the injection of ink. *The ant basal lobe is capable of invoking any pattern of fin. movement that occur. In normal life. The protraction and retraction of head but in addition the enorbasae lobe is able to cause the rotation of head. Ant basal lobe result* in reducing all component to change in direction such as *movement of funnel and associated move of head, eyes, arms, reproducing exactly normal behaviour of cattle fish.* When it turns to attack the prawn. Rising of Ist and 2nd pairs of arms to the position of attention then recurving the arms backwards around the head but this lobe is not responsible for longing the attack itself.

The interbasal lobe alone responsible for funnel of protrussion of tentacles to grasp the prey stimulation causes proctrussion of tips of tentacles out of the tentacles pits.

There must be extensive interreaction between other lobes of supraoesophageal mass for purpose of attacking, biting and swallowing the prey. The lateral basal lobe is responsible for expansion of chromatophores, creation of gill papillae on head arms, mantle. Efferent motor tracts pass from lateral basal lobes.

The sub padiculate lobes play a major part is sexual part of female octopus by inhibiting development of optic gland and thereby delaying the onset of sexual maturity until a body wt. of 1,000 gm has been reached.

Tests in males are delayed in a similar manner stimulation of *superior* and *inferior buccul ganglion* has intriculate and biting move of the jaws. The whole movement of eating is dependent on the interaction of these ganglion with medicine basal lobe with interior frontol lobe. *The peduncle and olfactory lobes stimulation cause general expansion of chromatophores* all over the body.

There are very good circuits for informations the tactile exploration by arms level of a system of tactile, learning and memory seems to be confined to octopodidae. The sub frontal lobe is divided internally into left and right valves between them there are numerous transverse connections. *The median inferior frontal lobe and vertical lobe plays do part in tactile learning process but it involves in distribution of tactile representation.*

3. *Optic lobes and correlation centres of supraoesophageal mass: The most useful information regarding the functions visual discrimination, visual learning using normal Octopuses and Octopuses which have suffered various branch legends. Neurons in substance of optic lobe tend to have dentrites pick up field oriented in plains corresponding to the horizontal or to the vertical plain in the visual field. Such neurons are tend to be fired by objects in visual field and oriented in particular base.* 'In this perhaps the other ways objects seen can be classified the vertical extent as compare to the horizontal extent.
   (1) This capacity optic lobes classify the object seen lay the foundation of for capacity of animal to discriminate such object.
   (2) To maintain representation of object seen until the object is handled and bitten when tactile and gustatory stimulli are also received.
   (3) To receive the both tactile representation from the skin and gurtatory representation form the mouth, to allow these representation to become dispersed through the substance of the optic lobe.

Classifying neuron summate the information received from these diverse sources on the basis of previous experience.

The result of analysis passed to the higher motor centre sites in the basal lobes of supraoesophageal mass.

OSMOTIC EQUILIBRIUM

A. Occurrence in Various Groups

Freezing-point measurements and chemical estimations have shown that the blood of marine molluscs of the major classes has the same total concentration of ions and other osmotically active particles as sea water. Thus the blood plasma is in osmotic equilibrium with sea water across the animal's permeable membranes, such "as the gills, at least within 1-2%. Table 5.1 shows most of the freezing-point data. To the

Table 5.1 : The Freezing-Point Depression (Δ) of the Blood Plasma of Marine Molluscs in Relation to that of the External Medium.

Species	Blood (Δ°C)	Sea water (Δ°C)	Equilibration period	Authority
Gastropoda				
Prosobranchia				
Archaeogastropoda				
Tugalia gigas	1.96	1.99	24 hr	Yazaki (1929)
Mesogastropoda				
Dolium galea	2.24	2.27	–	Bottazzi (1908)
Littorina littorea	1.99	1.97	2-21 days	Todd (1962)
L. littorea	1.42	1.42	4-7 days	
L. littorea	0.97	0.97	3-7 days	
L. littorea	1.06	0.48	3-7 days	
Opisthobranchia				
Anaspidea				
Aplysia fasciata	2.32	2.27	–	Bottazzi (1908)
Bivalvia				
Lamellibranchia				
Taxodonta				
Anadara inflata	1.95	1.95	24 hr	Yazaki (1929)
Anisomyaria				
Mytilus edulis	2.09	2.09	–	Potts (1954a)
M. edulis	1.54	1.54	–	
M. edulis	0.98	0.98	–	
M. edulis	0.58	0.58	4-5 days[a]	
M. dunkeri	1.93	1.98	24 hr	Yazaki (1929)
Modiolus capax	1.95	1.96	24 hr	Yazaki (1929)
Pecten maximus	1.91	1.91	–	Dakin (1909)
Ostrea circumpicta	1.95	1.98	–	Yazaki (1929)
O. circumpicta	0.92	0.85	21 days	
Heterodonta				
Venus merrenaria.76	1.76	–	Cole (1940)	
V. mercenaria	1.39	1.34	–	
Scrobicularia plana	1.89	1.90	2 days	Freeman and
S. plana	1.60	1.55	2 days	Rigler (1957)
S. plana	1.05	1.05	4 days	
S. plana	0.74	0.59	4 days	
Cephalopoda				
Coleoidea				
Octopoda				
Octopus vulgaris	2.30	2.27	–	Bottazzi (1908)
O. macropus	2.32	2.27	–	Bottazzi (1908)

[a]Valves propped open.

representatives of the seven orders it includes can be added all the species of Table 5.2, including members from five other orders, Neogastropoda (*Buccinum, Neptimea*), Notaspidea (*Pleurobranchus*), Acoela (*Archidoris*), Adapedonta (*Ensis, Mya*), and Decapoda (*Sepia, Loligo*); these have ionic concentrations 99-100% of those in sea water (98% in the case of *Sepia*). Isosmoticity within 1% for *Venus mercenaria. Mya arenaria* (Bivalvia), and *Busycon canaliculatum* (Neogastropoda). A few marine species, particularly among the bivalves, are euryhaline, withstanding abroad range of external salinities. Thus in the Gulf of Finland *Mya arenaria, Cardmm edule*, and *Mytilus edulis* can be found at, salinities of 4-5%, although much reduced in size.

There is no evidence that these species are other than isosmotic with the dilute environment. Bivalves often close their valves tightly when placed in dilute sea water or fresh water, and prolong their period of disequilibrium. *Ostrea circumpicta* placed in 50% sea water (Δ0.85°C) had not come into equilibrium completely even after 3 weeks (Table 5.1). But *Mytilus* in water of Δ0.58°C becomes isosmotic within 4-5 days when the valves are propped open; equilibrium is rapid, and is virtually complete in 4 hour in animals transferred from water of Δ2.08°C to water of Δ1.36°C. The slight hypertonicity of *Venus mercenaria* in slightly diluted sea water, blood Δ1.39°C against sea water Δ1.34°C (Table 5.1), is possibly only apparent as no details are given of the equilibration period; surface salinities may be lower than those of the water in contact with burrowing animals.

The intertidal limpet *Acrnaea limatula* (Archaeogastropoda) cannot regulate osmotically; its blood becomes isosmotic to the medium at low or high salinities over the range 17 50%.

B. Permeability to Water and Salts

In diluted sea water, increases in weight due to osmotic uptake of water occur in soft-bodied gastropods such as the opisthobranchs *Aplysia fasciata* (= *limacmo*) and *A. julicma*, and the marine pulmonate *Onchidium chamaeleon*. The explanation of the imperfect semipermeability of the molluscs' integument and gills is that, after the initial transfer of water, salts begin to be lost to the dilute medium, as has been shown in *Aplysia* by chloride analyses of the blood and external medium. The final drop in weight or *Aplysia* and *Onchidium* is probably due to muscular contraction with consequent forcing out of water through the tissues.

In concentrated sea water, *Aplysia* loses weight as water is abstracted by the hyperosmotic solution, but the theoretical values are not attained.

Further confirmation of permeability to ions is given by experiments with *Aplysia*, the nudibranch *Archidoris*, and *Onchidium* in various mixtures of sea water and nonelectrolytes, including urea, sucrose, and glycerin, and in sea water plus a solution of magnesium sulfate, all mixtures being isosmotic with sea water. In all these artificial solutions, the gastropods lose weight, the blood becoming viscous, although the loss is only slight in the urea-sea water solution in the Onchidium experiments. Bethe's interpretation of the loss in weight is outward diffusion of ions, particularly sodium and chloride, thus lowering the osmotic pressure of the blood from which water is then abstracted by the bathing solution, the two processes continuing since equilibrium cannot be achieved. The more rapidly penetrating urea reduces the imbalance, and the weight loss is smaller.

A physiologically more satisfactory demonstration of ionic permeability in *Aplysia* is obtained by keeping the animals in artificial sea waters of the same osmotic concentration but with different levels of ions. Thus in water with 0.48 mg Ca/ml, samples of blood from different *Aplysia punctata* gave values between 0.49 and 0.58 mg/ml, while in calciumfree water values of 0.15 and 0.16 were obtained after a stay of 6 and 24 hr, the animals themselves losing tonus; in calcium-rich water, 1.68 mg/ml, the blood of another specimen had a value of 0.70 after 5 hr, the animal showing increased tonus.

Similar experiments on *Aplysia fasciata*, in which the mouth was ligatured to prevent any possible ion transfer through the gut, showed that the body surface was permeable to calcium, magnesium, chloride, and sulfate ions.

C. Ionic Regulation of the Blood and Its Mechanism

Despite a presumed measure of permeability to ions, most molluscs have blood differing from sea water in the relative proportions of potassium, calcium, and sulfate. Ionic regulation may be defined as the maintenance in a body fluid of concentrations of ions differing from those of a passive equilibrium with the external medium. A measure of ionic regulation is obtained by comparing an analysis of one blood sample with another which has been dialyzed against sea water across a *collodion* membrane which is permeable to water and salts but not to protein. In such a passive equilibrium the concentrations (on a water content basis) of ions in the dialyzed blood differ slightly from those of the sea water, the differences depending on the concentration of the plasma proteins; the mean Donnan ratio, r, for univalent and divalent ions (except Ca^{++}) is, however, only 1.01-1.02 in prosobranchs and cephalopods where

$$r = \frac{Na_i}{Na_o} = \frac{\sqrt{Mg_i}}{\sqrt{Mg_0}} = \frac{Cl_o}{Cl_i}, \text{ etc. (i = blood, o = sea water)}$$

The calcium value is higher owing to the presence of a nondiffusible calcium-protein complex (*e.g.*, Ca_i/Ca_o is 1.05 in *Buccinum*, 1.19 in *Sepia*). Such experiments and analyses show that some or all of the ions of the blood are regulated in every mollusc tested (Table 2). In the gastropods and bivalves regulation consists chiefly in raised values of potassium and calcium. and lowered values of sulfate: sodium and chloride are virtually in equilibrium across the boundary membranes of these animals. Magnesium remains at the equilibrium value in practically all, being higher in the blood only in *Archidoris*. *Mytilus edulis* and *Ostrea edulis* show perhaps the least regulation. The Mediterranean *M. galloprovincialis* surprisingly accumulates sulfate.

In contrast, marked ionic regulation extending to all ions is shown by the three cephalopods. Noteworthy are the high potassium contents of the plasma, the slightly raised chloride, and the lower sodium and sulfate figures.

A progressive lowering of sodium in the series *Eledone, Loligo*, and *Sepia* is associated with a fall in sulfate and a rise in chloride, but the plasma remains isosmotic. These features have been interpreted in relation to the ideal laws of solutions. If sea water and body fluids are considered hypothetically as mixtures of isosmotic solutions of various salts, then the replacement of some of the Na_2SO_4 solution by the same volume of NaCl solution results in a lowering of SO_4^{--} and Na^+ ions, and an increase of Cl^- ions, while the whole solution retains the same osmotic concentration. Sodium ions are lowered because their activity is greater in association with Cl^- ions than with SO_4^{--} ions. Granted a marked decrease in sulfate as the result of active ionic regulation, then the chloride ions would rise to balance the cations, and the cations themselves decrease slightly. It can be calculated that in a synthetic sea water in which the sulfate is reduced to a fifth and the potassium doubled, the chloride would be 104.7% and the sodium 93.8% of the values in an isosmotic sea water of normal composition. As seen in Table 5.3 these values are approached in the analysis of *Sepia* plasma. With reduced sulfate but normal potassium, the sodium figure would be higher, 95.9% (Robertson, 1953).

Table 5.2 : Ionic Regulation in blood Plasma of Marine Molluscs[a]

Species	Concentrations in plasma as % concentrations in dialyzed plasma (prosobranchs and cephalopods) or in sea water (other groups) (water content basis)						Plasma protein (gm /liter)	Total ionic concentrations as % of that in sea water
	Na	K	Ca	Mg	Cl	SO_4		
Gastropoda								
Prosobranchia								
Neogastropoda								
Buccinum undatum	97	142	104	103	100	90	25.3	99
Neptunea antiyua	101	114	102	101	101	98	24.1	100
Opisthobranchia								
Notaspidea								
Plemabranchus membranaceus	100	117	112	99	100	102	0:3	100
Acoela								
Archidoris pseudoargus	99	128	132	107	100	96	0.4	100
Bivalvia								
Lamellibranchia								
Anisomyaria								
Mytilus edulis	100	135	100	100	101	98	0.3	100
M. galloprosincialis	101	121	107	97	99	120	0.8	99
Pecten maximus	100	130	103	97	100	97	–	100
Ostrea edulis	100	129	101	102	100	100	0.2	100
Adapedonta								
Ensis ensis	99	155	108	99	99	87	–	99
Mya arenaria	101	107	107	99	100	101	–	100
Cephalopoda								
Colenidea								
Decapoda								
Sepia officinalis	93	205	91	98	105	22	109	98
Loligo forbesi	95	219	102	102	104	29	150	99
Octopoda								
Elidone cirrosa	97	152	107	103	102	77	105	100

Data from Robertson 0949, 1953).

The mechanism of ionic regulation of the blood seems to be two fold: the elimination of a fluid which differs from an ultrafiltrate of the plasma in that its ions have been differentially excreted, and the replacement of this fluid by controlled absorption of ions and water in such a way that the blood remains isosmotic and approximately constant in ionic composition.

1. DIFFERENTIAL EXCRETION BY RENAL ORGANS

In marine molluscs the fluid excreted by the renal tubules has been analyzed in cephalopods. This fluid of low protein content (1/100 or less that of the plasma) is practically isosmotic with the blood in *Sepia* and *Eledone* (Table 3), and also in *Octopus vulgaris* (blood Δ2.30°C, urine Δ2.24°C). Wide differences from the plasma or a plasma ultrafiltrate are found with respect to the concentrations of almost all the ions. In *Eledone* there occurs a reabsorption of potassium, calcium, magnesium, and chloride and secretion of sulfate and sodium, tending to raise the level in the blood of the former ions and lower that of the latter; in fact, relative to sea water the blood does show higher levels of potassium, calcium, magnesium, and chloride, and lower levels of sodium and sulfate.

Table 5.3 : Plasma and Renal Sac Fluids of Cephalopods[a]

	Mg/gm water						
	Na	K	Ca	Mg	Cl	SO_4	Total mg-ions/kg water
Sepia officinalis							
Plasma	10.58	0.931	0.434	1.383	20.87	0.47	1145
Renal sac fluid[b]	8.35	0.465	0.302	0.936	20.85	1.01	1166
Sea water	11.31	0.409	0.432	1.364	20.38	2.845	1174
Fluid as % plasma	79	50	70	68	100	215	100
Plasma ultrafiltrate as % plasma	98	98	84	96	102	105	100
Eledone cirrosa							
Plasma	10.26	0.581	0.466	1.318	18.92	1.983	1081
Renal sac fluid	10.41	0.522	0.405	1.218	18.35	2.705	1073
Sea water	10.43	0.378	0.399	1.258	18.80	2.624	1082
Fluid as % plasma	101	90	87	92	97	106	99
Plasma ultrafiltrate as % plasma	99	99	92	98	101	102	100

[a]Data from Robertson (1949, 1953).
[b]Also 2.64 mg NH_4/gm water.

While the differences between the values in urine and plasma ultrafiltrate of potassium, chloride, and sulfate are in step with the composition of the blood in *Sepia*, that between the sodium concentrations is not: the low sodium content of the renal sac fluid, far below that of a blood ultrafiltrace, would tend to increase the sodium content of the blood, which, however, is only 93% of the value in sea water. This anomally is apparently the result of the large concentration of ammonium ions excreted in the fluid. Requirements of cation-anion balance and total osmotic concentration are met by reduction in sodium, the cation of greatest concentration. Ammonium forms 146 meq of the total cations, 613 meq/kg water, the chloride and sulfate anions coming to 609.

For an estimate of the water and ion turnover in cephalopods, estimates of the amount of urine produced and the volume of blood and other extracellular fluids are necessary. Figures are available for *Octopus bongkongensis*. Expressed in terms of volume as per cent wet weight, a mean value of 28.0% has been found for the extracellular fluid, based on the distribution of injected inulin. Of this, blood accounts for 5.8%, as measured by the dye T-1824 and colloidal mercuric sulfide (HgS). Urine output per 24 hr varies between 6.2 and 10.0% of the body weight, the lower value being obtained by tying off the renal sacs and measuring the increase in weight, the higher figure from catheterized animals. Thus the renal organs in this animal have a daily output equivalent to about 138% of the blood volume or 29% of the extracellular fluid volume.

In *Sepia* a minimal value for daily output of urine of 13% of the volume of extracellular fluid has born given by (Robertson, 1953).

2. CONTROLLED UPTAKE OF WATER AND IONS

Loss of water and ions through excretory tubules must be balanced by uptake of water and ions from sea water, through the gills and any other permeable portions of the integument, of via the gut. This uptake is indicated by the increases in weight of cephalopods such as the *Octopus* when the ureters are ligated. Absorption of pure water would alter the osmotic pressure of the body fluids. There is no evidence of this, and therefore, ions must be absorbed with the water in concentrations sufficient to make it isosmotic with the fluid it is replacing.

Unfortunately, there is no information on the absorption of ions such as has been obtained in "balance" experiments with the crab *Carcinus* (Robertson, 1960b), nor has any investigation been made with indicators, such as phenol red, of possible absorption of sea water via the gut.

An inference from the data of Table 5.3 that potassium, chloride, and possibly calcium are taken up against concentration gradients in the cephalopods may be correct, but a continuous loss of potassium in the excretory fluid without corresponding uptake, as in *Carcinus*, has not been excluded; sodium and sulfate may enter wholly or partly by diffusion, as their concentration gradients are favourable. In *Carcinus* the negative balance for potassium after 1 or 2 weeks' starvation may indicate that the crab normally obtains most of its potassium from food, not from ions absorbed from sea water. Its blood level of potassium rises after-feeding. But it is not so similar in case of most of the molluscs.

D. Ionic Regulation of Other Fluids

Fluids from the eyes and statocysts of cephalopods differ from blood in being colourless, cell-free, and almost protein-free. The eye fluids of Sepia, *Loligo* and *Eledone* have a total ionic concentration within 1% or so of the plasma, but show very considerable divergencies from it in composition: they are far from being dialysates or ultrafiltrates of the blood, but their mode of origin is obscure. Chief among the differences between the vitreous fluid (from the posterior eye chamber) and the plasma are consistently lower magnesium concentrations (9-19% on a water content basis) and higher sodium concentrations (112-115%). The aqueous humor-of *Sepia* (from in front of the lens) differs considerably from the vitreous fluid; in most respects it is intermediate in composition between the vitreous fluid and sea water, a finding consistent with the potential channel of communication with the exterior, the cornial pore (Robertson, 1953).

The statocyst of *Octopus* has both an inner and outer sac, so that as in the vertebrate labyrinth one can distinguish endolymph and perilymph. Whereas mammalian endolymph has a very low Na^+/K^+ molar ratio (<0.6), *Octopus* endolymph has a high ratio of 30, intermediate between that of the blood (17) and sea water +7). Perilymph has lower sodium and potassium concentrations than endolymph, but its ratio (33) is not markedly different.

Oceanic squid belonging to the family Cranchiidae have large coelomic spaces filled with slightly acid fluid (pH 5.2) containing a concentration of about 480 mg-ions ammonium, and 80 mg-ions sodium. Presumably these ammonium ions accumulate in the animals by secretion from the kidneys and perhaps other parts of the coelomic epithelium. The anions are almost exclusively chloride. This large volume of coelomic fluid, practically isosmotic with sea water and containing chiefly a solution of ammonium chloride, has a density lower than that of sea water and the heavier parts of the squid, thus bringing about buoyancy equilibrium in the animal.

III. OSMOTIC REGULATION

A. Definition

Osmotic regulation may be denied as the maintenance of the total particle concentration of body fluids at levels different from those of the external medium. The definition may be extended to terrestrial animals which control the concentration of their body fluids, although the source of ions and water is the more variable medium of ingested food and fluid. All fresh-water molluscs and probably a few brackish-water species show hypersomeric regulation, maintaining higher concentrations of ions in the blood than those of the medium. This type of regulation is a steady state in which energy is expended.

B. Brackish-Water Forms

Brackish water, that is, water of salinity 0.5-17‰, may contain two groups of molluscs as in the Baltic Sea: marine immigrants such as the bivalves *Mytilus edulis*, *Cardium edule*, *Macoma balthica*, and *Mya arenaria*; and fresh-water immigrants such as the pulmonate *Lymmaea peregra*, the prosobranchs *Bithynia tentaculata* and *Theodoxis fluviatilis*, and the bivalves *Anodonta cygnea* and *Dreissena polymorpha.*

In British waters the bivalves *Scrobicularia plana.* (Heterodonta) and the hydrobids *Potamopyrgus jenkinsi* and *Hydrobia ulvae* (Mesogastropoda) are typically brackish-water species, living in estuaries.

Little work has been done on those gastropods which might be expected to show regulation. The periwinkle *Littorina littorea*, collected intertidally at Millport, Scotland, is isosmotic in salinities of S 17-36$^{0}/_{00}$, but regulates when placed in dilute sea water (S 8.8$^{0}/_{00}$), maintaining a me-m blood A of 106°C (range 0.63°-1.50°) as against Δ0.48°C for the medium, after 3-7 days (Todd, 1962; Table 3). This species extends into the southern part of the Baltic, withstanding there a salinity of about 7-8$^{0}/_{00}$.

There is little information on the blood concentrations of the various Baltic pulmonates and prosobranchs of lacustrine origin which can tolerate low salinities, *Lymnaea peregra* tolerating 10-11$^{0}/_{00}$, *Bithynia tentaculata* and *Physa fontinalis* 6$^{0}/_{00}$ and *Theodoxus fiuyiatilis* up to 14%. *Lymnaea* is probably isosmotic with the medium at these salinities. Only Theodoxus has been studied experimentally Snails from a brackish-water population withstood a rise in salinity to 15‰ better than those from fresh water, and survived in 20$^{0}/_{00}$ if previously acclimatized to 6.5$^{0}/_{00}$. Another group kept at salinities of 3-25$^{0}/_{00}$ (about Δ0.15°-1.35°C) were slightly hyperosmotic to the medium over this range. In fresh water (Δ0.01°C) the blood of *Theodoxus* is hyperosmotic, Δ0.12°-0.18°, mean 0.15°C, the values being somewhat higher. Δ0.20°-0.21°C, in well-fed animals.

Unequivocal evidence of osmotic control in brackish-water bivalves is absent. *Mytilus edulis* seems to adjust to the medium; the blood passively followed changes in salinity down to the lowest compatible with life. They worked with animals from Stockholm (salinity 5.5$^0/_{00}$) among others, *M. edulis* acclimatized and 15 other marine molluscs to fresh water, after slow auction of the medium over a period of 5-8 months. A stage must come in this dilution process when the blood of the molluscs remains higher than the medium, since all fresh-water animals have mechanisms of active regulation. The salinity of the medium at this point may be quite low, since fresh-water lamellibranchs with large surfaces in contact with the external medium, and low metabolism, seem unable to maintain a high osmotic gradient between blood and water. The fresh-water mussel *Anodonta cygnea* comes into equilibrium with dilute sea water of Δ0.12°C, about S 2.3$^0/_{00}$. Thus in specimens of this species inhabiting Baltic coastal water of S 3-4$^0/_{00}$, the blood must be isosmoric with the medium.

The mussel *Dreissena polymorpha* is a fresh-water species found also in the Baltic [S 4.7$^0/_{00}$,] and in suitable salinities in the Black and Caspian Seas (about 10$^0/_{00}$ or less) where its blood must be isomeric with the medium. Like *Anodonta*, it comes into equilibrium with a medium of about 2.3$^0/_{00}$ (Rotthauwe, 1958).

Scrobicularia plane, an estuarine bivalve, cannot maintain an osmotic difference between blood and external medium in dilutions of sea water down to Δ1.05°C, if equilibrated for 48-110 hr (Table 3). Animals obeisant to remain open come rapidly into equilibrium, usually in 6-7 hr. But after a period of equilibration of 110 hr the blood of specimens in dilute sea water of Δ0.59°C had a mean Δ of 0.74°C. It is not clear whether in the latter case the animals are showing active control of the blood concentration or whether they have resisted final dilution to the outside level by keeping the valves closed.

C. Fresh-Water Forms

Fresh-water molluscs are faced with a corninuous osmotic inflow of water which they have to excrete, and with the problem of active absorption of ions from a dilute medium to replace those lost by outward diffusion and excretion. The level of blood concentration which they can maintain is low, and *Anodonta* has the distinction of having the lowest value recorded for any animal, Δ0.08°C, a concentration equivalent to

Table 5.4 : Concentration and Ionic Composition of Blood and Urine of Fresh-Water Molluses

Species	Osmotic concentration $\Delta°C$	As mM NaCl	Na	K	Ca	Mg-ions/kg water Mg	Cl	SO_4	Authority
Blvalvia									
Lamellibranchia									
Anodonta cymea			15.6	0.49	8.4	0.19	11.7	0.76	Potts (1954a)
Blood[a]	0.078	21.8	–	–	–	–	–	–	Picken (1937)
Urine	0.042	11.8	–	–	–	–	–	–	
Medium	0.01	3.4	–	–					
Hyrideita austratis				–	–	–	16.1[b]	–	Hiscock(1953a)
Blood	0.16	46.0	–	–	–	–	9.4[b]	–	
Urine	–	–	–	–	–	–	3.4[b]	–	
Medium	–	–	–						
Dreusseba oiktniroga				–	–	–	–	–	Rotthauwe (1958)
Blood	0.09	26	–						
Gastropoda									
Pulmonata									
Lymnaia peregra				–	–	–	–	–	Picken (1937)
Blood	0.25	71.5	–	–	–	–	–	–	
Urine	0.19	54.3	–	–	–	–	–	–	
Medium	0.01	3.4	–						

Species	Osmotic concentration Δ°C	As mM NaCl	Na	K	Ca	Mg-ions/kg water Mg	Cl	SO_4	Authority
Lyinnaea stagnalis									
Blood[b]	–	–	47.4	2.8	3.0	4.8	42.6	–	Huf(1934)
Blood[b]	–	–	31.2	1.2	7.1	2.1	27.2	–	Duchateau and Florkin (1954)
Planorbwws comeus									
Blood[b]	–	–	85.9	2.3	6.0	1.1	21.0	0.31	Florkin (1943)
Prosobranchia									
Theodoxus fluviattlis									
Blood	0.15	43	–	–	–	–	–	–	Neumann (1960)
Vmparus fasciatus									
Blood	0.21	60	–	–	–	–	–	–	Obuchowicz(1958)
Polamopyrms jenkinsi									
Blood	0.18	6051	–	–	–	–	–	–	Todd (1962)
Urine	0.15	43	–	–	–	–	–	–	

[a]Other anions, etc.: HCO_2^-, 13.6; CO_3^{--}, 0.056; CO_2, 0.92; HPO_4^{--}, 0.17; $H_2PO_4^-$, 0.03 (all values at pH 7.52).

[b]Mg-ions/liter.

about 4-5% sea water. Dreissend's value is only a little higher, and those of the pulmonates and prosobranchs investigated are about 2 to 3 times as high (Table 5.4).

Noteworthy in the analysis of *Anodonta* blood are the relatively high concentrations of calcium and bicarbonate ions. The internal concentration alters with experimental changes in the external medium, and different levels are sustained in hard and soft fresh waters [equivalent to 0.103 and 0.080% NaCl, respectively.

When *Anodonta, Hyridella*, and *Lymhaea stagnalis* are narcotized with ether or barbiturates, the animals increase in weight owing to osmotic uptake of water. In *Anodonta* and *Lynmaea* corresponding reductions in the concentrations of the principal ions in the blood occur, except for potassium which increases. During narcosis all the blood ions apparently begin to diffuse out into the medium, but cellular release of potassium more than counteracts the loss of this ion. Urine, formation also stops, as fluid practically ceases to filter across the heart wall into the pericardial cavity.

MECHANISM OF OSMOTIC REGULATION

The production of urine by initial filtration of fluid through the heart wall and partial reabsorption of ions from this filtrate by the excretory tubule (s) inevitably leads to the loss of salts from the body. By producing urine hypo-osmotic to the blood (Table 4), this loss is cut down to approximately 54-83% of what it would be if the urine contained salts equivalent to those in a blood filtrate. If nitrogenous substances of low molecular weight were secreted into the filtrate in *Anodonta*, the loss of salts would be less than that calculated proportionally from the osmotic concentration values.

The loss of salts, however, remains considerable in view of the large amount of urine produced, about 24% of the body weight (including shell) per day in *Anodonta*. The daily output of urine is equivalent in volume to 65% of the extracellular fluid.

Picken in 1937 suggested that ion absorption in *Anodonta* must take place at some part of the body surface, basing his proposal on the inadequacy of an estimate of the salt contribution from the food, and the knowledge that specimens remain alive for months in running tap water without serious fall in osmotic concentration, were kept mussels in running tap water for 22 months, and most specimens had normal freezing points until the twenty-second month, when blood values were about half that of normal.

Proof of ion absorption was supplied by Krogh (1939), who found, after depleting the salts of various molluscs by treatment with distilled water, that active uptake of chloride and sometimes sodium ions could be demonstrated in *Anodonta, Unio, Dreissena,* and the gastropods *Lymnaea stagnalis* and *Paludina.* Most of these could reduce the chloride of 0.01 Ringer solution or 0.7-1.1 mM NaCl solution to 0.1-0.15 mM chloride. Krogh believed that the renal organs of these molluscs must produce an almost salt-free urine when the animals are in distilled water; otherwise the process of depleting their salts would be much more rapid. (*Anodonta* was able to withstand the washing-out process for a month.) Undoubtedly, tight closure of the valves must reduce any osmotic uptake of water through the soft tissues, and consequently cut down excretion.

Shell closure in *Hyridella australis* enables this bivalve to resist desiccation when river margins dry out. Experimentally, it can survive at least 3 months in dried up mud; in two specimens the chlorinity of the blood was 27 and 34 mg-ions/liter, compared with a mean value of 18.7 (range 10-26) in tank specimens. Rise in weight of 11% and 13% in two other specimens from the dried mud when they were placed - in tank water indicated considerable dehydration during the enforced drought.

Differences in the she of the kidneys have been found in *Lymnaea peregra* from fresh and from brackish waters (S 6.5-7.0%g). These differences can be correlated with excretory activity: specimens from fresh water have larger kidneys in line with their presumed greater turnover of water

REFERENCES

1. Fox, D.L. (1966) In "Physiology of Mollusca" (K.M. Wilbur and C.M. Yonge eds) Vol-2 Academic Press. New York.
2. Goodwin T.W. (1972) In "Mollusca" (M. Florkin and B.T. Scheereds.) Chemical Zoology Vol-7, Academic Press, New York.
3. Grey Michell and Breat Hearts (1970) Shellfish Anatomy, Academic Press New York.
4. Mukherjee, H.K. MPEDA (1972).
5. Margaret E. Brown (1957). The physiology af. fisher.
6. Majumdar, J.R. CSc Publications Shellfish Physiology, Tata McGraw Hill Publication.
7. Swaminathans (1985): SEL Publication.

6

RESPIRATION

RESPIRATION

In marine mussels, the extraction of oxygen from the water occurs primarily at the gill surface. The gill therefore serves a dual function, in feeding and in respiration. The structure of the lamellibranch gill is well-suited to function in gas exchange, by virtue of its considerable surface area and rich supply of blood. Large volumes of water are passed through the mantle cavity, ensuring a supply of oxygen, and the diffusion distance for oxygen from water to blood is small. Normally, the efficiency with which the oxygen made available to the gill is utilised rather low (3-10%). Under certain circumstances this efficiency may be increased considerably, up to 30% or more and this provides one way in which a mussel may vary, its rate of consumption of oxygen in order to meet a variable metabolic demand. However, calculations suggest that a high proportion of the total oxygen consumed may be used to support the 'respiratory pump'; this and the high energetic costs of increasing ventilation rate impose an upper limit to the operation of the respiratory system.

Knowledge of the limits of respiratory functions are important for understanding the physiological adaptation of a species, since many features of aerobic metabolism can be studied indirectly by measurement of the rate of oxygen consumption by intact animals. This has led to a considerable number of investigations into the rates of respiration of mussels. As with many species, the rate of oxygen consumption has been found to vary with change in virtually any environmental variable. In this section, some of the factors known to craft: changes in rates of oxygen consumption by mussels are reviewed, in addition, since it is often impossible to discuss the intensity of respiration without at the same time considering the activity that contributes to the metabolic demand, we shall also review here some aspects of the activity of the individual.

Size

The relationship between body size (usually expressed as weight) and the rate of oxygen consumption by animals has received considerable attention. The relationship is most commonly expressed in the form of an allometric equation:

$$Y = aX^b,$$

based *tidal* phenomena. Rao (1954) claimed tidal rhythmicity in nitration rate in *Mytilus edulis* and *M. californianus*. The rhythm was independent of temperature between 9 and 20°C. and independent also of light. *M. edulis* transplanted over a distance of 5000 km gradually shifted their nitration rhythm over three weeks to correspond with the tidal pattern at the transplant site. Winter (1969) observed two maxima per day in (he feeding rate of *Modiolus* he suggested that this may indicate a tidal rhythm, but the data were insufficient to draw a firm conclusion. Later. Winter (1973) could find no evidence of a tidal feeding rhythm in *M. edulis. Salanki* (1966) observed an adduction abduction rhythm in *Lithophaga lithophaca* in Mediter ranean, and equated this diurnal changes in light intensity.

However, there has been no firm demonstration of rhythmic changes in the respiration rate of mussels. Thompson and Bayne (1972) recorded continuous feeding by *M. edulis* whenever food cells above a certain concentration were present; Bayne, Thompson and Widdows have been unable to detect a respiratory rhythm in *M. edulis.* Rhythmic activity in the digestive gland.

The effects of the rate of water flow on the respiration of marine bivalves has been neglected in most experimental studies. An increased growth rate in clams and oysters in flowing water was also recorded small oysters had a higher growth rate when kept at 183 ml min^{-1} flow than at 74 ml min^{-1}. On the other hand, found that growth of the scallop *Acquipecten* was inversely related to current speed until, at low current velocities, food became limiting. We have discussed the effects of water flow on nitration rate.

Nixon, Oviatt, Rogers and Taylor (1971 examined the influence of current speed on the 'metabolism' of a mussel bed *in vivo*. Dissolved oxygen was measured upstream and downstream of the bed, and the results corrected for diffusion to give an estimate of community respiration. Nixon *et al.*, (1971) also recalculated some earlier data of Kuenen's (1942) and they concluded that the relationship between respiration rate (R; g O_2 h^{-1} m^{-2}) and current speed (C; m s^{-1}) was described by:

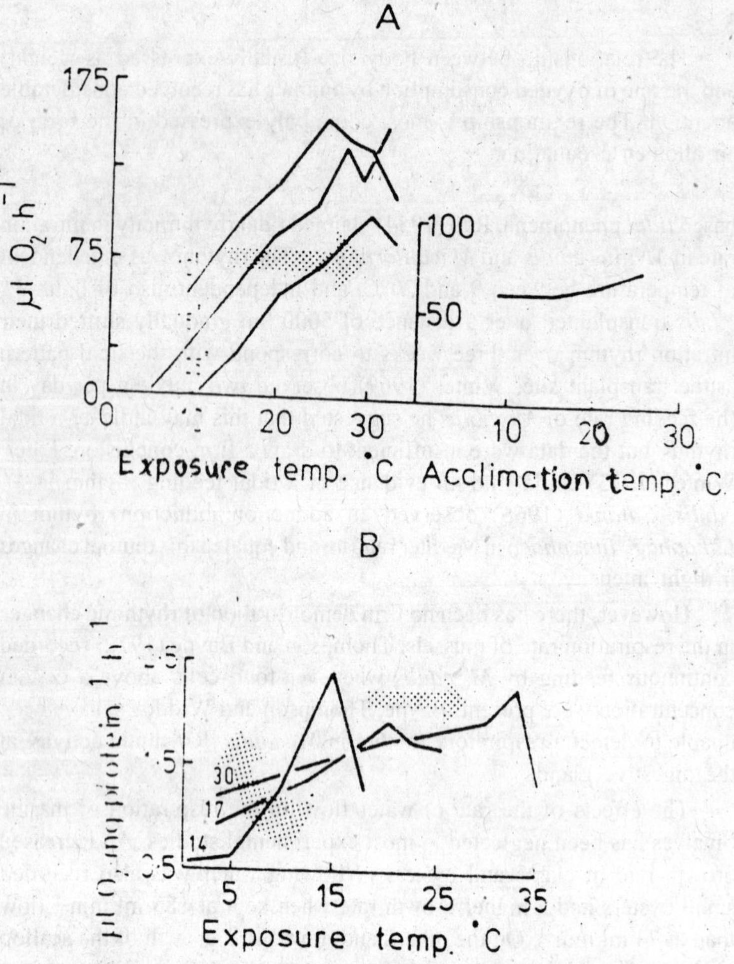

Fig. 6.1 : A. Left : Rate temperature Curves for the Routine Respiration Rate of *Crepidula fornicutu* of Standard 160 mg Dry Body Weight previously Acclimated at 10, 15, and 25°C. Acclimation shifts the curves along the abscissa (lateral translation). Right: the routine respiration at rate at the acclimation temperatures is almost constant After Newell (1979) (from, Newell and Kofoed, 1977). B. Rate-Temperature Curves to the Routine Respiration rate of *Littorina littorea* Collected from the Shore between Winter and Summer when the Environmental (Acclimation) Temperatures were 14, 17, and 30°C. Acclimation shifts the curves along the abscissa (lateral translation; end also rotates them towards the horizontal at higher temperatures. After Neweil (1979) (from Newell and Pye, 1971).

$$R = \frac{C}{0.006 + 0.35C}$$

over a range of current speeds from 0 to 0.6 m s^{-1}. The mechanism responsible for such an effect of current speed is unclear, and the relationships between current speed, nitration rate, particle concentration and respiration rate in bivalves remain to be elucidated.

Oxygen Tension

Shortage of oxygen is likely to be encountered by subtidal gastropods that burrow into sediments and by intertidal species that withdraw into their shells when desiccated at low tide. Gastropods that habitually experience low oxygen tensions might be expected to be 'oxy-regulators', increasing the efficiency of uptake as the oxygen supply decreases, whereas low-shore and surface-dwelling subtidal species would not be expected to do this being 'oxycon-formers'. Such is the case among chitons, close relatives of the gastropods in which aerial respiration is maintained over large decreases in oxygen tension in high-shore but not in low-shore species. Since oxygen enters the tissues by passive diffusion, 'oxyregulation' must involve physical changes of the respiratory system to facilitate oxygen uptake, but it is not known what these might be most gastropods are, however, likely to be oxyconformers (Fig. 6.2).

Prolonged anoxia will enforce anaerobic respiration and this may be of regular occurrence in some gastropods. Once oxygen becomes available again, the products of anaerobic respiration can be metabolised aerobically and the amount of oxygen required for this is the oxygen debt. *Littorina neritoides* extends to the splash zone on exposed rocky shores, where it may be kept dry for several weeks during calm, fine weather. Under such conditions *L. neritoides* closes the shell aperture with its operculum, remaining quiescent until wetted. Oxygen diffusion around the edges of the operculum must be minimal, and it seems likely that anaerobic respiration will be important during aestivation, although this has not been tested.

When experimentally *N. reticulatus* subjected to several hours of 'environmental anoxia'. *Nassarius mutabili; breaths* down the substrates phosphoarginine, glycogen and aspanate to form alanine and succinate (Gade *et al.,* 1984). *Bullia digitalis* also produces alanine under similar conditions (Brown, 1982). *Morula granulata* breaks down glycogen and to a lesser extent aspartate to form succinate and alanine when immersed in water of less than 15 per cent salinity. To avoid osmotic imbalance, the snail closes the operculum across the shell aperture, isolating itself from the surrounding water. The operculum closes more tightly as salinity decreases below 15 per cent and this is accompanied by a greater predominance of anaerobic respiration, reflected by a greater oxygen debt, as gaseous exchange is reduced. Compared with :he classical glycolytic pathway producing lactate, the succinate pathway has a higher ATP output per mole of glucose used.

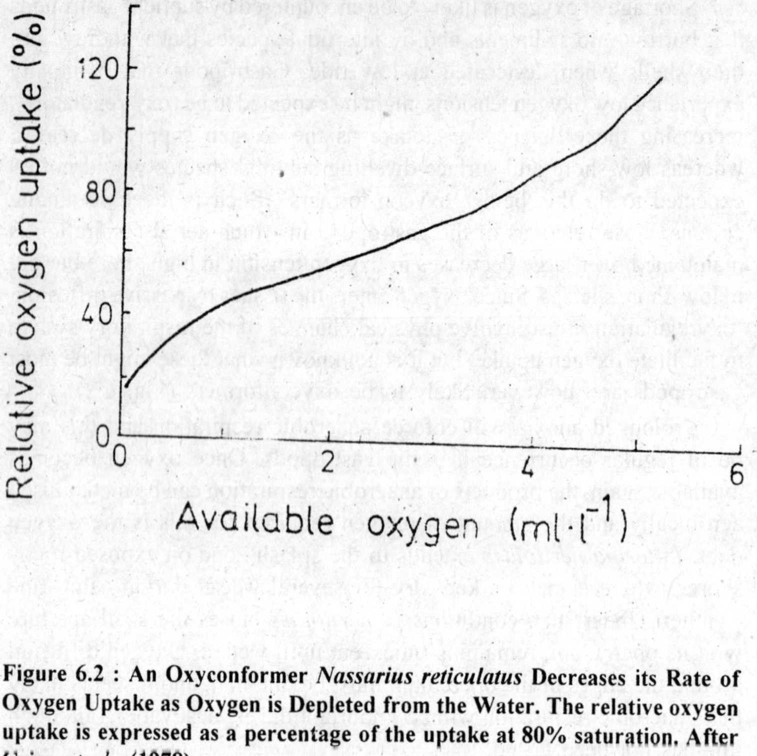

Figure 6.2 : An Oxyconformer *Nassarius reticulatus* Decreases its Rate of Oxygen Uptake as Oxygen is Depleted from the Water. The relative oxygen uptake is expressed as a percentage of the uptake at 80% saturation. After Crisp *et al.*, (1978).

Temporary anoxia occurs in muscles during sudden, strenuous exercise, and when this 'functional' or 'physiological' anoxia occurs in *Nassarius mutabilis*, arginine phosphate and glycogen are broken down to form octopine, as is the case among other invertebrates. Formation of octopine, or related products, apparently does not increase the yield of ATP per mole of glucose, but perhaps has other advantages such as stabilising the acid-base balance of the cell or preserving the regulatory properties of enzymes in the cell.

Thermodynamic considerations suggest that the maximum rate and maximum efficiency of anaerobic metabolism are mutually exclusive (*Gnaiger. 1983*) During physiological anoxia, energy is needed quickly because of the associated intense muscular activity, and thermodynamically grossly inefficient pathways sustain high metabolic rates for brief periods. But under long-term environmental anoxia a low, steady metabolic rate is achieved by thermodynamically more efficient pathways. These pathways are probably switched into operation by cellular changes of pH and adenylate phosphorylation potential associated with prolonged anoxia. Moreover, calorimetric studies directly measuring metabolic heat output indicate that unknown sources of energy may be important among marine invertebrates experiencing environmental anoxia.

Salinity

Estuarine and intertidal gastropods experience cyclic or intermittent salinity fluctuations either in the ambient seawater or in the mantle-cavity fluid, but since they are largely osmoconformers, their metabolic rate would not be expected to increase above normal when salinities fluctuate. On the contrary, the respiration rate of *Thais haemastoma* fails in response to lowered salinity and, since accompanied by withdrawal of the siphon, it is probably caused by decreased ventilation. Similarly, many other intertidal gastropods, including *Littorina littorea*, withdraw into their shells when exposed to low salinities, with a concomitant cessation of oxygen consumption. Anaerobic metabolism must be enforced during these conditions, the end products being metabolised aerobically when salinity returns to normal.

Respiration in Air and Water

Subtidal gastropods respire across exposed epithelia, especially over the gill which generates a respiratory current. Exposure to air causes withdrawal into the shell, collapsing of the gill and a decreased rate of respiration. Intertidal species maintain their oxygen consumption in air by ctenidial (branchial) respiration in the mantle-cavity fluid and by making greater use of vascularised mantle tissue Ctenidial respiration is more efficient in water, and diffusion across the mantle is more efficient in air.

The difference between the rate of aerial and aquatic respiration' varies among intertidal species, those from lower levels on the shore tending to have lower and those from higher levels greater aerial than aquatic respiration rates. Pairs of lower-shore and higher-shore species conforming to this trend include *Patella caerulea* and *P. lusitanica*

Notoacmea fenestrata and *Collisella digitalis* and *Lacuna vincta* and *Littorina littorea*. Exceptions do occur: West Indian species of *Nerita* from lower and higher shore levels have similar respiration rates in air and water.

Morphology of the respiratory organs may affect the differential effects of submersion and emersion on respiration rate. The high-shore limpets *Notoacmea pileopsis* and *Cellana radians* respire slower in air than in water, but the effect of desiccation is greater on *N. pileopsis*, which has a bipectinate gill, than on *C. radians*, which has secondary pallial gills.

An interesting, more complex pattern of variation in aerial and aquatic respiration rates occurs among South African limpets. *Patella cochlear* from the lowest shore levels has equal aquatic and aerial respiration rates. Mid-shore *P. oculus* and mid- to high-shore *P. granularis* vary in their response to aerial exposure according to body size, but in different ways. Small individuals of *P. oculus* live in rock pools and dam, crevices, respiring faster in water than in air, whereas large individuals occur on dry rocks and respire faster in air Fig. 6.3. Metabolic rate is therefore kept high by a reversal of the aerial/aquatic respiratory response as the limpets age, enabling them to maintain a high productivity by rapidly exploiting the plentiful food at mid-shore levels.

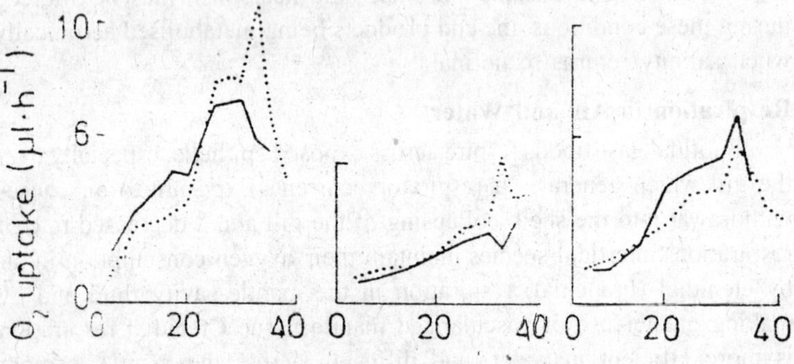

Fig. 6.3 : Respiration Rates in Air (Dottwd Lines) and Water (Solid Lines). Left, *Patella oculus*; middle, small *P. granularis*; right, large *P. granularis*. After Branch (1981).

On the other hand, *P. granularis* migrates up the shore as it ages, young individuals occurring lower on the shore and respiring faster in air than in water, older and therefore larger individuals occurring high on the shore but respiring faster in water than in air. This response to exposure and wetting is contrary to the general trend among high intertidal gastropods but, since tidal immersion is brief at these high levels on the shore, it will have the effect of lowering the average metabolic rate. Conservation of energy is probably important to *P. granularis* because food is scarce at the top of the shore.

Food Supply

Food intake, or lack of it, can profoundly influence metabolic rate. An increase in metabolic rate immediately following the consumption of food has been noted in a variety of animals and is often referred to as the 'specific dynamic action, perhaps associated with a surge of catabolism or anabolism. When given a meal of crab flesh after a period of starvation, *Buccnum undatum* excretes a pulse of ammonia during the following 24 h. signifying the deammarion of absorbed amino acids. There is a concomitant doubling of the rate of oxygen consumption, but this elevation is maintained at least for 50 h. so is unlikely to represent only the metabolic cost of protein catabolism.

Specific dynamic action is not usually distinguishable from other causes of increased respiration accompanying feeding. The rate of oxygen consumption of *Nucella lapillus* begins to rise while the dogwhelk drills its prey, continuing to a peak during ingestion Fig. 6.4. Although part of this respiratory increase must be associated with the metabolic costs of drilling and ingestion, some of it may be attributable to metabolic stimulation caused by the presence of food.

Gastropods are extremely sensitive to proteins emanating from their food: *Ilyanassa obsoleta* responds to a 10^{-10} M dilution of water-soluble glycoprotein, MW 120000, isolated from oysters *Elysia ccuze* responds to a 5×10^{-8} M solution of proteins isolated from the alga *Caulerpa* (Jensen, cited by Kohn, 1983). The mud-snail *Nassarius reticulatus* responds quickly to the scent of macerated crab or even of glycine. Within minutes, the rate of oxygen consumption rises 3-4 times above the routine level, sometimes, but not always, accompanied by increased activity. Recently fed snails show the respiratory response without everting their proboscis or showing any other signs of increased activity. When the olfactory stimuli are removed, the respiration rate quickly falls to the routine level. It also declines after 15-30 min of continuous exposure

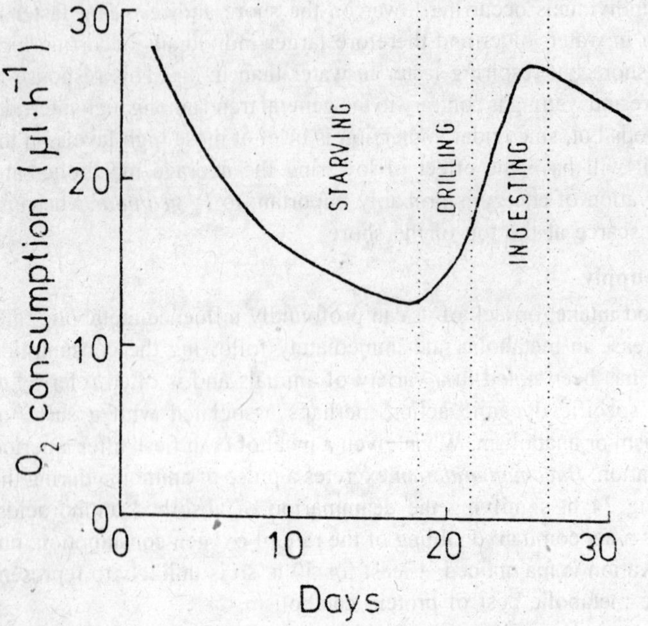

Fig 6.4 : When a Recently Fed *Nucella lapillus* a Deprived of Food, its Respiration Rate Declines until the Next Meal and then Rises Quickly to its Original Value during Drilling and Ingestion. The curve represents the respiration of a snail of 80 my dry flesh weight. After Bayne and Scullard (1978b).

to the olfactory stimuli Fig. 6.5A. This anticipatory increase of metabolic rate in response to olfaction may reflect physiological preparations of the digestive system, particularly enzyme production, in readiness for a meal, although this interpretation requires substantiation. When allowed to feed, the respiration rate of *N. reticulatus* gradually declines to the routine level over 3 or 4 days. This oxygen consumption (or the rate of fasting catabolism : Kleiber. 1961): metabolic processes associated with ventilation, movement and feeding; and gametogenesis. The last two of these contribute to the routine rate of oxygen consumption. Fig. 6.6 shows a simple schematic model that may help to account for some of the observed effects of temperature on the respiration rate of *M. edulis*. Gametogenesis, feeding/activity and the processes of maintenance contribute towards the observed rate of oxygen uptake. Temperature acts

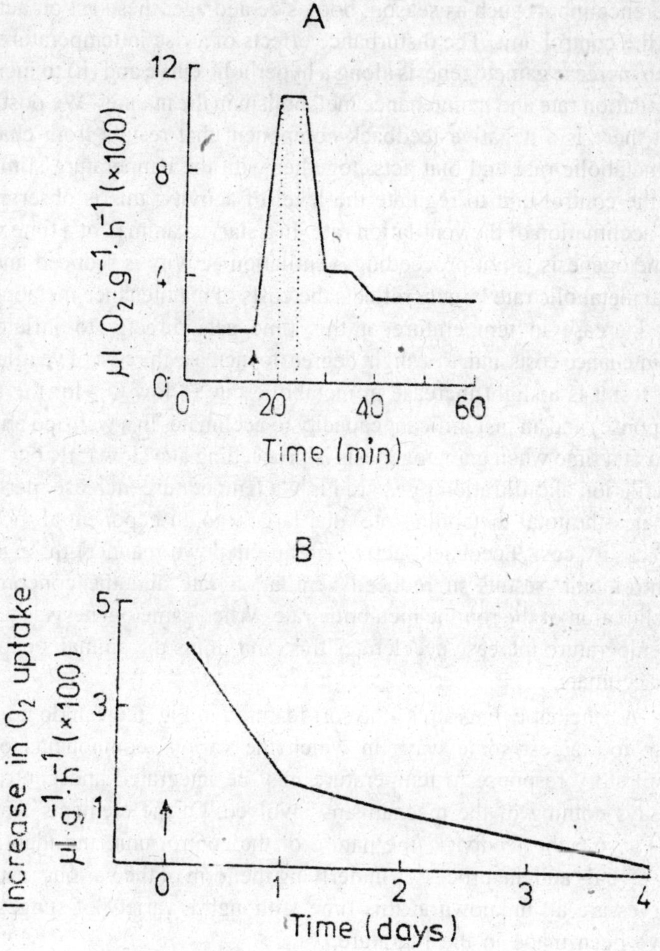

Fig. 6.5 : A. The Smell of Food increases the Rate of Oxygen Consumption by *Nassarius reticulatus*. Respiration rate per unit dry body weight declines after the meal. The arrow indicates the time when food was introduced. The dotted lines indicate the beginning and end of feeding. B. After a Meal. The Rate of Oxygen Consumption per Unit Dry Body Weight Declines over Several Days towards the 'Standard' Rate. The arrow indicates the time when food was introduced and eaten. After Crisp *et al.*, (1978).

directly on these processes as a 'disturbance' (Wieser, 1973) and also on a 'control unit' or temperature transducer as a 'stimulus'. Gametogenesis, feeding/activity and maintenance are influenced by reference inputs such as season, body size and age: these act on activity via the control unit. The disturbance effects of a rise in temperature are: (i) to increase gametogenesis along a hyperbolic curve and (ii) to increase ventilation rate and maintenance metabolism in the manner. We postulate that there is a negative feedback component that results from changes in metabolic rate and that acts, together with the temperature stimulus, on the control unit to regulate the level of activity; this is observed as the acclimation of the ventilation rate. In a starved animal, at a time when gametogenesis is not proceeding, ventilation activity is reduced and the total metabolic rate largely reflects the costs of maintenance metabolism. An increase in temperature at this time acts directly to inflate the maintenance costs and, to a slight degree, to increase the cost of ventilation; the result is a slight increase in metabolic rate (a low Q_{10} for the acute response) and an insignificant capacity to acclimate. In a well-fed animal, also at a time when gametogenesis is proceeding at a slow rate, but when ventilation and filtration rates are high, a temperature increase markedly inflates the total metabolic rate, due largely to an exponential increase in activity cost. Feedback, acting (in an unknown manner) through the control unit, results in reduced ventilation rate and the concomitant acclimation of the routine metabolic rate. When gametogenesis is active, a temperature increase accelerates this, and limits the animal's capacity to acclimate.

A schematic diagram of the sort featured in Fig. 6.6 can do no more than to suggest some ways in which the various components of the respiratory response to temperature may be integrated and controlled. It says nothing of the mechanisms involved. Of the elements featured in Fig. 6.6 the feedback, the nature of the control unit, the identity of the signals and the processes underlying the form of the various response curves are all unknown at this time, although a variety of suggestions have been made in the literature.

In attempts to systematise the considerable literature that has developed on the non-genetic adaptation, or acclimation, of metabolic rate to changes the muscles relax, the cavity decreases in size and the increase of pressure facilitates gaseous exchange with the blood in the veins in the roof. Then the pneumostome opens and air is expelled. Although pulmonate gastropods renew the air in their lungs at fairly frequent intervals, breathing is not as regular or as frequent as in vertebrates. Many fresh-water Basommatophora, such as species of *Lymnaea* come to the surface regularly to take in air. The varied conditions which prevail in these fresh-water pulmonates, ranging from air-breathing

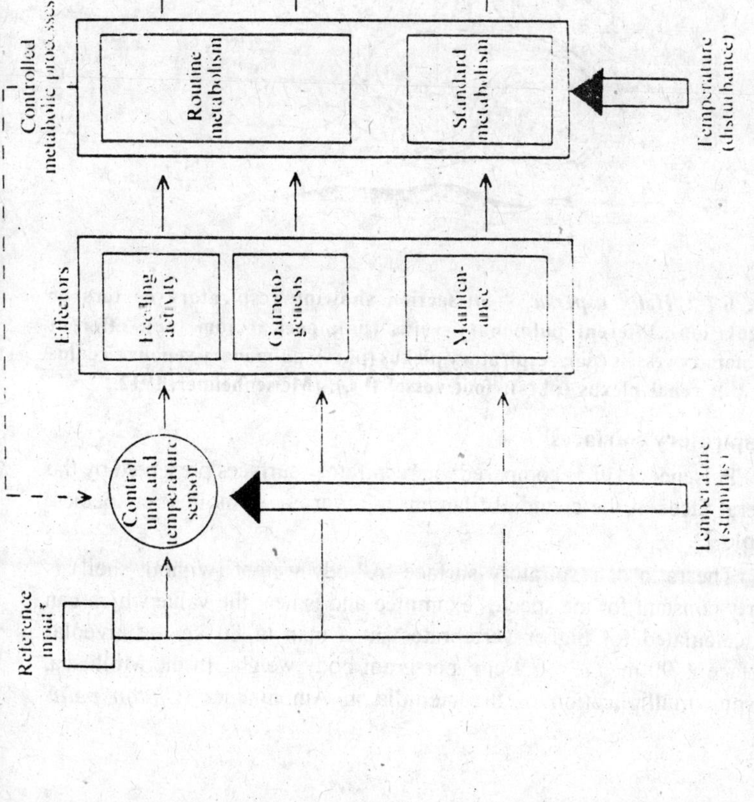

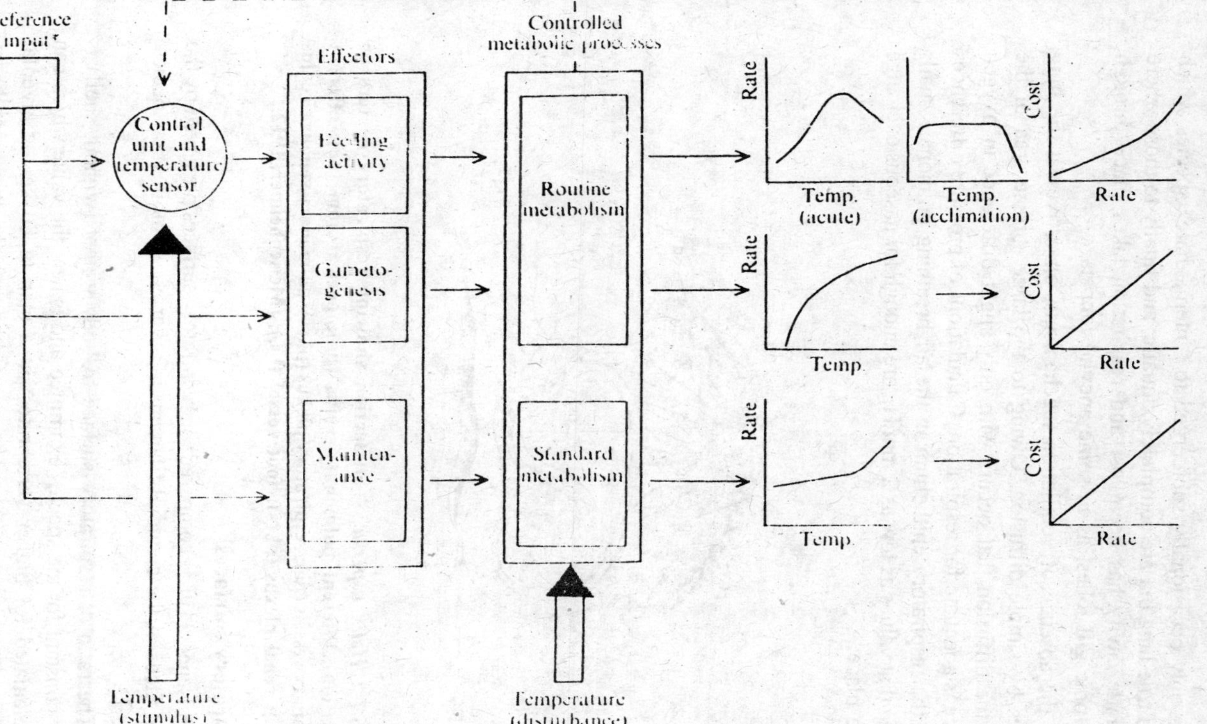

Fig. 6.6 : A scheme (based on Wieser. 1973) to illustrate some the effects of temperature on respiration and activity of *Mytilus edulis*. The form of the temperature response is shown as rate/temperature curves and, more speculatively, as curves of the 'energy cost' against rate.

animals only occasionally submerged to "intermediate" species which retain a true lung but are completely aquatic, and finally to those where the mantle cavity has been lost and respiration takes place through neomorphic gill lobes, as in some ancylid limpets.

In the absence of ventilation mechanisms, air arrives at the lung surfaces by simple diffusion. Owing to the large surface area of the lungs, the diffusion that occurs with only slight difference in oxygen pressure is adequate for respiration. Calculations of pressure differences across the respiratory epithelium in the air-breathing pulmonate snails indicate that values as low as 2 mm Hg are enough for gaseous exchange to take place.

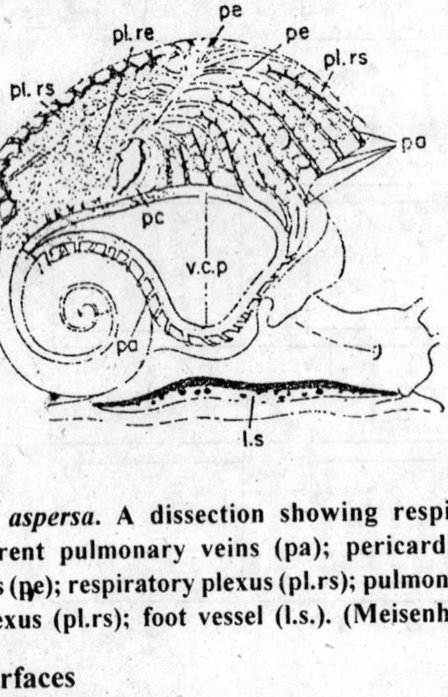

Fig. 6.7 : *Helix aspersa*. A dissection showing respiratory centers or circulation. Afferent pulmonary veins (pa); pericardium (pc); efferent pulmonary vessels (pe); respiratory plexus (pl.rs); pulmonary venous circulus (v.c.p.); renal plexus (pl.rs); foot vessel (l.s.). (Meisenheimer, 1912.)

Respiratory Surfaces

Pelseneer (1935) compared the respiratory surfaces presented by the lateral faces of the ctenidial filament in a variety of molluscan species. Table 1).

The ratio of respiratory surface to body weight (without shell) is fairly constant for the species examined and is near the value which can be calculated for higher vertebrates: in a man of 80 kg the alveolar surface is 90 m^2 *i.e.,* 10.9 cm^2 per gram body weight. In the Mollusca, despite multiplication of the ctenidia in Amphineura (*Chiton pellis serpentis* has 44 pairs), their duplication in *Nautilus* and their replacement by accessory organs (*e.g.,* pallial gills in *Patella*), the respiratory surface remains approximately constant. According to Yonge (1947) the values calculated by Pelseneer for *Mytilus* and *Cardium* are low. In fact, as a consequence of the elongation of the gill filament for feeding purposes, an increase of the frontal surface occurs and the lateral area is reduced in relation to the frontal length. A corrected estimation of the lateral faces gives a value of 13.5 cm^2 per gram for both *Mytilus* and *Cardium*.

Table 6.1 : Respiratory Surface in Adult Mollusca[a]

Class	Species	Body weight (gm)	Number of lamellae per gill	Total surface (Cm^2)	Cm^2/ gram wet meat
Amphineura	*Chiton pellis serpentis*	2.73	70	23.6	8.60
Gastropoda	*Trochus cinerarius*	0.35	225	2.55	8.64
	Patella vulgaia	7.26	272	68.0	9.36
	Buccinum undatum	20.0	256	158.7	7.94
	Purpura lapiilus	1.05	106	7.45	7.1
	Helix pomatia (lung)	13.0	–	107.5	8.3
Bivalvia	*Mytilus edulis*	12.0	–	108.0	9.0
	Cardium echinalum	11.8	122	107.5	9.12
Cephalopoda	*Nautilus macromphalus*	135.0	45-52	1257.0	9.3

[a]From Pelseneer (1935).

MECHANISMS OF RESPIRATORY EXCHANGES

1. Cephalopoda

Eledone, Octopus, Sepia, and *Loligo* have been the objects of most extensive studies into the mechanism of respiratory exchange in Cephalopoda.

(a) Ventilation

In Cephalopoda the respiratory movements—inhalation and exhalation, made possible by the rhythmic contraction of the mantle—bring water into contact with the ctenidia which are attached to the topographically upper (morphologically anterior) wall of the mantle cavity. The gills have a conical shape with the apex pointing upward and forward and, except for the side by which they are attached to the mantle, each gill protrudes freely into the cavity. The macroscopic and microscopic structure of the gills of cephalopods has been studied since detail by Joubin (1895).

In most cephalopods water is kept continually flowing over the gills by the coordinated activities of the mantle, of the funnel and of the inlet valves. A different respiratory mechanism exists in members of the family Cranchiidae. These deep water squids have a horizontal partition across the mantle cavity which is fused to the mantle wall and to the ventral midline of the enlarged coelom. In *Crcinchia scabra* the typical contractions of the coelom expel water through the funnel at the same time as water passes into the dorsal chamber. The water current necessary for respiration is, therefore, created by the enlarged coelom, not by the contraction of mantle muscle.

In *Sepia*, under normal conditions and when the animal is resting, the rate of respiration is about 55 inspirations per minute. In *Eledone*, during day time, the number of inspirations averages 12-14 per minute. Small specimens breathe rather more rapidly than larger ones. In *Octopus*, as shown in Table 2, the number of respiratory movements per minute decreases with increase in the body weight of the animal.

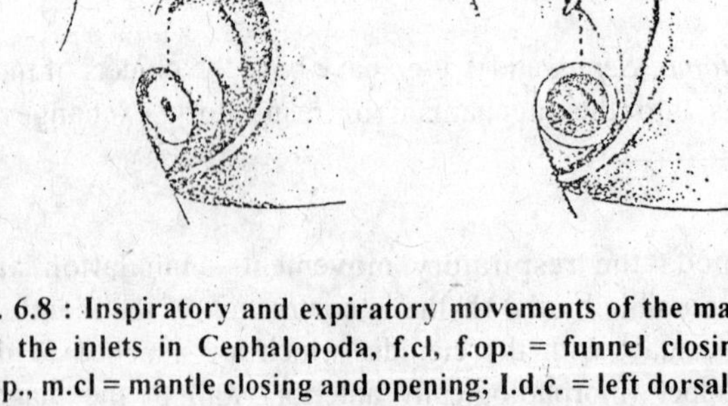

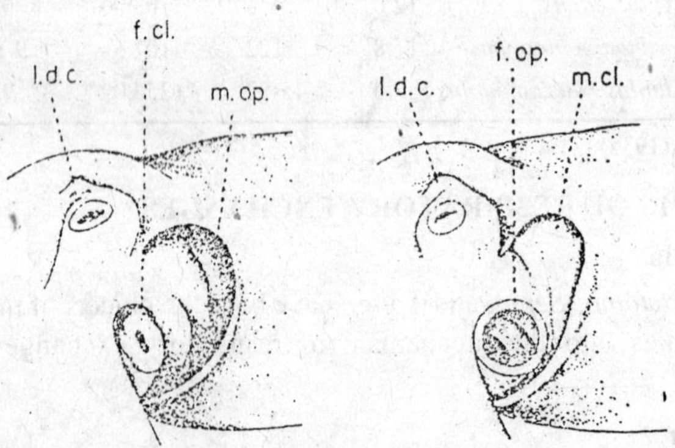

Fig. 6.8 : Inspiratory and expiratory movements of the mantle, the siphon, and the inlets in Cephalopoda, f.cl, f.op. = funnel closing and opening; m.op., m.cl = mantle closing and opening; l.d.c. = left dorsal cirrus. (Isgrove, 1909).

Table 6.2 : Respiratory Movements in Octopus vulgaris

Body weight (gm)	Number of inspirations per minute
2.5-3	51
5	45
9-23	40
40-50	36
100-150	30
200-450	27
500-500	26
900	25
1000-2000	24
2500-3000	23
3500	22
4000	18
8000	12

Generally the number of respiratory movements per minute decreases when the oxygen supply is diminished as well as when the blood flow to the brain is reduced or stopped by mechanically compressing the cephalic artery. Conversely, respiration greatly increases when the animal is in some way stimulated or excited: strong contractions of the mantle are seen, as during locomotion, and in *Octopus* these may even force the gills half way out the mantle cavity.

(b) Nervous Control

In Cephalopoda respiratory movements and the respiratory organs are under the control of nerves which run from the pallioviseral (posterior subesophageal.) lobe of the brain.

The paired pallial nerve has one main branch on either side running by way of the stellate ganglion—a roughly triangular structure giving off radiating nerves which may be seen inside the mantle aperture without dissection—to the musculature of the mantle. In Decapoda (*Sepia, Loligo*) besides many small fibers, each of the postganglionic branches of the pailial nerve contains one fast-conducting, so-called giant fiber. The postganglionic giant fibers are in synaptic contact with a single second-order preganglionic giant fiber from the palliovisceral lobe of the brain. According to Young, the mantle of the squid possesses a double mechanism of contraction: the rapid maximal contraction mediated by impulses in the giant fiber system and the graduated response mediated by the small, slow-conducting fibers. Wilson (1960), working with *Octopus* in which the giant fiber system is absent, found that the slow response facilitates with repetition of the impulses travelling in them. It is presumably the slow system which is normally in operation during respiratory movements not involving locomotion.

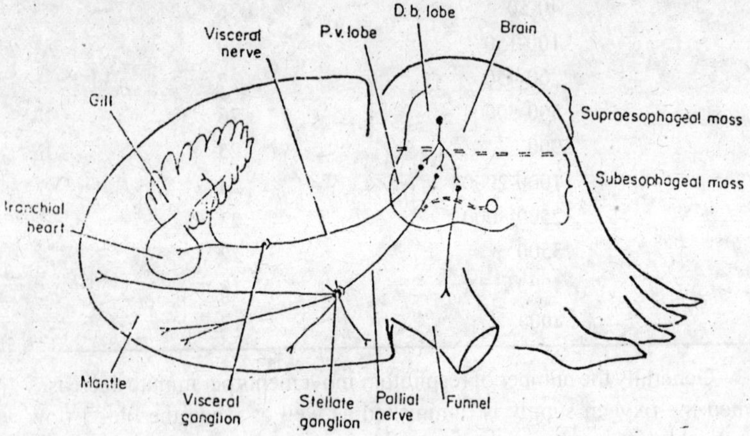

Fig. 6.9 : Generalized scheme of a Abranchiate cephalopod showing some of the efferent nerve pathways involved in respiration. Palliovisceral lobe (p.v. lobe). (From A. Packard, 1965).

The gills, the branchial hearts pumping venous blood to them, and also the auricle of the systemic heart which receives the oxygenated blood, are all innervated by end branches of the branchial nerve which itself is a branch of a ganglion on the visceral nerve. In Octopoda, in *Eledone* at least, the branch of the branchial nerve running to the gill of its own side swells to a ganglion at the level of each set of gill lamellae; from each ganglion fibers run to the corresponding lamellae. The paired visceral nerve arises from the palliovisceral lobe medial to the pallial nerve; presumably its branchial terminations serve in part to control flow of blood through the gills.

Besides these peripheral connections, the palliovisceral lobe is connected with supraesophageal (higher motor) centers which may modify the respiratory control set by the supraesophageal (lower motor) centers. The electrical stimulation of the medial basal lobe of *Sepia* produced rhythmical contractions of the mantle musculature and movements of water into and out of the mantle cavity as during normal inhalation and exhalation.

A supplementary respiratory control analogous to that of the Hering-Breuer reflex of mammals is present in Cephalopoda. Using a nerve-muscle preparation formed by the mantle and to nervous supply, it can be demonstrated that pressure on the gills causes contraction of the longitudinal fibers, whereas stretching of the mantle makes, the circular fibers contract.

(c) Chemical Control

The chemical control of respiratory activity has been studied in *Octopus*, who showed that carbon dioxide is able to increase the frequency and, to a lesser degree, the amplitude of the respiratory movements. Excess of carbon dioxide produces a tenfold increase in the volume of water pumped through the mantle cavity. Whether the effective respiratory stimulant is represented by hydrogen ions or by CO_2 per cent is not known, an whether these agents have a stimulatory action on the respiratory nerve center of the brain remains to be determined.

2. Bivalvia

The very complex and specialized ctenidia in this class are concerned with feeding as well as respiration. Their structure has been the subject of numerous investigation. Since the hypertrophy and specialization of these ctenidia is concerned with food collection, their detailed structure is discussed elsewhere. It should be noted that the great elongation of the filaments, by a corresponding increase in the lateral surface, produces a flow of water far greater than is required for exclusively respiratory needs. The frontal cilia originally concerned with cleansing here become the means of food collection, assisted in this by *laterochontal* cilia which act as strainers, throwing particles onto the frontal surfaces of the ctenidia. The general appearance of the ctenidia in *Cardium* (*Eulamellibranch*) and *Pecten* (*Filibranch*). There is clearly a great super abundance of respiratory surface.

(a) Ciliary Activity

The functional study of the ciliary currents has received the attention of numerous authors, but the mechanism of ciliary motion has still not been satisfactorily explained. It has long been a matter of debate as to whether or not ciliary activity is regulated by the nervous system. Previous workers held the view that the activity could be stimulated by the appropriate nerve, who studied the epithelium .of fresh-water mussels, came to the conclusion that ciliated cells are autonomous units and continue to beat in the absence of neural connections. The fact that any small piece of gill cut away from the body may be seen to continue beating with a metachronal rhythm, indicates that rhythmical activity is not controlled by one particular part of the tissue. The effect of various anesthetics and drugs on the ciliated epithelium of *Anodonta cataracta* and of several other species, the beating and the metachronism of the ciliary system are independent activities. It appears from this work that at least three control mechanisms underlie the activity of ciliated cells: one responsible for the ciliary beat; another which determines coordination within each individual cell; a third which regulates multicoordinated activity between the cells of the whole tissue.

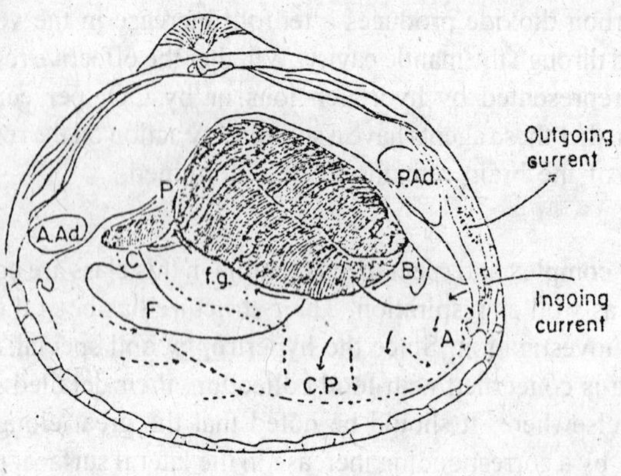

Fig. 6.10 : View of the mantle cavity of the common cockle, *Cardium edule*, to show the respiratory current and the currents connected with the mode of feeding. CP, Ciliated path on mantle which carries away the material rejected by the palps and that collected from the mantle; A, point at which heavier particles begin to drop out of main stream onto mantle, and also the region on the mantle whence the material collected by the ciliated path is finally shot out of the mantle cavity; B, gill-shield directing the ingoing current ventrally; C, point at which material is passed from the palps to the mantle; fg, food groove at the ventral edge of the inner gill lamella; P, lett outer palp, below the base of which lies the mouth; A.Ad,, anterior adductor; P.Ad., posterior adductor. The dotted arrows on the mantle and foot indicate the directions in which the cilia lash. The arrows on and at the edge of the gill indicate the paths of the food streams.

A detailed study of the central nervous system and of the branchial nerve in *Mytilus edilus* and *Modiolus modiolus* failed to reveal any innervation of the gills. Nerves do not enter the gill filaments even though contractile cells are present. In agreement with previous workers, Lucas concludes that the impulses responsible for coordination pass through the cytoplasm of the cell.

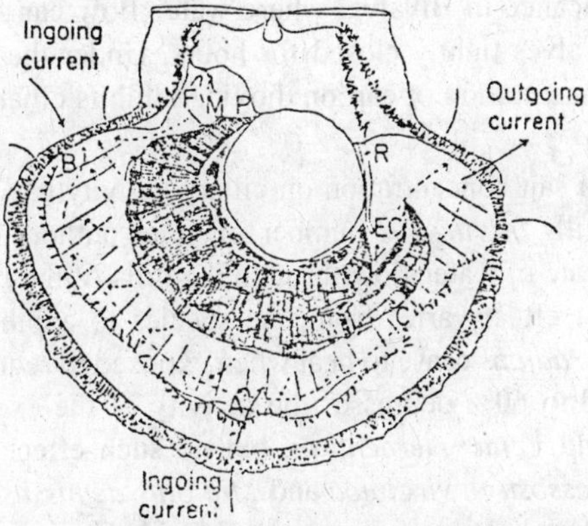

Fig. 6.11 : View of mantle cavity of the scallop *Pecten maximus* to show the food streams, seen from the lett side with the mantle supposed to be cut away. The dotted arrows indicate the directions in which the mantle cilia lash, and the dotted line on the ventral part of the mantle between A and B indicates the ciliated path. The small arrows at the edges of the gill lamellae and of the reflected filaments indicate the patlis of the main food streams which lead to M, the region of the mouth. The arrows at the proximal ends of the gills, as at EC, indicate the direction of the exhalant current. A. Point at which the heavier particles settle out of the main food streatus, B.C. the ciliated path on the mantle; F, toot; Al, region of mouth; P, left outer palp; R, rectum.

Temperature and hydrogen ion concentration have a profound influence on ciliary activity and, as a consequence, on the rate of flow of water produced by the gills and the respiration of the organism. In *Mytilus edulis* the frequency of beat of the gill cilia increases with rise in temperature in the range 0.33°C, although the amplitude of the beat remains constant. Between 34° and 40°C there is a marked fall in the amplitude followed by a reduction in frequency. At 40°C the cilia enter into a relaxed state and at 45°C they appear to be contracted. The resistance to high temperature varies with different molluscan species.

Vernberg *et al.*, (1963) found that, whereas the cilia of isolated gill pieces of *Aequipecten irradians* cease activity when exposed to 37°C, those of *Modiolus demisus* and *Crassostrea virginica* survive a temperature of 44°C. Generally the effect of temperature on the activity of cilia parallels that on other types of contractile protoplasm. At high temperatures time of exposure becomes important.

The response of ciliary activity to carbonic acid has a special biological significance in Bivalvia where water flow can be prevented by keeping the valves tightly closed for hours. Under these conditions the increased concentration of carbon dioxide inhibits ciliary activity on the gills.

The effect of salt concentration on ciliary activity has been studied on the excised gills *in vitro*. A number of other authors have studied its effect on the rate of water flow through the gills of the intact animal. The extent of the effect varies from one species to another. Gill cilia of *Aequipecten irradians* cease to beat when exposed to reduced salinity; sea water diluted to 60% depresses the activity of the excised gills of *Mytilus edulis* and *Venus mercenarin*, but no such effect can be seen in the gills of *Crassostrea virginica* and *Midiolus demissus*. Cilia of the excised gills resume beating if potassium chloride or veratrine is added to the medium. Magnesium ions have no effect. The velocity of the metachronal wave is affected by such physical factors as the viscosity of the surrounding medium.

The influence the activity of the lateral gill cilia as well as the relationship between frequency, velocity, and wavelength of the metachronal wave, found that the branchial nerve has a facilitatory role in *Mytilus edilus*. The same author has extracted a substance from the gill of *Mytilus* capable of restoring the activity of lateral cilia in the excised gill. This substance has an action like that of 5-hydroxy tryptarmine (5-HT) on the clam heart. Since 5-HT is able to activate and stimulate ciliary activity and since among all the cilioacceleratory substances described in the literature none is found to be so potent for bivalve cilia as 5-HT, it is possible that the activity of the branchial nerve, at least in the bivalve gill, causes the release of 5-HT or of a similar substance in the gill tissue and this activates the lateral cilia. According to these results, 5-HT would be a better candidate for the control of rhythmic ciliary activity than acetylcholine. The concept of a "local hormone" exerting its reaction in the same tissue which synthesizes and destroys it has already been applied to the autorhythmicity of the heart and to the ciliated epithelium of vertebrates.

Ciliary activity normally involves oxygen consumption; the oxygen uptake varies with the ciliary rate. In well-aerated conditions, between 0° and 30°C means within a given temperature range the rate was highest for the Arctic *P. groenlandicus*, intermediate for the boreal *P. vurius*, and lowest for the Mediterranean species *P. flexuosus.*

2. Salinity

The metabolic rate of aquatic Mollusca, as for other aquatic invertebrates, is generally affected by the osmotic pressure of the environment. Their response to salinity changes is, however, not uniform. *Theodoxus fluviatilis* and *Potamopyrgus jenkinsi* are species living both in brackish and fresh water. Whereas the former has the same oxygen consumption whether living in water of 11‰ salinity or in fresh water, individuals of *P. jenkinsi* from brackish water have a greater oxygen uptake than those from fresh water respiratory rates of specimens of *P. jenkinsi* from two fresh-water localities were different and that values for specimens living in brackish water were intermediate. The response to variations of salinity in the Australian fresh-water lake mussel *Hyridella australis* is very irregular. Oxygen uptake in this species decreases as the chloride concentration of the external medium rises from 0.25 to 5.0 mM.

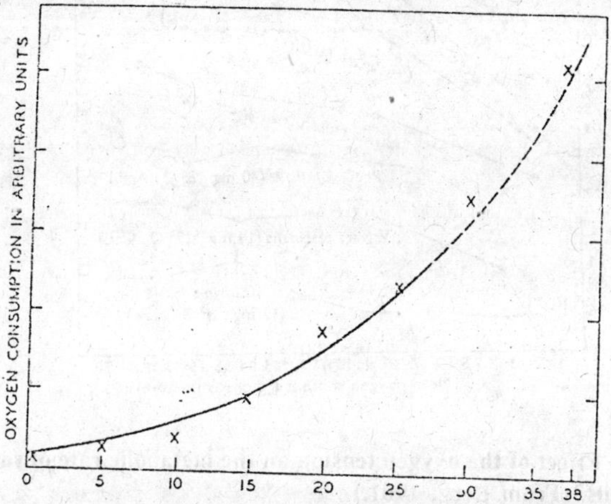

Fig. 6.12 : Relationship between temperature and oxygen consumption of Australorbis glabratus in the range 0.3-37°C. (From von Brand *et al.*, 1948).

Many osmoregulating animals respond to a decrease in salinity with an increase in the respiratory rate, but many nonregulating animals respond to a change in salinity with a decrease in respiratory rate. Although experiments of this kind suggest a causal relation between the osmotic demand and changes in metabolic rate, further consideration and experiments throw doubt on this simple hypothesis. Isolated tissues do not always show the same response to salinity changes as the whole animal. The gills of *Mytilus edulis* have a higher respiration rate in more dilute media, both hyper and hypotonic sea water decrease oxygen consumption. Isolated gills of the bivalve *Dreissena polymorpha* when transferred from tap water to sea water show an increase in respiration for about 2 hours, then the respiratory rate gradually returns to its initial value. Isolated gills of *Venus mercenaria* when transferred from sea water to isotonic and solution show an increased oxygen consumption. Apparently there is antagonism between sodium and divalent ions. The possibility that an increase in water content of the cells is responsible for the increase in oxygen consumption in lowered salinities was suggested by Schlieper. But experiments with crabs indicate that changes in respiration may not be the result of cellular hydration.

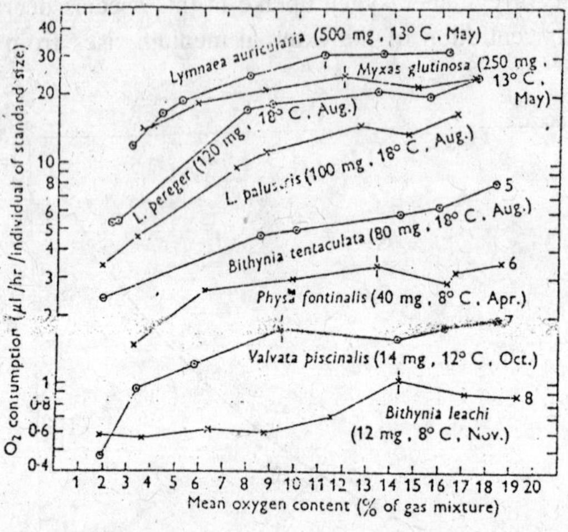

Fig. 6.13 : Effect of the oxygen tension on the metabolic rate of some fresh-water snails. (From Berg, 1961.)

3. Oxygen Tension

In molluscs all kinds of responses of metabolism to oxygen tension are found—from total or partial dependence to complete independence. In forms having little capacity for regulation, respiration is at least partially dependent on said varies linearly with external oxygen. This may be due either to incomplete saturation of the tissues with oxygen or to an increase in oxidative processes as a consequence of the increased oxygen tension in oxygen consumption, the maximum value being reached after one hour.

Table 6.3 : Influence of Oxygen Tension of the Oxygen Consumption of *Australorbis glabratus*[a]

	O_2 (ml/gm wet wt./hr)[b]	
O_2 (mm Hg)	30-40 mg	300-400 mg
760	288 (2.50-334)	144 (97-190.)
38	205 (186-221)	138 (99-178)
13	260 (234-279)	156 (110-204)
5	29.6(8.2-49.2)	12.5(3.8-17.3)

[a]From von Brand *et at.* (1948).

[b]Two size groups are shown. Figures are mean values and ranges at 30°C.

4. Seasonal Variations

In the same individual, the rate of oxygen consumption may change with the season. Berg and Ockehnann (1959) have compared two series of experiments with *Lymmaea palustris* and *Lymnaea peregra* and found that the Q_c of seasonally. There is also a seasonal variation of the oxygen content the limpet *Ancylus fluviatilis* in populations from both stagnant and running water. In spring and early summer Q_{or} is higher than in other seasons of the year. *Lymnnea pulustris* shows the highest respiratory rate in July and the lowest in November. It is most probable that the seasonal variations in oxygen uptake are intimately associated with some physiological activity of the animal such as reproduction in *Ancylus* or to concurrent biochemical changes of tissue composition as in *Mytilus*.

5. Starvation

During starvation only a slight decrease in oxygen utilization rate was observed in the fresh-water snails *Lymnaea peregra, Myxas glutinosa, Bithynia tentaculata*, and *Valvata piscinalis*. On the other hand, a distinct decrease shown by *Lymnaea palustris* and *Bithynia leachii*. Partial starvation had considerable influence on the rate of respiration of *Ancylus fluviatilis* fatal starvation caused a rapid decrease in oxygen uptake in this species: after 96 hours a decline to about three-fifths of the initial value was observed.

Similar, but delayed, effects have been observed in *Potamopyrgus jenkinsi* In pulmonate snails protracted starvation decreased the respiratory rate, at first rapidly, later slowly. The values for oxygen consumption by *Australorbis glabratus, Helisoma dtiryi, Physa girina*, and *Physa sp.* are shown in Fig. 6.14. Large deviations occurred only during the first days of starvation; in the following days the values fit very well to a single curve.

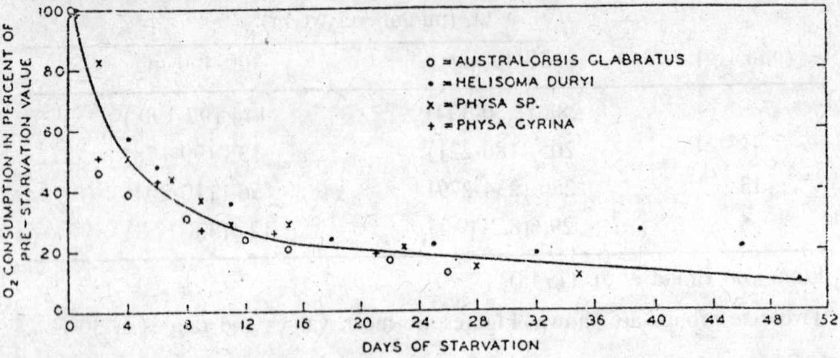

Fig. 6.14 : Effect of starvation on the rate or oxygen consumption of four species of pulmonate snails.

Effects of Physicochemical Factors

1. Temperature

Metabolic rate is generally related to temperature in Mollusca as in other poikilotherms. The relation between temperature and oxygen consumption in terms of a curve known as Krogh's normal curve. The respiration of *Australorbis glabratus* follows the curve and its extension to 37° (Fig. 6.15). However, deviations from Krogh's curve have been reported for a number of species: *Myxus glutinosa, Physa fontinalis, Bithynia leachii, Bithynia tentaculata,* and *Lymnaea peregra.* In *Mytilus edulis* respiration rose rapidly with increasing temperature and was directly proportional to temperature over the range 3-16°C. At higher temperatures, the respiratory rate passed through a maximum and decreased to the thermal death point. The maximal respiratory rate differs between species. For example, in *Mytilus edulis* the maximum was recorded at about 20° and in *Littorina littorea* at 35°. In individual species it is to be expected that the position of the maximum will be influenced by the temperature of acclimatization.

The measured the rate of oxygen consumption from very low temperatures to 30° for three species of *Pecten* from different. In fresh water pulmonate cutaneous respiration is chief means of respiratory exchanges. Aquatic habitat molluscs have a mantle cavity without ctendia which related to secondary function long time and is used in conditions when oxygen content of water falls significantly and oxygen drops below a definite value the animals to the surface of water to breathe, moving by floating or by travelling over the submerged objects. At the surface they orient the body in such a manner as to allow the resiphon like respiratory tube to be projected through the overlaying film. A definite relationship exists between the amount of air imposed and period of inspiration.

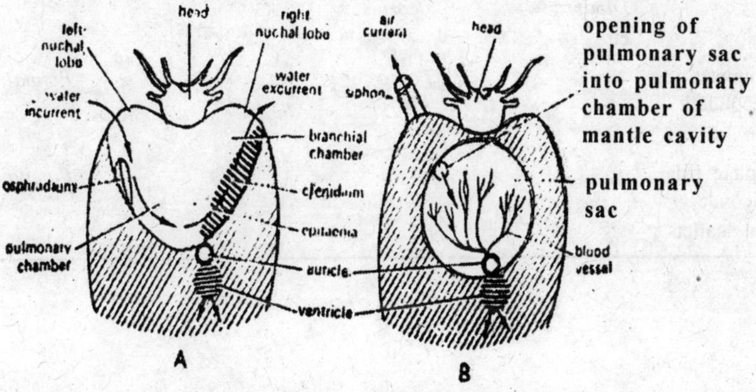

Fig. 6.15: *Pila*. Mechanism of respiration.

A Stereogram showing water current in aquatic respiration.

B– Stereogram showing air current in aerial respiration.

In short respiratory system comparise a series of small gills or ctendia. These are arranged in two rows, one in each pallial groove. The number of gill pairs varies from 6 to 80 in different species. The number of gill pairs even with in a single species.

According to Yonge (1999) the large number of gills in *Chitons* represents a multiplications of the original single pairs of the ancestral molluscs. The gills are inserted at the bottom of the pallial groove. Each gill consist of a thin central axis fearing an anterior and posterior row of smooth, delicate and more or less semicircular branchial leaflets or gill lamella, which one arranged like the leaves of book. The entire surface of the branchial is ciliated.

Table 6.4 : Diversity of respiratory arrangements in the Prosobranchia

Five genera in the Patellacea are marked with an asterisk (*), With the exception of the two general *Valuate* and *Adeorbis*.

	Two bi-pectinate ctenidia; shell slit or performated	One bi-pectinate ctenidiumn; water current left to right	Oe pectinibrach ctedium; water current left to right	No ctenidia
No secondary pallial leaflets	*Pleurotomaria* *Scissurella* *Emarginula* *Haliotis* *Diodora* etc.	Trochacea Neritacea **Patelloida* *Valvata* *Adeorbis*	+Mesogastropoda Neogastropoda	**Lepeta*
Incomplete ring of secondary pallial leaflets		**Lottia*		**Pctina*
Complete ring of secondary pallial leaflets				**Patella*

7

BLOOD, BODY CAVITY

In all molluscs but cephalopods the general body cavity is filled with blood brought back from outlying venous sinuses, the coelom comprises only the pericardium and the cavity of the gonad in *chitons*, gastropods and lamellibranches the head, foot and viscera are supplied directly with blood by close arteries from the ventricle. There are in molluscs no true capillaries. Except in the advanced and well-endowed cephalopods; in other classes blood seeps from the arteries into pseudo vascular spaces in the connective tissue .in gastropods peripheral blood from the head foot and much of mantle is returned into a central space known as the cephalopodan sinus. From the visceral mass it is brought back to a visceral sinus both these spaces discharge into a sub renal sinus lying near the columellar muscle at the base of the visceral mass. From the subrenal sinus blood is distributed in varying proportions through the respiratory organs and the kidney. Before returning to heart there is an extensive renal portal system into which all or part of the venous, blood may go passing thence either to gill or directly to the auricle by an alternative route the rectal sinus blood may be sent along the mantle roof straight to the gill without deploying through kidney re-oxygenated blood always returns by the efferent branchial vein directly to the auricle. Several memoirs describe in detail the arrangement of the molluscan vascular system in this book it can only depticted in the general course the circulation by the diagram. In the monotocardian prosobranch *Struthiolaria* an example not too untypical in its broad features of gastropoda. Some molluscs lack a heart, in the scaphopoda there are no well defined blood vessels of any sort, and pericardium is lost though paired kidneys remain, blood circulates between the organs largely by contractions of the body wall and re-oxygenated in simple transverse folds of the mantle roof in some sacoglossa. Among gastropods the blood is pumped not by a heart but by muscles of the body wall, especially—in *Alderia* and *stiliger*—by contractions of the club-shaped cereta. Even when present, the heart may play but a minor part in the shifting of blood changes in the shape of the body redistribute blood on a scale impossible for the ventricle, and in many gasteropods and lamellibranchs the aorta as it leaves the heart is valved against forcible backflow through the arteries. The presence of blood in the connective tissue spaces gives it another function besides its respiratory and trophic ones: it can now serve as a fluid.

BLOOD

The respiratory pigment of Gastropoda and Cephalopoda is haemocyanin, a copper-containing compound carried in the plasma, which in colour is faintly blue. This is similar oxygen-carrying properties to haemoglobin, though different molluscs vary widely in the oxygen pressures at which their haemocyanin becomes saturated. Thus the blood of most gastropods are saturated at low oxygen pressures, a factor fitting many Gastropoda to occupy habitats of poor aeration. On the other hand, cephalopod haemocyanin becomes saturated only at relatively high oxygen pressures, and these active molluscs are extremely sensitive to oxygen different lack. Dissolved oxygen never forms more than about 3% of the total blood volume in molluscs, except in cephalopods, where the percentage is 8 to 11, as compared with 20% in mammals. A few gastropods to live in especially poor oxygen conditions, have developed haemoglobin, which is here capable of working at very low oxygen pressures. In *Planorbis comeus* living in foul muds, for example, the haemoglobin is saturated in all parts of body and thus useless, in a medium containing more than 7% dissolved oxygen. With pressures from 7% down to 1% (where the animal dies). haemoglobin in useful; it is never so in molluscs for aerial respiration. The fast working muscles of the gastropod buccal mass frequently contain muscle haemoglobin, which may possibly work as an oxygen store during muscle contraction. It is not certain that any lamellibranch has haemocyanin, and in most species studied there is no oxygen carrier in the internal medium, the blood oxygen concentration being that of the outside water. Such a mechanism is able to supply sufficient oxygen to a sedentary animal by reason of the ample surfaces for oxygen uptake in the enlarged ctenidia and mantle. Some bivalves, particularly those living in poorly aerated substrata, possess haemoglobin, contained in non-amoeboid corpuscles. Examples include many of the Arcidae, and *Solen legunen* living in muddy sand in the Mediterranean.

The commonest blood corpuscles of molluscs are stellate amoebocytes which chister together in immense numbers in clotted blood. These cells pervade the whole molluscan body: they migrate out of the blood spaces and pass freely through the connective tissue, entering the mantle cavity and the lumen of the gut. They may aggregate in large depots near the wall of the gut, and appear to have a wide variety of functions, based upon their powers of phagocytosing fine particles. Haeckel first observed the uptake of carmine by molluscan amoebocytes, many years before the pioneer experiments of Metschnikoff on phagocytosis. Amoebocytes may serve as important vehicles of excretion, conveying particles into the lumen of the gut, into the pericardium and renal organ, through the outer body wall, or through special strips of permeable epithelium into the mantle cavity. In bivalves they may emerge from the ctenidial blood spaces on to the surface of the gill, where they take up food particles, then either retreating into the epithelium or being ingested at the mouth.

In the stomach of *Chitons*, gastropods and bivalves, the amoebocytes may be important agents of non-localised intracellular digestion. They freely engulf food particles such as diatoms, or artificially fed starch or fat or blood corpuscles, which they digest, afterwards re-entering the epithelium or cytolysing to yield their contents for absorption by digestive diverticula. Takatsuki has shown the oyster that the amoebocytes have a wide complement of digestive enzymes.

Wagge has demonstrated the importance of the amoebocytes of *Helix* in lime transport and shell repair. Calcium carbonate from the diet is stored in protein spheres in special lime cells of the digestive gland, whence it may be released to the lumen from time to time for the regulation of the pH of the gut contents. Amoebocytes and alkaline phosphatase are both active in the digestive gland cells and at the site of cell secretion. The amoebocytes have access both to the lime stored in the digestive gland and that laid down in previously secreted pats of the shell. Especially in land pulmonates where lime may be in the short supply, repair materials if not provided with the food - may be withdrawn from elsewhere in the shell. No lime is stored in the secreting edge of the mantle. At a broken part of a shell, amoebocytes densely collect, arranging themselves in a sheet to form a continuous organic membrane, which is then calcified. The mantle epithelium plays little part either in forming the membrane or in laying down the deposit of calcium carbonate.

FUNCTIONING OF THE VASCULAR SYSTEM OF CEPHALOPODS

The blood is pumped round the body in an elaborate vascular system with true cappillaries linking the arterial and venous sides. There are three hearts, a systemic one centrally placed in a capacious pericardial space in the visceral mass which receives oxygenated blood from the cetindia and dispatches it to the body and two branchial hearts, one at the base of each gill and also in the pericardial cavity which receive the deoxygenated blood from the system and pump it through the gills.

Circulation is further added by the contractibility of the walls of many main vessels. In *Cephalopods,* the main veins to the branchial hearts pass on their walls. There is no renal portal system as in those groups.

The vascular system in *Bivalves* is like that of *Gastropods* in including considerable haemocoelic spaces around the viscra and in the foot.

The heart consists of a ventricle which comes to lie around the rectum during development and into which two auricles discharge. Blood is discharged both anteriorly and posteriorly from the ventricle but the vessels (anterior and posterior aortae) can not be homologous with the two aortae of gastropods. The anterior passes forward and supplies blood to the head, the foot and the visceral mass while the posterior goes backwards and takes blood to the mantle and siphons. Blood is returned from the anterior part of the mantle to the viscera and the foot into a vessel which takes it to the kidney, thence to the gills and back to the heart.

Circulation

The heart is situated on the mid-dorsal line behind the posterior termination of the hinge, where it is enclosed in the pericardial cavity. The heart is composed of a median ventricle and two lateral auricles. Blood leaves the ventricle by the single anterior aorta from which it is distributed to the body through five channels:

(a) pallial arteries, which supply the mantle and the pallial muscles;
(b) the gastro-intestinal arteries which supply the stomach, intestine, posterior retractor muscles, posterior adductor muscles and mesosoma;
(c) the pericardia! artery, which carries blood to the pericardium, intestine and nearby genital glands;
(d) hepatic arteries which go to the digestive gland; and
(e) The terminal arteries which distribute blood to the anterior parts of the body.

The vencus system collects blood into three main sinuses; the pallial, the pedal and the median ventral sinus. From these sinuses the blood is carried to the kidneys and thence to the gills, and finally back to the heart via the oblique vein. It should be noted that there is not a complete branchial circulation in mussels. Blood can be returned from various organs to the heart, without passing through the gills.

The Haemolymph

The blood, or haemolymph (the terms will be used synonymously), consists of cells within a colourless plasma. There is no respiratory pigment and the oxygen-carrying capacity of the haemolymph of *Mytilus edulis* is equivalent to that of seawater. The blood volume of bivalves is large; Martin, Harrison. Huston and Stewart (1958) have reviewed this topic, and they recorded a blood volume for *Mytilus californianus* of 50.8% of the wet body weight excluding the shell.

Blood Cells

The blood cells of molluscs have been variously referred to as haemocytes, amoebocytes, leucocytes, lymphocytes and phagocytes. Cheng and Rifkin (1970) consider these terms to be synonymous; we shall use the generic term haemocytes. The haemocytes are cells capable of independent movement, and they play an important role indigestion, nutrient transport, excretion, regeneration and repair, and internal defence. The haemocytes are distributed throughout the vascular system, and because this system is an open one they are also found within the tissues.

The classification of cell types present in molluscan blood has been the subject of much speculation, but there is still no general agreement recognised seven cell types in *M. edulis* and other bivalves based on the presence or absence of cytoplasmic granules, and on the cells different staining affinities. However. Galtsoff (1964) stated that there are only two cell types in the oyster *Crassostrca virginica* viz., granular and hyaline cells. The granulocytes are recognised by the large numbers of granules present in the cytoplasm, although the hyaline cells are not completely devoid of cytoplasmic granules. Moore and Lowe (1975) have recently described three morphological types of haemocyte in *M. edulis*; lymphocytes, macrophages and granulocytes. The agranular lymphocytes and the macrophages are believed to form a developmental series. maturing from small, RNA-rich lymphocytes to the larger, phagocytic macrophages. The granulocytes probably represent a separate cell line.

Moore and Lowe (1975) suggest that the lymphocytes represent circulating stem cells; they are small (4-6 μm), spherical, with little cytoplasm and a spherical nucleus (Fig. 7.1 a). As the cytoplasm increases in volume the cells assume an irregular appearance, with pseudopod-

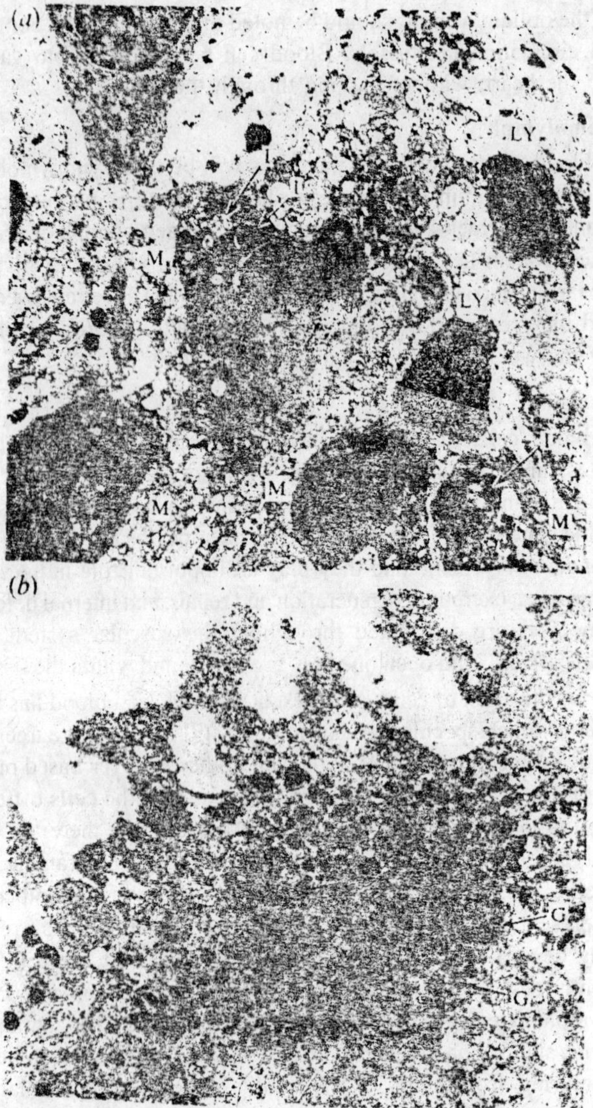

Fig. 7.1 (a) : Haemocytes from the mantle connective tissue of *Mytilus edulis*. This group of cells comprises two a granular lymphocytes (LY) with large nuclei surrounded by a small volume of cytoplasm; and a number of larger macrophages (M) with irregular nuclei, and containing numerous cytoplasmic inclusions (I) which are believed to be phagosomes and phagolysosomes. Uranyl acetate and lead citrate staining, (b) A granular haemocyte (granulocyte) from the mantle connective tissue of *Mytilus edulis*. The small spherical nucleus differs from the nucleus of an agranular cell in the distribution of the intensely stained. The cytoplasm is packed with electron-dense membrane-bound spheroidal.

like formations and vacuoles in the cytoplasm. These macrophages (7-10 mm in diameter: Fig. 7.1 b) are phagocytic, and observed high activity of lysosomal hydrolase enzymes, associated with vacuoles, suggesting the presence of an intracellular vacuolar digestive system. The cytoplasm of the granular leucocytes is filled with spherical granules (0.5-1.0 μm) which range from neutrophilic to acidophilic in staining properties. The degree of acidophilia of these granules apparently increases with increasing size and Moore and Lowe were unable to use the staining properties of the granules to make any functional distinction between different granulocytes.

Haemopoiesis

The origin of the haemocytes in molluscs is unknown although Wagge (1955) has suggested that they originate in the epithelial and connective tissues of the mantle and the digestive gland.

Digestion and storage

Haemocytes play an important role in the digestion and transport of material in the alimentary canal. They ingest directly particles of nutritional value which are too large to enter the cells of the digestive diverticulum. These haemocytes are enzymatically equipped with lipases hemolymph shows great variation from one group to another, ranging from about 9-10% in *Eledone moschata* and *Octopus vulgaris* to 1-4% in *Helix ligata* and *Cryptochiton stelleri* Closely related species do not present great differences in the concentrations of hemocyanin in their bloods. During the life span of an animal the concentration of hemocyanin seems to remain fairly constant under normal physiological conditions. In most cases molluscan hemocyanin is by far the most important if not the only protein present in the blood. Hemocyanin amounts to 98% of the total protein in *Octopus*, 95% in *Helix*, and 90% in *Busycon*; the remaining few per cent consist of proteins of low molecular weight which can be separated easily from the hemocyanin by physical or chemical methods.

Nothing is known about the biosynthesis of hemocyanin, in which organ it is formed, its precursors, or the pathways of its metabolism. Recent studies on the biosynthesis of hemocyanin *in vivo* show that copper metabolism is involved. Distribution studies have indicated that the hepatopancreas is the molluscan organ richest in copper. Nutrition experiments on *Octopus vulgaris* fed on normal and copper-deficient diets clearly indicate that (1) all the copper of the animal is of dietary origin; (2) the alimentary copper is stored in the hepatopancreas; and

(3) the metal is utilized for the biosynthesis of hemocyanin throughout life. Using a very sensitive immunological method, it has been demonstrated that in *Sepia officinalis*, hemocyanin is absent in the unfertilized egg. It makes its first appearance as a copper-protein complex ready to function as respiratory pigment, at stage XIV, about one week before hatching

The main function of hemocyanin in molluscs is to transport oxygen from the external medium to the tissues. Accessory functions are also known and pure hemocyanin is able to act as a pseudophenolase and pseudocatalase. These enzymic activities are due to the presence of copper in the protein molecule and are inhibited by cyanide and copper reagents.

During the oxygenation of hemocyanin one molecule of oxygen combines stoichiometrically with two atoms of copper; this ratio has been unequivocally established for all hemocyanins whatever their origin. The respiratory function of hemocyanin has been firmly demonstrated in cephalopods and in some gastropods (.*e.g.*, *Busycon canalicidatum*). The oxygen capacity of hemocyanin-containing molluscan blood is rather low, ranging from I to 5 volumes per cent and is related to die activity of the animal. Up to 90 of the oxygen is removed from the pigment as it passes through the tissues in Cephalopods. As with other respiratory pigments, the saturation of molluscan hemocyanin with oxygen increases with the partial pressure of oxygen. Loading and unloading tensions are generally dependent on salt concentration, pH, and temperature. The oxygen dissociation curves vary in shape from hyperbolic to sigmoid; undulatory curves have also been observed. The dissociation curve of purified Busycon hemocyanin is accurately described by Hill's equation,

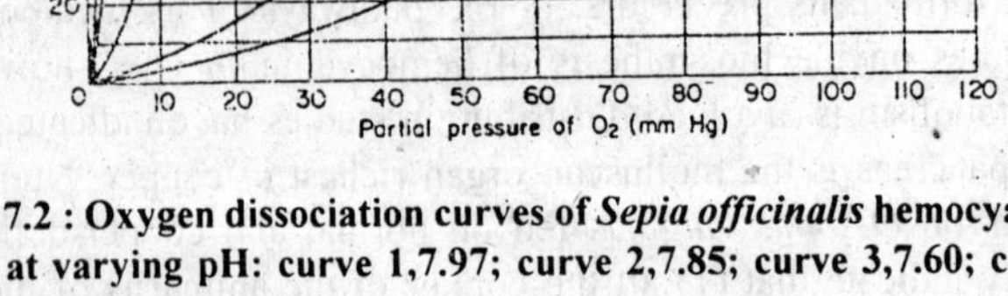

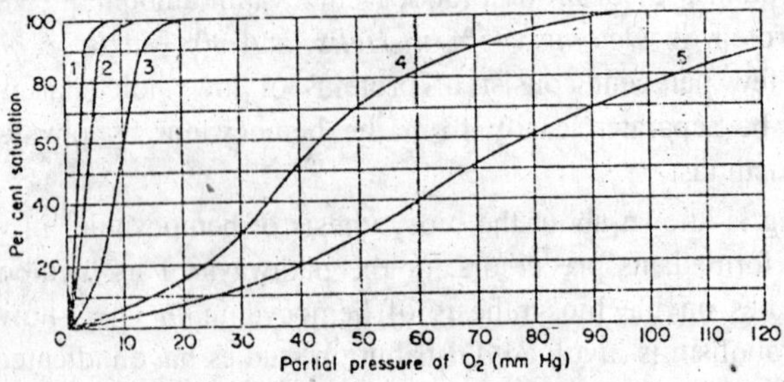

Fig. 7.2 : Oxygen dissociation curves of *Sepia officinalis* hemocyanin at 14°C and at varying pH: curve 1,7.97; curve 2,7.85; curve 3,7.60; curve 4,7.35; curve 5,7.24,

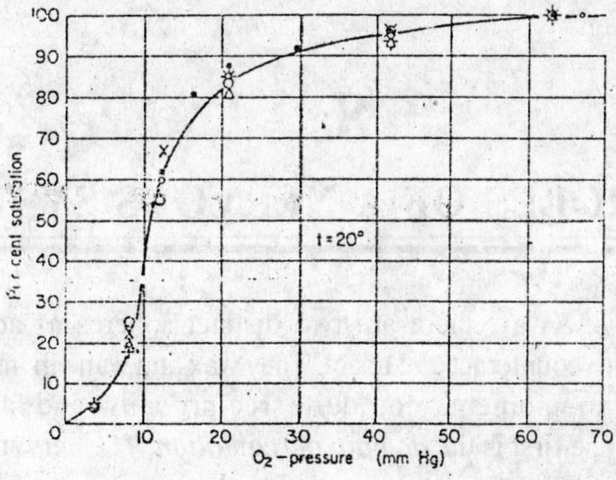

Fig. 7.3 : Oxygen dissociation curve of *Helix pomatia* hemocyanin at different seasons; temperature, 20°C and pH, 7.65.

taking n = 1. The blood of *Helix pomatia* displays a sigmoid dissociation curve at pH 8.2 and a hyperbolic curve at pH 7.0 same blood after dialysis gives a hyperbolic curve over the pH range 4.0 to 9.0; for a sigmoid oxygenation curve the presence of calcium and magnesium ions is required.

The influence of pH on the oxygen equilibrium is also important.

8

BASIC ROLES OF A NERVOUS SYSTEM

Viewed in one way, there are two distinct spheres of activity of nervous system; to counteract and to act. They wax and wane in importance from moment to moment, not quite reciprocally and not quite independently. The first is the *domain of regulation.* The nervous system operates as one of the mediators, probably the main one, of reactions that preserve the animal's status quo. Certain of the events or states (stimuli) that tend to displace some functional feature of the organism call forth a compensatory response resulting in the reduction of the displacement and restoration of the norm. These regulatory reactions are mostly simple, prompt, short lasting, involving only a part of the body or only visceral effectors, "automatic" in the meaning that "volition" or the highest centers are not required-in short, reflexes. By no means are all reflex, however, for even some very complex sequences are both compensatory and stimulated by a need.

The other sphere of functions of nervous systems embraces actions, which do not preserve but alter the status quo; it may be called the domain of initiation. The system does not simply wait for stimuli and react to cancel their effect; it also initiates or deviates from that status. Especially in higher animals one phase or mood replaces another; exploratory or appetitive behaviour is prominent. These can be regarded in a long view as regulatory, but the difference between this domain and the merely compensatory is not only one of time scale. Many of the activities of animals have been found, upon experiment, not to be responses to need, in the sense that stimuli representing needs—trigger the action, but spontaneous in the proper meaning of the word. For example the initiation of movements of various kinds of polychaetes worms was not to explicable by accumulation of wastes, lack of food, or oxygen-though their movements normally reduced such conditions. Nest building, migration, and even eating and drinking have been shown in given cases not to be performed "in order to" achieve a desirable result, but simply to satisfy an internally arising state (drive) that demands the motions (sometimes even less) which would in nature bring about an adaptive end.

These generalizations may appear doctrine or forced but should not be discounted for that reason alone. They do at least serve to frame some of the questions that exemplify the state of the subject and provide points of departure for new work.

Closely related to speed, with the consequent possibility of many steps in a given response, is the role of the nervous system in providing small and hence numerous units of information handling. These are conveniently developed from the cells, 'invented' phylogenetically on other ground. But relative to other organ system, also composed of many cells can act discretely and can influence each other specifically. We shall see, as a major conclusion of our survey, that the units and their properties are essentially alike from the lowest to the highest animals with nervous systems. But the number of units, especially of those in between receptor and motor neurons, increases greatly in higher forms; the number and profusion of their branching processes, together with the differentiation of shapes and connection does likewise. The complexity is the structural background, which provides for complex manipulation of the signals representing internal and external events; they are coded in a special way åssuring high temporal resolution, with many steps per second and with faithful propagation for long distances in the body.

DEFINING FEATURES OF A NERVOUS SYSTEM

If these functional generalities are kept in mind, it may be easier to define our object of study. A nervous system may be obvious and easily identified, as in higher group, at least in its central parts; or it may be extremely difficult either to recognize or to deny, as in the multicellular sponges and the highly coordinated ciliate protozoan. A nervous system may be defined as an organized constellation of cells (neurons) specialized for the repeated conduction of an excited state from receptor sites or from other neurons to effectors or to other neurons. This formulation automatically excludes any coordinating system that may exist in unicellular or a cellular form. Protozoan are because of the intrinsic interest in any specialized conducting organelles that there may be, even though they may have no genetic relation-to the systems of cells in metazoans.

The relevance of the definition for sponges, and any other metazoans in which a nervous system is to be demonstrated, is that it provides anatomical and physiological criteria to be satisfied by any true nervous system. Thus presumptive nerve cells must be shown to connect receptor sites with effectors or to connect with each other. Cells that send prolongations to end in open tissue or spaces or on spicules are unlikely to be nervous, unless other arguments clearly exist. Cells which reach from a superficial position, presumptively receptive, to a deep position in the neighbourhood of shape-changing effectors, can be regarded as indifferent, supporting, or epithelial as well as nervous, unless physiological evidence of specialization for repeated conduction of excitation is adduced. Since the property of excitability is probably general for living material and since any collection of like cells can be called a system, it is the combination of connectiveness and specialization for propagating an excited state that we must look for in a nervous system.

The other problem requiring treatment is simply how to recognize a nerve cell! Granted that it has been shown that a nervous system exists in a given animal, satisfying the criteria above, it may be difficult sometimes to decide whether a certain cell or cell type should be regarded as a nerve cell. For example, how can we decide whether a neurosecretory cell is or is not a nerve cell? The ideal criteria are like those for a nervous system: specialization for conduction and connection with other nerve cells or with receptors or effectors. These criteria may be difficult to use, even apart from the technical obstacles to adequate histological and physiological demonstrations. First, neuroglia, epidermal cells, gland cells, vascular or other elements may make apparently intimate contact with nerve cells and thus reduce the ease of application of the criterion of connection. Second, neuroglia and possibly other types of cells have been reported to exhibit a transient change of membrane potential, like a show action potential.

Cytological criteria such as the presence of Nissil granules or neurotubules may occasionally be helpful but often are not; similarly, the possession of long processes like dendrites and axons may not be decisive; a unipolar nerve cells, for example, may have a much reduced stem process associated with a neurosecretory function. Some connective tissue and neuroglial cell have long, branched processes. It must be admitted, finally, that no specific criterion is certain. Like the question of nervous mediation, we can only build up a case for reasonable identification by accumulating suggestive evidence. Two signs are perhaps the most basic, even where the axon is greatly reduced:

(a) electron microscopic or oscillographic evidence of real functional contact with other cells, themselves clearly,

(b) a reasonably brief action potential and refractory period, permitting repeated signals at moderate frequency. But other indications should also be sought, such as an evolutionary series in related species, suggesting homology with typical nerve cells, ontogenetic stage, and possibly histochemical diagnostic features.

COMMON FEATURES OF NERVOUS SYSTEMS

To set off the next section on major trends and contracts in the organization of nervous system, it will be useful to add to the defining features now before us some notice of other features that are either universal or nearly so.

The neuron and its main properties are universal, within the limitation just discussed. The neuron doctrine, which we own chiefly to Cajal but also to many other workers, some of whom contributed by vigorous opposition, maybe stated thus: all nervous systems consisting essence (whatever other, non nervous element may be present) of distinct cell called neurons, which are specialized for nervous functions and which produce prolongations and branches. Each nerve fiber is a cytoplasm-filled tube, which extends from the cell body of the neuron of which it is a part. This means that cellular units, and hence embryological and atrophic units, comprise the nervous tissue. Like other cell, neuron are generally discrete and separated by cell membranes, though special cases of fusion are perfectly possible and compatible, probably always secondarily derived. Each nerve fiber grows out of a nerve cell and remains organically connected to it. Functional union among neurons is ordinarily by contact or continuity, rarely and secondarily by continuity of cytoplasm although, as Cajal emphasized, occasional finding of continuity is not fatal to the general doctrine. The opposing view is reticularism, which holds that all nerve processes are continuous protoplasmically, soneuropile and central gray matter are without discrete neurons and excitation has complete freedom of spread from any point to any other. Interaction by more remote influence between neurons that do not make contact is not excluded. The neuron doctrine makes no statement as to the direction of normal transmission between neurons; that is the subject of a separate generalization-also due to Cajal is doctrine. There is no necessary implication that the neuron always acts as a whole, though such a supposition has frequently been made his a sense the neuron must be a functional unit, being a single cell, but the term is difficult to define further; we know that the whole neuron does not always act when part of it acts. An important contemporary concept of the neuron is that it consists of several or many regions of different functional capacity, facultatively interacting in complex ways. Some of these functionally diverse regions correspond to the anatomically distinct parts of the cell.

A corollary of the neuron doctrine is that any process severed from the nucleated portion of the neuron (cell body, soma, perikaryon) can be expected to die. This is the essence of Waller an degeneration but, beside the fragmentation of the axon, that phenomenon includes also the deterioration but, besides the fragmentation of the axon, that phenomenon includes also the aeterioration of the Schwann cell membrane now known to form the myelin sheath. The reaction of the uninjured Schwann cells when the axon is injured is a especially sensitive example of Tran cellular dependence. Retrograde reaction of the soma or proximal axon and transynaptic degeneration are known to occur in varying degrees but are not predicted by the doctrine.

The properties common to neurons, besides their connections to other neurons, can only be stated on the basis of a small sample studied physiologically. It is not necessary that these properties be universal, and some reasons exist to suspect otherwise. All neurons so far examined are capable of an all-or-none, brief (0.3-10 msec) membrane change called the nerve impulse. Of repeating from dozen to hundreds of their per second and of propagating it without decrement along process called axons. Since many neurons which have never been properly examined physiologically, have only short processes, it is not clear whether they have all impulse. All neuron properly examined a still smaller sample, a sample form 4 major phyla (annelids arthropods, molluscs and chordates)-exhibit a different form of response called postsynaptic potential. When adequate impulse in other neurons called presynaptic.

THE COMPOUND EYE

As we proceed phylogenetically from the protozoa with their eyespot and flagellum photoreceptor system to coelenterates, flatworms, roundworms, and segmented worms, we find a variety of photoreceptor structures ranging from eyespots to photosensory cells to simple "pin-hole" eyes. In arthropods and molluscs we find, in addition to these more primitive receptors, imaging compound eyes and refracting eyes. We can see that in the course of evolution invertebrates have used every known optical every known optical device for light detection and for forming an image.

The molluscs are by far the most numerous of the invertebrates, included among the molluscs are clams, oysters, nautiluses, squids, and octopi.

Almost every kind of imaging eye is found among them, from the simple pin-hole eye of the Nautilus to refracting eye of the *Octopus*.

THE CENTRAL NERVOUS SYSTEM

The ganglia are characteristically distinct node like clusters of nerve cell bodies, communicating by cell-free connectives and commissures. (The pedal ganglion of some lower families forms an exception). The ganglia are few in number and conservative-that is homologous ganglia can be found, generally speaking, through otherwise highly divergent forms and novel ganglia are not conspicuous, though they occur perhaps more often than in any other group of animals, with the exception of one class of molluscs, the cephalopods.

(a) *Gross Anatomy of Ganglia, Connectives, and Commissures :*

The Cerebral ganglia, Sub cerebral Commissure, Labial ganglia, Subradular ganglia, Buccal ganglia, Pleural ganglion. Pedal ganglion. Intestinal ganglia. Parietal ganglia, Visceral ganglion, Accessory ganglia.

(b) *Microscopic Anatomy :*

The ganglion cells, the inclusions of the ganglion cells. Nerve fibers and trunks. Central ending, synapses. Neuroglia cells, Neurosecretory cells.

THE NERVES AND THE PERIPHERAL NERVOUS SYSTEM

There are a staggering number of descriptive accounts of the nerves of particular species of gastropods. Only a beginning has been made in the attempt to homologize nerves and to consolidate nomenclature; this is largely due to the fact that most authors have not accurately determined the area of distribution of the respective nerves and that the areas innervated vary in size and position in evolution, having few fixed landmarks. Here it will only be possible to set forth in outline the anatomy of a few species, without considering them as necessarily typical of their subclasses.

Some further aspects of the peripheral nervous system are of importance here; for example, do receptor neurons send axons directly into the central nervous system- and only three-or are there relays, or convergences, or peripheral reflex possibilities? Similarly, are there peripharal relays of efferent fibers? Are there peripheral connection with

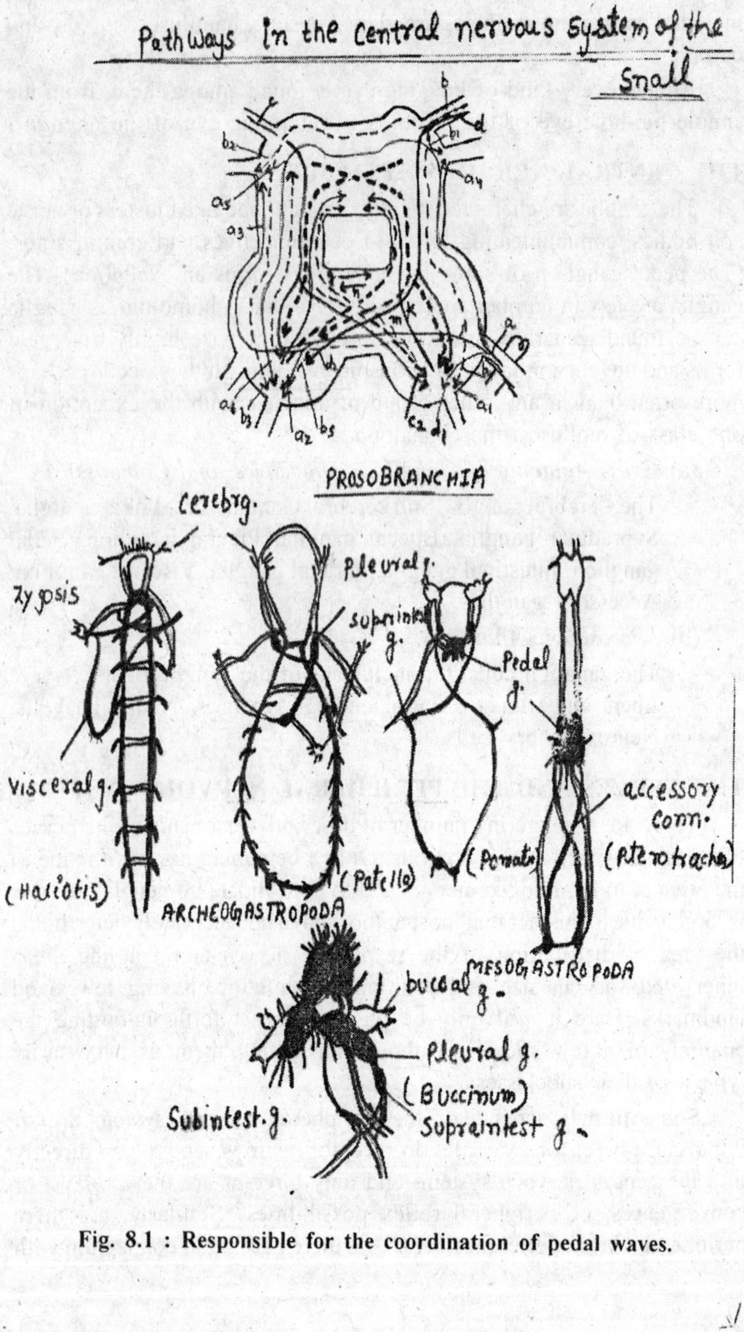

Fig. 8.1 : Responsible for the coordination of pedal waves.

sensory neurons? Although these questions cannot really be answered, the numerous reports will be surveyed and evaluated.

Function of the Pedal Ganglion and the Sole Plexus

In gastropods, unlike pelecypods and cephalopods, it appears that the pedal ganglia are the principal center for locomotion and major body movements. The brain plays only a moderating and directing role and the pleural and visceral ganglia play still less of a part. Destruction of the pedal ganglia leads to complete locomotor paralysis in *Helix* and *Aplysia* and nearly complete in pterotrachea, but not in *Limax* or *Arion.* It will be convenient of treat locomotor and other phasic movements separately from muscle tone; there has been little effort to establish relationships between the two.

Function of Other Ganglia

Electrophysiological studies on the visceral ganglia of *Aplysia* with respect to excitability to synaptic transmission. Many cells have been penetrates chiefly in the visceral ganglion of *Aplysia* but also of *Helix* and of the marine pulmonate onchidium (*Hagiwara*). Several physiological types of cells are distinguishable and often-in characteristic places. However, there is also a pronounced lability and changes in behaviour of cells may last for hours after minimal stimulation. Seasonal differences in behaviour of certain cells are also present, as well as activity characteristic of a certain time of the day.

Function of the Peripheral Nervous System

The peripheral plexus of the foot de *Helix* and of the parapods of *Aplysia* has been dealt with above; Crozier and Arey have reported reactions, which they attribute to such a structure in the gill crown, tentacles, rhinophore, and pharynx of *Chromodoris*. In general, the conclusion of Hofmann still appears tenable that three is no nerve net in the sense of ability to spread excitation for some distance. The isolated sole of *Limax*, which can still conduct waves, may prove this conclusion wrong, but this conduction may be either muscular or a series of local stretch reflexes. The plexus does seem able to mediate local reflex response—at least pieces of mantle, gill crown, and the like can respond to mechanical stimuli by local contraction; the plexus is not proved to be a mediator. The problem recalls that in annelids and other groups also, where peripheral ganglion cells, scattered along nerves or in a plexus, have not yet been assigned a function.

On the basis of Alien's work, four classes of inter neurons can be distinguished.

1. Cell body in the brain or in a ventral ganglion axon descends to the posterior end of the cord having arborization in all or must of the ganglia on its path.
2. Cell body in a ventral ganglion; axon descends to the brain.
3. Cell body in a ventral ganglion: axon ascends or desired to the adjacent ganglion.
4. Cell body in one ventral ganglion; axon ascends going off collateral fibres in the adjacent ganglion and ends in the subjacent one.

Thoracic Motor Neuron

Each thoracic ganglion gives rise to 2 main nerve bundles the anterior mainly to the limb, the other mainly to the body wall. The anterior nerve is ready double and soon divides.

Nervous System of Pila

Cerebral Commissure	→	This connects the 2 cerebral ganglion
Cerebral Ganglion	→	To the brain, supply impulse to eye and tentacle
Buccal Ganglion	→	To the buccal mass
Pedal Commissure	→	Connect the pedal Ganglion, Nerve goes to foot
Pleural Commissure	→	To the some visceral organ
Visceral Ganglion	→	To Visceral organ, Heart, liver and digestive system
Pedal Cords	→	To the posterior region
Cerebral connective	→	Connection between cerebral and pleural ganglion
Ospharadial	→	To the right portion of the body
Ospharadial	→	To the left portion of the body.

MOLLUSCA-CEPHALOPODS

Cephalopods are among the most active invertebrates live by hunting prey and have a very highly developed nervous system. A varied repertoire of behaviour, recognition of complex objects, rapid muscle, colours and luminescent responses and rapid learning are characteristics, in the decapoda but not in the octopoda, the stellar nerves also contain giant fibres which are only activated when the animal is alarmed, and which initiate maximal contraction of the pallial muscle fibres. The giant fibre system comprises 3 series of neurons; there are 2 giant fibre of first order.

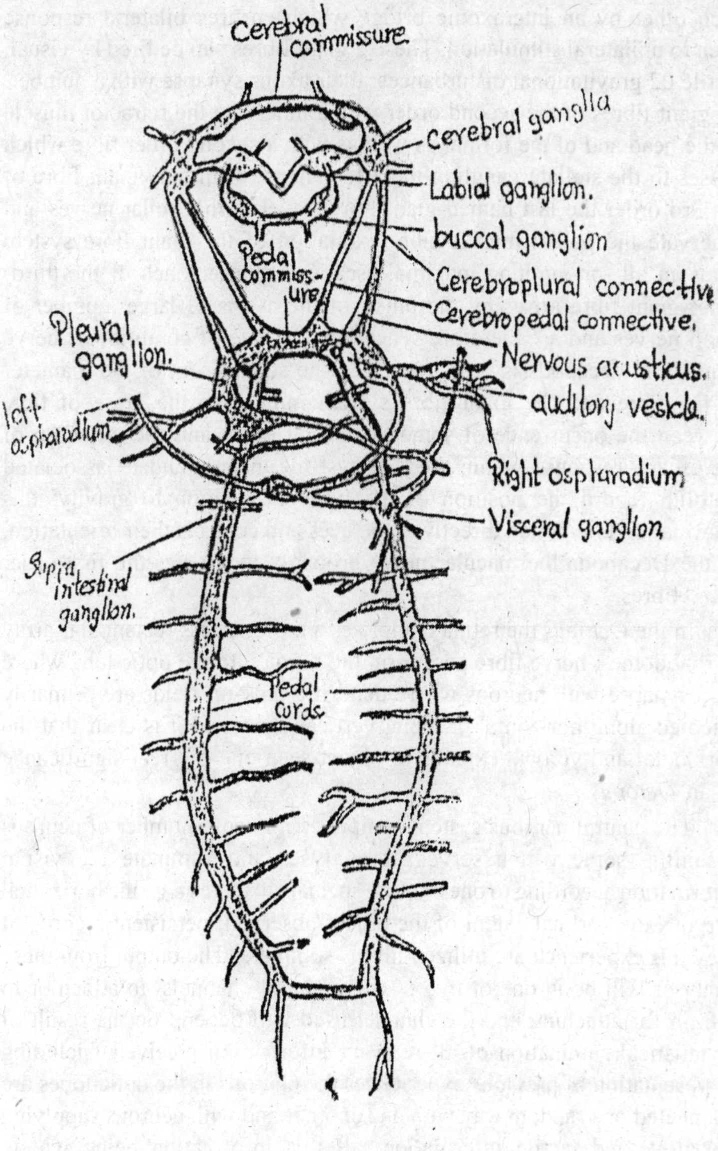

Fig. 8.2 : Nervous System.

One on each side lying in the central neuron system and connected to each other by an interaxonic bridge which ensures bilateral response even to unilateral stimulation. These 2 giant fibres can be fired by visual, tactile 02 gravitational disturbances; their axons synapse with a number. Of giant fibres of the second order which innervate the retractor muscle of the head and of the formal as well as with a second order fibre which passes to the stellate ganglion there to synapse with may giant fibre of the 3rd order the last named giant fibre travel in the stellar nerves and innervate the pallial muscle fibre excitation of the giant fibre system elicit an all -or- nothing maximal escape response. Each of the third-order-giant fibre arises by the union of the axons of large .number of small nerves and are therefore syncytial. The rate of condition of nerve impulse in decapods is proportional to the square root of the diameter of the fibre and the giant fibre system minimizes the lapse of time between the occurrence of some disturbing event and the initiation of the escape response within the statocyst the macula and its associated statolith record the position of the body in relation to gravity; this information is supplied directly to the eyes and controls their orientation. In the Decapoda the macular nerve probably innervates the first order giant fibres.

In the *Octopus* the retina comprises a more or less rectangular array of rhabdomes, nerve fibre pass form the retina into the optic lobe where they synapse with neurons where dendritic pick-up fields are primarily oriented along horizontal or along vertical axes, and it is clear that the horizontal and vertical extends of object seen one of great significance to an *Octopus*.

The central nervous system comprises immense number of neurons including some which serves as analyses and summate the visual information according to one or more special attributes *e.g.*, the horizontal nervous the vertical extent of the object observed; persistent records of previous experience are utilized in this summate. The output from these analyses will be in one of two possible neural channels -to attach or to refrain for attaching and the channel used will depend on the result of a statistical summation of all relevant information received, including representation of previous experiences the neurons in the optic lobes are connected in a random way with one another and with neurons supplying gustatory and tactile information, all this information being widely distribution, throughout the optic lobes serve lesion to the supra-oesphageal brain centres leaving the optic lobes undamaged, will destroy any long-term memories. A short-term memory presumably comprise lf-excitatory reverberations in cyclical series of neurons in the optic bes, while a long - term memory may become established if by the petition of the situation these reverberating circuits are extended to clude ports of the supra-esophageal brain mass, especially the vertical be.

The learning process concerns the imprinting of changes in the ·obablity of attack under given circumstances. These changes cannot oncern a single series of neurons, but must be widely distributed through ıe substance of the brain, the more firmly is the memory established. he·long-term memory probably involves rhythmical reinforcement of ıe cyclical reverberation passing through neurons in the vertical, sub-ertical and optic lobes.

The functions of the various parts of the brain have been investigatec ›y observation of the effect of stimulation by the application of electrode ıt various points. On the brain, and by observation of the behaviour o: ›pecimen after brain lesion of known extent.

MOLLUSCA-GASTROPODS

The most diverse class of mollusca and one of the most diverse classes of animals, in level of organization and habit of life, the gastropods present a particular challenge in the study of nervous systems.

No nerve fibres ending among the ganglion cell bodies of the end. A few indication of ending in the cell bodies occur in the works of *Havct* and of *Dreyer*. It world be necessary to doubt the existence of synapses on neuron stomata in gastropods were it not for the single report of Veratti (1990), His success with golgi impregnation in gastropods in outstanding and no readily assignable artifact or misidentification for *e.g.*, the interaction between orthodromically elicited synaptic potentials and intrasomatically applied electrical stimuli. But Jauc (1960) has shown that synaptic transmission persists after paralyzing the soma by passing a large current through it via an internal electrode.

In number and volume far and away the main synaptic field is the neuropile of the core. The most specialized synaptic field is there found in the dense fine texture neuropile of the procerebrum by higher pulmonates where very many short process of small nervous end in a luxurious tangle of tightly packed small-caliber twigs.

Smaller and slower conducting impulses are elicited in one pedal nerve by electrical stimulation of a contralateral one than by direct stimulation of the same nerve. The former are also activated by stimulation of the contralateral cerebropedal connective and this can make the fina common path reflectory to stimulation of the contralateral pedal nerve, stimulation of the contralateral pedal nerve can preoccupy the common moto neurons to the exclusion of response to stimulation of the contralateral connective. The larger and faster pedal efferent actually requires excitation convergent from more than one *sonru* for *e.g.*, contralateral pedal nerve and isolateral or contralateral connective. These large fibers branch and distribute their action potentials to several divisions of a pedal nerve as are seen by stimulating one and recording from another.

Somatic reflex movements due to impulse from the viscera are demonstrated by bilateral response of the body mediated through the peripharyngeal ganglia and of the visceral connectives. Movement of the pharynx and esophagus occur at the same time if the buccal ganglia are intact.

Neurogila Cells

Are abundant and quite similar to those in annelids and arthropods. Just as in the articulates, there are indications that neuroglia are better developed in higher species or larger species as well as non nervous cell are much better represented in helix and Arian man in *Lymnae* and *Planorbis*.

Local Neurons

Many of the chromatic cells so similar to the globuli cells of arthropods in form and arrangement are apparent by intramural, mating connection in the neuropile of this segment of the brain without sending process elsewhere.

The functional significance of the procerebrum is different to guess, since it is not know to have efferent connections with the rest of the brain. The cell are functionally connected and microscopically similar to the second-order afferent nervous in that ganglion, and it seems reasonable to say that the procerebrum is a higher order sensory association center especially for tentacle receptors, which are probably chemoreceptive.

The neuronal composition is not as elaborate informs lacking well-developed distal testicular ganglia.

The most important topographical fibers of the neuron from the physiological point of view is the position of the dendrites and axon arborization in the 3-dimenson selection to those of other *redseo* the effects of calcium on central synapses, although there have been numerous attempts to correlate the over-all results with those on motor end plates in muscle so thoroughly investigated over the past two decades. We have studied these effects in single abdominal ganglion cells of *Aplysia californica* and have attempted to correlate our results with those mentioned above in peripheral nerve and, to a lesser extent, to correlate the changes observed in EPSP's and IPSP's to those described for end-plate potentials.

Our results may be divided into four parts.

(a) The effects of high calcium concentration in the external medium on membrane potentials when the magnesium concentration remained normal.

(b) The effect of low external calcium with normal magnesium.

(c) The effects of altered calcium concentrations when magnesium was absent from the external medium.

(d) The effects of altered calcium concentration on the responses to light stimuli of certain ganglion cells.

Chloride in Nerve Cells

Keynes (1963) measured the chloride concentration in squid axoplasm and found that it amounted to 108 mM. This value suggested that the Cl^- concentration was not controlled by the membrane potential but by a Cl^- pump taking the ions inward. This was confirmed by showing that the inward flux of Cl^- was reduced one half by the addition of 0.2 mM DNP. Quabain had no effect on the inward flux, suggesting that the movement was not linked to the Na^+ pump. The outward Cl^- flux was not affected by DNP.

In the snail there are two types of cells: H cells which are hyperpolarized by the addition of ACh (the cells that show IPSP's are H cells) and D cells which are depolarized by ACh Fig. 8.3. Strumwasser (1962) showed that in *Aplysia* one intermediate neurone could drive both an H cell and a D cell, the transmitter (ACh?) causing a simultaneous hyperpolarization in the H cell and depolarization in the D cell.

There is thus a problem. To what is this difference in the action of acetylcholine due?

Oomura (1964) suggested that there is a difference in the internal chloride concentrations of the H and D cells and that the action of ACh is to allow the cells to reach their chloride equilibrium potential. We developed glass electrodes sensitive to the Cl^- concentration. These electrodes were made by the chemical deposition of silver inside low-resistance glass electrodes. This "chloride electrode" and a normal sodium acetate filled glass electrode were both inserted into a known cell in the snail brain and the potential between them measured. The chloride electrode gives a good linear response between the external Cl^- concentration and the electrode potential. By injecting Cl^- into the nerve cell we were able to show that the electrodes were most likely measuring the internal Cl^- concentration and that the E_{ACh} and E_{IPSP} and the chloride electrode potential all changed in a corresponding manner. It was also possible to test the electrode by measuring the Cl^- concentration in the large algal cell *Nitella*, whose Cl^- concentration can also be determined by direct chemical means. The two methods give similar results for the internal chloride concentration.

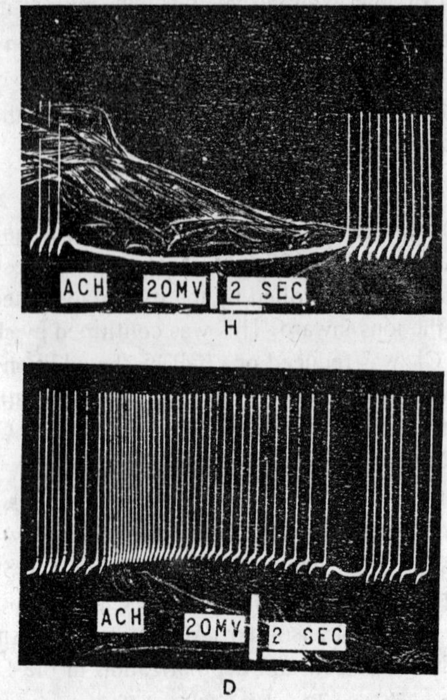

Fig. 8.3 : H and D cells in *Helix*. When acetylcholine is added to some specific nerve cells there is a hyperpolarization (H), whilst other specific cells show a depolarization (D). (From Kerkut and Meech, 1966.)

If we assume that the activity of the Cl^- in the cytoplasm is the same as that of the Cl^- in the Ringer solutions that we used to calibrate the electrodes, we can estimate the Cl^- concentration in the nerve cell.

The value for the internal Cl^- in the H cells was 8.7 ± 0.4 mM ($n = 25$). The value for the internal Cl^- in the D cells was 27.5 ± 1.5

mM (n =25). The effect of acetylcholine on the H cells is to bring it close to the E_{Cl}. The action on the D cell is more complex, and we think that A_{Ch} also allows an increased permeability to Na^+.

The point that should be emphasized here is not the relationship with the E_{ACh} but instead the fact that the H cells and the D cells have different internal Cl^- concentrations. Although it has been generally accepted for some time that different cells can give off different transmitters, this chemical heterogeneity has not been extended to include the ionic composition of the cells. The fact that two types of cells in the snail brain normally have different Cl^- concentrations and that this is reflected by differences in the cell's responses to drug action extend the concept of chemical heterogeneity in the CNS.

Glutamate as a Transmitter

The possible role of glutamate as a transmitter in the mammalian CNS has been discussed by Curtis (1965). One of the difficulties has been that glutamate is normally considered as a metabolic chemical and as not at all implicated in any transmitter system.

There are many invertebrate nerve-muscle junctions where the nature of the transmitter is not known investigated the nature of the possible transmitters in the snail brain and found that certain brain fractions contained no ACh but brought about a marked contraction of the pharyngeal retractor muscle. Later, purification showed that the most active fraction always ran on thin-layer chromatograms at the same Rp as a ninhydrin-positive substance (NPS) which appeared to be glutamate.

We then set up a series of nerve muscle preparations from the *snail*, cockroach, and crab (*H. aspersa, P. americana, C. maenas*) and perfused them. We improved our techniques of recovering small quantities of NPS from large volumes of perfusate and found that when the nerves of the preparation were electrically stimulated a NPS appeared in the perfusate. The amount of NPS was proportional to the number of stimuli given to the nervous system. The NPS had the same R_F as glutamate in six different solvent systems. The evidence for the NPS being the transmitter for the *Helix* muscle is summarized.

The application of glutamate to the functional region of the crayfish muscle brought about a depolarization. Glutamate will make the cockroach and bivalve foot-leg contract—the threshold to L-glutamate was 10^{-7} g/ml whilst that for D-glutamate was 10^{-6}. More recently we have entrance of K^+. The relative effect of Cl^- and K^+ is shown in Fig. 8.4. Cl^- has a quicker effect. If we provide a solution so that the resulting changes in E_{Cl} and E_{-K} will oppose each other, we can then see from the E_{IPSP} and E_{ACh} which ion is the more dominant. Fig. 8.5 shows that the Cl^- has the dominant effect on the E_{ACh} after such a change of solution. It has a similar dominant effect on the E_{IPSP}. From the values of the reversal potentials we estimate that during the IPSP the increase in the membrane permeability is such that 90% of the current is carried by the Cl^- and 10% by the K^+ Fig. 8.4.

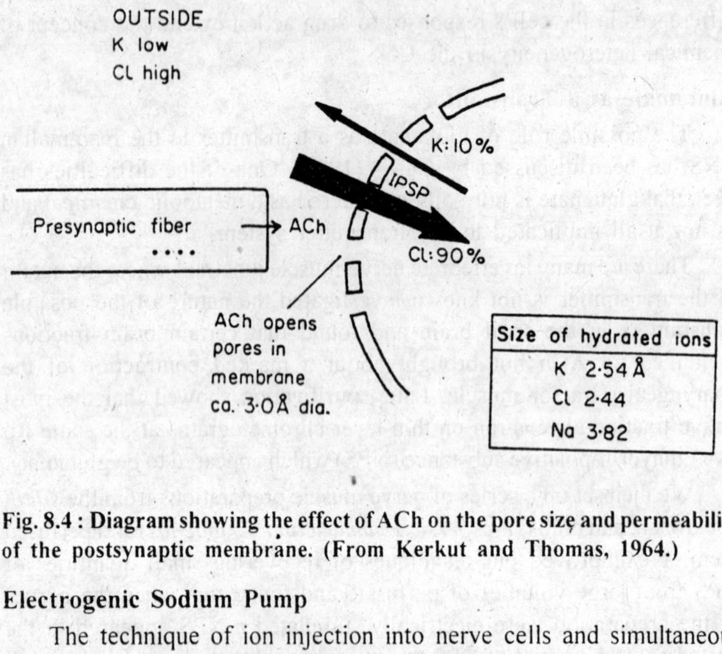

Fig. 8.4 : Diagram showing the effect of ACh on the pore size and permeability of the postsynaptic membrane. (From Kerkut and Thomas, 1964.)

Electrogenic Sodium Pump

The technique of ion injection into nerve cells and simultaneous alteration of the external solution has also been applied by Kerkut and Thomas (1965) to a study of the resting potential in snail nerves. Certain snail nerve cells increased their resting potentials with the rise in internal Na^+ concentration. This effect was absent for K^+, but a temporary hyperpolarization followed injection of lithium ions Fig. 8.5. The sodium-induced hyperpolarization could be reduced by the addition of quabain, parachloromercuribenzoate (PCMB), or by reducing the external K^+ concentration Fig. 8.6. This sodium induced hyperpolarization is interpreted as due to the presence of an electrogenic Na^+ pump in these nerve cells. There is some evidence that the pump is sensitive to the presence or lack of oxygen, and this fits in well with the studies of

Arvanitaki and Chalazonitis on the changes in resting potential in *Aplysia* cells under anoxia and high oxygen concentrations.

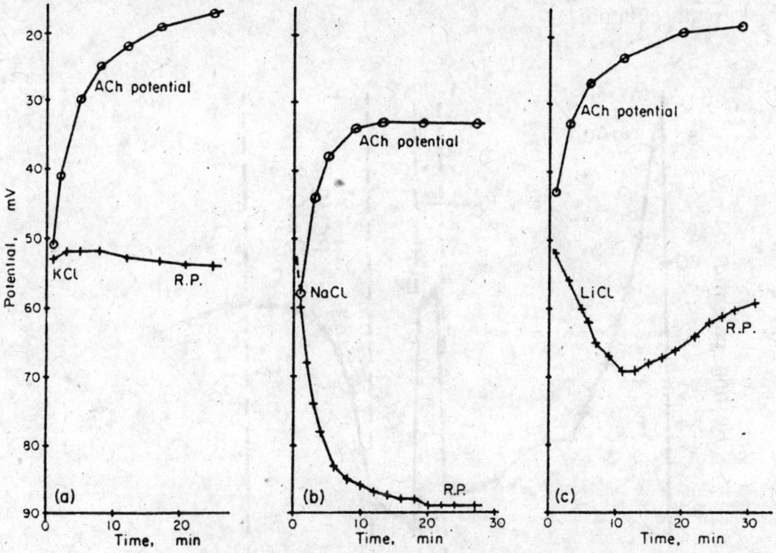

Fig. 8.5 : The effect of injecting KCl, NaCl and Li on the membrane potential of nerve cells. The reversal potential of ACh indicates the rate that the chloride ions diffuse into the cell. In all cases the chloride ions diffuse into the cell. There is only a prolonged hyperpolarization in the case of Na^+ and a short one in the case of LiCl. (From Kerkut and Thomas, 1965.)

EFFECTORS AND MOTOR CONTROL

Octopuses move in a mysterious way. Being flexible the movements that they make are often difficult to specify and correspondingly difficult to investigate. The literature does not contain a description of octopod walking comparable with descriptions of the six-legged, tripod gait of insects, or the stereotyped locomotor patterns of snails or pelychaetes. Descriptions of posture run into very similar difficulties and perhaps partly because of this, research into motor control in cephalopods has proved a less attractive proposition than research on sensory analysis and learning.

Certain peripheral elements of the control system have been examined and described. The control of the heart and blood vessels (and of respiratory movements) is discussed in the control of ingestion and digestion in while the question of sense organ innervation from the CNS is covered in each case when dealing with the organs concerned, mainly in. Here we are concerned with the hierarchy of centres controlling the movements of the arms and suckers, and with the chromatophores and skin musculature.

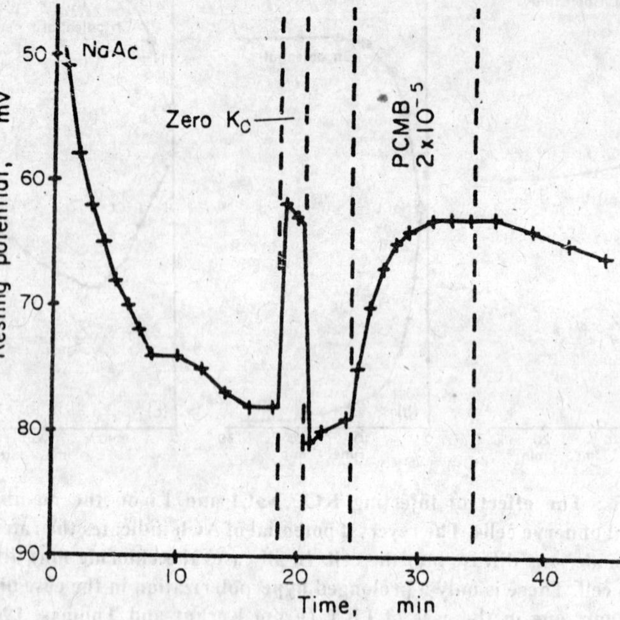

Fig. 8.6 : The sodium-induced hyperpolarization is inhibited by removing the external potassium ions and by adding parachloromercuribenzoate. (From Kerkut and Thomas, 1965.)

Mapping Brain Function

An Outline of Nervous Anatomy

The nervous system of *Octopus* is a very diffuse system, by vertebrate standards. Most of the neurones are outside the brain, in a series of peripheral ganglia. Of these, the arm nerve cords alone contain more than twice as many nerve cells as the central supra and suboesophageal ganglia; 3.5×10^8 compared with 1.7×10^8 in the brain (Young, 1963a). The cell bodies of the sensory neurons throdughout the body are, of course, peripheral even to these outlying ganglia. There is an extensive plexus, including both sensory and motor elements around the gut and many of the blood vessels.

In overall control of the activities of these outlying parts are the thirty or so lobes of the central sub- and supraoesophageal brain. Some of these control the patterning, though often not the details, of movement. Others seem to be concerned only with sensory analysis or with learning, a separation of sensory integrative and motor executive operations that has enormous potential convenience for anybody-trying to relate structure and function.

A great deal of information is available on the anatomy of the brain and about some of the peripheral ganglia. The brain of *Octopus* is certainly the best-mapped of all invertebrate brains, and is quite possibly as completely known as that of any vertebrate at the present time. Young's 'Anatomy of the nervous system of *Octopus vulgaris*' is a far larger book than this and elegantly demonstrates how much can be deduced from structure, granted an anatomist with first hand knowledge of the behaviour of the system one is studying. It is clearly quite out of the question to include anything but the most superficial summary of this work here.

Very broadly speaking, then, the brain consists of a trio of suboesophageal lobes, the branchial, pedal and palliovisceral, which house neurons that include motor units forming the penultimate link between brain and effectors. Together with these are regions of cells with motor neurones that run directly to the chromatophores and to many of the blood vessels. All these suboesophageal lobes are plainly motor control centres, and they include a proportion of very large neurons, as one might expect in any final common pathway Table 8.1.

Connected to the suboesophageal lobes and thus only indirectly to the periphery are a further series of centres of complex structure, the magnocellular and basal lobes around and above the gut. The former has a small number of neurons with nuclei in the 15-20 μm diameter class but in general these regions lack the very big nerve cells characteristic of the suboesophageal brain. The same is true of the superior buccal lobe, which controls the mouthparts and salivary glands via the inferior buccal ganglion.

In addition to these regions, which it would be reasonable to classify as motor control centres on the grounds of their anatomy and connections alone, there are several supraoesophageal lobes of an entirely different

Table 8.1 : The numbers and sezes of nerve cells in certain lobes of the *Octopus* brain (after Young, 1963a). Figures show numbers in thousands and nuclear diameters in a brain from an animal of about 450 g. Nerve cell numbers increase considerably at the animal grows (Packard and Albergoni, 1970).

	Total nuclei	Diameter of nerve cell nuclei % < 5µn	% 5-10µm	% 10-15µm	% 15-20µm	% 30-25µm	Probable function
Suboesophageai lobes, lower and intermediate motor centres							
Brachial, pedal and palliovisceral	692	26	44	24	5	1	Final or penultimate pathways to arms, guts and body musculature
Vasomotor	1307	18	74	8%	–	–	Control of the circulation
Chromatophore	526	19	50	27	3	–	Colour and pattern changes
Circumoesophageal and Supraoesophageal lobes							
Superior buccal	150	23	53	21	1	–	Mouthparts and eating
Magnocellular	581	44	54	21	1	–	Relay to lower centres, defence and escape reactions
Higher motor centres							
Anterior basal	380	28	71	2	–	–	Head and arm movements
Medial basal	245	37	56	8	–	–	Swimming, respiration; Dorsal part–visceral activities

	Total nuclei	Diameter of nerve cell nuclei % < 5μn	% 5-10μm	% 10-15μm	% 15-20μm	% 30-25μ	Probable function
Lateral basal	127	43	54	2	–	–	Chromatophores, relay from optic lobes
Sensory and learning centres							
Inferior frontal	1085	91	7	2	–	–	Distribution and sorting of tactile input
Subfrontal	5303	100	–	–	–	–	Tactile memory
Superior frontal	1854	96	2	2	–	–	Distributes input from the optic lobes and touch centres
Vertical	25066	100	–	–	–	–	Visual and tactile learning
Optic	128940	95	5	–	–	–	Visual analysis and learning

structure, characterized by relatively enormous numbers of very small neurones. Some of these, the inferior frontal and optic lobes, receive sensory inputs directly from the arms and eyes. Other like the superior frontal, are relay centres, passing inputs onwards from the regions that receive sensory inputs, to the small-cell areas of the subfrontal and vertical lobes. These lobes are only quite remotely connected to the periphery on either the sensory or the motor output side; most of their tiny neurons are *amacrines*, cells without an obvious axon, connected to their neighbours but lacking processes to lead outside their parent lobe. Brain lesion work, to be reviewed in that they are concerned in learning while having no detectable effect on motor performance.

The appearance of the brain seen from the side and from above. Fig. 8.7 : is a drawing of a longitudinal section through the supraoesophageal lobes, while Figs. 8.8 and 8.9 summarize some of the internal connections and the sources or destinations of peripheral nerves, shows a longitudinal section through the whole brain. Some further details of connections within the supraoesophageal.

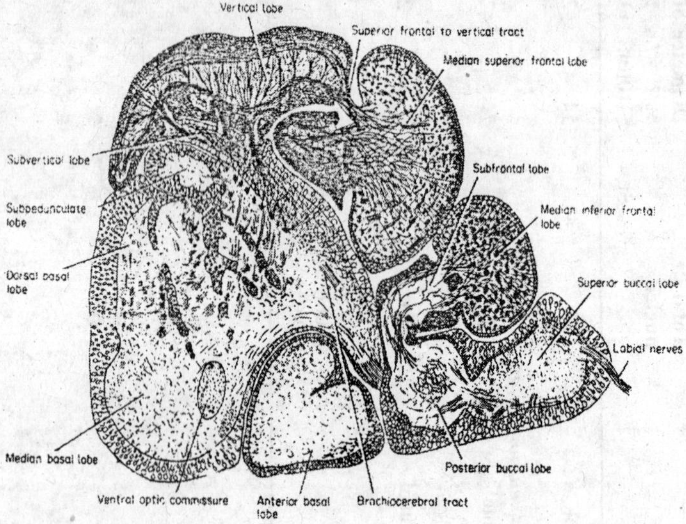

Fig. 8.7 : Sagital section through the suprabesophageal lobes, slightly to one side of the mid-line (from Young, 1971).

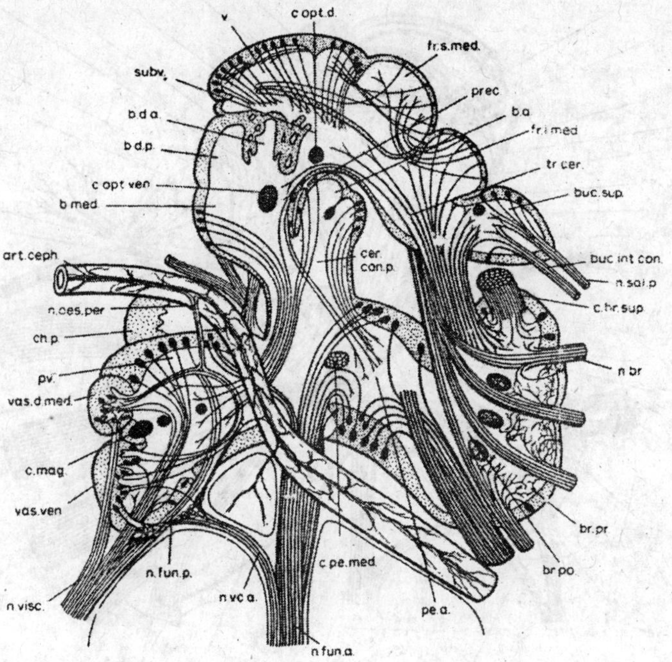

Fig. 8.8 : Connections in the brain of *Octopus*. The diagram represents a section taken to one side of the mid-line. The cephalic artery branches and passes through the suboesophageal brain between thc palliovisceral and pedal lobes. 1., lobe; n., nerve; art.ceph., cephalic artery; b.a., anterior basal lobe; b.d.a. and b.d.p., dorsal basal anterior and posterior basal ls.; b.med., median basal 1.; br.po. and br.pr., postbranchial and prebranchial is., buciticon, interbuccal connective; buc.sup. superior buccal 1.; c.br.sup., suprabuccal commissure; c.mag., magnocellular commissure; c.opt.d. and c.opt.ven., dorsa and ventral optic commissures; c.pe.med., medial pedal commissure; cer.con.p., posterior cerebral connective; ch.p., posterior chromatophre l.;fr.i.med. and fr.s.med., median inferior and superior frontal ls:; n.br., brachial n.; n.fun.a. and n.fun.p., anterior and posterior funnel nerves; n.oes.per., perioesophageal n.; n.sal.p., posterior salivary gland n.; n.vc.a., anterior vena cava n.; n.visc., visceral nerve; pe.a., anterior pedal l.;prec., precommissural l.;pv., palliovisceral 1.; subv., subvertical 1., tr.cer., brachiocerebral tract.; v., verti-'al 1.; vas.d.med and vas.ven., median dorsal and ventral vasomotor ls. (from Young, 1971).

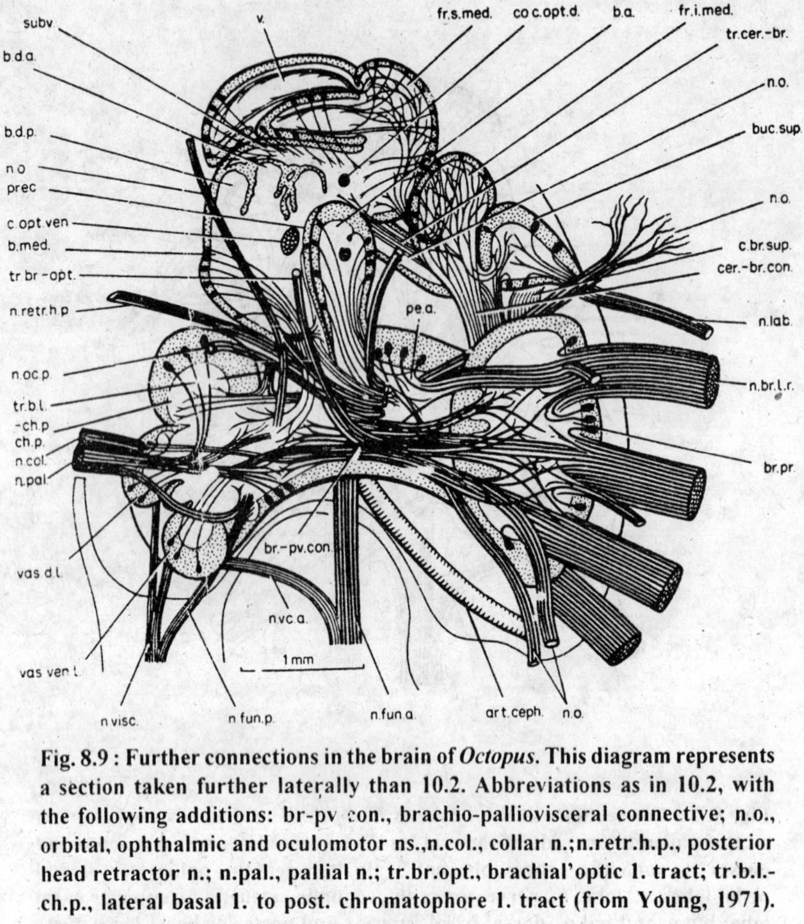

Fig. 8.9 : Further connections in the brain of *Octopus*. This diagram represents a section taken further laterally than 10.2. Abbreviations as in 10.2, with the following additions: br-pv con., brachio-palliovisceral connective; n.o., orbital, ophthalmic and oculomotor ns.,n.col., collar n.;n.retr.h.p., posterior head retractor n.; n.pal., pallial n.; tr.br.opt., brachial'optic 1. tract; tr.b.l.-ch.p., lateral basal 1. to post. chromatophore 1. tract (from Young, 1971).

Electrical Stimulation of the Brain

If the brain of a cephalopod is exposed and stimulated through a wire electrode, movements of the arms and body, and chromatophore changes, can be elicited. This method was being used to map the parts of the brain concerned in motor control as early as 1867 when Bert attempted Faradic stimulation of the brain of *Sepia*, found the supraoesophageal lobes electrically inexcitable-he called them 'silent' areas-and concluded that they were not involved in motor control. Von Uexkull (1895) used similar methods on the brain of *Eledone*, and showed that Bert was only partially correct. Some of the supraoesophageal lobes are truly 'silent', stimulation of others, including the optic lobes, can produce motor responses.

Since then, a number of stimulation experiments have been made with *Octopus*, mainly by Boycott and by Young. More detailed accounts of the effects of stimulating particular systems within the CNS are included in Altman (1968) and Messenger (1965).

Boycott (1961) classified the lobes of the cephalopod brain into five categories, distinguishable by their anatomy and the effects of stimulation. These categories begin with the *Lower motor centres*. These house the neurons that supply the muscles directly, without further intermediate synapses. Most such centres are peripheral. They include the arm nerve cords, the inferior buccal and gastric ganglia and the ganglia controlling the hearts. Within the brain the chromatophore and vasomotor lobes appear to include at least some neurons with axons that run directly to the muscles they control so that these two are also, by definition, lower motor centres. Electrical stimulation (1-6 V, 3ms pulses at 60 Hz) of these final motor pathways will cause contraction of isolated groups 01 muscles,, expanding the chromatophores in a limited area, for example, but never evoking integrated patterns recognizable as components of the normal behaviour of the animals.

The lower motor centres receive most of their input from a superimposed level of command, the *Intermediate motor centres*. These do not, in general, include final motor pathways and their stimulation excites whole groups of muscles; stimulation of the brachial lobe, for example, produces contraction in all of the arms at once; individual arms can be moved independently only by placing the electrodes close to the point of exit of one of the brachial nerves. The pedal lobe seems also to contribute to arm control but is probably divided into several distinct functional regions since stimulation here can also evoke head, funnel and eye movements; it forms the link between statocyst inputs and compensatory eye movements. The palliovisceral region, perhaps better thought of as the 'posterior suboesophageal mass' (Young, 1971) because it plainly consists of a number of structurally and functionally distinct lobes, is a mixture of lower and intermediate motor centres. Some nerves run directly to muscles in, for example, the funnel and collar, while trios'- controlling mantle contraction have synapses in the stellate ganglia. Stimulation of the central part of the posterior suboesophageal mass may evoke contraction of the mantle and/or funnel and collar movements, but never the sort of co-ordinated action of all three necessary for the normal aspiratory rhythm of the animal. At the intermediate motor centre level, one is dealing with regions that can organize recognizable elements of the movements that intact animals make, but these are always very simple components of behaviour.

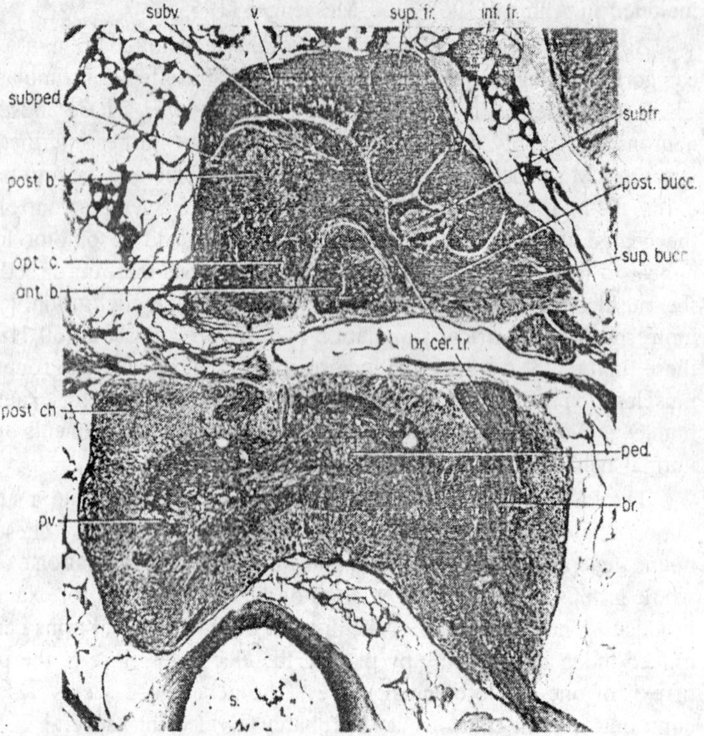

Fig. 8.10 : A longitudinal vertical section taken slightly to one side of the midline through the brain of *Octopus*. The supra- and sub-oesophageal pans of the brain are separated by the oesophagus. Lobes of the brain: ant. b.-anterior basal; br.-brachial; inf. fr. -inferior frontal; ped. -pedal: post. b.-posterior basal; post. bucc.-posterior buccal; post. ch.-posterior chromatophore; pv.-palliovisceral; subfr.-subfrontal; subped.-subpedunculate; subv.-subvertical; sup. bucc.-superior buccal; sup. fr.-superior frontal; v.-vertical, br. cer. tr.-branchiocerebral tract; opt. c.-optic commissure; s.-statocyst. (Photograph, J.Z. Young.)

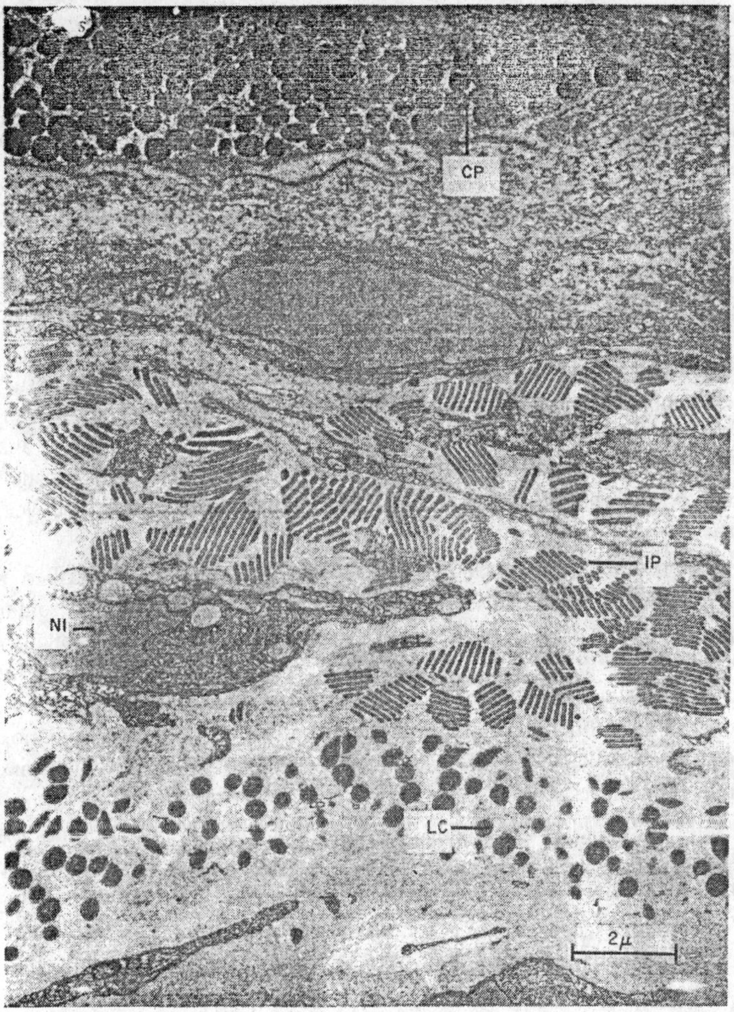

Fig. 8.11 : A section through the dorsal skin of *Octopus*. CP-part of the pigment containing sac of a chromatophore. Below this are the iridiophores, containing stacks of reflecting lamellae (IP): N1 is the nucleus of an iridiophore. A second, lower, reflecting layer is formed by the leucophores, which contain granules of a white opaque material-LC. (Photograph, D. Froesch.)

The *Higher Motor Centres* form the third tier in the hierarchy of motor control. These are all supraoesophageal, the basal lobes of the brain. Stimulation of these regions can produce well-co-ordinated responses, such as a regular respiratory rhythm, with associated movements of funnel and collar. The median basal appears to control respiration and the more violent contractions of the mantle that occur when the animal swims. The anterior basal controls head, arm and eye movements-an *Octopus*, stimulated in this region, will walk using all the arms-while stimulation of the lateral basal evokes changes to the chromatophores and skin papillae extending all over the isolateral side of the body. Subdivision of function within the major lobes was not detectable.

The Silent areas, where stimulation has no visible motor consequences, include the inferior and superior frontals, the vertical and the subfrontal lobes. Brain lesion work, reviewed in quite clearly that these parts are concerned in sensory analysis and in learning. Also silent are the dorsal basal, subvertical and precommissural lobes lying between the vertical lobe and the higher motor centres.

On the input side are a series of lobes perhaps better described as *Receptor analysers* than as *Primary sensory centres*. The inferior frontal/ posterior buccal region handles inputs from the arms, and there is abundant evidence that these have already been processed very extensively in the arm nerve cords. The optic lobes must carry out correspondingly elaborate operations upon the visual input. The receptor analysers are for the most part electrically inexcitable; only the inner regions of the optic lobes (and the output fibres from the inferior frontal system) will produce responses if stimulated. These responses may be quite complex; stimulation of the base of the optic lobe can produce patterned chromatophore activity and may evoke any of the locomotory responses seen as a result of stimulating the basals.

Electrical stimulation by wire electrodes is a crude way of exciting a system as elaborate as that of *Sepia* or *Octopus*. Effects are liable to be generated at a considerable distance from the point of stimulation if tracts rather than cell bodies are excited, so that a fairly detailed knowledge of brain anatomy is required before reliable interpretations can be made. Boycott (1961) points out several instances where the results reported by previous workers were almost certainly attributable to this sort of spread of excitation, and for this reason alone it is best to be wary of findings from stimulation experiments that are not backed by brain lesion experiments.

The results are nevertheless useful in showing several general features of the organization of the brain. The most obvious are:

(1) There is plainly some sort of hierarchy of motor control, with regions superimposed upon one another. The lowest are final motor pathways, the highest organize whole patterns of activity in space and time.

(2) Certain areas, on the sensory input side of the higher motor control regions, seem to play no part in the organization of locomotion.

(3) Within lobes, there is rarely any indication of a topographical subdivision of function. The neurons controlling, say, respiratory movements, seem to be distributed throughout the lobes that play any part in their organization.

Brain Lesions, Posture and Movement

An alternative, and almost equally crude means of investigating the function of different parts of the CNS is to destroy regions and see what effect this has on behaviour. The usefulness of this approach depends upon a rather precise specification of what has been destroyed and this necessitates the examination of serial sections from damaged brains. It is inevitably a slower way of mapping brain function than electrical stimulation, but at least it is possible to define what way done.

Lack of detailed maps of lesions made renders much of the early literature less valuable than it could have been, so that experiments like those of Fredericq (1878), von Uexkiill (1895) or Buytendijk (1933) are difficult to interpret. Buytendijk, for example, claimed that octopuses are unable to climb out of their tanks after removal of the superior frontal and vertical lobes, and this is plainly untrue; he must have damaged the basal lobes as well, but there is no way of checking this now. So far as brain lesion work is concerned, it is safer to rely on work done from 1950 onwards when it became accepted as routine to section the brain and check the lesions in all the experimental animals used.

Even so, there are problems of interpretation, since the effect of a lesion changes with time. Behaviour within a few hours of operation may be influenced by discharges from damaged tissues, while that of the same animal several weeks later could be complicated by regeneration.

The regeneration in the mantle nerves at summer temperatures of 20-25°C, the breakdown of axons separated from their nerves normally takes between one and three days. Regeneration from the central stump proceeds at 7-18mm h^{-1}, a rate comparable with that of vertebrates at a similar temperature; functional connections can be established if the regenerating nerves proceed through the scar tissue and into the peripheral stump. A return of central control of chromatophore changes within two to four months after section of the mantle connective on the side concerned. The situation following surgical removal of parts from the brain is less certain. Degeneration studies have been very extensively used in mapping the brain. Regeneration certainly occurs as well and has been observed repeatedly when examining serial sections from animals with lesions but no evidence for the re-establishment of functional connections has ever been obtained.

In the account that follows, the effect of brain lesions on the control of movement and posture is considered. The results of a variety of experiments confirm and extend the impressions gained from preliminary anatomical and electrophysiological mapping; sensory and motor regions of the brain are separated and there is a clear hierarchy of levels at which motor responses are controlled and elaborated. The special case of chromatophore control is considered separately.

The Control of Locomotion

Monocular Vision and Interocular Transfer

The visual fields of the two eyes overlap (Heidermanns, 1928), but the *Octopus* rarely, if ever, approaches objects that it sees using both eyes to fixate the target. Normally the head is held sideways. Objects must, nevertheless, pass from one visual field to another as the animal, or the object, moves about, and it would seem inevitable that some mechanism exists to ensure that the experience accumulated while the animal is using one eye can be applied when the same image is focussed on the other retina.

Muntz (1961a, c) has examined this matter, taking advantage of the monocular approach to train animals to discriminate using one eye (by presenting visual figures always to the same side) and then testing them using the other. Interocular transfer is rarely complete; in transfer tests, in training and in extinction (unrewarded responses) the performance using the untrained eye is always a little poorer. The effect is most marked for difficult visual discriminations. Transfer was close to 100% only after prolonged training to discriminate between horizontal and vertical rectangles, a task known to be very easy for octopuses. Cutting the links between one side of the brain and the other after training on one side only, Muntz (1961b, c) was able to show that the memory trace is bilateral; once the animal is trained even the trained side optic lobe can be removed without abolishing subsequent correct responses by the untrained side.

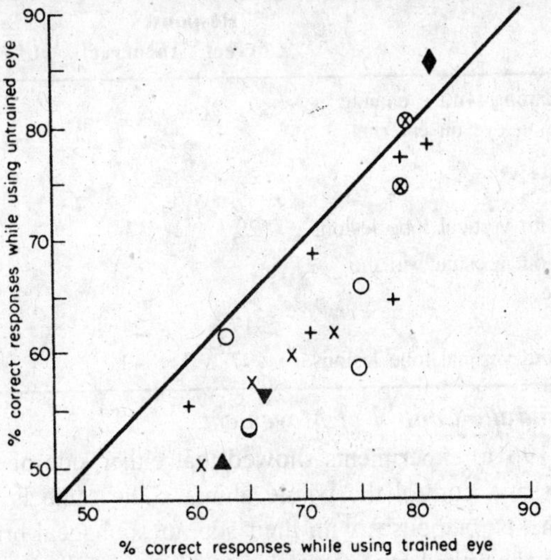

Fig. 8.12 : Interocular transfer. Octopuses were trained using one eye and tested using the other; a summary of the results of 3 series of experiments. In series 1, 2 groups of animals were trained to discriminate between 10 × 2 rectangles, shown horizontally and vertically (results shown. A series of transfer tests, with figures that ranged from a 4.5 × 4.5 cm square, through rectangles to a 10 cm row of three 2 × 2 squares, was run alternately with further training trials (results shown O). In a second series animals were trained on the rectangles (♦), T shapes (Δ) and diamond vs triangle (Δ). In a third series, animals were trained on the rectangles and then subjected to a series of unrewarded tests, extinguishing their responses; plots + and × show the scores at different stages in learning and extinction. Except for the very simplest discriminations the performance using the untrained eye always lags behind that of the trained (from Muntz, 1961a).

The octopuses will det our through the apparatus in order to reach crabs seen through one or other of two alternative windows. In the course of these runs octopuses often changed from leading with one eye to leading with the other, without any consequent drop in the proportion of correct runs made Table 8.2. This again indicates that interocular transfer can be complete, given favourable circumstances.

Table 8.2 : Eye changes and errors made; an analysis of the results of completed runs through the apparatus (from Wells, 1967).

	Responses Correct	Responses Incorrect	Proportion of errors (%)
(1) Trials beginning with a change in the leading eye on entering corridor			
Controls	50	4	8
Animals with vertical lobe lesions	29	7	24
(2) Trials without a change in the leading eye			
Controls	317	24	8
Animals with vertical lobe lesions	197	42	21

Split Brains and the Control of Movement

Muntz's (1961c) experiments showed that either side of the brain could, if necessary, control the whole animal. The same is found in detour experiments; octopuses with their supraoesophageal brains split by a longitudinal vertical cut move about quite normally most of the time, guided apparently by whichever eye happens to be leading at the time. There is no evidence that one side is normally dominant, in this or in any other experiment so far carried out; the *Octopus* brain, so far as we can discover, is functionally symmetrical. Split brain animals, however, run into difficulties in cases where the input to the two eyes leads to conflicting demands by the two sides of the body. In detour experiments it is not unusual to observe the arms on one side of the animal reaching forward towards a crab, while those on the other are clutching at the brickwork of the 'home', apparently attempting to gull the animal in the opposite direction. One consequence is an unusually high proportion of abortive runs-the *Octopus* sees the crab, moves into the corridor, and then returns home without completing a detour Fig. 8.13. The decision as to which way to move is presumably made as a result of comparing the inputs in relation to some inbuilt scale of priorities, the result of the past genetic and personal history to the animal. A crab remembered is evidently less effective than a crab seen in generating commands to move onwards, so the signals arising from the attacking side of the body become stiffed by the competing output from the contralateral command centres, operating on the principle that the animal should not leave the shelter of its home without good cause. The conflict can be resolved by blinding the splitbrain octopuses in one eye, after which there is a marked reduction in the proportion of abortive runs (Table 8.3).

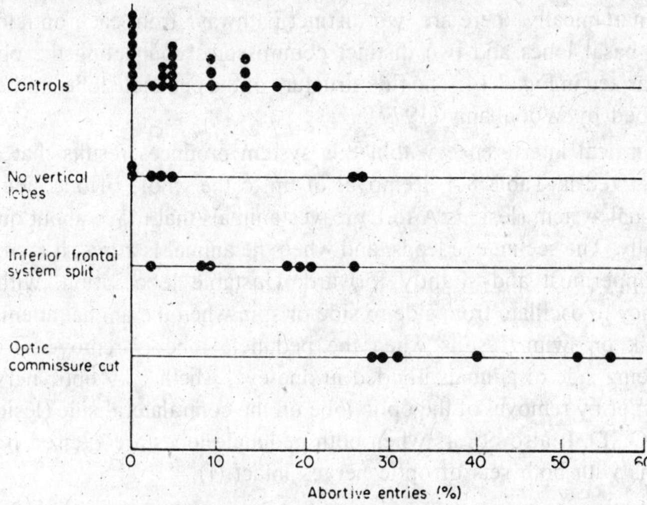

Fig. 8.13 : Showing the proportion of trials at which animals with various lesions made only abortive entries into a detour apparatus. Each point shows the result obtained with a single individual having the stated type of lesion (after Wells, 1970).

Table 8.3 : Split brains and abortive runs (after Wells, 1970).

Animal	Before blinding in one eye			After blinding in one eye		
	Completed	**Abortive**	**%**	**Completed**	**Abortive**	**%**
N1	39	15	28	59	2	3
N3	7	14	67	14	15	52
N20	41	9	18	30	1	3

The Peduncle Lobes

In the intact animal, conflicts of command between the two sides of the brain must be resolved at supraoesophageal level, since there is evidently no mechanism for doing so in the suboesophageal brain. Octopuses can be taught to make visual discriminations after removal of their vertical and superior frontal lobes. The sensory inputs that initiate attack or retreat must come in through the optic lobes on their way to the higher motor centres in the basal lobes and it would seem reasonable to search above the basal lobe level for the mechanisms that resolve conflicts between the two sides of the brain.

Anatomically, there are two distinct pathways from each opticlobe to the basal lobes and two distinct commissures connecting the optic summarized in Fig. 8.14. The fine structure of the peduncle lobe has been described by Woodhams (1977).

Surgical interference within this system produces results that are summarized in Table 8.4. Removal of up to the whole of one side of the visual system (lesions A to E) leaves animals that move about quite normally. The seeing eye leads, and when the animal is at rest it is often held uppermost and slightly forward. Unstable locomotion, with a tendency to oscillate from side to side or spin when the animal attempts to walk or swim, results when the peduncle lobe is removed from the seeing side of animals blinded in one eye, whether by optic nerves section or by removal of the optic lobe on the contralateral side (lesions N, 0, Q, T). It also occurs when both peduncle lobes are excised from animals with both sets of optic nerves intact (I).

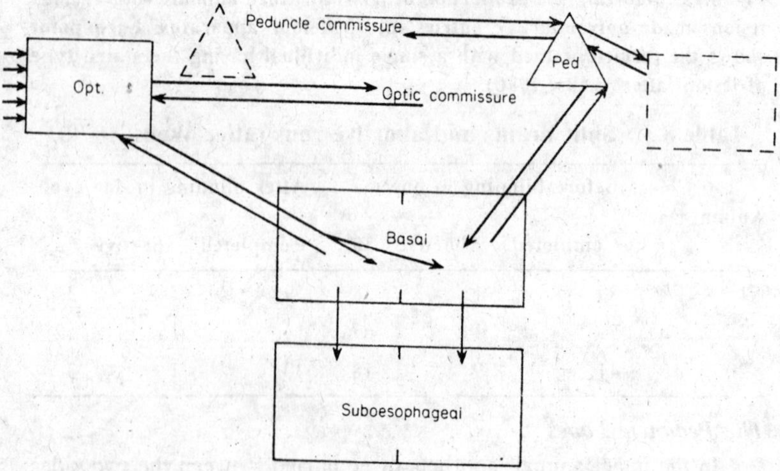

Fig. 8.14 : Diagram showing the functional connections between the optic, the peduncle and the basal lobes. For clarity the optic lobe only is shown on the left and the peduncle lobe only on the right (from Messenger, 1967b).

Table 8.4 : Lesions of the visual locomotor control system (after Messenger, 1967b;.

Lesion	Posture	Unstable movement	Forced circling
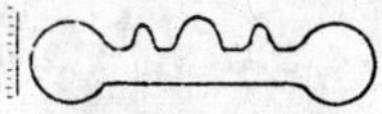 A $n=3$	Normal. intact eye leads	–	–
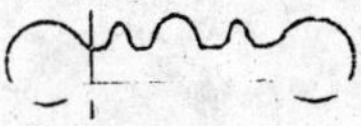 B $n=11$	Operated side forward. higher	–	–
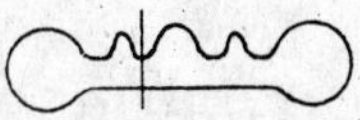 C $n=20$	Normal		In dark only. side without p.l. walks forward
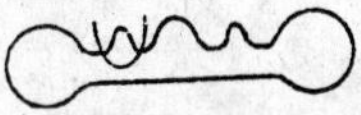 D $n=16$	AsB	–	–
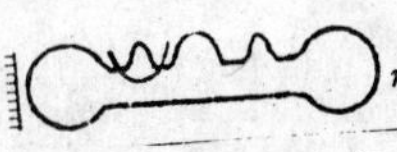 E $n=4$	AsB	–	–
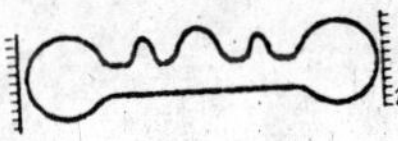 F $n=14$	Normal	–	–
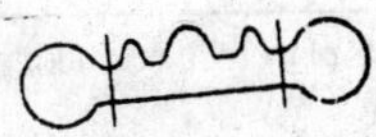 G $n=11$	Normal		
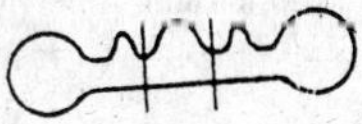 H $n=11$	None. immobile. limp		
 I $n=20$	Normal. but may sit with head up. down or sideways	Yes	–
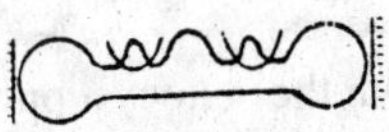 J $n=12$	Normal	–	–
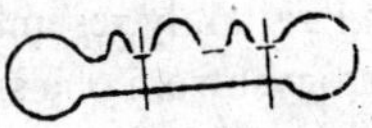 K $n=17$	Side with peduncle lobe down	Circles	Yes. side without p l. walks forward

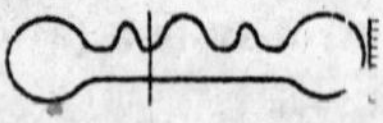	L $n=8$	As K	Circles	As K
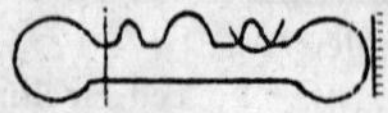	M $n=15$	As K	Circles	As K
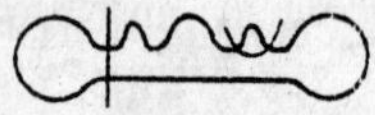	N $n=6$	Seeing side up, forward	Yes	–
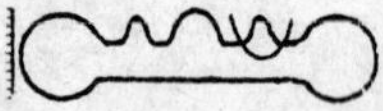	O $n=($	As N, but less marked	Yes	–
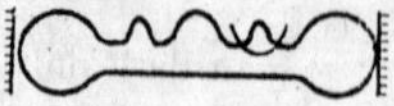	P $n=7$	Normal in walking, P.I. side leads	–	–
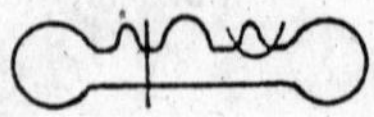	Q $n=6$	Normal, sometimes head down	Yes	–
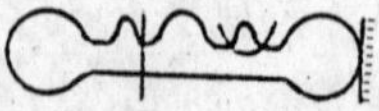	R $n=5$	Intact optic lobe up	–	–
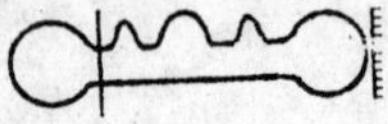	S $n=7$	Normal	–	–
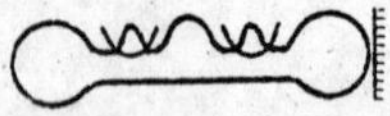	T $n=4$	Seeing eye up, forward	Yes	–

'Unstable, movement' = Oscillations, particularly in the rolling plane, when walking or swimming.

'Normal' = head up if on side of tank, eyes up if on bottom, symmetrical, tone similar on the two sides.

p.l. = Peduncle lobe

Standard diagram and lesions:

Peduncle lobe removed

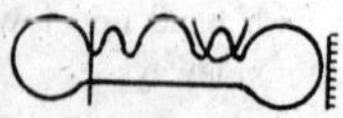

optic tract cut   optic nerves cut

Each peduncle lobe receives a direct input from the statocyst on the same side of the body Fig. 8.15, Hobbs and Young, 1973) and the effects of peduncle lobe removal on walking and swimming certainly resemble those of removing the statocysts. Removal of the statocyst from one side and the peduncle lobe from the other produces severe disturbances in locomotion, resembling the effects of bilateral peduncle lobe removal.

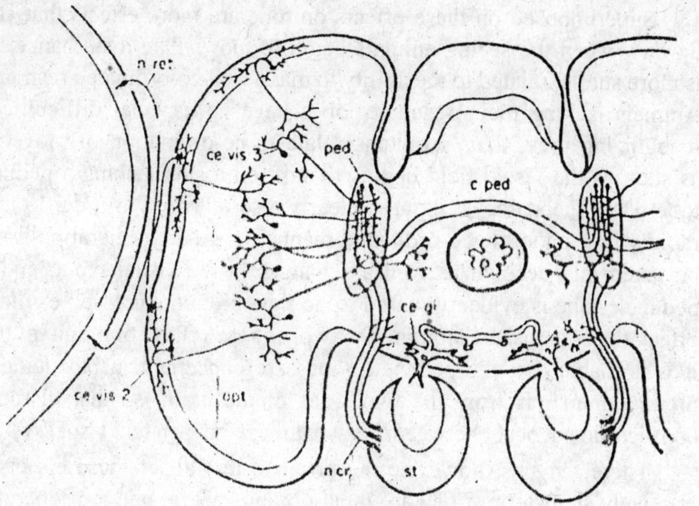

Fig. 8.15 : Diagram of connections of the peduncle lobe of *Loligo*, seen in transverse section; the condition in *Octopus* is believed to be similar. The axis lies in an oblique vertical plane, ce.gi., first order giant cell; c.ped., peduncle cominissure; ce.vis.2, ce.vis.3., second and third order visual nerve cells; n.cr., crista nerve from the statocyst; n.ret., retinal nerve fibre; opt., optic lobe; ped., peduncle lobe; St., statocyst (after Hobbs and Young, 1973).

The instability produced by interference with the peduncle lobes is not, however, attributable to cutting the statocyst input, since it is only found in animals that can see. Operations J, P and R Table 8.4 produced blind, but stable animals, unlike their seeing counterpart, O and Q.

The peduncle lobes seem to have two sorts of effect upon locomotion. One arises from their effect upon muscle tone; each lobe appears to have an excitatory effect on the musculature on the opposite side of the body, though the mechanism may in practice be inhibition of inhibition by the contralateral basal lobes. The matter is complicated by the direct output from the optic lobes, which again seems to increase muscle tone, though here the effect is apparently bilateral, since removal of one optic lobe does not produce asymmetry. The full effect of peduncle lobe removal on muscle tone is only seen in bilaterally blinded animals that lack one or both of their optic lobes. These animals (operations K, L and M in the Table 8.4 series) circle around the side with an intact peduncle lobe. The same circling was observed in seeing animals with one optic/peduncle lobe complex removed (operation C) when visual cues were eliminated by switching out the lights.

Superimposed on these effects on tone are more effects that show up only when the seeing animals begin to move. The disturbance here is more subtle, related to the ability to make well co-ordinated responses. Animals lacking their peduncle lobes have no especial difficulties in recognizing prey, their problems relate to co-ordination of movement as soon as the visual field begins to shift on a side lacking a peduncle lobe. Here the statocyst input is clearly also relevant; in order to move rapidly in a smoothly co-ordinated manner, angular acceleration it must be taken into account as well as changes in the visual input, and the peduncle lobe is evidently wired to do this. From the crossed-excitation effect, the commissures joining the two, and the effects of brain-splitting, it is arguable that the peduncle lobes are concerned in balancing the motor commands from the two sides of the brain, so that disruptive conflicts do not occur between the two largely independent visual systems.

There remains, however, the problem that all of these effects are relatively short term. They are most obvious within hours of operation, but decline progressively so that within a week or ten days the postural and locomotor effects of the lesions figured in Table 8.4 have disappeared. This transience suggests that the optic and peduncle lobes may not themselves organize patterns of movements at all. Rather, they provide the sensory information to trigger patterns organized within the basal lobes, where, in contrast, lesions have permanent effects on locomotion. The fact that electrical stimulation of the base of the optic lobes can sometimes produce better-patterned locomotion than stimulation of the basal lobes themselves is a reflexion of the crudity of the means used to excite the basal lobes directly—we cannot produce an adequate pattern of stimuli in this way. By exciting the tracts running from the optic lobes we are causing discharges in nerves that normally carry signals summarizing specific patterns of input (from the 'classifying' cells in the optic lobes, Sections 8.18 and 12.2.1) and these signals, in contrast to those generated by direct stimulation of the motor neurons in the basal lobes, may constitute an adequately patterned input. *Ex-hypothesis*: the transient disorganizing effect of damage to the visual pathways will be due both to the sudden withdrawal of normal sensory inputs and to the presence of irregular signals from wounds rather than to removal of a part of the motor system. Muscle tone, similarly, is lost with the normal source of exciting sensory input, but becomes replaced in time as the threshold of response to alternative, tactile, inputs drops. It is, on this view, wrong to describe the optic/peduncle lobe system as a 'visuo-motor' control system. After it is removed, the motor control system remains, and produces as smooth movement as before, but based on an alternative set of inputs.

Computing the Visual Attack

Octopuses that are used to attacking crabs in their tanks will carry out a relatively stereotyped series of movements when they see prey at the far end of their aquaria. The *Octopus* emerges from the home, walks forward, gathers itself together and makes a jet-propelled leap to

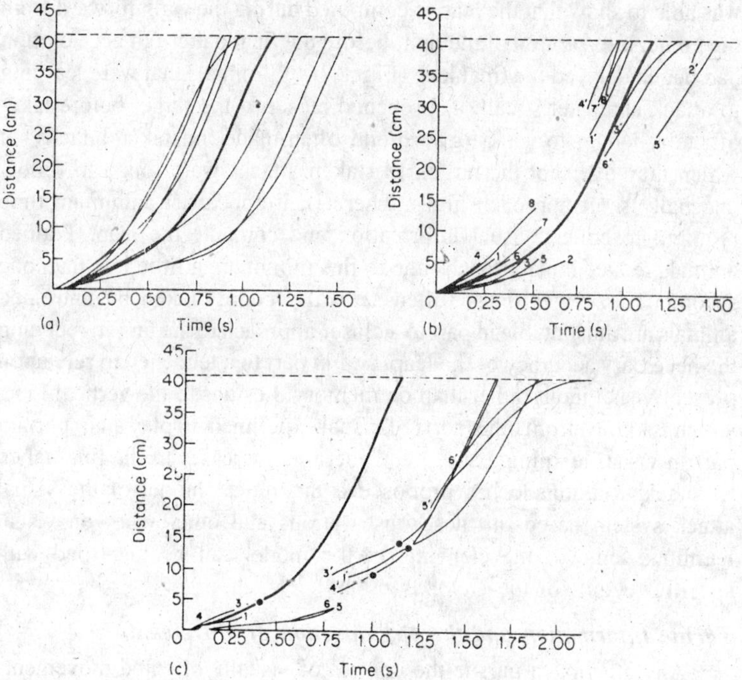

Fig. 8.16 : Time and distance plots of the courses of attacks on crabs, (a) Course of seven attacks at trials with illumination throughout. Heavy line: average course. The curves are cut off at 41 cm from the home, the distance where tactile information may begin to play a part. (b) Course of 8 attacks at trials with periods during which the light was interrupted. Heavy line: average shown in (a). Numbers 1 to 8 stand for the instant at which the light was interrupted; numbers with primes stand for the instant at which the light was switched on again, (c) Course of 6 attacks in which the light was interrupted as the octopus was walking forwards from its home. Heavy line: average course from (a). A dot on curves 1, 3 and 4 indicates the beginning of final pattern of acceleration (from Maldonado, 1964).

cover the crab with its web. Detail analysis of the time course of attacks can be used to establish which parts of this sequence depend upon visual feedback analysed cine film to establish the normal course of acceleration and deceleration during the leap and compared this with the pattern shown by animals in cases where the lights were extinguished for a period after the beginning of the jump. As Figs. 8.16a and b show, the leap is unaffected by interrupting the light, so it must be fully programmed before take-off. Tracing the attack back to its earlier phases Maldonado was able to show that the leap is computed during the walk forward from the home. Interruption of the light before the final pattern of acceleration had begun delayed the final leap (Fig. 8.16c). Animals that were learning to attack characteristically approached closer to the target before take-off, took longer to walk forward and often made 'mistaken attacks' in which they overshot the target. Mistaken attacks were found to follow unusually short approach times. There is, it appears, a minimum time required to collect visual information and compute the leap. Trained animals reduce the approach time to this minimum, a little less than one second. The longer times (often several seconds) taken by untrained animals are attributable in part to a closer approach to the target, reducing the necessary accuracy of the leap, and in part to a tendency to rely upon present visual input rather than on memory. Lesions to the vertical lobe, which is known from other sorts of visual experiment to play an important part in visual learning, caused considerable increases in the time taken to attack. Maldonado has proposed a theoretical model of the visual attack system based on these observations and other work on visual learning; some of the elements in the model can be identified with specific regions of the *Octopus* brain.

Tactile Information and the Organisation of Movement

Any attempt to pursue the control of visually oriented movements into the basal lobes using behavioural and brain lesion techniques eventually founders because the necessary operations cut off the sensory input that is driving the system. Further analysis of the control of movement can, however, be made using tactile stimuli and observing the performance of the arms. Blinded animals, Octopuses with their optic lobes removed and even Octopuses with their optic tracts cut centrally to the peduncle lobes will walk and swim stably if they are allowed a long enough period to recover from the short-term effects of the operations. Since the only sensory inputs now come from below from the mouthparts via the superior buccal lobe, and via the suboesophageal lobes from the aims and mantle-the effect of lesions to all parts of the central supraoesophogeal brain can be investigated.

Destruction of the superior frontal and vertical lobes has no detectable effect on movement, whether the animals can see or not. Damage to the subvertical, dorsal basal and precommisural lobes similarly seem to have little effect (though this matter has never been considered in detail). Extension of the damage into the anterior or posterior basis, in contrast, has gross and apparently permanent effects on locomotion and posture.

Basal Lobes and the Inferior Frontal System

Removal of all parts of the brain behind and above the median inferior frontal and subfrontal lobes produces an animal that lies on the floor of its aquarium in a tangled and rather flaccid heap in time, muscle tone improves a little, and the animal may move slowly about if prodded, apparently as a result of small stepping movements by the suckers. These animals can feed (indeed they can be trained to make tactile discriminations; They will live and grow for many weeks. Apparently spontaneous activities increase with time-the animals are found to have moved between periods of observation—but there is never any return to normal walking or swimming. So far as locomotion is concerned an *Octopus* with the basal lobes destroyed resembles one with the whole of the supraoesophageal brain removed. It exhibits certain reflexes. It will, for example, move arms to search in approximately the right position if it is scratched on the back of the abdomen. If the suckers of one arm are stimulated by contact, they will grip and hold on. If an attempt is made to withdraw the probe, neighbouring arms will come across to the area of contact, and if a pull continues, the arms on the opposite side of the-body may align themselves along the direction of the strain. The whole preparation may then pull convulsively. Handling a preparation often initiates 'writhing' or 'cleaning' movements in which the arms are curled back into tight spirals and then twist back and forward from the base placed upside down, the animal lacking basal lobes may right itself; but it is difficult to tell whether this is achieved 'deliberately' or simply because the rather feeble grip of the suckers holds the arms right way up once they have by chance achieved this position. Basal lobe animals are not infrequently found upside down, mouth and suckers upwards with the arms folded back and applied to the floor-a 'defence' position that is again, perhaps, achieved by chance.

If the front part of the supraoesophageal lobes is removed leaving the basal lobes intact, *Octopus* remains capable of walking and swimming.

So far as locomotion is concerned, these animals are indistinguishable from 'normal' blinded Octopuses with the whole of the supraoesophageal brain intact. Differences can only be detected when their behaviour is considered in detail. Thus lesions to the inferior frontal system are found to interfere with learning to discriminate by touch, and elimination of the whole of this region prevents touch learning altogether. Destruction of the superior buccal interferes with motor control of the mouthparts so that the animals cannot chew food that they will still take with the arms; There are also changes to that arm reflexes; extensive damage to the posterior buccal and subfrontal lobes leaves animals with 'sticky suckers', which seem to have difficulty in letting go of objects that they have grasped. The effect is most marked when any attempt is made to withdraw an object in contact with the suckers. It does not, however, interfere with normal locomotion, so that one can sometimes observe the apparently paradoxical situation of an animal in training running away but evidently unable to shake off an object that it has grasped. This implies a dual control system. The posterior buccal/sub rental region can inhibit the grasping reflex. So can the locomotor orders from the basal lobes. The latter, by themselves, are ineffective only when very strong stimulation of the grasping reflex is experienced.

Suboesophageal Lobes and the Interbrachial Commissure

The behaviour of animals with the whole of the supraoesophageal brain mass removed is predictable from the combined effect of removing the basal lobes and the inferior frontal system. These animals do not walk or swim, and they cannot learn to recognize objects by touch. In training experiments, to be reviewed later, they err mainly by taking nearly all small moveable objects that are presented to them. With the elimination of the inferior frontal system, suboesophageal animals have 'sticky suckers', with the result that responses dependent upon adhesion, such as the aligned pull reflex, are easier to demonstrate; the animals will hang on where preparations with basal lobe lesions let go and they can, for example, be lifted out of the water by the grip of their own suckers. Perhaps because of this their posture, sucker-side-down, tends to be a little more reliable than that of basal lobe animals, particularly during the first few days after operation.

There is one further level at which integration of the activities of the individual arms might be achieved. The interbrachial commissure links the arms just centrally to the last of the nerve cell bodies that form the ganglia of the arm nerve cords; central to this point there are only fibres running to or from the arms. The nerves in the interbrachial commissure are all centripetal; they arise from longitudinal fibres running towards the brain, and from the neuropil of the first ganglion in each arm linking the arm with its immediate neighbours and probably directly or indirectly with all the other arms. Stimulation of the commissure between two arms produces sweeping and stepping movements of the arms on either side, as does stimulation of the arm nerve cord distal to the interbrachial commissure. With continued stimulation (3 V, 0.7 ms pulses at 100 Hz) the effect spreads to further arms on either side. The spread of excitation is prevented by cutting the commissure but not by sectioning the arm nerve cord on the brain side of the commissure.

Movement Control at Arm Nerve Cord Level

The bulk of the muscle in the arms is enclosed in a tubular connective tissue sheath to form a solid mass surrounding an inner sheath around the axial nerve cord Fig. 8.17. Within this core of intrinsic musculature there are longitudinal and transverse fibres, and a complex system of oblique fibres arranged in three series that differ in orientation. Outside this core are the extrinsic muscles controlling the movement of the skin and suckers.

The nerve cords in the eight arms are together estimated to contain between two and three times as many neurons as the brain. The bulk of these are to be found in the axial cords, but there are also four intramuscular cords in each arm, lying parallel to the axis Fig. 8.17, and a subacetabular ganglion immediately above each sucker. Each of these elements contains motorneurons and receives sensory input from outlying nerve cells. The axial cord includes swellings, alternately to each side, related to the individual suckers. These are generally and conveniently referred to as ganglia, though the term is a little misleading since the nerve cell body wall of the axial cords is in fact continuous up to the level of the interbrachial commissure; the ganglionated appearance seen in sections arises because the cord flexes from side to side as it runs down the arm.

In view of the very considerable complexity of the arm nerve cord system it is not surprising to find that the behavioural responses even of isolated arms tend to be elaborate. Attempts have been made to list arm 'reflexes' but the results are less than satisfying because the responses observed and reported are clearly not an exhaustive list, and because they are commonly overlain by movements that may be truly spontaneous, or longterm responses to amputation and manipulation.

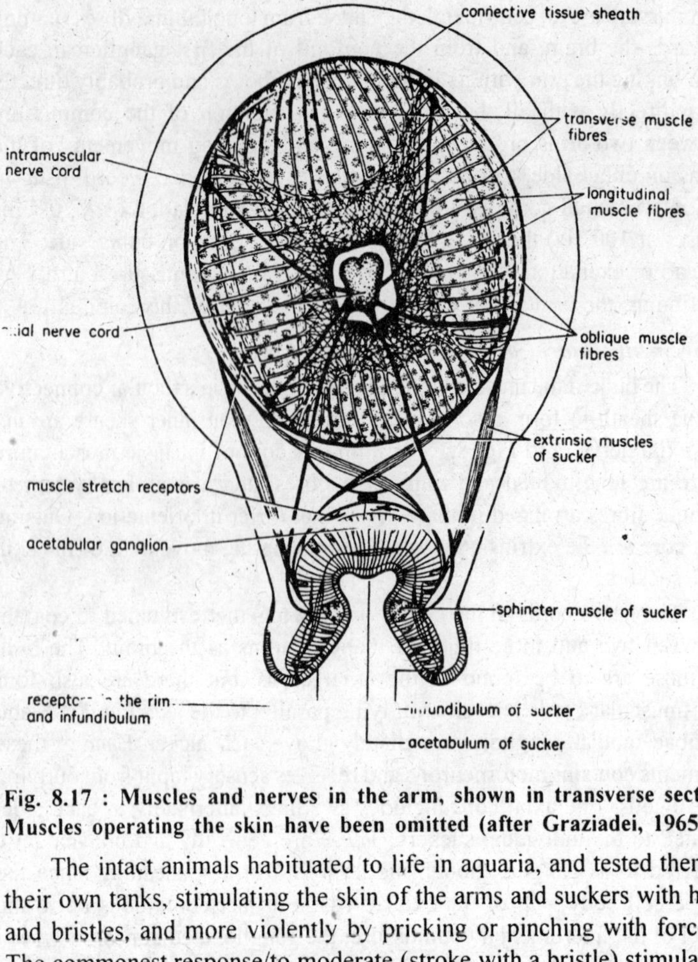

Fig. 8.17 : Muscles and nerves in the arm, shown in transverse section. Muscles operating the skin have been omitted (after Graziadei, 1965b).

The intact animals habituated to life in aquaria, and tested them in their own tanks, stimulating the skin of the arms and suckers with hairs and bristles, and more violently by pricking or pinching with forceps. The commonest response/to moderate (stroke with a bristle) stimulation of a sucker or its vicinity, was slight withdrawal followed by extension of the sucker disc towards the point of origin of the stimulus. Stronger stimulation caused withdrawal followed by flexing of the arm and extension of several suckers towards the stimulus. If the extending suckers touched anything, they grasped it.

The same series of responses could be elicited from isolated arms; 'The reflex movements of the suckers ... do not differ in any way from the reflexes of intact animals'. He proceeded to analyse the matter further and concluded that three levels of reflex are could be detected in the arms. The first of these must operate through the subacetabular ganglion, since the rim of the sucker would still bend inwards and the disc still contract if touched after cutting all the peripheral nerves from the arm cord. Suckers cut off below the level of the subacetabular ganglion did not respond. A second level, with a much lower threshold, operated through the arm nerve cord ganglion immediately overlying the sucker concerned. Stimulation in the centre of the sucker disc led to flattening and expansion, stimulation of the more peripheral parts to extension and grasping. Cutting the cord on either side of the ganglion did not alter these responses. Still longer arcs could be traced in cases where stimulation led to recruitment of suckers, all turning towards the source of stimulation. Since stripping the intervening part of the arm nerve cord of its ganglion cells and peripheral connections did not prevent the spread of such effects. Ten Cate argues that these must be due to simple reflex arcs with long neurones connecting the more distant respondents to the stimulus source.

Rowell (1963) repeated some of Ten Gate's experiments, with essentially similar results. He failed, however, to observe the turning and grasping movements noted by Ten Cate (1928) and von Uexkull (1894). This, as he points out, is hardly surprising since he was wording with direct electrical stimulation of isolated arms.

In tactile training experiments, the arms of blinded Octopuses regularly twist and grasp objects touched against the suckers or upper surface of the arms. A ring of arms isolated from the rest of the body by a cut below the level of the buccal mass will do the same and, when fresh, can be relied upon to grasp, flex and pass pieces of fish in towards the base of the arms, where the mouth once was. Less commonly, such preparations will reject inedible objects. Similar results (and most of the reflex responses described above) can be obtained with denervated arms, left attached to the *Octopus*. These chronic preparations differ from excised arms only, it seems, in their muscle tone; they tend to be flaccid until excited and in general require more vigorous stimulation if they are to perform like freshly amputated limbs. The implication is that the brain has an excitatory effect in the intact animal that is mimided by wound trauma in the acute preparations.

Recordings from the Arm Nerve Cords

Recordings from the nerves entering the arm cords from the suckers are dominated by the discharges of rapidly adapting mechanoreceptors.

Inside each arm ganglion, Rowell (1966) found units that responded to touch on the ganglion's own sucker, rim, cup or skin, and units that responded only to tactile stimulation of other suckers. Thus, for example, there were units that responded only to stimulation of either of the two suckers distal and contralateral to the recording point. A further series of neurons fired only when a sucker was moved, or chose to move itself. These presumably represented discharges from proprioceptors in the muscles.

Table 8.5 : Take and reject responses by preparations with increasing amounts of the tactile system intact (from Altman, 1971).

	Sardine			Sardine soaked in quinine		
	Accept	Reject	Refuse	Accept	Reject	Refuse
Isolated ring of arms n =9	15	–	2	1	10	8
Denervated arm, chronic preparation n = 7	11	–	3	–	–	13*
Suboesophageal brain only (n + 4, 4 sets of tests)	32	–	–	23	1	8
Inferior frontal system intact (n = 4, 4 sets of tests)	32	1	2	7	22	1

2 Tests with each food were given by touching the pieces against an arm.

* but on 7 of these occasions the arm, having failed to grasp the bait. bent towards the mouth in a typical accept reflex.

Moving centrally, to record from the longitudinal tracts of the axial cords, it is evident that a very great deal of peripheral processing takes place before the input from the suckers is relayed to the brain. One would expect this on anatomical grounds, since the number of cells in each axial ganglion (about 1.3×10^5) is overwhelmingly greater than the number of axons in the peripheral nerves (about 5000). Rowell (1966) found interneurones that responded to areas as small as the rim of a single sucker, and as large as the whole tactile field distal or proximal to the recording point. Units were found that would respond to stimulation of the skin, or of the suckers only. There was a considerable range in threshold.

The great majority of interneurons in the axial cords were fast adapting, though a few would remain active for several seconds after the end of any obvious stimulation. Of the fast adapting nerves, many were apparently connected to detect differences in stimulation, adapting to a repeated stimulus pattern but immediately aroused by any change; it appears that these units habituate rather than fatigue, so that there is a measure of learning even at this level in the *Octopus* CNS. In addition to the 'novelty' units there were interneurones that distinguished between contacts made by the experimenter and those arising as a result of the animal's own movements (Rowell, 1966).

Stimulation of the Arm Nerve Cords

The complexity on the sensory side is matched by what is plainly a complex motor output system. Von Uexkull (1894) was the first to investigate this and more recent reports can be found in Altman (1968) and Rowell (1963); Fig. 818, is taken from Rowell's summary. Almost any movement of which a sucker is capable can be evoked by suitable stimulation of the ganglion controlling it. Widespread effects, with progressive recruitment of neighbouring suckers can be evoked by continued stimulation. Interestingly, Rowell (1963) was unable to excite nerves leading to the release of sucker grip, although from behavioural studies it is quite clear that the animal must have a detailed sucker-by-sucker control of this; the animals can hold small objects and move about, or pass objects from sucker to sucker in the course of take or rejection movements.

Stimulation of the middle region of the axial nerve cords can produce contraction or extension, twisting or sideways movements of the arm. Again repeated stimulation at threshold levels can lead to a slow build-up and progressive spread of excitation, indicating a range of polysynaptic pathways. The intramuscular nerve cords seem to control only the extrinsic dermal musculature (Rowell, 1963), which is perhaps surprising since they receive proprioceptive inputs from receptors buried in the axial musculature (Fig. 8.17).

Hierarchic Control, an Attempt to Summarize

Behavioural studies show that most of the apparently complex things that an *Octopus* does with its individual arms can be organized at arm nerve cord level. The function of the brain in relation to tactile discrimination, for example, would seem to be limited to triggering the sets of movements that result in taking or rejection. In locomotion all the detail of adapting the arms to the substrate over which the animal is walking could be dealt with at arm nerve cord level.

The fact that it might be does not prove that it normally is. The brain could override and control the details of what the arms are doing. Two lines of evidence argue against the possibility. One is the nature of the sensory information that is relayed from the arm nerve cords to the brain. Rowell (1965) found no proprioceptive component to the messages recordable from the brachial nerves, while the number of fibres in these nerves (about 30 000 compared with some 4×10^7 neurones in the axial cord of each arm) itself implies that only the most summary statements about conditions in the arms can ever be relayed to the brain. The second block of evidence that argues against central control of motor details is behavioural, arid has been outlined in very briefly, Octopuses never seem to learn tasks that would require a detailed central control of their arm movements.

Integration of the movements of the individual arms seems to be determined at several levels. The interbrachial commissure plays a part, as does the suboesophageal part of the CNS. But walking does not seem to take place unless the basal lobes are also present. Again, it is possible that the results obtained in the rather crude experiments so far carried

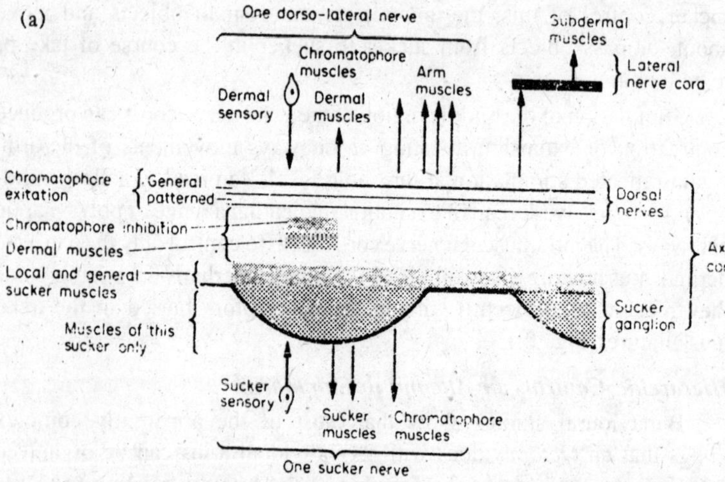

Fig 8.18(a)

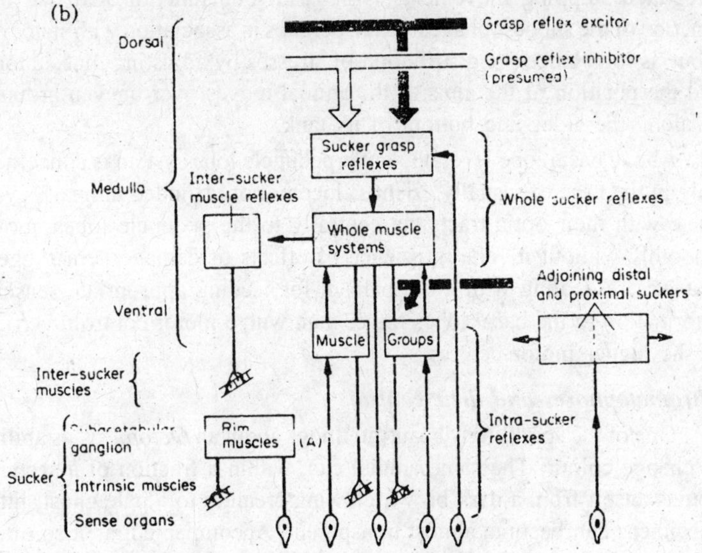

Fig 8.18 (a) : A stimulator's guide to the brachial medulla of *Octopus*. A short length of medulla with one sucker 'ganglion' is shown and the effects obtained by electrical stimulation of different areas and of the nerves leaving it are indicated. Areas giving contraction of the dermal muscles and inhibition of the chromatophore muscles are small and poorly defined; the diagram shows only their dorso-ventral level on medulla, and not their lateral extent which is "discontinuous. Unshaded areas control the general musculature of the arm. (b) Diagram of the nervous organization in the medu'h affecting the muscles of one sucker of the arm. Anatomical levels arc shown on the left-hand side. The two reflexes mechanisms labelled (A) and (B) are respectively those described by Ten Gate (1928) and von Uexküll (1894). The label 'whole nuscle systems' indicates for example all the longitudinal muscles of the sucker, while 'muscle groups' would mean a small patch of the longitudinal muscles (from Rowell, 1963).

out are misleading and that in fact, walking could be organized by the suboesophageal lobes, given an appropriate set of triggering stimuli. But suboesophageal preparations, and those with the inferior frontal system intact and the basal lobes missing will live for several weeks. This is more than enough time for the recovery of posture and locomotion by Octopuses with their optic tracts cut where, too, the effects at first seem to be attributable to removal of a higher level of motor control. On balance, it would seem reasonable to assume that the organization of integrated stepping movements is a basal lobe function and that the function of the suboesophageal brain, perhaps in association with statocyst input, is to balance the distribution of stresses by adjusting muscle tone and the position of the aims as the animal moves over uneven ground, or along the sides and bottom of its tank.

Above basal lobe level, the optic/peduncle lobe system is concerned only in the case of visually oriented locomotion. Blinded animals, even those with their optic tracts cut centrally to the peduncle lobes, move smoothly without it. The transience of effects of damage here suggest that we are dealing with a mechanism for feeding appropriate sensory information to the basal lobes rather than with a motor control override of the higher motor centres.

Chromatophores and their control

One of the sparklingly beautiful things about an *Octopus* is its ability to change colour. The same animal can, within a fraction of a second, convert itself from a dark brown glowing creature to a pale ghost; little specimens can become almost transparent. Accompanied, as it so often is, by a change in skin texture, so that the animal is at one time smooth and immediately afterwards prickly with extended papillae, the capacity for colour change can both camouflage the animal directly, and baffle would-be predators simply by virtue of its changeability. Search image is disrupted. If, as is also often the case, the retreating *Octopus* discharges a puff of mucus-bound ink as it switches from dark to pale and jets away, a decoy is added that can regularly confuse even the most sophisticated human observer, who knows exactly what is going to happen.

Octopuses use their capacity for colour change in a variety of ways. For *Crypsis*, as above; for displays between themselves for alarming potential predators (the 'dymantic' display) and possibly for flushing prey that would otherwise remain motionless.

Until quite recently, observations of chromatophore structure and electrophysiology were limited by the resolution of the light microscope and the absence of microelectrodes capable of penetrating the very small muscle fibres that surround the bag of pigment. In the absence of unequivocal observations, there was much argument about the nature of the innervation of the muscles, about cell elasticity (radial muscles pulled it out, but how did the chromatophore contract again?) and about a whole range of problems arising from the behaviour of chromatophores in recently excised or denervated skin.

Chromatophore Structure

Each chromatophore consists of a bag filled with granules of pigment, held in place on a framework of fibres. The bag is probably elastic, but it is also possible that the filaments around it are contractile; in *Octopus*, extracellular channels penetrate to the bag surface in a manner strongly reminiscent of the T-system in muscle cells (Froesch, 1973a). Attached to the bag is a ring of radial muscles, linked one to another by tight junctions, so that the possibility of electrical interaction between the muscles exists (the fact that the muscles sometirr JS pulsated in synchrony led some of the older workers to believe that the fibres formed a syncytium, which is now plainly not the case). The upper and lower surfaces of the pigment bag are enclosed in a single cell, the surface of which is thrown into a series of folds in the contracted state. Fig. 8.19a is a reconstruction of a pigment cell, based on electronmicrographs, while 8.11b shows the approximate relative dimensions of the cell when it is contracted and expanded.

The radial muscles are obliquely striated, with large and small (presumed myosin and actin) filaments surrounding a central core of mitochondria. There is no trace of large, paramyosin filaments which is perhaps surprising, since the muscles can apparently remain contracted for long periods and a molluscan 'catch' mechanism might have been expected. Along each muscle there is a nerve, covered in ganglia on its exposed surface. Each nerve contains between one and four axons, lying in a groove along the surface of the sarcolemma and forming 2 continuous synaptic junction along the whole of the length of the muscle. In the contracted state the nerve shakes along the muscle fibre in the manner shown in Fig. 8.19a in *Loligo*, in *Octopus*, the neuromuscular junctions are more localized (Fig 8.19c).

Innervation and contraction

The expansion of a chromatophore reflects the combined activity of the radial muscles around it. Fig. 8.20 shows the effect of increasing the frequency of stimulation through a nerve of several fibres; the muscles twitch, and the twitches combine to give a tetanus. Reducing the voltage produces a stepwise reduction in the strength of contraction; the chromatophores contract, but irregularly as one after another of the radial muscles drops out. Fig. 8.21 traces the outlines of a group of chromatophores during progressive expansion caused by progressive increases in the voltage applied at 20 pulses s^{-1}; full expansion in this

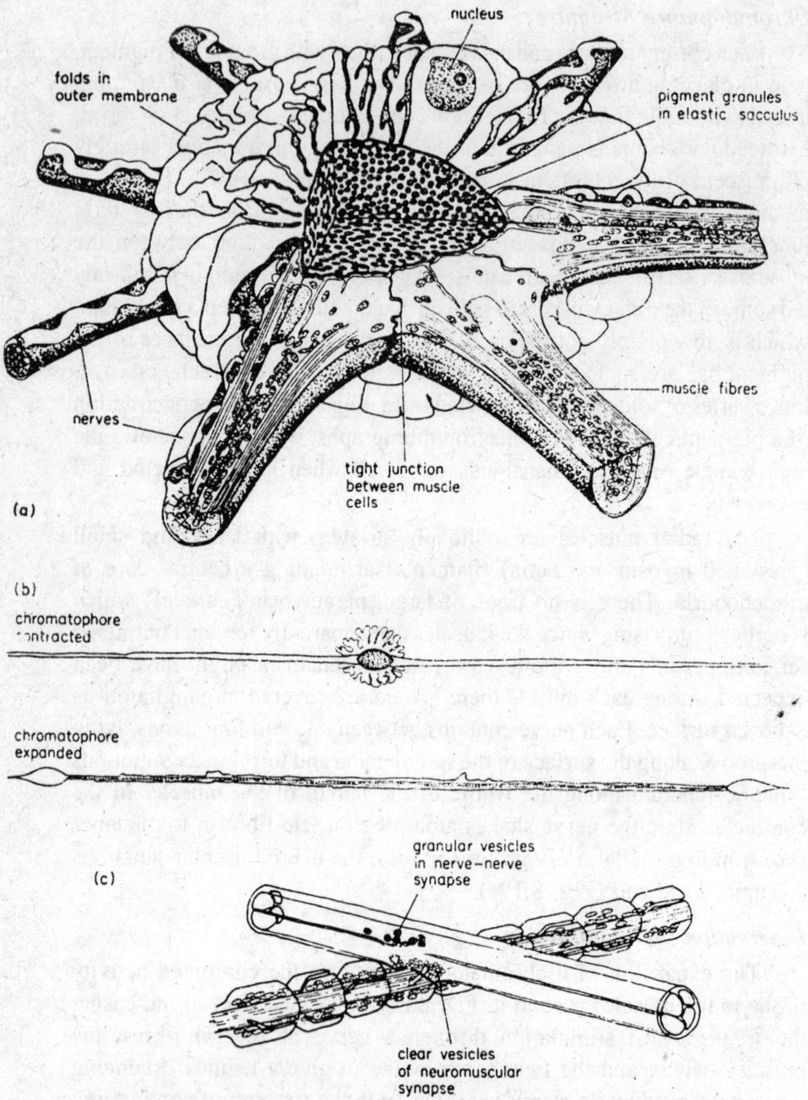

Fig. 8.19 : Chromatophore structure, (a) Chromatophore of *Loligo* opalescens with the muscles relaxed. In this condition, the pigment sac would have a diameter of about 50 p.m. (b) The same in section showing the relative dimensions with the chromatophore muscles relaxed and contracted, (c) Chromatophore muscle and nerves from *Octopus vulgaris*. Here the mitochondria lie around a central core of muscle fibres. The nerves do not lie along the fibres but synapse where they cross. As well as neuromuscular synapses there are presynaptic nerve junctions, (a) after Cloney and Florey, 1968; (b) after Florey, 1969; (c) after Froesch 1973a)).

instance follows recruitment of three separate motor units, each of which innervates some of the muscles from each of the chromatophores.

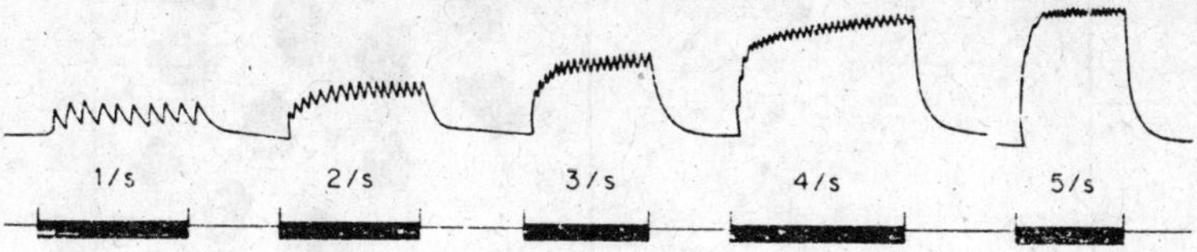

Fig. 8.20 : Tracings of contractions of chromatophore muscle fibres resulting from stimulation of their motor axons. The upper channel records contraction, the lower channel the stimuli. The numbers indicate the frequency of stimulation in pulses pers (after Flqrey, 1969).

Intracellular microelectrodes inserted into individual muscles show that the effect of nerve stimulation is to produce local excitatory postsynaptic potentials. In *Loligo*, at least, there is no sign of spike generation, at whatever voltage. Instead, there is a stepwise increase in local postsynaptic potentials with increasing voltage that presumably indicates polyneuronal innervation; the smaller potential changes show a slower rise time and a longer latency than the larger.

Isolated areas of skin, denervated on the otherwise intact animal, or removed and pinned out in seawater, are at first pale but later become coloured by waves of chromatophore contraction. The contractions appear spontaneously in one or a small group of chromatophores, spread outwards and return in a series of 'wandering clouds' which may continue for many hours *in vitro* or until nerve regeneration occurs in the intact animal. Muscles from neighbouring chromatophores appear to form bridges (Plate 2d) so that direct mechanical excitation is the most likely explanation of the spread.

A notable feature of the 'wandering clouds' contraction is that all the radial muscles in a given chromatophore may contract together. In chronic preparations, at least, it is obvious that this cannot be achieved nervously; the nerves have degenerated. Microelectrodes inserted into the spontaneously active muscles now reveal a change in physiology. The muscles generate spikes and can stimulate one another, presumably through the tight junctions already noted above.

The chromatophore muscles in isolated skin often show tonic contractions that may last for minutes on end examined muscles in this condition and found that tonic contraction was associated with prolonged showers of miniature potentials, apparently due to spontaneous release

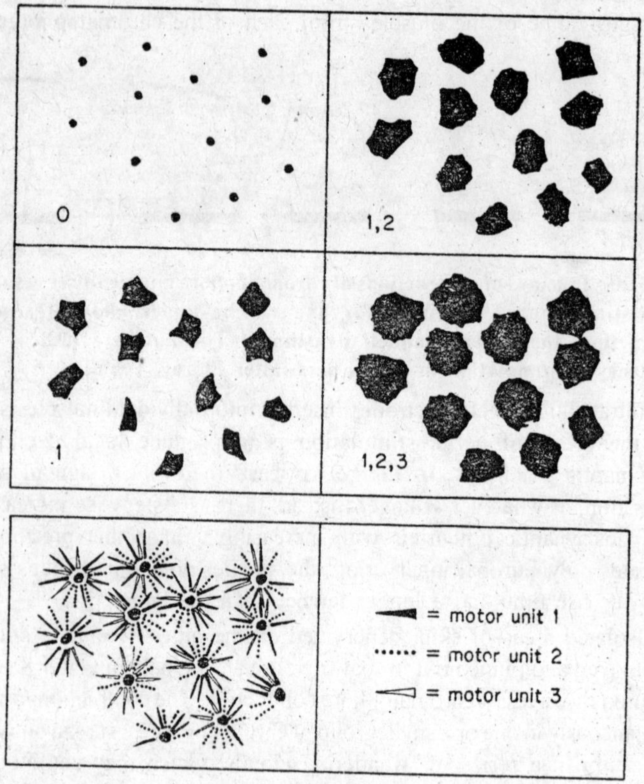

Fig. 8.21 : Stepwise expansion of chromatophores due to recruitment of additional motor units. The upper set of four diagrams shows the appearance of the same group of brown chromatophores in the absence of stimulation (0), and when 1, 2 or 3 motor axons of the appropriate nerve bundle are stimulated at the same frequency of 20 pulses s^{-1}. The lower set of diagrams illustrates the distribution of the motor units. The diagrams are based on a series of photomicrographs (from Florey, 1969).

of transmitter by the chromatophore nerves. Transmitter release was not affected by tetrodotoxin, which would eliminate nerve impulses, so there is no possibility that the tonic effects are due to continuing activity in the nerves of the excised skin. Once set up, tonic contractions can persist despite the disappearance of the miniature potentials; the muscle stays contracted until something is done to release it, a molluscan 'catch' muscle without the paramyosin fibres often associated with the 'lock-on' capacity Fig. 8.22.

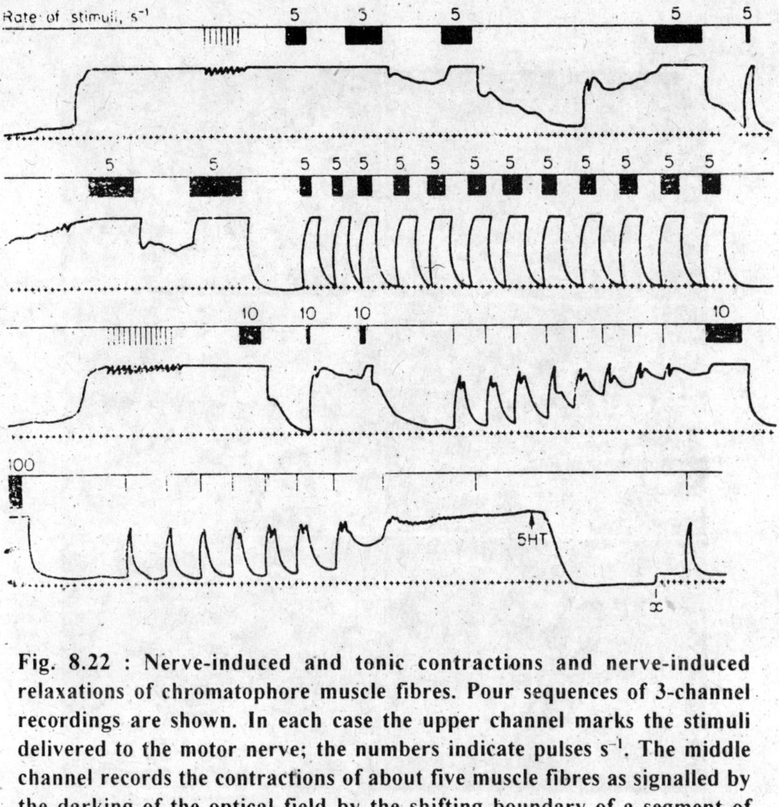

Fig. 8.22 : Nerve-induced and tonic contractions and nerve-induced relaxations of chromatophore muscle fibres. Pour sequences of 3-channel recordings are shown. In each case the upper channel marks the stimuli delivered to the motor nerve; the numbers indicate pulses s^{-1}. The middle channel records the contractions of about five muscle fibres as signalled by the darking of the optical field by the shifting boundary of a segment of the pigment mass of a single chromatophore. The lower channel provides time marks at 1 s intervals. Near the end of the fourth (lowest) sequence, 5-HT (10^{-6} g ml^{-1}) was applied to the preparation resulting in almost immediate loss of muscle tone. At x the trace was adjusted (after Florey and Kriebel, 1969).

The actions of acetylcholine (ACh) and 5-hydroxytryptamine (5-HT) strengthen the analogy with catch muscle. ACh induces tonic contraction and 5-HT supresses it. Pharmacological experiments of this sort, the observation that excitation of the chromatophore nerves can cause relaxation in muscles tonically contracted Fig. 8.22, and the existence of clear and dense-cored vesicles in separate nerves innervating the muscles all suggest a double innervation with excitatory and inhibitory components; the obvious implication is that ACh is the excitatory and 5-HT the inhibitory transmitter. Internal electrodes, however, never

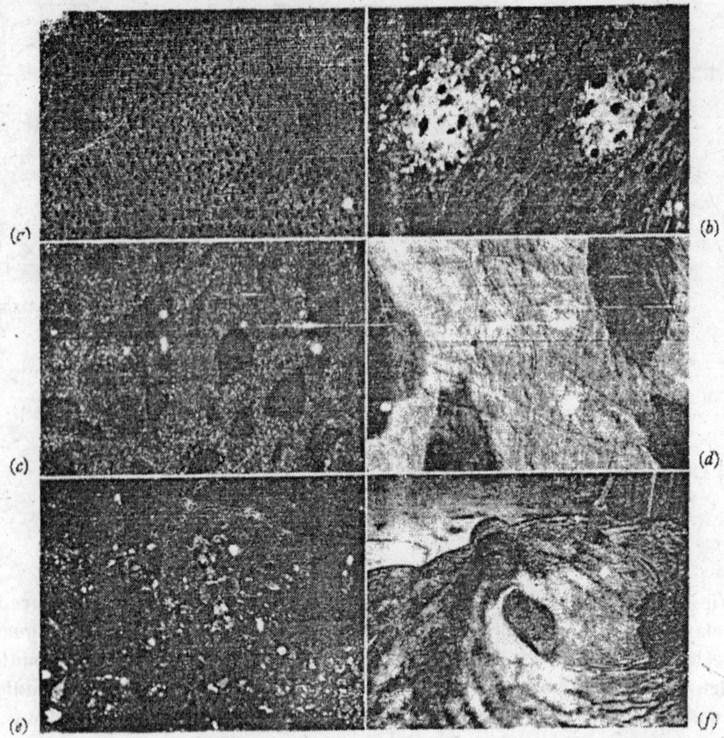

Fig. 8.23 : Polygonal unit with the chromatophores largely contracted; the white leucophores can be seen against a greenish background provided by the reflecting iridiophores. Incident white flash. Frontal head skin. × 20. (b) Chromatophores find leucophores from the web. Dark field, x 25. (c) Chromatophores and iridiophores from the iris, where there are no leucophores. Incident white flash, × 50. (d) Chromatophores and their muscles in the web. Interference contrast. x 120. (Photographs by D. Froesch.) (e) Small octopus with the skin papillae raised. (Photograph from Packard and Sanders, 1969.) (f) A larger animal with the skin smooth, dark hood and partial dymantic. Animal with tube to blood pressure recorder (Photograph by M.J.W.).

show hyperpolarization as a result of nerve stimulation or drug application. ACh enhances and 5-HT reduces the frequency of miniature potential changes, but the effect seems to be presynaptic, altering the rate of 'spontaneous' release of transmitter. If either substance were a transmitter itself, one could expect it to have an effect on the permeability of the muscle membrane. In the absence of any such effects, Florey, and Kriebel (1969) propose that the continued release of transmitter from the nerves of ageing preparations is attributable to elevated free calcium levels, a failure to return to binding sites within the sarcoplasmic reticulum. A motor nerve impulse, or the application of 5-HT, reactivates the calcium transport system, reduces 'spontaneous' transmitter release and allows the underlying muscle to relax. Additional support for this view comes from an additional action of 5-HT on the muscle itself, which contracts and relaxes faster when 5-HT is present-again, exhypothesis, a calcium binding effect.

These experiments are interesting in relation to the physiology of muscle contraction in general. They are not immediately relevant to conditions in the intact cephalopod where, so far as we know, the chromatophore muscles never show tonic contractions. They appear always to be under direct nervous control, their state of contraction or expansion being attributable to variations in the frequency of a continuous shower of nerve impulses. Observed closely, the chromatophores of a living cephalopod are in perpetual shimmering movement, even when the animal, seen at a distance, remains a steady colour.

Hormonal effects, if they occur at all, are attributable to the action of substances at CNS rather than at chromatophore level.

Reflecting elements in the skin; colour and tone matching

Octopus vulgaris hatches with about 70 red-brown chromatophores. As it grows the number increases steadily, so that an adult weighing 500-1000 g will carry between two and three million of these tiny organs, at a density of 100-200 mm^{-1} (Packard and Sanders, 1969).

By the time the little animals settle out of the plankton, the population of chromatophores has become subdivided into two groups, a very dark set, which can vary from black or red-brown, and a paler series which appears red when contracted and pale orange-yellow in extension. Plainly, these two alone cannot possibly account for the full range of colours that an octopus. The explanation lies in the possession of two sorts of light-reflecting structure, the iridophores and the leucophores, lying beneath the chromatophores (Plates 4 and 5 and Packard and Hochberg, 1977).

Iridophores contain stacks of platelets of a chitinous material spaced out at about one quarter of the wavelength of visible light, acting as broad-band reflectors of all the colours in the environment around the cephalopod (Denton and Land, 1971). Leucophores are irregular in shape, lie below the iridophores and the chromatophores, and are filled with a creamy guanine-like substance. They are opaque, and reflect white light. The ultrastructure of cephalopod iridophores has been described by Froesch (1973a- *Octopus*) and by Mirow (1972b-*Loligo*); nobody has yet reported on the fine structure of the leucophores.

All the available evidence suggests that *Octopus* is colour-blind. It nevertheless manages to match the colours of the plants, animals and stones among which it lives to a quite amazing degree. Reflection is clearly a part of the explanation, and it is reflection rather than the chromatophores that is responsible for the cryptic coloration of an *Octopus* at rest in the sea or matching blue and green colours in its aquarium. Reflection is passive. What the *Octopus* can and does do actively is match the intensity of the light that it reflects to the intensity reflected from its background. The chromatophores can be expanded to screen the reflecting elements to any desired extent and the seeing animal clearly does this; a blinded *Octopus* at rest.

Colour and tone patterns

Chromatophores; iridophores and leucophores are not uniformly distributed. At a whole-animal level, there is a well-marked pattern of

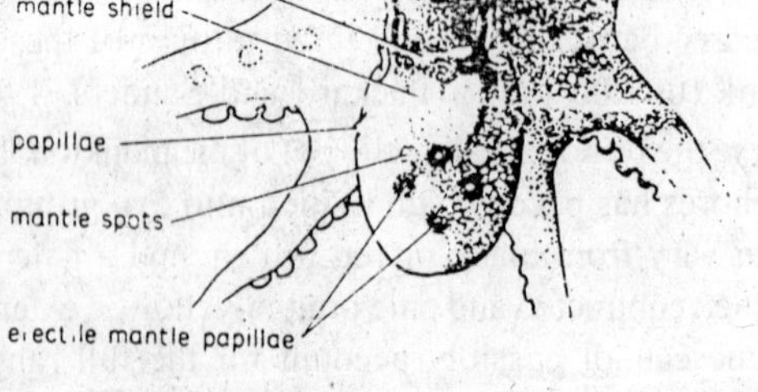

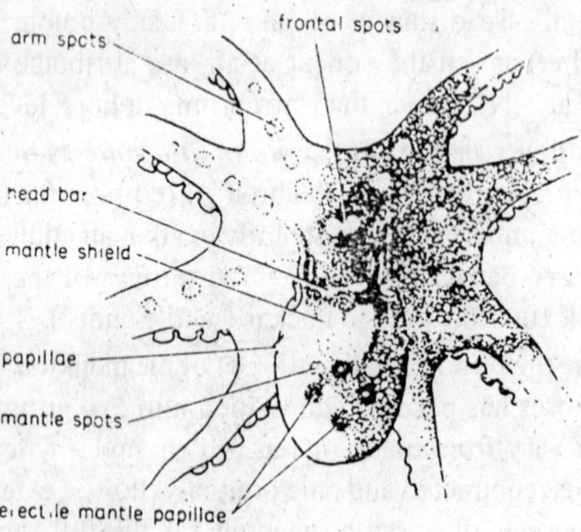

Fig. 8.24 : Permanent white components of the patterns shown by *Octopus vulgaris*. These are always present and visible except when shielded by expansion of the chromatophores. The frontal spots are particularly conspicuous, and nearly always shown. The four long mantle papillae are included as landmarks (after Packard and Sanders, 1969, 1971).

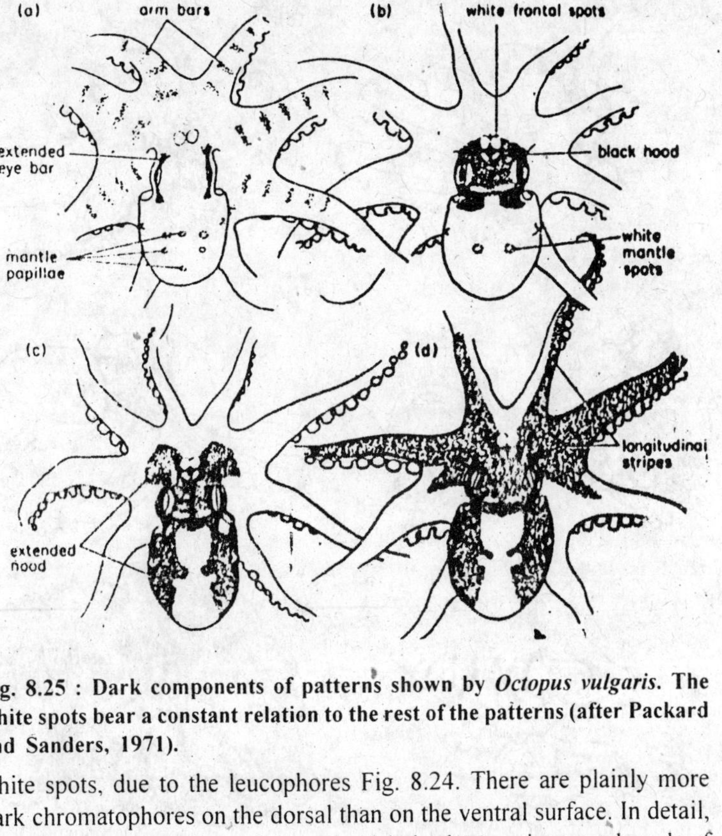

Fig. 8.25 : Dark components of patterns shown by *Octopus vulgaris*. The white spots bear a constant relation to the rest of the patterns (after Packard and Sanders, 1971).

white spots, due to the leucophores Fig. 8.24. There are plainly more dark chromatophores on the dorsal than on the ventral surface. In detail, the whole upper surface of the animal is broken up into patchwork of polygons, each with a pale whitish centre (leucophores) surrounded by a reflecting region (iridophores) rimmed by a dark groove the groove colour is due to chromatophores, concentrated by the folds around the edges of the polygon; there are relatively few iridophores here. White spots and the polygonal network are permanent. As the animal grows each patch gets larger; the precise pattern may be as characteristic of the individual as a fingerprint, with no two Octopuses quite alike. Permanent, too, are the locations of the erectile skin papillae on head and back and arms Figs. 8.24, 8.25 and 8.26. These can affect colour because contraction of the underlying muscles concentrates and reorients the iridophores. The animal, indeed, can hardly move without in some way affecting the colours that it shows to the world, so that the different

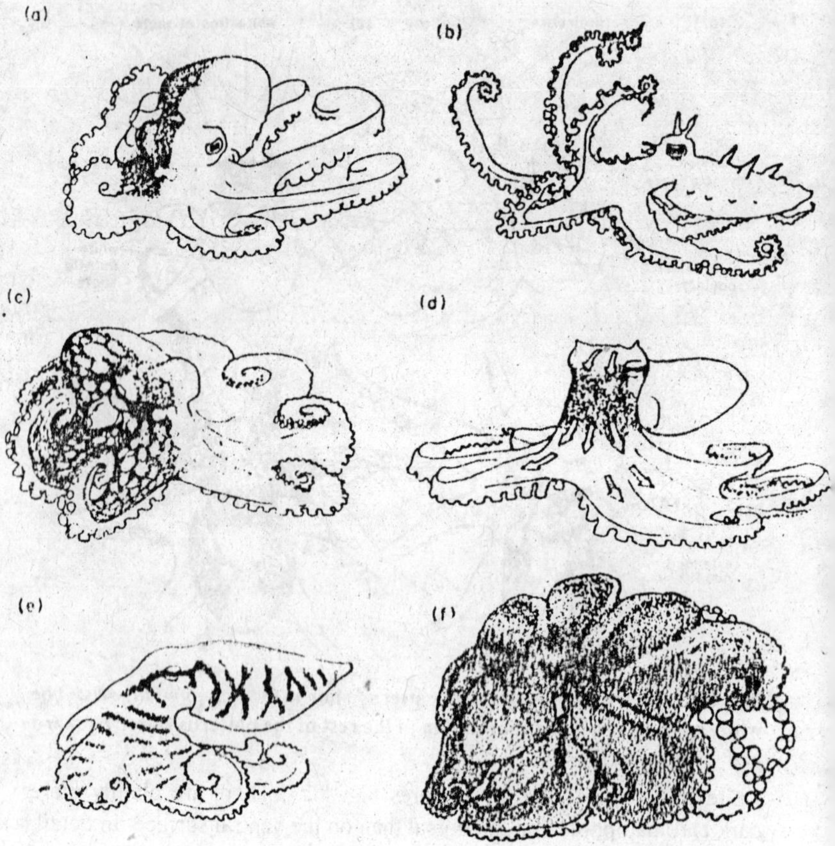

Fig. 8.26 : Postures and patterns associated with particular situations, (a) Unilateral extended hood displayed 111 response to an approach from the animal's right side. (b) The flamboyant posture adopted by small (< 5 g) animals. This is probably cryptic. The colour is uniformly dark in specimens of < 1 g, and later becomes replaced by black hood, broad mottle and lateral stripes. In animals larger than 5 g, the flamboyant response to disturbance is replaced by the dymantic (Plate 2 and (b) in Frontispiece). (c) Broad 'conflict' mottle here displayed unilaterally towards another *Octopus* approaching from the animal's right side. (d) Moving flush, shown when a small object approaches to one side; this display may serve to startle stationary prey into movement, (e) Zebra crouch by a young animal in the presence of another *Octopus*, (f) Uniform dark red of an animal that is pushing back another in fighting or an attempt to copulate (after figures and photographs in Packard and Sanders, 1969, 1971).

postures and forms of locomotion that it adopts all play a part in determining the patterns that it displays (Packard and Sanders, 1969, 1971; Packard and Hochberg, 1977).

Chromatophore expansion is directly controlled from the central nervous system and here to there is evidence of permanent patterning. What at first appears to be an endless reportoire proves on analysis to be built up from a strictly limited number of components, which recur regularly but in differing combinations and at differing intensities to give a wide variety, but by no means an infinite range of displays.

In an attempt to draw up an exhaustive list of the components of patterns of chromatophore expansion, Packard and Sanders (1969, 1971) have listed the following: Arm bars; eye bar and extended eye bar; eye ring, and dark edges to the suckers (perhaps always occurring together, as in the dymantic response the dorsal network of polygons (see above), which can give a range of mottled effects; black hood and extended hood; longitudinal stripes; transverse stripes.

These components are shown in Figs. 8.25 and 8.26 and can be recognized in some. They are of course superimposed upon the permanent white patches.

The light areas and their dark overlays are used to build up patterns that may be sustained for hours or days, or switched on only for a matter of seconds. Sometimes specific components are associated with particular postures, as in the dymantic display. Others seem to occur only in specific situations; the centrifugal flush shown in Fig. 8.26d for example seems only to occur when a smallish object moves close to the octopus; the display is evidently a transient version of the uniform dark colouring that is switched on in fighting or when the *Octopus* is about to pounce upon its prey. Mottled patterns, produced by varying degrees of expansion of the chromatophores in the dorsal network, are typically associated with conflict situations, when the *Octopus* is uncertain whether to attack or retreat; they are often shown only on the side of the eye viewing the problem to be assessed (Fig. 8.26c).

Although a start has been made in this direction, it is quite clear that a great deal more work will be required before the full significance of many of the patterns shown by *Octopus* is understood. Different species show different patterns, and may or may not use similar patterns for the same purposes. *Octopus cyanea*, for example, has a sexual display in which a dark longitudinal strip is shown down each of the arms-*O. vulgaris* only produces longitudinal stripes down arms 1 and

2 and never, so far as we know, displays these to advertise its sex (Wells and Wells, 1972a). A detailed study of the *repertoire* from a range of Octopuses could well yield information about phylogeneric relationships, quite apart from its interest in relation to visual signalling.

Central nervous control

The chromatophores are controlled through ten of the thirty or so nerves that fan out from each side of the brain to the periphery. Section of any one of the ten de-enervates the chromatophores in a specific area of skin (Fig. 8.27; Froesch, 1973b).

There is no overlap in the fields controlled by different nerves, and it if obvious that most of the patterns shown in Figs. 8.23 and 8.24 must be determined by fibres in several nerves - there is no question of patterns arising from the activation of whole fields at a time since the maps of the projection areas Fig. 8.27 cannot readily be correlated with the maps of chromatophore expansion.

Inside the brain, on each side, there are three regions that appear to be more or less exclusively concerned with chromatophore and skin texture control. The two lower centres, in the suboesophageal lobes, serve no other muscles. Mechanical stimulation of the walls of the posterior chromatophore lobe can produce expansion of the chromatophores in limited areas of the mantle. The cell bodies, in this lobe at least, seem to be laid out so that the motor neurons forming the cortex constitute a topographical projection of a part of the animal's surface (Boycott, 1961). The anterior chromatophore lobe controls the chromatophores in the arms. Again, the control appears to be direct, with no further synapses between the brain and the muscles.

The third, supraoesophageal chromatophore control centre lies in the lateral basal lobe, to the side of the hind part of the brain. The neurones here connect only indirectly to the periphery, via the anterior and posterior chromatophore lobes (Boycott, 1953).

Electrical stimulation of any of the three chromatophore lobes produces a dark flush on the side concerned, which may spread across to the other side of the animal; the suboesophageal centres are connected by commissures. Skin musculature is also affected; the chromatophore lobes control the erection of skin papillae and the state of the folds in the polygonal patchwork (see above) as well as the muscles of the chromatophores themselves.

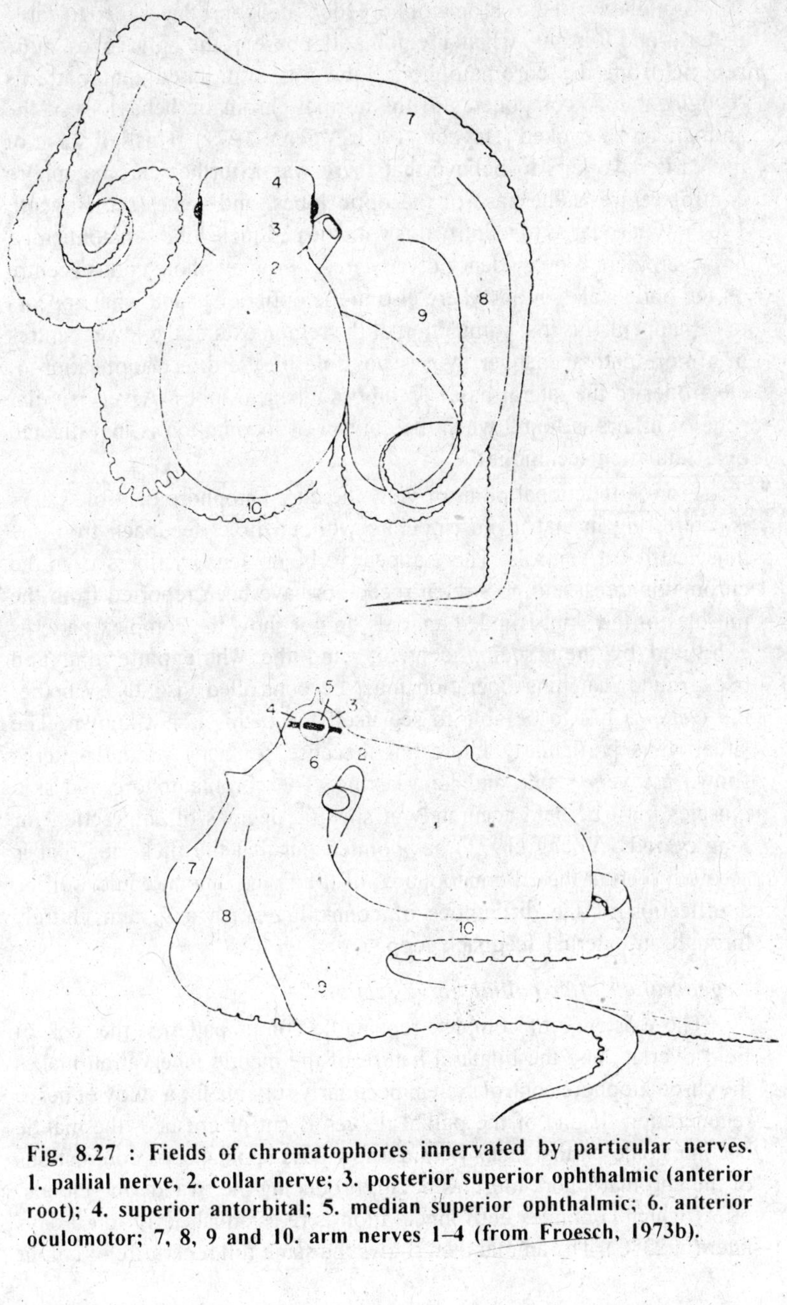

Fig. 8.27 : Fields of chromatophores innervated by particular nerves. 1. pallial nerve, 2. collar nerve; 3. posterior superior ophthalmic (anterior root); 4. superior antorbital; 5. median superior ophthalmic; 6. anterior oculomotor; 7, 8, 9 and 10. arm nerves 1–4 (from Froesch, 1973b).

Stimulation of the chromatophore lobes never produces recognizable patterns and it is only when a region at the base of the optic lobes, quite remote from the chromatophore lobes, is stimulated that patterns recognizable as components of the normal chromatic behaviour of the animal can be evoked (Boycott, 1961; Young, 1971). This led some of the earlier workers to believe that there was a further chromatophore control centre at the base of the optic lobes, and some (see Boycott, 1961) went so far as to identify this with the peduncle lobe. Anatomically, however, there is no evidence of a discrete chromatophore control centre on the optic stalks or anywhere else in the optic lobes and what appears to be happening is that stimulation of the region excites the lower centres in a more natural manner than is possible by the direct application of electrodes to the lateral basal or suboesophageal lobes. A very similar state of affairs is found when the control of locomotion is investigated by stimulation techniques.

From a functional point of view the chromatophore control system is interesting in that it must operate without direct feedback from the muscles that it controls. There appear to be no sensory fibres from the chromatophores, and no stretch receptors have been reported from the muscles of the skin. Blinded animals do not show the complex patterns exhibited by their seeing controls, and the whole patterning and background matching operation must be controlled visually. Whether the *Octopus* has to be able to see itself to do this is not known. The situation is particularly interesting because so many of the patterns shown are very stable and long-lasting. The chromatophore and skin muscles must be held accurately at specific degrees of contraction for long periods. Young (1971) has pointed out that the flickering that is so often seen in the chromatophores of the living animal could well be a reflection of the difficulties of controlling such a system entirely through an external feedback loop.

Regeneration after pallial nerve section

The existence of complex repeatable colour patterns, the lack of field overlap, and the bilateral nature of the mantle innervation makes the chromatophore control system peculiarly suitable for a study of nerve regeneration. If one of the pallial nerves is cut or crushed, the mantle becomes pale on that side. Within a few days spontaneous contractions of the chromatophore muscles begin, producing the 'wandering clouds' effect which continues until innervation is re-established 40 to 60 days later (at 23°C). The animal then shows the same full repertoire of colour patterns on the control and the operated sides, a state of affairs that can only mean that the regenerating nerves have re-established functional connections with the same groups of chromatophores that they innervated before section (Sanders and Young, 1974).

In this system it should also be possible to study regeneration after removal of areas of the skin, and just possibly after grafting in new regions. It will be most interesting to observe how regenerating nerves incorporate new chromatophores into the patterns that they reestablish; an octopus can more than double its weight in 40-60 days and will presumably have formed many more chromatophores in the interim. The skin of the mantle clearly has much to offer us as a preparation for studying the means by which nerves dentify and home in on their muscular targets.

All in all nervous systems are collections of secretory cells, differing biochemically from each other in their abilities to produce activating substances that may serve as transmitter agents (neurohormones) and also activating substances that may have other uses in the body. There later products are distinguished as neurohormones and also as appeared to neurohormones. Transported through a fluid tissue they exert their effect at a distance from their sites of origin. Cells producing neurohormones are generally designated neurosecretory, or more specifically, neuroendocrine, cell, and their products are sometimes called incretions rather than secretions. In invertebrates as in vertebrates some nonneural cells synthesize and discharge hormones but in invertebrates, neural origin, on present evidence, seems more greater than nonneural?

The criteria for neurosecretory cells, in this somewhat limit use of the world, are drived both from cytological and physiological data. The cells have morphological attributes of a neuron, with a cell body and a extended processes with neurofibrils, one of which is axon-like, their end often in swellings, in close association with some area of a body fluid frequently a blood vessel or space, the combination constituting a neurohaemal organ. This cytoplasm contain membrane-bounded vesicles or granules.

Neurosecretory cells have been described in a wide variety of molluscs, including cephalopods, several lamellibranchs, many gastropods. A cephalopod earliest cytological reports of them in invertebrates was from the opisthobranchs *Aplysia* and *Pleurobranchaea* in lamillibranchs and gastropods the presumptive neurosecretory cells lie in ganglia, usually in particular areas in most of them, but in some species only in particular ganglia. There is little experimental evidence for the role of neurohormones in the organisms in which such cells or organs have been found, but what there is relates it principally to maintenance of water balance and reproduction.

Characteristic neurosecretory cells have been found in the cerebral, visceral and pedal ganglia of the snail *Vivipara vivipara,* with numbers reaching a peak in June and July, the natural breeding reason in *Crepiduton* they have been shown in the cerebral ganglia only; the stainable granules in them are maximal in individuals at the interses stages. The have been found in cerebral, *pleural*, pedal and buccal ganglia, and buccal ganglia of the nudibranch *Hominala,* in which a possible role in water regulation has been shown by operative removal of the different ganglia. A similar function has been attributed to the neurosecretory cells of *Patella* for when the animals were immersed in hypotonic solution there cells were empty of granules, but become full of them when the limpets were in hypertonic solution. The Cerebral, visceral and paretal ganglia of pulmonates are especially rich in neurosecretory cells. In *Lymnaea staynalis,* they are grouped in the cerebral ganglia in the lateral lobes and under the mediodorsal and latero-dorsol bodies. These alocotions correspond to those also found in *Planorbarius.* A junction in water balance and reproduction has been attributed to the neurosecretory cells in *Lymnaea* observation of the cyclic changes in them led to the presumption that the medio-dorsal and latero-dorsal cells function in spermatogenesis, while and then group, located in the caudal-dorsal part of the ganglion, function in ovipostion in extirpation experiments it was found that removal of pleural ganglion resulted in an increase of water in the body. Right pleural ganglion responsible for recreation of the agent controlling water balance.

In *Helix* neurosecretory cells have been identified in all of the ganglia, including the buccal, and a neuroscretory-endocrine complex described in the opic tentacles. Amputation of the optic tentacles of the slug *Arion* is reported to result in a significant increase in the production of eggs, while injection of an extract of from the tentacle reduced this to ordinary levels, in injection of an extract of the brain into infact animal resulted in increased production of eggs.

The hypothesis has, therefore been proposed that neurosecretory cells in the brain produce a hormone that stimulates eggs production. Apparent neuropaemal organ have been observed in some of pulmonates. In *Lymnaea* and *Planorbarius,* processes from the medio-dorsal and latero-dorsal neurosecretory cells ends in finy bulbs in the perineurium of a nerve next to a blood space.

Among lamellibranchs, neurosecretory cells have been found in the cerebropleural and visceral ganglia of *Mercenaria mercenaria.* Neurosecretory cells have also been reported in the pedal ganglia of this clam, as well as in those of *Unio.* Altogether they have been found in representative of four orders in this class of mollusc.

In cephalopods, cells with the histological characteristic of cells recreating hormone have been found in optic glands, tissues of the vena cava, and in *Octopus vulgaris*, then tissue around the infrabuccal ganglion and the blood trinus. The optic a glands common to all cephalopods except *Nautilus,* are highly vasculrised and innervated bodies zying in the optic stalks this recretion has been shown to function in sexual maturation of the animal, stimulating development of the reproductive organ and this gland. The activity of the glands is suppressed by impulses transmitted to them by nerves originating in the brain. In the Octopuses *Eledone Cirrosa* and *octopus vulgaris* nerve originating in the visceral lope of the brain run to the vena cara. Their ferminations from a neuropil under the epithelial lining, which, because of uneveness in thickness of the neuropil, is ridged in the area a where the nerves end.

In *Sepia officinalis* the ridges extend into other vein adjacent to this area. Typical neurosecretory vesicles can be demonstrated in the neuropil, indicating the neurosecretory function of the nerves that constitute it in O. Vulgaris has reveculed that neurons of juxtaganglionic tissue is filled with membrane bounded, electrondense granules similar to those in cells with established neuroendocrine function. The oxon of there neurons, also often filled with granules, end in direct apposition to the basement membrane of the buccal rives. Although, nothing is known of the nature of the recretion of this neurons, or of its effects, both this ultrastructure and this anatomical relationship are strongly suggestive of a neuro-endocrine function.

SENSE ORGANS AND THEIR FUNCTION

In no other phylum except the chordate does the evolution of nervous system cover such a wide span as in the mollusca. An early mollusca such a *chiton* is a slow-moving creature with a nervous sys. best compared with that of a flatworm. But the highest production of the mollusca. The cephalopod brain and senxe organs are revilled only among the vertebrates themselves.

The nerve cells in the longitudinal coards soon become localized in distinct ganglion. In the earliest gastropods we find already a pain of peural ganglion concetrated at the sides of the nerve signs at the head of the pallial cords. The cords themselves are reduced to connectives, forming along visceral loop with a parietal ganglion becoming distinct on either side and a single or sometimes paired visceral ganglion where the loop crosses the gut behind. Haliolis or trochus may be chosen to illustrate this early condition with no distinct cerebral ganglion and the pedal ladden still concentrated with the cerebral ganglion are the next to appear, while in most mesogastropods the neurones of the pedal cords have been drown into the nerve rings as a pain of ganglia, while the cords as such disappear. Thus we find the typical gastropods nerve ring, with two cerebral ganglion dorsolly and two pleural and two pedal ganglion below. The cerebral and pedal pairs are linked in the middle line by cerebral and pedal commissures, and at either side the cerebrals are linked with pleurals and pedals, and the pleurals with pedals by short connectives.

From the pleural ganglia springs the visceral loop, connecting to the nerve being the parietal ring ganglion, from which nerves pags to the pallial organs. During the torison of gastropods visceral loop is twisted into a figure of the condition known as *chiasmaton* every. The right particle ganglion now crosses above the gut to lie upon left being called the supraintestinal ganglion (through it really lies near the oesophagus). The original left parietal ganglion passes and sheath the oesophagus to lie on the right as the subintestinal ganglion.

In the Montocardia, the supraintestinal ganglion takes on the innervaltion of the remaining left side of the mantle cavity including the surviving gill. It also supplies the osphradium, either directly or through outlying osphradial ganglion. On either sides the pleural and parietal ganglion may established a direct link by a connectives known as zygonevery.

In the higher gastropods all the ganglion finally come to lie in the nerve ring and the visceral loop is lost. First in the higher prosobranchs such as *Buccinam* the subintestinal ganglion is withdrwan across to its original side to lie close against the left pleural ganglion. Then the supraintestinal gargli on moves back over the oesophagus to becomes attached to the light pleural membe.. In the pulmonates, the whole visceral loop is shortened with the incorporating of the visceral ganglion in the ring with now consists of nine large ganglia with the sonalles buccal ganglia still attached by connectives. Torison of the visceral ganglion mass has not been abolished in the pulmonates, but the evidence for it in the twisting of the visceral loop-can no longer be clearly, and in *helix* the nerve ring is very compact with its ganglion in direct contact.

The opisthobranchia, however have completely reversed tosion. During their phylogency the mental cavity has moved backwards along the right side towards its original site and the visceral loop actually twisted. All the ganglion lie in the nerve ring thing, usually above the oesophagus and unliked only by the pedal commisure below the early opisthobranch *Acteon* and the primitive pulmonate chilina are interesting transitional firms in the reduction of the visceral loop.

Aplysia

Aplysia californica has a pair of eyes at the base of the rhinophores; the eyes are about 600 μm in diameter and can be withdrawn into folds in the skin. The eye is again of the closed vesicle types with a large lens abutting on the retina so that, as in *hermissenda*, it seems unlikely that a sharp image can be formed.

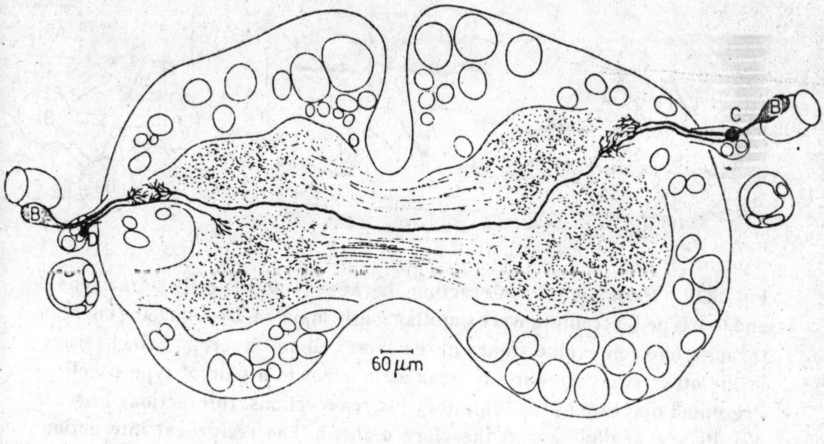

Fig. 8.28 : *Hermissenda brain,* showing second-order visual cell projecting contralaterally. Cell C receives input from isolateral type B cells a near the some and contralateral type B cells on its terminals; it inhibits contralateral optic ganglion cells (D) (Alkon, 1973a).

The functional organisation of the retina is not fully understood. In contrast to *Hermissenda,* there are about 3,600 receptor cells, as well as about 1,000 "secondary cells," situated at the periphery of the retina; there are also pigmented cells and glia. Some of the receptor cells bear loosely organised microvilli but, in *A. punctata*, other are "ciliary" in that they contain equal numbers of cilia and microvilli (Hughes, 1970a). The receptor cells may be presumed to make the major contribution tot he optic nerve, but some must derive from the secondary cells. The proposal, made on electrophysiological grounds (Audesirk, 1973), that the receptor cells are coupled by electric junctions to the secondary cells and that it is their axons alone that contribute to the optic nerve has not been substantiated by a recent Procion Yellow study (Jacklet, 1976).

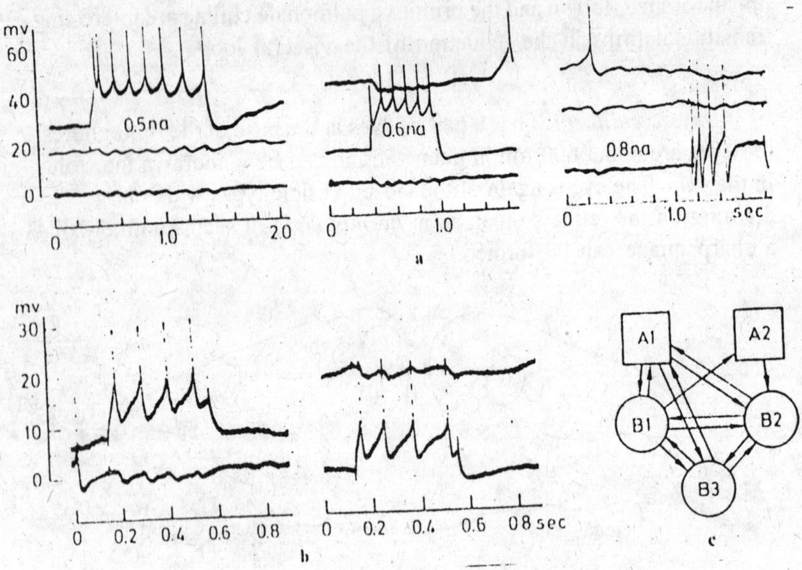

Fig. 8.29 : *Hermissenda:* interactions between A and B cells a One type A and two type B receptors were simultaneously impaled. Depolarizing currents through one cell evolke spikes in the same cell and hyperpolarizing waves in the other two receptors. b Some as in a but in a pair of type B cells. c Presumed diagram of the inhibitory interconnections. Interactions between A cells are negligible and therefore omitted. The reciprocal interactions between all B cells are well established (Alkon and Fourtes. 1972).

It is the retinal neuropil that has proved refractory to analysis. The synaptic organisation is not clear, and the situation is complicated by the presence of two classes of vesicle, one of which may contain dopamine (Luborsky-Moore and Jacklet, 1976a, 1977; Lon and Jacklet, 1977), the other a polypeptide that may have a neurosecretory role in the regulation of circadian-driven locomotor behaviour (Harf *et al.*, 1976; Rotiman and Strumwasser, 1976).

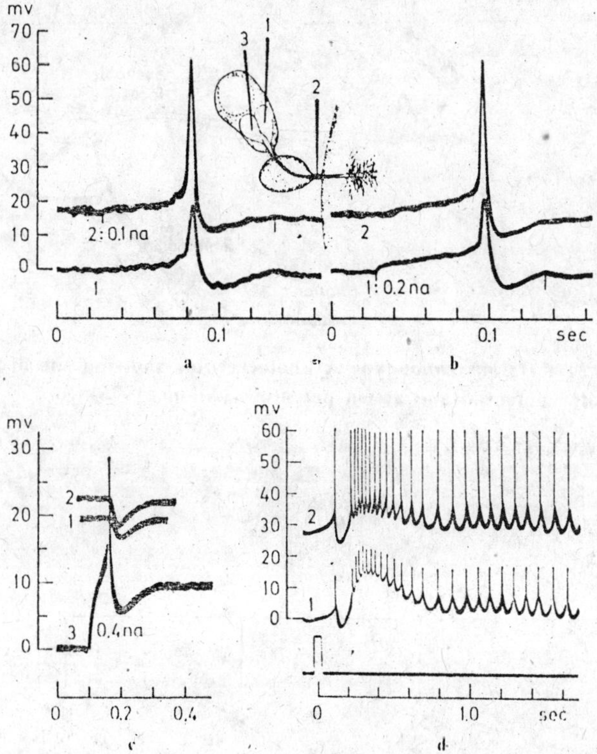

Fig. 8.30 : *Hermissenda*: responses recorded from some and axon of same cell. The distance between the two electrodes was 80-100 μm. a Respones to a step of depolarizing current through axonal dectrode. b Response to a step of current through the some electrode. In either case the spike is larger and earlier in the axon. c Two electrodes were inserted in the some and axon, respectively of one receptor, and a third electrode was inserted in the some of a second receptor. A step of depolarizing current through this third electrode evoles a spike in one receptor and a hyperpolarizing synaptic potential in the other. The synaptic potential is larger and has shorter time to peak in the axon. d Responses to a flash of light. The generator potential is larger at the some (*lower trace*) while the spikes are larger at the axon. Downgoing bars in a and b indicate timing of current steps (Alkon and Fourtes. 1972).

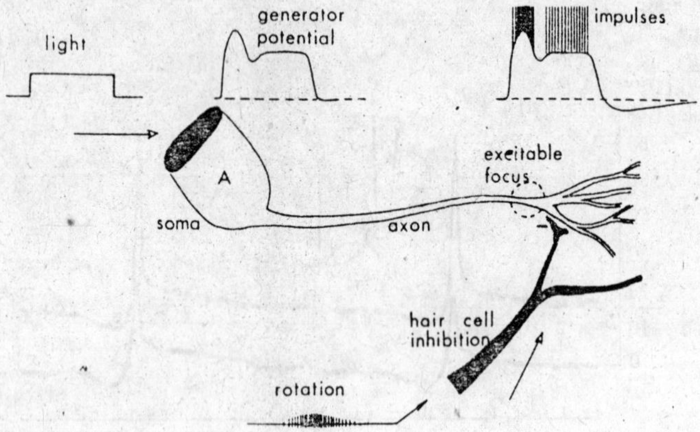

Fig. 8.31 : *Hermissenda*: type A photoreceptor showing site of origin of generator potential and action potentials (Alkon, 1976).

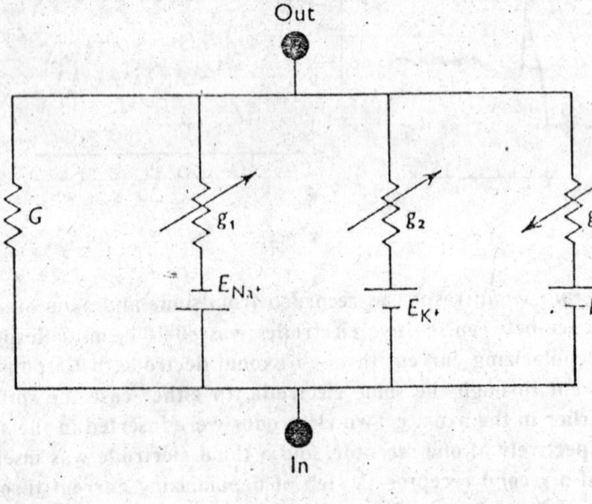

Fig. 8.32 : Scheme of the electric circuit that must operate in the *Hermissenda* photoreceptor. There are three light-sensitive conductances g_1, g_2, g_3 and a fixed conductance (i. light increases g_1 and g_2 but decreases g_3. The batteries represent equilibrium potentials for either K+ or Na^+ (Detwher, 1976).

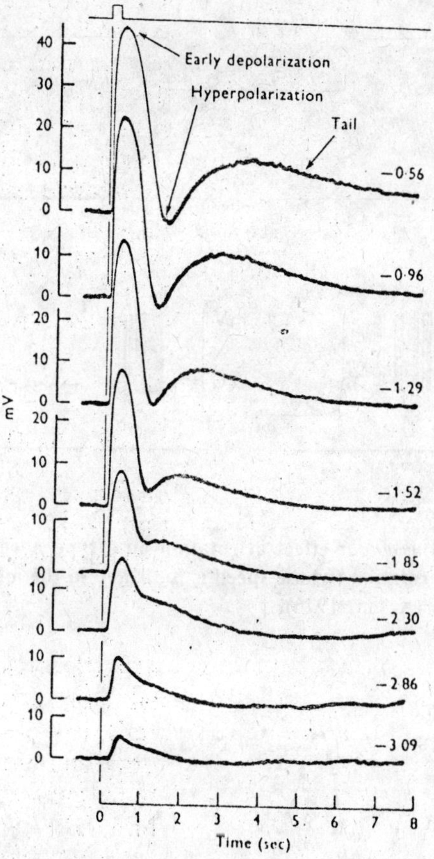

Fig. 8.33 : *Hermissenda*: intracellular records of photorespones to flashes of increasing intensity delivered at 63-s intervals. Zero on left-hand scale represents - 50 mV; stimulus intersities on right (Detwher, 1976).

There is a typical ERG, which changes polarity as the recording micropipette penetrates of the receptor layer and which is not affected by tetrodotoxin (TTX). Intracellular recording reveals a resting potential of 40-50 mV, yielding graded potentials of up to 55 mV on illumination. In the dark-adapted retina there are discrete bumps on the graded response to brief flashes. Other cells in the eye respond to light with action potentials that are in synchrony with the extracellularly recorded compound action potentials (CAPs). Yet others are tonically active and are inhibited when illuminated (Jacklet, 1969b).

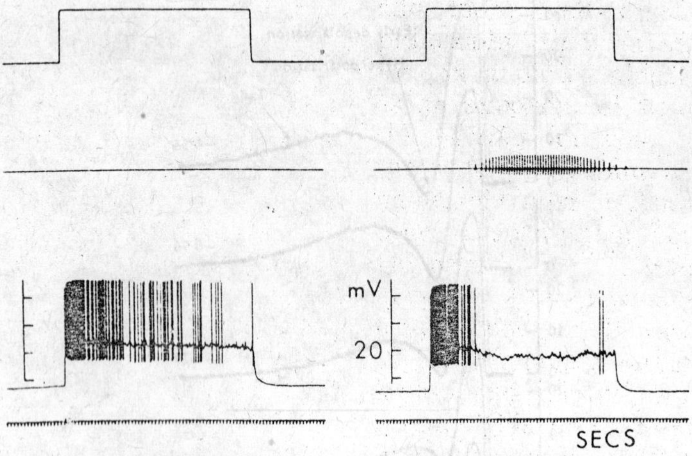

Fig. 8.34 : *Hermissenda*: effect of rotation on a type A cell. Left: alone. Right: light and rotation (whose speed is indicated in the middle traced by monitor signals) (Alkon, 1976).

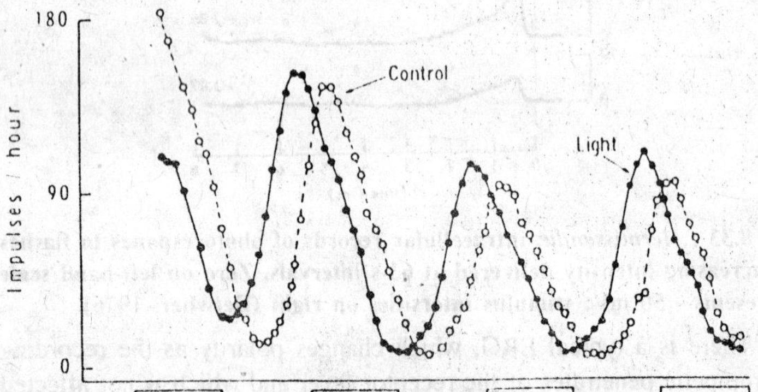

Fig. 8.35 : *Aplysia*: cireadian rhytham of optic nerve discharge in the isolated eye. The experimental eye (●) was exposed to 6-h light (black bar) and then kept in constant darkness as was the control eye (O) throughout. It will be seen that the light has produced a phase advance in the experimental eye. The open bars at the bottom represent the projected (12 h) light portion of the cycle (Eskin, 1971).

There is evidence (Luborsky-Moore and Jacklet, 1976b) that the optic nerve contains deferents (perhaps cholinergic), and it has been suggested that these act upon the secondary cells, whose function remains to be established (but see Jacklet, 1976). It could be that they are involved in the production of circadian rhythms, for the most remarkable feature of the *Aplysia* eye is that if it is exclosed and isolated in continual darkness and at constant temperature, it will fire spontaneously for up to 2 weeks and in its frequency and amplitude changes, exhibit the clearest possible circadian rhythm (Jacklet, 1969a). There is a considerable literature on the spontaneous discharge of the isolated *Aplysia* eye but, as most workers have considered it primarily as a system for examining circadian rhythms, we can only touch upon this here.

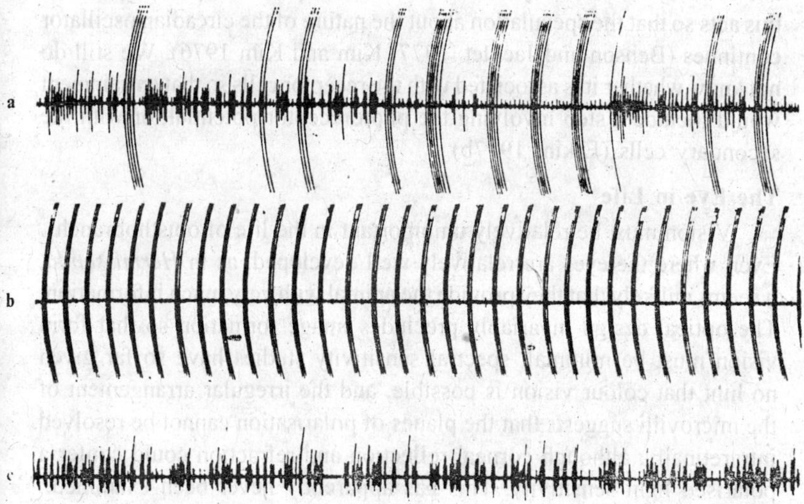

Fig. 8.36 : *Aplysia*: influence of efferent fibres on the circadian rhythm. a Record from eye attached to cerebral ganglion; b Concurrent recording from eye isolated from the ganglion; c Activity in optic nerve attached centrally with eye removed (Eskin, 1971).

How the activity of an isolated eye alters during a 25-h period, with a peak production of CAPs near "projected" dawn. Eskin (1971) found that it is possible to alter the period of the rhythm (*i.e.*, to "entrain" it) by exposing the eye to a light "pulse." *in vivo*, a 13-h advance of the light/dark 12:12 cycle can complete entrainment after only one exposure of the animal tot he new regime. Yet four or five exposures are required *in vitro* before entrainment is complete, suggesting there must be centrifugal regulation of the rhythm in the infact animal Fig. 8.36. Eskin subsequently showed that electrical stimulation of the thinophore or tentacle nerve, prior to a light flash, decreases the response latency, but the function of this efferent activity is not clear.

Eskin (1974) has also shown that phase shifts in the circadian rhythm can be elicited *in vitro* by "pulses" of high K^+, so that evidently the mechanism underlying the circadian rhythm is coupled to membrane properties. There is good evidence for the Na-K pump being involved (Eskin, 1977a; Eskin and Corrent, 1977), as well as Ca^{2+} and Mg^{2+}. Recently it has been shown that the circadian rhythm can be phase shifted, *in vitro*, by low concentration of 5-hydroxytryptamen (5-HT) and that 5-HT is present in the eye in high concentrations (Corrent *et al.,* 1978). Unfortunately, however, we do not known precisely where this acts so that the speculation about the nature of the circadian oscillator continues (Benson and Jacklet, 1977; Kim and Kim 1976). We still do not know whether it is associated with the receptor cells so that entrainment would include a step involving the photoreceptor potential, or with the secondary cells (Eskin, 1977b).

The Eye in Life

Vision must be relatively unimportant in the life of opisthobranchs. Even where the eyes are relatively well developed, as in *Hermissenda*, it seems unlikely that they provide the animal with very much information. The optical design invariably precludes image formation so that form vision must be minimal; spectral sensitivity studies have so far given no hint that colour vision is possible, and the irregular arrangement of the microvilli suggests that the planes of polarisation cannot be resolved intraretinally, although corneal reflection and refraction could confer a polarised light sensitivity. This has apparently never been examined.

Despite this, light must play a considerable role in the life of most opisthobranchs. Thus, Rahat and Monselise. (1979) have described the complex behaviour of *Elysia*, which harbours symbiotic algal chloroplasts in its parapodia. According to the intensity of the light, these are either exposed or concealed, but this response only partially involves the eyes; it largely depends on extra-ocular photoreccptors.

Extra-Ocular Photoreception

In *Aplysia* there are photoreceptors in the mantle skin (Lukowiak and Jacklet, 1972), in the oral tentacles (Block, 1975, cited in Lickey *et al.,* 1976), and in identified neurons in the cerebral ganglion (Block and Smith 1973) and the abdominal ganglion (Arvanitakiand and

Chalazonitis, 1961). Those that have received most attention are the photosensitive neurons, which (through the work of A.M. Brownand his colleagues) have yielded important data about photoreceptor processes.

In the abdominal ganglion, three neurons have been studied: R2. R15, and 1,10 (Kandel., 1976). The R2, or giant neutron, contains β-carotenes in membrane-bound cytoplasmic pigmented granules learned lipochondria, which also contain calcium (Krauhs *et al.*, 1977; Baur *et al.*, 1977). Intracellular injections of Ca^{2+} simulate the photocurrent, while ethylene glycol tetra-acetic acid (EGTA) blocks it and, in an elegant paper, Brown *et al.*, (1977) suggest, on the basis of several features, that calcium ions may act as an intermediary in the establishment of a current. The light-evoked potential in R2 has been measured under voltage clamp (Brown and Brown, 1973) and evidence amassed that it can be attributed to an outward flow of K^+. Presumably, the effect of light is to act on the pigment, releasing Ca^{2+}, which, in turn, increases the K^+ conductance. Although unusual by vertebrate standards, photoresponses utilising the K^+ channel occur elsewhere in molluscs.

R15 the bursting pace-maker neuron, also responds to light by a slow hyperpolarisation associated will an increase in membrane conductance. The spectral sensitivity curve of this cell has a single peak al 485 mm. close to the value obtained for the K 2 cell (490 nm). This cell itself exhibits a circadian rhythm that may he mediated by the eyes.

The ventral photosensitive neuron. L. 10. has been the subject of two very important recent papers. This cell has a response threshold 1000-fold lower than that of the R 2 cell, but its properties are otherwise similar. Again, increases in membrane conductance are involved, though the light-activated K^+ conductance is neither time nor voltage dependent. Again stress the role of diffusion of Ca^{2+} in the photoresponse of this neuron and develop an important model of the photoreceptor process (*rf.* the model developed for the turtle photoreceptor by Baylor *et al.*, 1974).

The function of the various photosensitive neurons is not known, but Brown *et al.*, (1977) have data showing that the skin overlying them is sufficiently transparent to the ambient light likely to be encountered in the sea. Some neurons may regulate the locomotor rhythm, which has been shown to exist in *Aplysia* by Lickey and his collaborators (1977), who, unlike most recent experimenters on *Aplysia*, disposed of the eyes and examined the animal. Intact *Aplysia* are diurnal in habit with an onset of activity just before dawn. Eyeless *Aplysia* respond on the whole like intact animals, even at low levels of illumination, although there are behavioural differences, notably the great reduction of pre-dawn activity. Other workers have obtained results slightly different from these, probably for methodological reasons. But, if substantiated, the findings of Lickey *et al.*, would indicate that the eye cannot be the only photoreceptor coupled to locomotor rhythm nor the only site of a circadian oscillator. Their results are a salutary reminder of the difficulties of assessing tie role of visual input in the life of molluscs, many of which are replete will) extra-ocular photoreceptors.

To emphasise tilis point, consider again *Onchidium*. with its paired cephalic eyes and 46 "dorsal eyes". Fujimoto *et al.*, (1966) have shown that if all these dorsal and cephalic eyes are removed, the animal *still gives a shadow response.* Perhaps this is mediated by one of the light-sensitive neurons in its CNS that have been examined physiologically by Gorow (1975).

Closely related to the scallops are the file shells, and one of these, *Lima scahra*, has been examined morphologically and electro-physiologically by Bell and Mpitsos (1968) and Mpitsos (1973), who report many similarities to the *Pecten* eye. McReynolds (1976) has confirmed and extended this work using intracellular recording: there may be synaptic interaction in this retina Mpitsos (1973) has noted the existence of a second pigment (λ max 520 nm) whose function may be to increase the sensitivity of the "off" response, and Gorman and Cornwall (1976) now have complementary data for *Pecten.*

Extra-Ocular Photoreception

We have seen that the loss of cephalic eyes has been followed by the acquisition of pallial eyes. Yet, despite their being more recent in evolutionary terms, and despite their complexity, they are not always able to provide all the photic information needed by the animal. Thus Mpitsos (1973) has good evidence for "off" responses from dermal receptors in *Lima*, even after the eyes have been extirpated like cephalic eyes, pallial eyes have also been supplemented by extra-ocular photo-receptors.

There is a widespread dermal light sense in bivalves, often mediated by diffuse, unspecialised single cells, as in *Mya's* siphon. It usually serves a variety of protective renexes, such as siphon withdrawal or shell closure.

Kennedy (1960) has examined in detail the shadow response of *Spisula solidissima,* which has a typical "off" neuron, in the pallial nerve, that is inhibited by while light. Adaptation to strong blue light, however, causes an "on" response to red light kennedy (1960) interpreted this as evidence for two opposing mechanisms, but it could be the result of a single pigment giving different electrical responses to light of different wavelength, as in the distal retina of *Pecten (Cornwah* and *Gorman,* 1976). No photoreceptor has been identified morphologically in this *Spisula* nerve, and it is thought that a restricted part of the axon must itself.

Single axons in the siphonal nerve of the clam *Mercenaria mercenaria* also exhibit "off" responses very like those of *Pecten* distal photoreceptors. Wiederhold *et al.,* (1973) give a very full account of this response which may be mediated by "ciliary" receptors. Their paper emphasises the long lime contains of "off" responses in bivalves: The "off" response to constant intensely stimuli continues to grow with stimulus duration for over 8 minute.

Summary

Cephalic eyes, homologous with those of gastropods and cephalopods are found in very few genera; in *Mytilus,* it is claimed that the adult cephalic eyes may be rhabdomeric.

Pallial eyes have been developed (probably independently) in several groups. Very elaborate eyes are found that project to a special part of the viscero-parietal ganglion, emphasising their non-cephalic nature. In *Pecten* the eyes have a double retina, one rhabdomeric and one ciliary; the receptors in these have different functions, partly related to the optics. Many authors have attempted to relate these to the microvillar or ciliary organisation.

Table 8.6 : Size and cell numbers of molluscan eyes[a]

Genus	Diameter	Receptor cells (approx. number)
Onithochiton	50 μm	100
Strombus	1.2 mm	50,000
Hermissenda	80 μm	5
Tritonia	250 μm	5
Aplysia	800 μm	4,000-5,000
Helix	1 mm	4,000
Pecten	1.5 mm	10,000
Nautilus	15-20 mm	4×10^6
Octopus	25-30 mm	20×10^6

[a]Based on data from various authors cited in the text

Considerable attention has been paid to the hyperpolarising "off" cells in the distal retina of *Pecten*. The elegant experiments to *Gorman* and *McReynolds* have shown that this photoresponse is due to an increase in the membrane permeability to K^+, perhaps resulting from an increase in Ca ion permeability.

It is appropriate here to summarise the major morphological features of representative mollusc eyes (Table 8.6). Data on receptor-cell numbers alone serve to mark off the "lower" (but still so complex!) molluscs from the vertebrate-like cephalopods, to which we now turn our attention.

Cephalopoda

Living cephalopods fall into two groups, the Coleoidea containing the cuttle fish, the squids, the octopuses and the vampyromorphs (Voss. 1977), and the Nautiloidea a relict group represented by a single genus, *Nautilus*. There are extensive and fundamental differences between nautiloids and colcoids and, as these include the nature of the visual system, it is convenient to treat *Nautilus* separately.

Nautilodiea

The large, paired eves of *Nautilus* are unusual in that they are open have no cornea or lens and appear to function like a pilihole camera. The pupil is generally round (2 mm diameter) or elliptical, but it may become crescent-shaped (Hurley *et al.*,. 197H). The vertical line running from the pupil remains vertical as the animal moves, for there are compensatory eye movements in the pitching plane at least (Packari), personal communication).

Nautilus has a well marked pupillary response that is unusually slow, even by cephalopod standards; after a decrease in light intensity, the pupil takes 50s to open fully but, after an increase in light, constriction requires no less than 90s (Hurley *et al.,* 1978). Although illumination of one eye is followed by constriction contralaterally, the time course of this constriction is different.

The chamber of the eye, which is open to the sea, is filled with a jelly-like substance. There are receptor cells and large supporting cells, both *μm*; and bear microvilli distally. These are not as regularly arranged as in the colcoid retina, but they form an extensive array of tubules, many of which have their long axes perpendicular to the incident light. In the inner segment of the receptor cell, mycloid or somal bodies are extraordinarily well developed, but whether this is because there is retinochrome in the *Nautilus* eye is not known. There appears to the little synaptic interaction in the layer below the receptor cells, where axons leaves the receptor cells to form the optic nerves. This contrasts markedly with the situation in *Octopus* and suggests that there may be little sharpening of contour in this eye. Though no precise estimates are available it is clear that there are a large number of cells in the retina. An approximate estimate based on data in *Merton*, and *Barber* and *Wright* and on *Gilpin-Brown's* photograph Fig. 8.37 yields a Fig. of 4×10^6 cells per retina. This figure may be regarded as conservative but, even so, it is of a different order of magnitude from that of the other molluscs we have so far considered Table 8.6, though lower than that of the *Octopus* retina. The optic nerves differ from those of nearly all coroids in that they do not exhibit a dorsoventral chiasma (*Young*. 1965). They may be up to 10 mm long, and they project to an optic lobe, whose organisation appears to be far simpler than in *Octopus* or *Loligo* (*Young*, 1965).

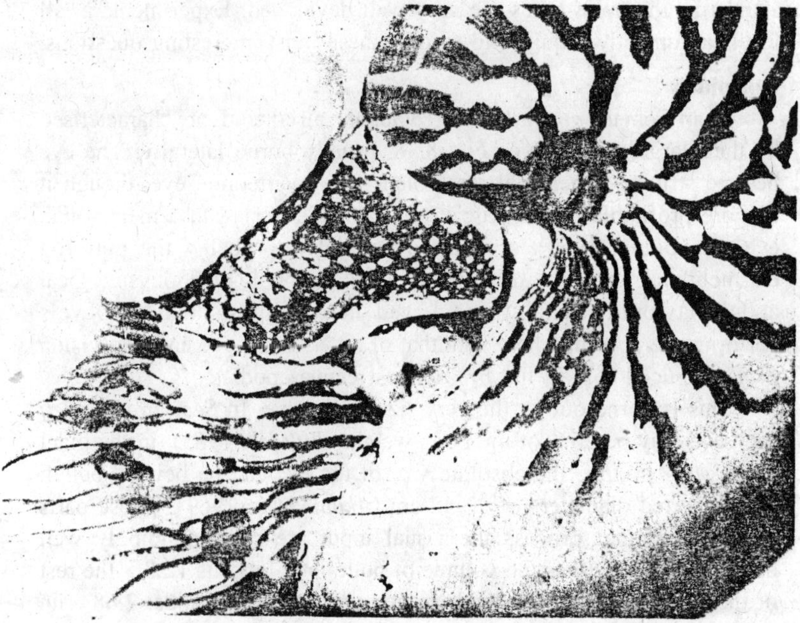

Fig. 8.37 : *Nautilus*: eye of a living animal; the dark line running from the pupil is normally maintained vertically eye length approx, 25 mm (Fig. kindly supplied by Dr. J.B. Ghpin-Brown).

There is much we need to know about the *Nautilus* eye. Are there regional differences in the retina? Are there two photopigments? Is there polarised light sensitivity? What of light and dark adaptation? Are there deferents in the optic nerves? Are there extra-ocular photoreceptors? So little is known about the behaviour of *Nautilus* that the role of the eye in life remains speculative. Is the eye poor at form vision, as is generally supposed (*Young*, 1965a), or capable of reasonable resolution by virtue of a pupil acting as a low pass filler (*Hurley et al.*,. 1978)?

Nautilus has well-developed olfactory receptors (*Young*, 1965), and the undoubted simplicity of the eye, the pupil size, the lack of dioptic apparatus, the absence of a chiasma, and the small and more simply organised optic lobe have led to the assumption that their vision is relatively poor. This is true only when they are compared with the coleoid cephalopods. When compared with the rest of the molluscs, it can be seen that eye is an advanced organ Table 8.6. There are something like lour million closely packed photoreceptors per eye and, by gastropod standards, the optic lobes are very well developed. Experiments at all levels argumently required to resolve these most interesting questions.

Colcoidea

Apart from the *anomalous Cirrothauma*, all coleoids are characterised by their large, paired eyes Fig. 8.38, usually borne laterally. The eye bears a striking superficial resemblance to a vertebrate eye, though it is more profitable to compare it with a fish's than with a mammalian eye (*Pumphrey*, 1961); the closed parallel is probably the pure-rod elasmobranch eye (*Packari*, 1972). The morphological physiological and behavioural evidence reviewed here suggests that the eye's performance is comparable with that of a vertebrate eye and that vision plays a crucial role in the life of most cephalopods.

This is borne out by the very size of the eye. In *Sepiola rondeleti* each eye may be 25% of the body weight (*Bullock*, 1965); in the giant squid, *Architeuthis,* the absolute size of the eye can be beyond belief, with reported diameters of 37-40 cm (*Akimushkin*, 1963). Those parts of the brain that process the visual input are correspondingly well developed. In *Octopus* the volume of both optic lobes is 1.28 × the rest of the brain, and this fig. lends to he even larger in squids: 2.68 × in *Loligo vulgaris* and 4.04 × in *Pyroteuthis margaritifera* (Wirz, 1959).

One very interesting feature of cephalopods is that many deep-water forms (*e.g.*, *Vampyroteuthis* and *Bathyteuthis*) possess large eyes though they occur at depths that in fish would be associated with a reduction in eye size (*Packari* 1972).

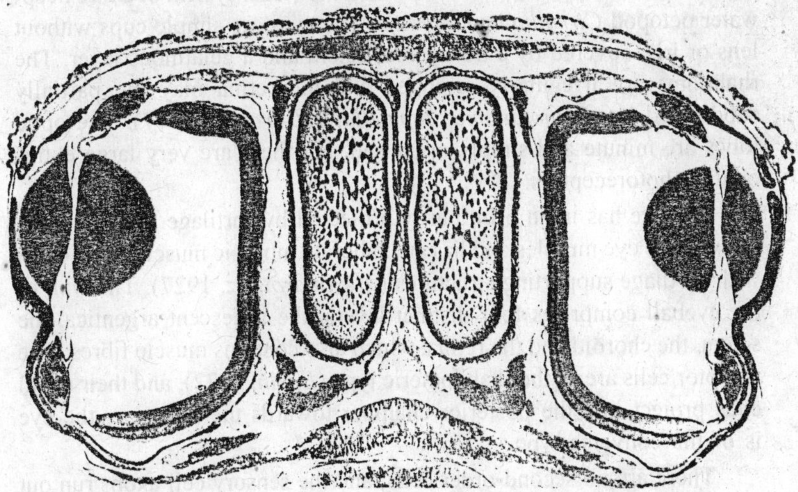

Fig. 8.38 : The eyes and optic lobes of a young *Sepiola* in transverse section. The rectilinear from of the dorsal retina is presumably a fixation artifact. Cajal silver X 12 (Original).

Cihin (1910) has magnificent drawings of may fascinating cephalopods eyes: in squids like *Corynomma, Toxeuma* and *Bathothauma,* there are stalked eyes. *Sandalops* has curiously shaped eyes pointing obliquely downward; the pelagic octopod. *Amphitretus,* has a pair of tubular eyes oriented directly upwards like twin telescopes. The bathypelagic squid, *Bathyteuthis (= Benthoteuthis),* must also stage vertically upwards, for it has a remarkable ventral "fovea," where the closely relates eye form to age and to habit in *Sandalops* and *Taonius.*

One particularly bizarre cephaloped is the squid *Histioteuthis* (= *Calliteuthis*), which has one eye very much larger than the other. The lens of the larger eye is bright yellow, but the small lens is transparent down to 310 nm, *Muntz* (1977b) has suggested that the large lens function to "break" the ventral luminescent countershading system of squids or fish, for a yellow lens would attentuate light from the photophores less than light from the surface so that the animal producing it would appear brighter. In support of this theory is the observation that while the small eye is directed down-wards, the layer eye is directed upwards and so would gaze at the undersides of such bioluminescent animals (*Young,* 1975a).

Aldred et al., (1978) have examined the visual system of a rare deep-water octopod *Cirrothauma murrayi.* The eyes are simple cups without lens or iris, covered by a transparent sclera and a gelatinous layer. The rhahdoms are irregularly arranged and the retina may the partially subdivided (*ef. Bathotanama, Japetella;* J.Z. *Young*, 1977). The optic lobes are minute and simply organised, but there are very large extra-ocular photoreceptors.

The eye has in an orbit partly protected by cartilage. It bears a set of extrinsic eye muscles and has an elaborate intrinsic musculature, often with cartilage supporting structures (*Alexandrowicz*, 1927). The wall of the eyeball comprises four distinct layers, the iridescent argentica, the sclera, the choroid and the retina, which also contains muscle fibres. The receptor cells are of the rhabdomeric type (*Eakin*, 1972), and their distal ends project into the posterior chamber towards the lens; *i.e.,* the eye is of the converse type.

There are no second-order cells, and the sensory cell axons run out of the back of the eyeball in the optic nerves, which after undergoing a dorsoventral chiasma, enter the optic lobe of the brain to terminate in the "deep retina" of *Cajal.* The optic nerves contain centrifugal fibres, some of which end in the retina. Others, however, serve the intrinsic eye muscles and those of the ciliary ring, which hears the lens and which may he important in accommodation. They are also said to carry the important pathways for controlling the iris. The iris has separate muscles for constriction and dilation: its fine structure has been examined by *Froesch* (1973).

The posterior and anterior chamber (*Boycott* and *Young* 1956a) are well defined. Both contain fluids with a higher K^+ concentration than sea-water. There is an anterior chamber organ that may function to regulate pressure in the eye. The so-called cornea is not continuous with the sclera; it is essentially a transparent "window" in the head skin. It is vary then and, as in fish, cannot play a very important part in image formation by refraction. There are differences, once held to be of great taxonomic significance, between the arrangement of the corneal folds in different groups of coreoids.

The Lens and the Pupil

The large spherical lens is composed of crystalline proteins very similar to those found in the vertebrate lens, and it produces a sharp (though reduced) image anywhere on the retina. Cephalopod lens shows almost no spherical aberration and must therefore be optically non-homogeneous. Presumably, different proportions of the constituent protein fractions are laid down in the concentric layers formed as the lens grows (*Arnold,* 1967), so that there is a gradation in refractive index from high at the centre (1.53) to low at the periphery (1.33). *Brahma* (1978) has considered the development of the lens immunologically and *Delamere* and *Duncan* (1977) its membrane potential.

In surface squids, such as the flying form *Onychoteuthis banksi,* the lens is bright yellow and absorbs strongly in the blue and ultraviolet, while in deep-water forms such as *Spirula* and *Chiroteuthis,* the lens is transparent down to 310 nm.

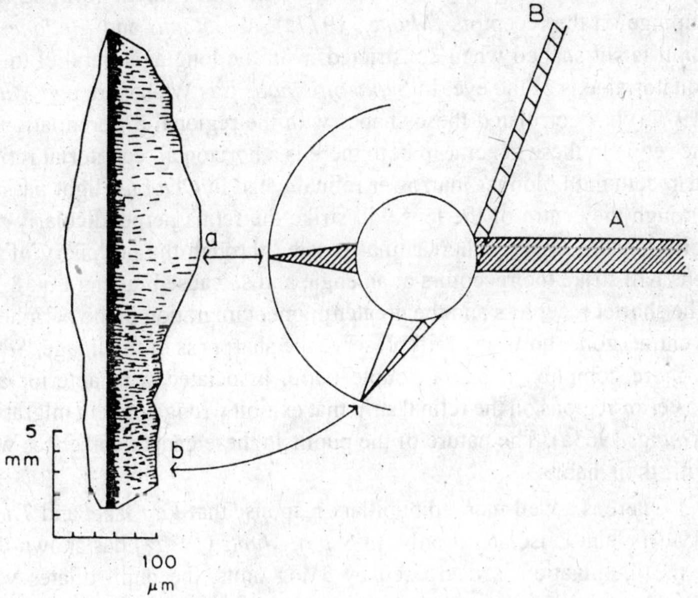

Fig. 8.39 : *Octopus*; pupil and retina. On the left is depicted a straightened-out vertical section through the centre of the light-adapted retina. The actual length of the retinal cells (inner and outer segments) is shown, as well as the position and density of the sereening pigment. On the right, the vertical section through the constricted pupil shows that light reaches the centre of retina (a) perpendicularly but the dorsal and ventral (b) regions obliquely (*Muntz.,* 1977).

Accommodation is not fully understood. *Alexandrowicz,* (1977) showed that electrical stimulation can after the shape of the enclosed eye, contraction of the muscles in the retina and sclera protruding the lens outwards, contraction of the ciliary muscle pulling it inwards. But there is no evidence that this system is used in the living animal; the abundant musculature could have another function, for example to cushion any distortion of the eyeball arising as a result of acceleration or other forces (*Alexandrowicz,* 1978). Certainly the short focal length of the lens (2.5 × lens radius; Matthiessen's ratio) and the great length of the outer segments (200-300 μm) imply that accommodation needs must be minimal in cephalopods.

Because the lens obeys Matthiessen's ratio, the design of the pupil will affect the quality of the image found on different parts of the retina by affecting the brightness and the angle at which image-forming rays impinge on the receptors (*Muntz.,* 1977a). In *Octopus* and *Eledone* the pupil is slit shaped when constricted, with the long axis parallel to the equatorial axis of the eye. In *Sepia officinalis* it is W shaped, and *Muntz* (1977a) has correlated these shapes with the regional differentiation of the retina in these genera. In both there is a horizontal, equatorial retinal strip containing longer, narrower retinal cells. In *Octopus*, light passing through the centre of the lens will strike the retina perpendicularly and form a sharp image, whereas that passing through the periphery of the lens will strike the receptors at an angle and so cause blurring Fig. 8.39. The shorter receptors and the greater pigment migration in the peripheral (ventral) zone, however, will preserve the sharpness of the image. *Sepia* is more complex; it has a double pupil, associated with anterior and posterior regions on the retinal strip that exhibit wrong pigment migration (*Young*, 1963a). The nature of the pupils in these genera correlates well with their habits.

There is a well-marked pupillary response that *Van Weel* and *Thore* (1936) claim is isolateral only. In *Sepia, Muntz* (1977a) has shown that if the illumination is decreased by 3 log units, the pupil dilates very slowly, taking 30 s to reach full size; later, the pupil contracts again, perhaps because of pigment withdrawal. On increasing the illumination (by the same amount), the pupil closes within 5s, which is still slow by mammalian standards. The reasons for this are not clear. In several genera (*e.g., Octopus*) the pupil can be dilated to give the eye prominence during threat or courtship (*Wells* 1966); this is true of *Nautilus* incidentally (*Hurley et al.,* 1978). The pharmacology of the iris muscles has never been systematically explored *Chichery* and *Chanelet* (1972) have some evidence that the constriction pathway may be cholinergic.

Eye Movements

The cephalopod eye has a complex set of extrinsic eye muscles, and eye movements are very obvious in captive animals. The muscles themselves have been described by several early workers. In *Octopus* there are only six muscles, arranged as four "recti" and two "obliques" (*Packard*, 1972), but in *Sepia* there are as many as 13. These muscles can turn the eye in any direction. They con contain many motor endings, these being the terminals of fibers originating in the lateral pedal lobe (*Young*, 1971, 1976); this is a distinct "oculomolor centre" and receives input from the statocyst. In *Octopus* there are some 3,000 nerve fibres (distributed among six nerves) running to the eye muscles on each side (Young, 1971).

The eye moves as a result of visual and mechanical influences. Visually induced eye movements include Saccadic and pursuit movements, convergence movements, and reflex movements (or optiokinetic mystagmus). Saccadie and pursuit movements have been observed in *Octopus* (*Messenger*, 1965) and *Sepia* (*Messenger*, 1968). Such movements imply that there may be a specialised area within the retina, a "fovea". Convergent eye movements are especially well defined in *Sepia*, which fixates its prey binocularly prior to attacking it (*Messenger*, 1968). During visual fixation the influence of the statocyst on the eye's position can be completely overridden.

Reflex (or compensatory) eye movements can be elicited experimentally by the nystagmus apparatus (*Packard*, 1969; *Messenger*, 1970; *Messenger et al.*, 1973). Stripes differing in brightness elicit nystagmus and optomotor responses. The mystagmus, which shows the characteristic slow and fast components, persists after the statocysts are removed (*Dijkgraaf*, 1961); *Messenger*, 1970). In *Sepia* the oculomotor response has been quantitatively studied by *Collewijn* (1970) Fig. 8.40; who shown that eye movements are maximal when both visual and rotatory influences are present.

Mechanically induced eye movements arise as a result of input from receptors in the statocyst that detect linear and angular acceleration. *Budelmann* (1970a, b, 1975, 1977) has made a detailed study of statocyst ultrastructure and physiology in *Octopus* and *Sepia* and has shown unequivocally that the counter-rolling of the eyes during enforced rotation is determined by gravity receptors on the macula. He concludes that the macula's sensory epithelium controls the counter-rolling eye movements on the basis of the direction of shear forces.

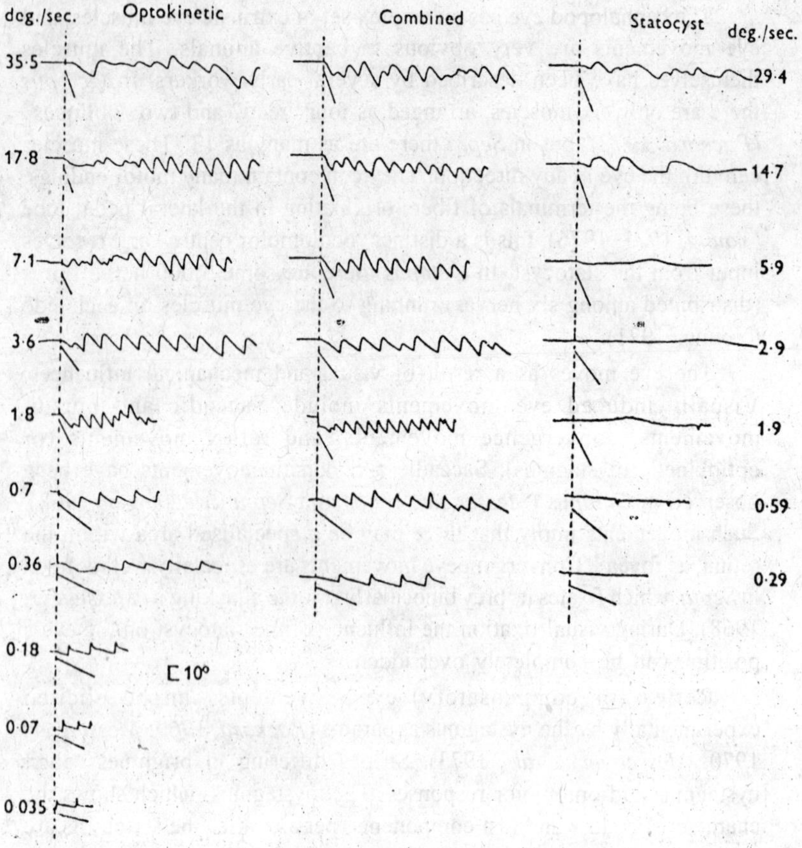

Fig. 8.40 : *Sepia*: records of eye movements elected by drum rotation to the right (left column), and passive rotation of the animal to the left in the dark (right column) and in the light (centre). Calibration of horizontal time axis varies (*Collewijn*, 1970).

The influence of gravity in keeping the eye correctly orientated was elegantly demonstrated by *Wells* (1960). If the statocysts are bilaterally removed from an *Octopus*, the horizontality of the eyes is lost and the pupils may be at 45° to the horizontal, or even more. This not only leads to postural and locomotor disturbances (*Boycott*, 1960), but it also destroys the frame of reference for the visual perception of orientation. Octopuses that could previously discriminate between horizontal and vertical rectangles now fail to do so, unless the rectangles are presented parallel or orthogonal to the pupil axis as it lies in the orbit (*Wells*, 1960).

The influence of angular acceleration on the eye was first shown by *Dijkgraaf* (1961), who observed post-rotatory eye movements when octopuses revolving on a turntable were abruptly halted. *Sepia*, too, shows after-nystagmus; *Dijkgraaf* (1963) and Mnssi;NGiiR (1970) demonstrated that this persists in the absence of visual cues though it is suppressed when these are present. As in *Octopus*, removal of the statocysts abolishes after-nystagmus.

No one has yet described very fine eye movements in cephalopods (tremor, drifts and saccades). It seems likely that they occur, however, not only on general grounds, hut also because extremely rapid adaptation has been demonstrated physiologically in the isolated cephalopod retina. *Lettvin* (cited in Young, 1971) has suggested that the function of the intraretinal muscle fibres could he to maintain line tremor during relaxation.

The Retina

(a) Organisation of the Receptor Cells

The retina comprises the receptor cells themselves, sometimes termed retinular cells but here referred to as retinal cells; the supporting cells; and the retinal glial (or epithelial) cells (*Young*, 1971), which lie in the deepest part of the retina (common usage describes the photoreceptive end of the cell as the "outer segment," even though it is directed towards the lens). Thus, it is convenient to recognise three layers in the cephalopod retina: a long outer (or distal) segment separated by a basal membrane from a short inner (or proximal) segment and, below this, the plexiform layer Fig. 8.41. Beyond the plexiform layer, there are choroid and scleroid layers (with blood vessels and muscle fibres), while external to the outer segments there is a definite limiting membrane separating the receptors from the posterior chamber.

The retinal cells are enormously elongated and thin; in *Octopus* their diameter varies between 4 urn and 10 μm and they may be even narrower in *Loligo* (*Cohen*, 1973a). Their nuclei lie below the basal membrane, and their distal processes extend above it for about 200 μm in *Octopus* and 320 μm in *Loligo*. Within the main axis of the retinal cell there are pigment granules; this pigment moves distally in the light and retracts in the dark to the base of the outer segment.

The distal segment bears two sets of parallel tubules or microvilli, at right angles to the main axis and opposite each other Fig. 8.43; they

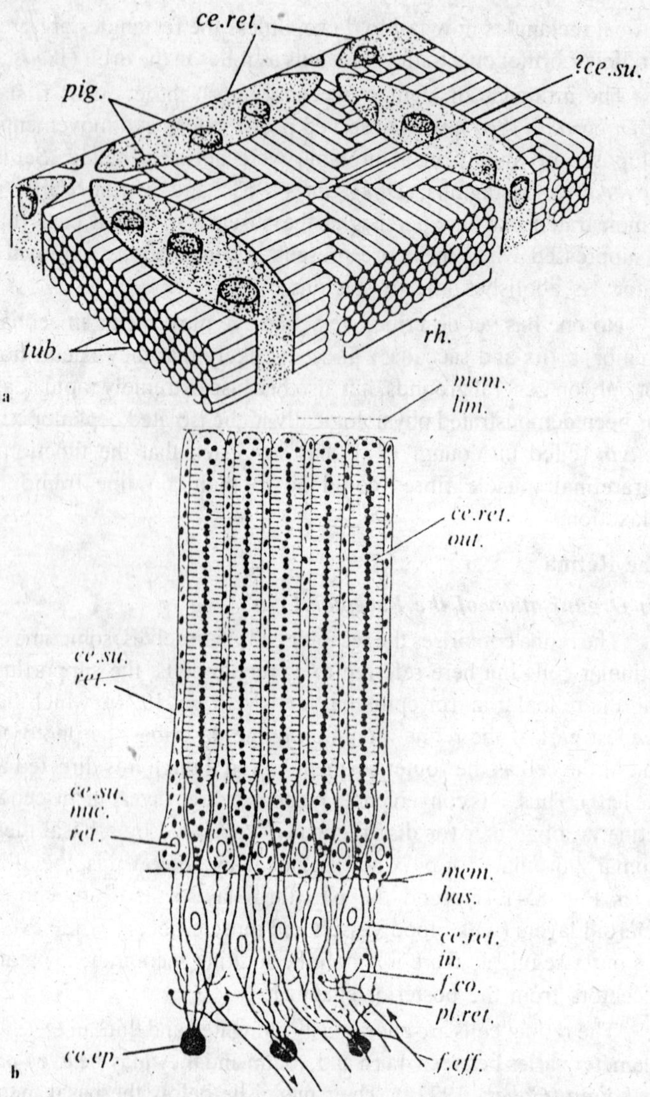

Fig. 9.41 : *Octopus* retina. a "ideal" rhabdom, formed from the rhabdomeres (rho of four retinal cells (Moody and *Parriss,* 1960). b Vertical section through retina, diagrammatic. note the inner and outer segments and retinal plexiform layer *ce. ep.*, epithelial cell of retina; *ce. ret.*, retinal cell; *ce. ret.* in., proximal portion of retinal cell; *ce. ret. out.*, distal portion of retional cell; *ce. su. nuc. ret.*, nucleus of retina-supporting cell; *ce. su. ret.*, supporting cell of retina (whose processes may extend to the space marked? *ce. cu.*); *f. co.*, collateral fibre of retinal cel; f. *eff.*, efferent fibre to retina; mem. bas., basal membrance; *mem. lim.*, limiting membrane; *pig.*, pigment granules; *pl. ret.*, retinal plexus; rh., rhabdomerc; tub., tubules or microvilli (Young, 1962a).

thus lie in the tangential retinal plane, perpendicular to incident light. The set of tubules on each side of a retinal cell is termed a rhabdomere. *Zonana* (1961) estimated that there are between 200,000 and 700,000 microvilli per rhabdomere in *Loligo*. Each tubule has a diameter of about 60 nm and the longest may reach 1.2 µm (*Moody* and *Parriss*, 1961; *Zonana*, 1961; *Cohen*, 1973 a).

They are regularly packed as evidence from low-angle X-ray diffraction bears out (*Worthington et al.*, 1976). These tubules which are derivatives of the cell membrane have a high lipid content [over half the photoreceptive membrane is lipid according to *Moson et al.*, (1973)] and, although there are important differences in the composition of the phospholipids compared with those in a vertebrate, this is a highly suggestive finding when we recall that the visual pigment is a rhodopsin (*Hubbard* and *St. George*, 1958). *Moreover, Mason* and *Fager* (1974) have demonstrated that light increases the amount of phospholipid exiractable from rhahdoms. *Saibh.* (1978) has found that actin is a major component of the distal segments and suggests that it may he associated will the photoreceptor membranes.

Moody and *Parriss* (1961) suggested that the rhodopsin molecules might be oriented with respect to the plane of the tubule membranes, and although there is no direct evidence for this, the idea is still current (*Mason et al.*, 1973). It gains support from *Moody's* (1962b) Findings that the outer segment layer shows weak birefringence (the fast directions lying in the tangential plane of the retina) and weak dichroism (light being preferentially absorbed when its electric vector lies again in the tangential retinal plane). *Moody* and *Parriss* (1961) assumed that the molecules lie parallel to the membrane around the tubule, bill are disposed randomly within it. However, *Hagins* and *Liebman* (quoted in *Moody*, 1964) measured the dichroic ratio (tangential to radial planes) in the squid and found it to be 6, which suggests that the molecules are axially aligned within the plane of the subtle membrane. Incidentally, the dichroism disappears on exposure to bright light, implying that this regular orientation is lost.

One attraction of the theory that the rhodopsin molecules are specifically oriented in the tubule membrane is that it would explain the polarised light sensitively of cephalopods. If each rhabdomere acted as a dichroic analyser of polarised light, maximum absorption would occur when the e-vector was parallel to the axes of the microtuhules.

It seems likely that there is only one type of retinal cell in the coleoid retina although there is variation in size (Young 1971). *Zonana's* (1961) suggestion that the squid retina contained two types of receptor has not stood up to scrutiny. It is important to note, however, that the retinal cells are organised rather precisely with respect to their neighbours and that there are differences in these associations that could have a functional significance.

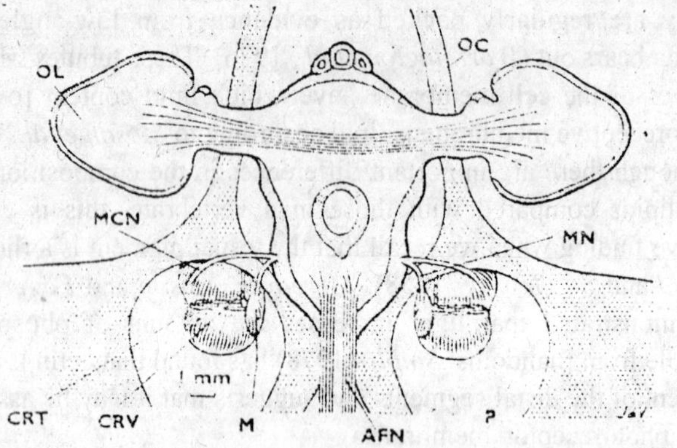

Fig. 8.42 : Transverse section through the head of an *Octopus*, seen from the anterior side, showing the positions of the statocysts in relation to the horizontal plane—which is indicated by the optic commissure (OC) joining the two optic lobes (OL). For interpretation of other lettering. (Originally published in Young, 1960. *Proc. roy. Soe.* B. 152. 5, fig. I.)

The statocysts arise as ectodermal invaginations which become enclosed in hard cartilaginous capsules, and in a moderately large *Octopus* this capsule can be felt as a hard lump beneath the posterior border of each eye. The original ectodermal invagination persists in the adult as a slender blind canal, known as Kolliker's canal, which no longer communicates with the surface. In decapods the statocyst lies in contact with the surrounding cartilaginous capsule, but in octopods the sac is suspended from the cartilage in a space which is filled with perilymph. The wall of the octopod statocyst is mainly thin and membranous, but it is supported in places by cartilaginous thickenings; its shape is maintained by the action of scattered muscle fibres in the wall of the sac against the internal fluid pressure. Scattered fibres and blood vessels cross the perilymph to support the statocyst. The sac is firmly attached to the cartilaginous capsule only at one point, at which a number of separate nerves pass from the statocyst to the brain.

There are three types of sense organ in the statocyst: the crista, which serves to record angular accelerations in different planes; the macula, which records the bodily position in relation to gravity; scattered "hair" cells which may the endolymph which fills the lumen of the statocyst.

The crista is a continuous ridge which is arranged in three planes on the inner surface of the statocyst. The statocyst wall is supported externally by a system of cartilages which form a continuous framework external to the crista and the macula. The transverse limb of the crista corresponds with the horizontal plane joining the two eyes: another limb of the crista is also horizontal, but at right angles to the first while the third limb of the crista lies vertically and parallel with the sagittal plane. In comparison with this arrangement, in the Vertehrata the inner ear possesses two semi circular canals in vertical planes at right angles to each other but both set obliquely with reference tot he principal axes of the animal and only one semicircular canal in the horizontal plane. The Cephalopoda and the Vertehrata are the only animal groups to have developed their statocysts as organs for the resolution of angular accelerations in different directions.

The sense organ which comprises the crista is not continuous, but it consists of nine units or which three occur in each limb of the crista. These nine units are not identical, there being two types which differ slightly and which are arranged alternately. Each limb of the crista therefore possesses two units of one kind and one of another kind. Each of these units consists in principal of either a double or a single row of sensory "hairs" which together form a flap projecting into the lumen of the statocyst. Each limb of the crista possesses its own nerve, by means of which afferent impulses are transmitted to the brain Fig. 8.43.

In gastropods the sense organs are of small and in simple order contain one stage statocystor after natively several smaller statocyst. Such statocyst are unadequate needs.

1. *Crista :* It serves to record regular acceleration in different planes.
2. *The macula :* It records the body position in relation to gravely.
3. *Scattered hair cells :* It may detect movements in the endolymph fills the lumen of the statocyst.

The sense organ comprises the crista is not continuous but it consists of a units of 3 occur in each limb of crista. These units are not identical. They are being 2 types each differs fragmently and are arranged alternatively. Each of these units in principle of either a double or single row of sensory hairs which are together from flap projection into the lumen of statocyst. Each limb of crista posses its own nerve by means of the afferent impulses are transmitted to brain. Removal of one/both statocyst, was found that the ability of animal to move changes in some cases animal would only swim backwards. After bilateral statocyst removal animal would not swim unless force do so and was unable to maintain direction.

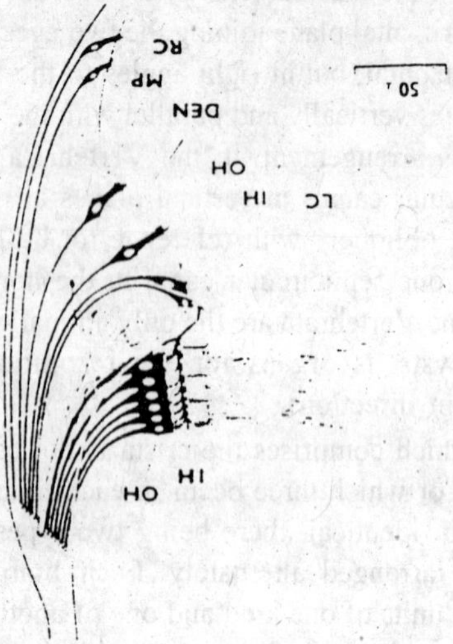

Fig. 8.43 : A diagrammatic section through a single-row portion of the crista of *Octopus vulgaris*. In such a portion of the crista a row of large hair cells (LC) is flanked both above and below by inner and outer rows of hair cells (IH. OH). Other receptors scattered over the inner surface of the wall of the statocyst are probably associated with multipolar cells (MP), as shown. For interpretation of other lettering, see p. 448. (Originally published in Young, 1960, *Proc. roy. Soc.* B, 152, 17).

It swims in spiral sometimes performing summersault and shows rolling pitching, yawning. After bilateral statocysts removal the animal seemed unable to distinguish both horizontal and vertical surface the visceral mass displaced by gravity the normal animal always move with the head permanent or relative to slow moving animal in most cephalopods they are highly specialised as animal fast moving.

There is no evidence to suggest that *Octopus* can hear and any responses to vibration can be made equally well by individual which have had both statocysts removed.

- Macula lies exactly in the vertical plan multipolar nerves cells passed from sensory epithelium to macular nerve. The sensory epithelium includes concentric rings of neurosecretory hair cells.
- Disturbances in water causes upset the in equilibrium of decapods will be detected by macula and afferent impulses passing in the macular nerves may result in firing the 1st order giant cells and so involved in the rapid escape deduction.
- The correct orientation of both eyes is maintained even after the loss of one statocyst nerve fibre from the state compensatory eye movements depend upon the integrity of the statocyst and correct interpretation of visual information is wholly depended upon the proper function of the at least one of the statocyst.

Visual Functioning : An image of the an object in an area is from on the retina of each eye and information is being passed from the majority of retinal receptors to optic lobe. In the absence any significant movement there is no feature is the seen around the after the animal and only background information is being passed from optic lobes into the centre co-relation in the brain which comprise vertical superior; frontal and sub vertical lobes. With the movement of the prey the retinal receptors, deep seated neurons those one classified for occurrence of movement and information will be pursued to the collectively detect these classifiers must be individually capable of either of two outputs. One such output promotes attack while the other suppressed attack. The optic lobes also receives tactile and gustatory information.

Taclite discrimination : When Octopoda diverged from their pelagic ancestory and adopted for benthic made of life they began to use the arms fro tactile exploration and this naturally through an additional burden on the central. Nervous system. There one majority of sensory nerves passes without synopses through the lower centres of brain and continue to supraoesopheal lobes.

Olfactory Senses : Various herbevore, Gastropods can detect the presence of suitable sea weeds and move in correct direction to feed on these.

The piscivorous spp. of *Conus* detect presence of their brain by olfactory sence. In *Bulia* (opisthobranchiata) the immediate responses come in the presence of chemical trimethylamine and tetra methyl ammonium salts which are the natured products of decomposition of food substances. In *Patella* sensory epithelium on sides of food detect the differences in salinity.

To summarise, any sudden external event in the vicinity of a decapod –*e.g.*, the appearance of a potential predator, or a major physical disturbance of the water—will probably evoke an escape reaction by the decapod. The reaction may be initiated by a visual or a tactile disturbance or by a disturbance in the animal's stability, and is transmitted by an integrated system of giant nerve fibres which ensures "instantaneous", maximal. bilateral response even to unilateral stimulation. The first pair of giant fibre cells doubtless require fairly massive stimulation to ensure that the escape reaction is not invoked unnecessarily. The interaxonic bridge between these two cells ensures that even a unilateral esternal disturbance evokes a full bilateral escape response. Withdrawal of the head and of the tunnel will compress the mantle cavity and so augment the escape reaction and there is only one synapse on the giant fibre pathway which mediates this withdrawal. The escape reaction depends on close co-ordination of contraction of all the circular muscle fibres of the mantle; this is achieved by the branching of the third-order giant fibres, so that no more than ten or eleven of these fibres command the entire array of circular pallial muscle fibres. There are only two synapses on the giant fibre pathway to the circular pallial muscle fibres: there is minimal consumption of time in transmission of impulses through synapses, while the very extensive nature of. the synaptic contacts ensures that there will be no failure to excite the neurones distal to the synapse.

The giant fibre system aids the escape reaction in one further respect, which is especially worthy of mention. It has been shown that the rate of conduction of a nerve impulse varies according to the diameter of the nerve fibre, conduction being faster in the larger fibres. The rate of conduction increases approximately in proportion to the square root of the diameter of the fibre. The presence of giant fibres, and the utilisation of these when the escape reaction is invoked, results in a great saving in "reaction time". Due to possession of a giant fibre system an alarmed decapod can get under way far more rapidly than it could otherwise. Furthermore, it has been shown that in the series of third-order giant fibres, the longer stellar nerves contain giant fibres of greater diameter, and the shorter stellar nerves contain giant fibres of lesser diameter. In consequence there is a more rapid delivery for impulses which have a greater distance to travel, and the time of contraction of the pallial muscles is a little more closely co-ordinated than would be the case if all nerve impulses travelled at a uniform speed. The cxhalant jet of water will be expelled with slightly greater vigour, and the animal will dart backwards at a slightly higher speed than would otherwise be the case.

Structure and Function of the Slatocysts

The stalocysis of gastropods and of bivalves are comparatively small and simple organs which may contain one single statolith or alternatively several smaller statoliths. Such statocysts are adequate to the needs of these sedentary or relatively slow-moving animals. In the dibranchiale Cephalopoda, which are highly specialised as fast-swimming predacious carnivores, the Slatocysts differ in oeing relatively large and very much more complicated in structure. The statocysts in the Dibranchiata evolved in response to the requirements of nektonic animals and there was no significant alteration in their basic mechanism or in their utilisation in the Octopoda when these diverged and adopted a benthic mode of life. This explains certain limitations in the organisation and the capabilities of the octopods as revealed bv experimental investigation. Attention is directed here to the structure and mode of utilisation of the stalocysts in *Octopus vulsaris* (Fig. 8.44).

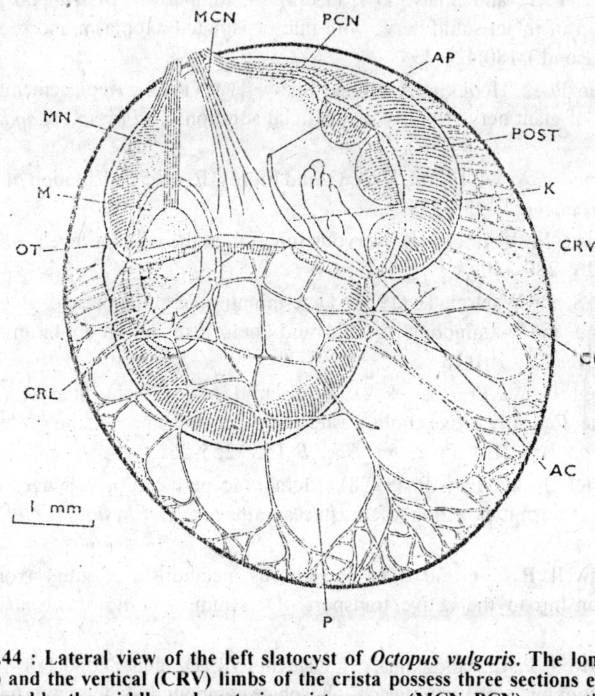

Fig. 8.44 : Lateral view of the left slatocyst of *Octopus vulgaris*. The longitudinal (CRL) and the vertical (CRV) limbs of the crista possess three sections each; they are served by the middle and posterior erista nerves (MCN, PCN) respectively. The otolilh (OT) lies in the macula (M) which is served by the macular nerve (MN). The statocyst lies suspended in a wide space filled with perilympn (P). For interpretation of other lettering, (Originally published in Young, 1960, Proc. roy. Soc. B, 152, 5, fig. 2.)

REFERENCES

Araki, T. M. and Oscarsson, O. (1961). Anion permeability of the synaptic and nonsynaptic motoneurone membrane. *J. Physiol. (Lond.)* 159: 410-435.

Arvanitaki, A. and Chalazonitis, N. (1961). Excitatory and inhibitory processes initiated by light and infra-red radiations in single identifiable nerve cells. In *Nervous inhibition*, ed. E. Florey. pp. 194-231. Oxford: Pergamon Press.

Arvanitaki, A. and Chalazonitis, N. (1965). Oxygen control of the neuronal activity of identifiable *Aplysia* nerve cells. In *Proc. XXIII Int. Congr. Physiol. Sci., Tokyo*. p. 95. Amsterdam: Excerpta Medica.

Awapara, J., Landua, A. J., Fuerst, R. and Seale, B. (1950). Free γ-aminobutyric acid in brain. *J. biol. Chem.* 187: 35-39.

Baker, P. F. (1965a). Phosphorus metabolism of intact crab nerve and its relation to the active transport of ions. *J. Physiol. (Lond.)* 180: 383-423.

Baker, P. F. (1965A). A method for the location of extracellular space in crab nerve. *J. Physiol. (Lond.)* 180: 439-447.

Baker, P. F. and Shaw, T. I. (1965). A comparison of the phosphorus metabolism of intact squid nerve with that of isolated axoplasm and sheath. *J. Physiol.* (Lond.) 180:424-438.

Baker, P. F., Hodgkin, A. L. and Shaw, T. I. (1962). Replacement of the axoplasm of giant nerve fibres with artificial solutions. *J. Physiol. (Lond.)* 164: 330-354.

Bazemore, A. W., Elliott, K. A. C. and Florey, E. (1957). Isolation of Factor I. *J. Neurochem.* 1: 334-339.

Bennett, M. V. L. (1964). Nervous function at the cellular level. *Ann. Rev. Physiol.* 26: 289-340.

Berl, S. and Waelsch, H. (1958). Determination of glutamic acid, glutamine, glutathione and γ-aminobutyric acid and their distribution in brain tissue. *J. Neurochem.* 3: 161-169.

Bissett, G. W., Frazer, J. F. D., Rothschild, M. and Schachter, M. (1960). A pharmacologically active choline ester and other substances in the garden tiger moth *Arctia caja* (L.). *Proc. my. Soc. B* 152: 255-262.

Boistel, J. and Fatt, P. (1958). Membrane permeability changes during inhibitory transmitter action of crustacean muscle. *J. Physiol. (Lond.)* 144: 176-191.

Caldwell, P. C. (1960). The phosphorus metabolism of squid axons and its relationship to the active transport of sodium. *J. Physiol. (Lond.)* 152: 545-560.

Caldwell, P. C., Hodgkin, A. L., Keynes, R. D., and Shaw, T. I. (1960). The effects of injecting "energy-rich" phosphate compounds on the active transport of ions in the giant axons of *Loligo*. *J. Physiol. (Lond.)* 152: 561-590.

Carlsson, A., Faick, B. and Hillarp, N-A. (1962). Cellular localization of brain monoamines. *Acta physiol. scand.* 56 (Suppl. 196): 1-28.

Chalazonitis, N. (1959). Chemopotentiels desneuroncs geants fonctionellement differences. *Arch. Sci. physiol.* 13: 41-78.

Chalazonitis, N. (1961). Chemopotentials in giant nerve cells (Aplysia fasciata). In *Nervous inhibition*, ed. E. Florey. pp. 179-194. Oxford: Pergamon Press.

Chalazonitis, N. and Arvanitaki, A. (1956). Chromoproteides et succinoxydase dans divers grains isolables du cytoplasme neuronique. *Arch. Sci. physiol.* 10: 291-319

Chalazonitis, N. and Gola. M. (1964). Analyses microspectrophoto-metriques relatives quelques catalyseurs respiratories dans Ie neurone isole *(Helix pomatia)*. C. R. Soc. *Biol. (Paris)* 158: 1908-1914.

Chalazonitis. N. and Takeuchi, H. (1964). Variations de l'excitabilite directe somatique, en hyperoxie (neurones geants d'*Aplysici fasciaia* et *Helix pomatia*). *C. R. Soc. Biol.* (Paris) 158: 2400-2408.

Coombs, J. S., Eceles, J. C. and Fatt, P. (1955). The specific ionic conductances and the ionic movements across the motoneuronal membrane that produce the inhibitory post-synaptic potential. *J. Physiol. (Lond.)* 130: 326-373.

Crescitelli, F. and Geissman, T. A. (1962). Invertebrate pharmacology: selected topics. *Ann. Rev. Pharmacol.* 2: 143-192.

Curtis, D. R. (1961). The effects of drugs and amino acids upon neurons. In *Regional neurochemistry*, eds. S. S. Kety and J. Elkes. pp. 403-422. Oxford: Pergamon Press.

Curtis, D. R. (1965). The actions of amino acids on mammalian neurones. In *Studies in physiology* presented to J. C." Eceles, eds. D. R. Curtis and A. K. Mchtyre, pp. 34-42. Heidelberg: Springer-Verlag.

Curtis, D. R. and J. C. Watkins (1960). The excitation and depression of spinal neurones by structurally related amino acids. *J. Neurochem.* 6: 117-141.

Curtis, H. J. and Cole. K. S. (1942). Membrane resting and action potentials from the squid giant axon. *J. cell. comp. Physiol.* 19: 135-144.

Diamond, J. (1963). Variation in the sensitivity to gamma-aminobutyric acid of different regions of the Mauthner neurons. *Nature (Lond.)* 199: 773-775.

Dudel, J. and Kuffler. S. W. (1961). Presynaptic inhibition at the crayfish neuromuscular junction. *J. Physiol. (Lond.)* 155: 543-562.

Eceles, J. C. (1964). *The physiology of synapses*. Berlin: Springer-Verlag.

Erspamer, V. (1961). Recent research in the field of 5-hydroxytryptamine and related indolealkylamines. *Progr. Drug Res. (Fortschr. Arineimittelforsch.)* 3: 151-367.

Erspamer, V. and Benati, O. (1953). Isolierung des Murexins aus Hypobranchialdrusenextrakten von *Murex trunculus* and seine Identifizierung als β-[Imidazolyl-4(5)]-acryl-cholin. *Biochem. Z.* 324: 66-73.

Faick, B. (1962). Observations on the possibilities of the cellular localization of monoamines by fluorescence method. *Acta physiol. scand.* 56 (Suppl. 197): 1-26.

Florey, E. (1961). Comparative physiology: transmitter substances. *Ann. Rev. Physiol.* 23: 501-528.

Florey, E. (1962). Comparative neurochemistry: inorganic ions, amino acids and possible transmitter substances of invertebrates. *In Neurochemistry*, eds. K. A. C.

Elliott, I. H. Page and J. H. Quastei. pp. 673-693. Springfield: Charles C. Thomas.

Florey, E. (1965). Comparative pharmacology: neurotropic and myotropic compounds. *Ann. Rev. Pharmacol.* 5: 357-382.

Frontali, N. (1961). Activity of glutamic decarboxylase in insect nerve tissue. *Nature (Lond.)* 191: 178-179.

Frontali, N. (1964). Brain glutamic acid decarboxylase and synthesis of γ-aminobutyric acid in vertebrate and invertebrate species. In *Comparative neurochemistry*, ed. D. Richter. pp. 185-192. Oxford: Pergamon Press.

Furshpan, E. J. and Potter, D. D. (1959). Slow post-synaptic potentials recorded from the giant motor fibre of the crayfish. *J. Physiol. (Lond.)* 145: 326-335.

Gelder, N. M. van (1965). The histochemical demonstration of γ-aminobutyric acid metabolism by reduction of a tetrazolium salt. *J. Neurochem.* 12: 231-237.

Gerschenfeld, H. M. and Lasansky, A. (1964). Action of glutamic acid and other naturally occurring amino-acids on snail central neurons. Int. *J. Neuropharmacol.* 3: 301–314.

Greenberg, M. J. (1960a). The responses of the Venus heart to catecholamines and high concentrations of 5HT. *Brit. J. Pharmacol.* 15: 365-374.

Greenburg, M. J. (19606). Structure-activity relationship of tryptamine analogues on the heart of *Venus mercenariu. Brit. J. Pharmacol.* 15: 375-388.

Grundfest, H., Reuben, J. P. and Rickles, W. H., Jr. (1959). The electrophysiology and pharmacology of lobster neuromuscular synapses. *J. gen. Physiol.* 42; 1301-1323.

Hagiwara, S., Kusano, K. and Saito, S. (1960). Membrane changes in crayfish stretch receptor neuron during synaptic inhibition and under action of gamma-aminobutyric acid. *J. Neurophysio.* 23: 505-515.

Hodgkin, A. L. and Huxley, A. F. (1945). Resting and action potentials in single nerve fibres. *J. Physiol. (Lond.)* 104: 176-195.

Hodgkin, A. L. and Keynes, R. D. (1955). Active transport of cations in giant axons from *Sepia* and *Loligo*: *J. Physiol. (Lond.)* 128: 28-60.

Hyden, H. and Lange, P. (1961). Differences in the metabolism of oligodendroglia and nerve cells in the vestibular area. *In Regional neurochemistry*, eds. S. S. Kety and J. Elkes, pp. 190-199. Oxford: Pergamon Press, to M., Kostyuk, P. G. and Oshima, T. (1962). Further study on anion permeability of inhibitory post-synaptic membrane of cat motoneurones. *J. Physiol. (Lond.)* 164: 150-156.

Kerkut, G. A. and Cottrell, G. A. (1962). Neuropharmacology of the pharyngeal retractor muscle of the snail *Helix aspersa. Life Sci.* 1: 229-231.

Kerkut, G. A. and Cottrell, G. A. (1963). Acetylcholine and 5-hydroxytryptamine in the snail brain. *Comp. Biochem. Physiol.* 8: 53-63.

Kerkut, G. A. and Meech, R. W. (1966). Microelectrode determination of intracellular chloride concentration in nerve cells. *Life Sci.* 5: 453-456.

Kerkut, G. A. and Price, M, A. (1961). Histamine content of tissues from the crab *Curcinus naenas. Comp. Biochem. Physiol.* 3: 315-317.

Kerkut, G. A. and Thomas, R. C. (1963). Acetylcholine and the spontaneous inhibitory post synaptic potentials in the snail neurone. *Comp. Biochem. Physiol.* S: 39–45.

Kerkut, G. A. and Thomas, R. C. (1964). The effect of anion injection and changes in the external potassium and chloride concentrations on the reversal potentials of the IPSP and acetylcholine. *Comp. Biochem. Physiol.* 11: 199-213.

Kerkut, G. A. and Thomas, R. C. (1965). An electrogenic sodium pump in snail nerve cells. *Comp. Biochem. Physiol.* 14: 167-183.

Kerkut, G. A. and Walker, R. J. (1961). The effects of drugs on the neurones of the snail *Helix aspersa. Comp. Biochem. Physiol.* 3: 143-160.

Kerkut. G. A. and Walker, R. J. (1962). The specific chemical sensitivity of *Helix* nerve cells. *Comp. Biochem. Physiol.* 7: 277-288.

Kerkut, G. A. and Walker, R. J. (1966). The effects of L-glutamate, acetylcholine and GABA on the miniature end plate potentials and contractures of the coxal muscles of the cockroach *Periptaneta americana. Comp. Biochem. Physiol.* 17: 435-454.

Kerkut, G. A., Sedden, C. B. and Walker, R. J. (1966). The effect of DOPA, α-methyldopa and reserpine on the dopamine content of the brain of the snail, *Helix aspersa. Comp. Biochem. Physiol.* 18: 921-930.

Kerkut. G. A., Shapira, A. and Walker, R. J. (1965). The effect of acetylcholine, glutamic acid and GABA on the contractions of the perfused cockroach leg. *Comp. Biochem. Physiol.* 16: 37-48.

Kerkut, G. A., Shapira, A. and Walker, R. J. (1966). The liberation of labelled glutamate from the snail nerve muscle system. *Comp. Biochem. Physiol*, 16, 154-166.

Kerkut, G. A., Thomas, R. C. and Venning, H. B. (1964). A transistorized linear sweep circuit for determining reversal potentials in nerve cells. Med. *Electron. Biol. Enynig.* 2: 425-430.

Kerkut, G. A., Leake, L. D., Shapira, A., Cowan, S. and Walker, R. J. (1965). The presence of glutamate in nerve-muscle perfusates of *Helix, Carcinus*, and *Periplaneta. Comp. Biochem. Physiol.* 15; 485-502.

Keyl. M. J., Michaelson, I. A. and Whittaker', V. P. (1957). Physiologically active choline esters in certain marine gastropods and other invertebrates. *J. Physiol. (Lond.)* 139: 434-454.

Keynes, R.D. (1963). Chloride in the squid giant axon. *J. Physiol. (Lond.)* 169: 690-705.

Koechlin, B.A. (1955). On the chemical composition of the axoplasm of squid giant nerve fibers with particular reference to its ion pattern. *J. biophys. biochem. Cytol.* 1:511-529.

Kravitz, E. A. (1962). Enzymic formation of gamma-aminobutyric acid in the peripheral and central nervous system of lobsters. *J. Neurochem.* 9: 363-370.

Kravitz, E.A. and Potter, D.D. (1965). A further study of the distribution of γ-aminobutyric acid between excitatory and inhibitory axons of the lobster. *J. Neurochem.* 12: 323-328.

Kravitz, E. A., Kuffler, S. W. and Potter, D. D. (1963). Gamma-aminobutyric acid and other blocking compounds in Crustacea. III. Their relative concentrations in separated motor and inhibitory axons. *J. Neurophysiol.* 26: 739-751.

Kravitz, E. A., Kuffler, S. W., Potter, D. D. and Gelder, N. M. van (1963). Gammaaminobutyric acid and other blocking compounds in Crustacéa. II. Peripheral nervous system. *J. Neurophysiol.* 26: 729-738.

Krnjevic, K. (1965). Actions of drugs on single neurones in the cerebral cortex. *Brit. Med. Bull.* 21: 10-14.

Krnjevic, K. and Phillis, J. W. (1963). lontophoretic studies of neurones in the mammalian cerebral cortex. *J. Physiol. (Lond.)* 165: 274-304.

Kuffler, S. W. (1960). Excitation and inhibition in single nerve cells. Harvey Lectures, 1958-1959, pp. 176-218. New York: Academic Press.

Kuffler, S. W. and Edwards, C. (1958). Mechanism of gammaaminobutyric acid action and its relation to synaptic inhibition. *J. Neurophysiol.* 21: 589-610.

Lewis, P. R. (1952). The free amino-acids of invertebrate nerve. *Biochem. J.* 52: 330-338.

Loveland, R. E. (1963). 5-Hydroxytryptamine, the probable mediator of excitation in the heart of *Mercenaria (Venus) mercenaria. Comp. Biochem. Physiol.* 9: 95-104.

Lowe, I. P., Robins, E. and Eyerman, G. S. (1958). The fluorimetric measurement of glutamic decarboxylase and its distribution in brain. *J. Neurochem.* 3: 8-18.

Mollusca-An introduction to their from and function by—J.E. Moston

Young (1982) Biology of Mollusca

Octopus Physiology and Behaviour of an Advanced Invertebrates M.L. Wells, 1978. Champman and Hall London.

Physiology of Mollusca, Karl M. Wilbur and C.M. Yonge Acad. Press 1964. Vol I.

9

REPRODUCTIVE SYSTEM

INTRODUCTION

In primitive molluscs opposite sexes gametes were discharged into the sea water via the nephridia, fertilization were random, and the small eggs developed into trochophore larvae. The reproductive function of adult was limited to the emission of large numbers of gametes, and no protection or special provision was made for the developing embryos. This condition is retained in various modern lineages. In other lineages there are specializations of one kind or another.

In the monoplacophora two pairs of gonads gametes discharge via the third and fourth pair or Nephridia, and fertilization is presumably at random in the sea water. It is suggested that the mollusca are metamerically segmented animals, which, except for the Monoplacophora, have only one genital segment.

In the Scaphopoda and in most of the Polyplacophora there is only one, unpaired gonad, the gametes are shed into the sea and development is external. Most of the Aplacophora are specialized as hermaphrodites which discharge gametes via gonadal ducts into the mantle cavity and hence into the surrounding water, the fertilization and development are external.

The gastropoda show considerable diversification as regards there reproductive systems. In the prosobranchia some members, if the Archaeo-gastropoda, retain the primitive ancestral condition; in more advanced types the right kidney has aborted safe for the renal contribution to the genital duct and the genital duct is carried across the floor of the mantle cavity either as a ciliated and glandular groove or by the enclosure of this groove to form the pallial gonad duct.

In most of the higher prosobranchs the male has a penis and after copulation the fertilized eggs receive nourishment from the albumen gland at the inner end of the pallial oviduct and a protective coat from the capsule gland which comprises the distal part of the pallial oviduct. As regards the complexity of their reproductive systems, the neritacea, are more advanced than other members of the Archaeogastropoda, and this raises a difficult question in connection with their systematic position.

Phylogeny of the Molluscs

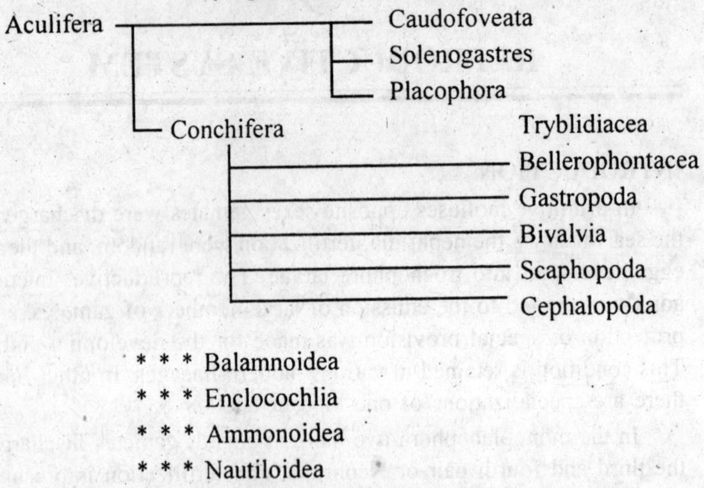

Fig. 9.1

THE PRIMITIVE MOLLUSCAN REPRODUCTIVE SYSTEM

It is suggested as a basis for discussion that the sexes were separate, that there were no accessory sexual organs and no special genital ducts, and that the gametes were discharged via the nephridia into the sea water. The eggs were numerous, small and lacking in yolk, cleavage and gastrulation occurred rapidly while drifting in the sea, and the organism hatched as a planktonic trochophore larva. The parental generation was slowly concerned with the emission of large numbers of gametes for which no protection was provided.

The Reproductive System

Neopilina galatheae (monoplacophora) exhibits metamerism in many of the organ systems including the muscles the nerve connectives, the nephridia, the auricles of the heart, and the gonads. In the male there are clearly two pairs of testes, which pass their products by two gonad ducts into the third and fourth pairs of Nephridia and thence to the exterior. The female system is probably identical, but the division between the first and second pair of aviaries has not been established with certainty.

The problem is to decide whether this metamerism has been derived from a segmental plan of construction possessed by the remote, common pre-molluscan ancestor, or whether it is a secondary specialization confined to the Monoplacophora. This remains unresolved, but I myself think the former theory to be simpler, and therefore to be more readily acceptable. On this basis it judged that the gonad of the remote ancestral mollusc was probably a paired, segmental structure, confined to an exceedingly small number of reproductive segments. Possibly all early

The Scaphopoda retain a primitive reproductive system, there being a single unpaired gonad in the roof of the mantle cavity. A single gonad duct leads from the anterior end of the gonad into the right excretory organ- a condition which also occurs in the more primitive prosobranch gastropods. The animals have separate sexes, the gametes are shed into the water, and development is external.

In the Polyplacophora the sexes are separate. The gonad is paired in one genus, *Nuttallochiton*, but in the remainder there is a single median gonad which is thought to have been produced by fusion of the left and right gonads. The gonad lies anterior to the pericardium and communicates with the lateral pallial groove by means of paired gonad ducts which do not join the nephridia. In some genera the eggs are fertilized in the lateral pallial groove– presumably by spermatozoa which have been drawn in with the inhalant water current- and are retained in the pallial groove during the early stages of development, but in other cases the fertilised eggs are laid single or on strings, and are abandoned.

The majority of the Aplacophora are hermaphrodite, the ovotestis discharging gametes into the pericardium- which is, of course, a remnant of the coelom. The gametes are voided from the pericardium by the nephridia, which are somewhat specialized as adjuncts of the reproductive system. A caecum on the proximal wall of each nephridium may serve as a sperm reservoir; the distal ends of the two nephridia fuse to from median pouch with glandular walls where the fertilised eggs are enclosed in a capsule; copulation is known to occur in some species. *Chaetoderma* is unsexual in being unisexual, with a single median gonad, and with simple, unmodified nephridia.

The reproductive system remains simple and unspecialized in most of the Bivalvia. Typically the sexes are separate and the gametes are discharged into the mantle cavity by gonad ducts which open very close to the excretory pores. In the majority of cases the eggs are fertilised by random contact with spermatozoa in the water, and develop into free-swimming bi-valved veliger larvae. Certain bivalves show specialized featured such as functional hermaphroditism, sex reversal, incubation of developing embryos and young larvae within the ctenidium, and even development of a part of the ctenidium into a special brood plouch or marsupium. These specializations will be discussed individually under appropriate headings later in this chapter.

The sexes are separate in the Cephalopoda; there bring a single median gonad at the posterior end the visceral mass. Ova or sperm are passed into the perivisceral coelome, and are discharged into the mantle cavity by a single gonad duct which lies on the left side of the visceral mass. In the male the gonad duct is extremely complicated and is the site of manufacture of spermatophores. In the female the ventral surface of the visceral mass bears paired nidamental glands and paired accessory nidamental glands, secretions from which provide secondary coverings to the fertilised eggs. The male has no penis, and yet mating occurs, the spermatophores being transferred by one of the arms of the male.

Finally, we have to consider the Gastropoda. Here in the Archaeogastropoda we find Patella and in Haliotis the gonad duct opens directly into the kidney, while in other Docoglossa the gonad duct opens into the ureter. The gametes therefore pass through a duct which is of dual origin, the proximal part being derived from the gonad and the distal part from the nephridium (Frettter, 1946). Retention of this condition would seem to hinder the inception of internal fertilization by copulation; development of the accessory structures necessary for internal fertilization can only occur after the genital pathway has been isolated from the products of excretion.

In the Mesogastropoda and in higher gastropoda the right excretory organ has disappeared except for that part of it which contributes to the genital duct-the renal oviduct of the female-the renal vas deferences of tee male. In the Mesogastropoda a further relic exeretory organ may be found in the gono-pericardial duct, which occurs in the female only, and which links the renal oviduct with the pericardium. Its purpose is not known, but it may serve some essential function since it appears at sex reversal from the male to the female phase in *Calyptraes and in Crepidula.* After dissociation of the excretory and reproductive systems in this way there is increased prospect of the emergence of internal fertilization by copulation.

In the Mesogastropoda the gonad duct of the female has been extended across the floor of the mantle cavity alongside the similarity extended rectum by the development of a third component to the genital duct. This distal extension of the female duct was probably effected, firstly, by the development of a ciliated and glandular grove crossing the floor of the mantle cavity and, secondly, by the fusion of the ridges bounding this groove, so converting it into a closed ciliated and glandular duct. Since the distal component of the female duct has presumably been derived in this way from the floor of the mantle cavity, it is generally known as the pallial oviduct. Thus the female duct comprises firstly the oviduct *sensu stricto*, then the renal oviduct which is now the sole remnant of the right nephridium, and, finally, the pallial oviduct.

In the Mesogastropoda the ova are fertilised at the inner end of the pallial oviduct by spermatozoa which have been conveyed to this point in a variety of ways, and the main functions of the pallial oviduct are to surround the fertilised eggs with a supply of nourishing material and then with a protective coat or capsule. The inner end of the pallial oviduct is specialized to from an albumen gland for the former purpose, while the distal three quarters is developed into a capsule gland. In some, *e.g.*, *Littorina*, and the Neogastropoda, there is a bursa copulatrix adjacent to the orifice of the pallial oviduct, for the reception of spermatozoa from the penis of the male. The spermatozoa are then transported inwards along a ciliated grove in the floor of the pallial oviduct to a receptaculum seminis sited at the inner end of the pallial oviduct. Spermetozoa are liberated from the receptaculum as and when required fro the purpose of fertilization of eggs arriving from the renal oviduct.

Pedal glands may participate in the reproductive process; in the neogastropoda Nucella lapillus is the egg capsule is passed to the ventral pedal gland, by which it is moulded into its final shape and firmly fixed to the substratum; in the pedal opisthobranch Onchidella celtica Thedoxus fluviatilis. The inner end of the vagina passes to the receptaculum seminis, which communicates with the inner end of the pallial oviduct.

In some mesogastropoda the growth from of the shell results in lateral compression of the mantle cavity to such an extent that the introduction of a penis during copulation would occlude the mantle cavity and interface with the processes of respiration and feeding. In such examples the male lacks a penis, the pallial oviduct of the female is econdally simplified into an open ciliated and glandular grove and the spermatozoa are drawn into this groove via the inhalant water stream. Prosobrancha may have two kinds of spermatozoa. Secondly, there are oligopyrene spermatozoa of various forms, which may reach enormous size. An oligopyene sperm transport thousands of eupyrene sperms which are oriented on its tail with their heads embedded in the cytoplasm.

Such compound structures are called spermatozeugmata. These can transverse considerable distances and they obviate the need for copulation as they can swim into the open pallial oviduct of the female, *e.g.*, Ceritiopsis and Clathrus. Lanthinajanthina has no locomotory organs and no copoulatory organs, and Wilson and Wilson (1956) judge that fertilization is effected by such spermatozeumata which swim across the relative large spaces between individuals snails in the swarm. Spermatozeugmata swim to the female and are transported up the glandular pallial groove to the site of fertilization, *e.g.*, in Turritella communis (Fretter, 1946, 1951, 1953). Other examples in which the shell is coiled in a tight spiral, the mantle cavity constricted, and the pallail oviduct is secondarily open throughout its length, include *Cerithiopsis Triphora, Clathrus, Bittium,* and *Balcis.*

The genital duct of the male similarly consists of two sections in the Archaeogastropoda, but of three sections in the Mesogastropoda and the Neogastropoda. The proximal section, derived from gonad, serves as a vesicular seminalis and the spermatozoa are stored, here. The second section is the renal vas deference, and this commences with a sphincter muscle which only permits spermatozoa to leave the vesicular seminalis at the appropriate time. In the Archaeogastropoda the spermatozoa are released from the renal vas deference into the mantle cavity and are emitted in the exhalant water stream. In the Mesogastropoda and Neogastropoda, due to the development of a closed pallial oviduct in which the eggs are encapsulated, internal fertilization is necessary and in the male a penis has been developed on the right side of the head. In the most primitive condition the seminal fluid is conveyed from the mouth of the renal vas deferens to the penis by means of a ciliated groove on the floor of the mantle cavity and up the side of the penis. During copulation the ridges bordering this ciliated grove will approximate above the groove, so converting it temporarily into a closed tube, *e.g., Trichotropis, Littorina, Crepidula, Capulus*, and even the *Cypraeid Simnia* (Fretter, 1955; Graham, 1954).

In Littorina, and in the secondarily open groove of Turritella, glandular strips on either side of the groove provided prostatic secretions. In the more advanced Mesogastropoda, and in the Neogastropoda, the ridges bordering this groove in the floor of the mantle cavity have fused above the groove, with the formation of a permanent tubular pallial vas deferens which penetrates the penis and opens at its tip. The pallial vas deferens has glandular walls and apyrene and eupyrene spermatozoa are produced; sperm and prostatic secretion is emitted in a cloud which does not disperse much, and this is sucked in through the inhalant siphons of females in close proximity to the spawning males. Only the eupyrene spermatozoa have been found in the female reproductive system (Fretter, 1951).

Some Members of the Archaeogastropoda resemble the Mesogastropoda in processing a closed prostatic tube crossing the mantle cavity, and a penis on the right side of the head, *e.g.,* the Neritacea, including Nerita and Theodoxus. A similar parallelism is found in the females system of *Calliostoma zizyphinum* which processes a closed pallial oviduct.

Some prosobranchs exhibit protandric hermaphroditism and the reproductive system is naturally complicated by the development of both male and female components. In *Crepidula* and *Calyptraea*, *Alyptraea*, for instance, the gono-pericardial duct begins to develop during the preparations fro sex reversal from the male to the female phase. Omalogyra atomus is a minute protandric hermaphrodite with quite remarkable modifications to the reproductive system. I the spring young specimens pass through a male phase when the gonad duct serves as a vesicular seminalis. There is no penis. A bursa copulatrix opens into the prostatic pallial vas deferens, and the opening of the bursa copulatrix opens into the prostatic pallial vas deferens, and the opening of the bursa is developed into a long slender, Muscular tube which lies within the pallial vas deferens, pointing inwardaly. It is thought that this provides a devices whereby the bursa can become filled with native spermatozoa during the male phase. The method of copulation is not known, but it seems possible that the long slender tubular orifice of the bursa may be reserved in position so that it can be protruded through the genital aperture and used as a penis. It is not at all clear how this reversal of the tube could be effected. The organs of the female system are present at the same time as the male organs, but do not reach maturity until later. Distally the pallial oviduct and vas deferens join to from a her maphrodite duct. In the summer the ovary is greatly developed and the animal how enters the female phase. Individuals which develop from eggs laid by summer spawners fail to develop the mail system fully, although a testis with ripe spermatozoa is produced, but pass quickly into a functional female phase. There is a succession of generation in the summer months and self-fertilisation may possibly occur.

In some Mesogastropoda and Neogastropoda special areas of the reproductive system have acquired the capacity to destroy superfluous or senile spermatozoa; these are ingested by the epithelial cells which processed to digest the spermatozoa intracellularly. This is known in the male system in *Omalogyra atomus, Ocenebra erinacea*, and *Buccinum undatum*. It occurs in the female system in *Cerithiopsis tubercularis, Trivia monacha, Nucella lapillus, Ocenebra erinebra*, and *Buccinum undatum,* and in the last-named species the ingesting area can also ingest superfluous yolk. The engulfing of spermatozoa may also occur in the Opisthobranchia *e.g., Onchidella celtica.* It is curious that species within the same genus may differ in processing or in lacking this capacity, thus it is present in *T. monacha*, but absent in *T. arctica*. Senile spermatozoa might be able to penetrate the ovum thereby blocking entry by and other spermetozoon, and yet be unable to initiate normal development. If this were true then the capacity to destroy superfluous spermatozoa would individuals, for in Aplysia the last individual in the chain does not become fertilized. The achievement of functional hermaphroditism in the Opisthobranchia and Pulmonata provided the climax to the development of the reproductive system. Much diversity in forms has been revealed in published descriptions of the reproductive system. Much diversity in form has been revealed in published description of the reproductive systems of many different genera, *e.g.*, the parasitic Pyramidellidae, the tectibrancha *Philine, Actaeonia, Onchidella*, the *Ellobiidae*, *Oxychilus*, etc. It does not seem necessary to investigate such diversity. The development in some examples of three distinct genital ducts to the exterior. These are the vas deferens which passes to the penis, the oviduct by means of which fertilised eggs are laid, and the vagina which has become separated off form the oviduct for reception of the penis and passage of foreign spermatozoa to the head of the pallial oviduct. This "triaulic" condition has already been encountered in the *Neritacea, e.g., Nerita* and *Theodoxus*, which as members of the archaeogastropod Prosobranchia are unusually advanced in this respect.

Nevertheless, it is interesting to note the occurrences of hypodermic impregnation in *Limapontia capitata*, in which the bursa copulatrix lies immediately below the body wall but does not open to the exterior. During copulation hypodermic impregnation occurs at this point, the slender, gently curved penial style being stabbed through the body wall and into the bursa of the partner, and spermatozoa with prostatic secretion are deposited in the bursa. It is striking that *L. depressa* differs in possessing an open bursa copulatrix, and here the penial style is adapted for reataining its position in the bursa during copulation, having a distinctly angular contour, and bearing four short lateral spines. The occurrences of notable differences in the genital armature in two closely related species is doubtless of importance in establishing and maintaining the two separate species. In the sacoglossan opisthobrach *Elysia maoria* there is no vagina and fertilization is by hypodermic impregnation in spite of the fact that the penis is not armed with a style. The turgor pressure of the penis evidently sufficient to ensure penetration of the body wall of the partner. Prostatic secretion has been found in the haemocoel subsequent to copulation, but no spermatozoa were found there and it is not clear how the spermatozoa injected into the haemocoele pass thence into the bursa copulatrix (Reid, 1964). Hypodermic impregnation is also known in *Alderia modesta.*

The two British cowries *Trivia monacha* and *T. arctica* are very closely related, and were originally grouped together as one species, T. europaea. One of the few deferens between the two species is the size and shape of the penis, that of T. arctica being larger and having a broader, leaf-like shape. This deferens is clearly of great importance in preventing crossbreeding, and so maintaining these two species. A somewhat similar condition is found in the Pulmonata. in the genus Cepaea. The geographical ranges of *C. horrtensis* and *C. nemoralis* overlap considerably, and populations of the two species are sometimes found occupying the same habitat. Under natural conditions the two species are prevented from cross-breeding by small deferens in their courtship behaviour, and also by structural differences. These pulmonate snails possess a dart which is used for reciprocal act.

A sexual reproduction never occurs in Mollusca but the animals may be either monoecious or diocious. Though some of them produce only a few eggs, other produce large numbers. It is stated for example that millions of eggs may be laid by a single oyster. Some snails and some freshwater bivalve molluscs are viviparous.

Molluscan eggs are holoblastic but undergo unequal cleavage. Development typically includes a trochophore stage, but this, is not represented in the development of land and freshwater molluscs. The cephalopods have a larvae that develops the egg or within the egg.

The trochophore larvae is top shaped with a ring of cilia about the margin of the expanded upper portion and with an eye spot at the apex of the body. The mouth opens near the ring of cilia and the anal opening is below at the tip of the body.

Since a trochophore larvae appears in the development of bryozoans and annelids as well as molluscs and since the resembles in a certain

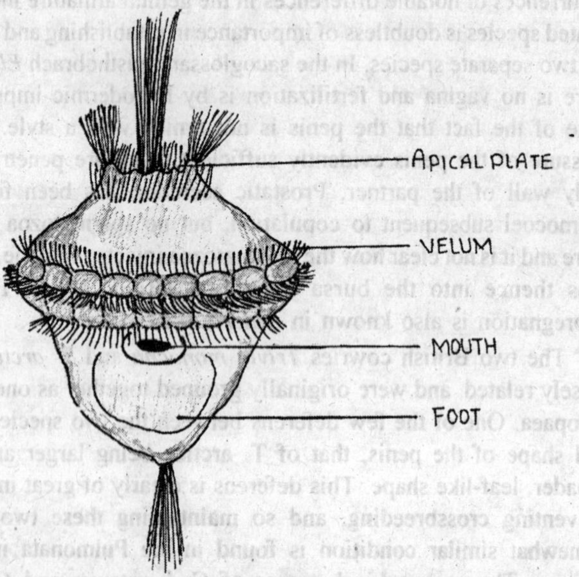

VENTRAL SIDE OF TROCHOPHORE OF LARUAE.

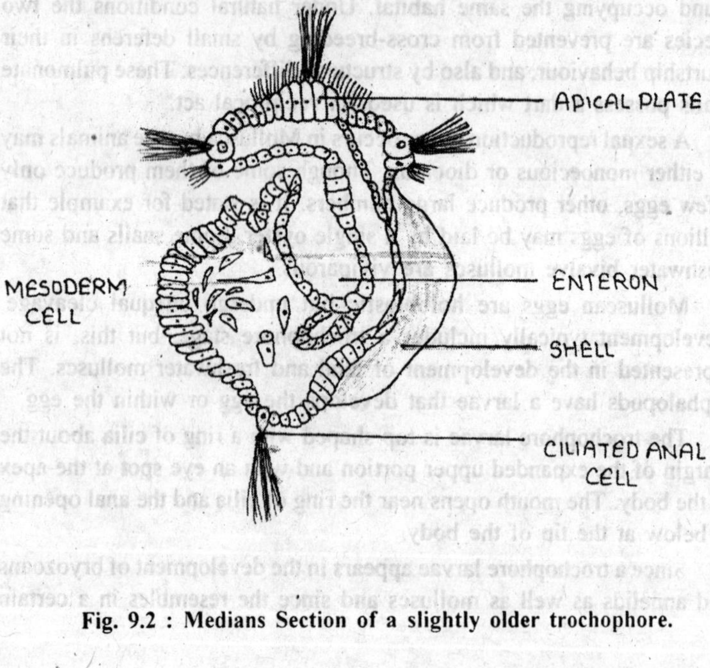

Fig. 9.2 : Medians Section of a slightly older trochophore.

degree some rotifers, the suggestion has been made that these groups may have all descended from a common ancestral from called as a trochozoa.

Regeneration of lost parts and repair of injuries are general among the molluscs, but they do not include replacement of ganglia, the loss of which causes death.

As it is to be expected in such a large group, there is variation in the life history, corelated in part with the varied habitats of the adults. Close relationship with the annelids is seen, however, in the occurrence of a trochophore stage in many forms, particularly in the lower gastropods, or archaeogastropods (*e.g., Patella*) and in the lamellibranches; it is found also in the development of *Chitons* (Amphineura) and *Dentalium* (Scaphopoda). This larva is rapidly transformed into a more complex stage, the veligar larva, which is particularly characteristics of the gastropods and bivalves. In this the prototroch is drawn out into a pair of ciliated lobes, an arrangement that considerably increase the support given to the larva and makes for a more vigorous and controlled locomotion. The reason for this development is apparent in the advanced state of differentiation reached by the veliger.

In the larval life of gastropods there is another complication that is more difficult to explain. This is the process known as torsion, which is a twisting of the viscera through 180° relative to the rest of the body; it brings the originally posterior mantle cavity to the anterior end, and leaves the originally left side of the pallial complex on the right. This transformation, which is effected very quickly in some spp., in few minutes is brought about through the contraction of an asymmetrically arranged retractor muscle. This runs from the right side of the shell to be inserted on the left side of the foot and head.

Torsion, which could conceivably have resulted from a genetic mutation promoting asymmetrical muscular development, facilitates the with drawl of the head and velum, because the mantle cavity is now in a more favourable position. This would have had sufficient survival value to ensure persistence of this new feature, and thereby to establish the fundamental distinguishing character of the class gastropoda. Again in Garstang's words,

Predaceous foes, still drifting by in numbers unabated where baffled now by tact is which there dining plan frustrated. There prey upon alarm collapsed, but Promptly turned about, with tender morsal safe within and the horny foot without —"Sir Garstang"

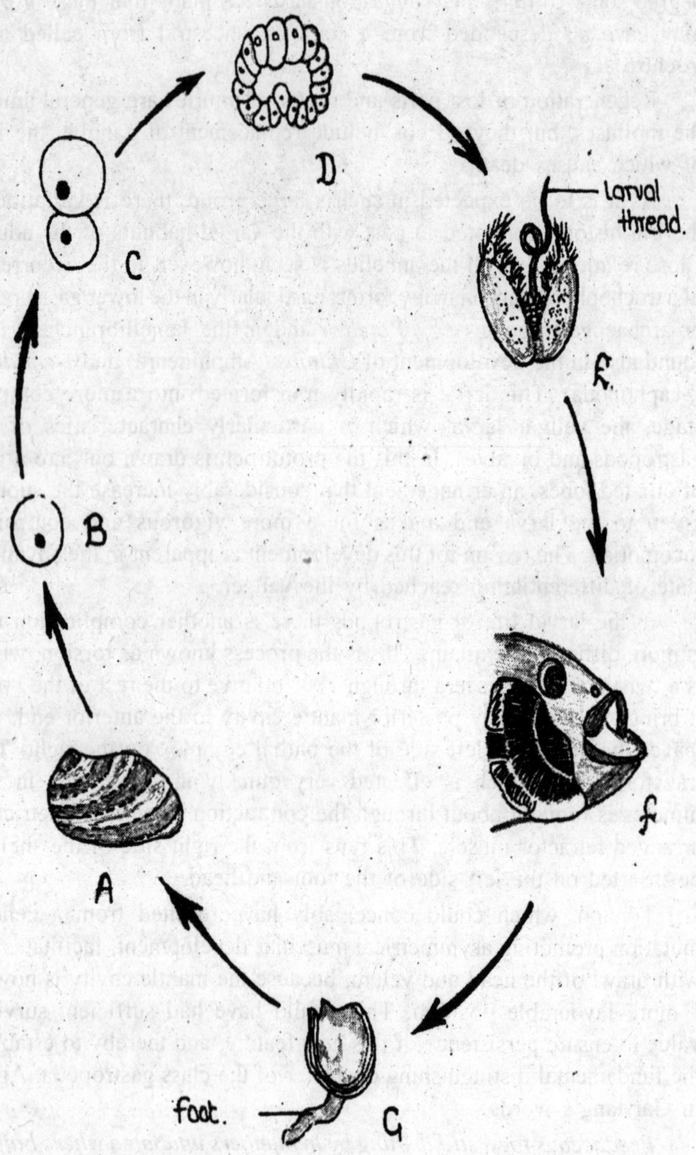

Fig. 9.3 : Illustration of the life history of a fresh water mussel.

A	→ E	:	are of union complanatus.
F	and G	:	of lampsilis Ligamentina
A	→		The adult × about 1/4
B	→		The egg × 55
C	→		Two cell stage, showing unequal cleavage × 55
D	→		Section of the Gastrula × 122
E	→		Glochiclium × 92
F	→		Head of rock bass with operculum cut away to show the glochiclia attached to the gills
G	→		Young mussel one week after leaving the fish

Reproductive Pattern of Mollusca

(Pelecypoda)

Venus merenaria, the hard-shell clam, is also known as the "little-neck" or quahog. It burrows in mud or sand flats near the lower part of the intertidal zone. The quahog is common at Woods Hole, Mass., and can be dug from the mud off Devil's Foot Island. Another clam, *Callocardia*, obtained in the same areas, is similar in appearance, but has a smooth margin to its shell, it is considerably smaller than the mature *Venus*.

Venus is "partially hermaphroditic," but is never self-fertile. It is usually protandrous until the second year. When the animal either remains male or reverses its sex.

Mid-June to mid-August, with a peak period extending from late June through early July. Individual batches of animals will shed for about 20 days only. The temperature of the sea water must be above the critical level of 23 to 25°C.

A. Care of Adults : These animals will remain in good condition indefinitely in the laboratory, if they are kept in large aquaria provided with flowing sea water.

B. Procuring Gametes : Since all attempts to prepare cultures by cutting up the gonads have proved unsuccessful, it is necessary to use naturally-shed gametes. Animals kept in the laboratory have been known to spawn three different times during a single season, extruding the gametes through the excurrent siphon in a fine stream, stated that mature quahogs spawn at night when the water temperature reaches 24.5°C., and do best when they are not covered with sand. To induce shedding, place a beaker containing: an adult into a waterbath, raising the temperature rapidly to 32-33°C. Spawning usually occurs when the animal is cooled slowly. If the normal sea water temperature is above 20°C.. the procedure is not effective until the animal has been pre-treated by chilling in an ice box for 24 to 36 hours.

C. Preparation of Cultures : The ova should be inseminated immediately. Add four or five cc. of sperm suspension to a fingerbowl of eggs and mix well. In a few minutes pour the contents through a gauze 80 stainless steel sieve with 177-micron openings, into a fingerbowl of sea water. This will remove large debris. Repeat, using a finer sieve, which will retain the eggs and allow excess sperm and other small particles to be eliminated. Place the eggs in large battery jars of sea water and aerate continuously. Leave them undisturbed until the early veliger stage, then change the sea water every second day. Add a small quantity of mixed microplankton to the cultures daily; shells should be added to older cultures, to provide surfaces for attachment at metamorphosis. Using this method, Loosanoff and Davis (1950) reared cultures through metamorphosis without difficulty. This kind of experiments were tried in Pantnagar by author also.

D. Extension of the Loosanoff and Davis (1950) have conditioned quahogs so that they spawn in winter. They are taken from their natural beds at a temperature of about 0°C., and placed in trays of running sea water having a temperature of approximately 5 to 7°C. At intervals of three to five days, the temperature is increased by several degrees, up to 20 to 22°C. This conditioning period takes about three weeks. Actual spawning may now he induced by the method outlined above. Not all egg batches will have the same vitality, but with experience, poor batches can he recognized and eliminated.

A. The Unfertilized Ovum : Due to pressure within the ovary, the newly-shed egg is often irregular in shape, and one axis may be longer than the other. When spherical, the egg measures about 70 to 73 microns in diameter. It is surrounded by a gelatinous membrane which swells markedly on contact with sea water the diameter across this membrane measures 167 to 170 microns. The ovum is very yolky and appears white and opaque. It is shed in metaphase of the first maturation division. The germinal vesicle apparently will not break down in sea water.

B. Fertilization and Cleavage : A fertilization membrane is formed, cleavage is spiral and unequal. Poorly defined polar lobes are developed. Gastrulation is by epiboly, followed by invagination.

C. Time Table of Development The following schedule of development at about 22°C. is presented by Loosanoff and Davis (1950). Time is calculated from insemination.

Stage	Time
first cleavage	45 minute
second cleavage	90 minute
ciliated blastula	6 hours
early gastrula	9 hours
trochophore	1.2 hours
straight hinge velliger	24-36hours
fully formed velliger	12 days

D. Later Stages of Development and Metamorphosis : The gastrula elongates to form a top-shaped trochophore, which has no apical flagellum, but which is propelled through the water in a spiral direction by a crown of cilia at its anterior end. A mouth and shell gland are formed. The shell of the early veliger is bivalved and has a straight hinge line. Two adductor muscles are present. The mouth, stomach and small intestine are clearly visible, but the mantle is inconspicuous. The velum is elliptical and well ciliated. In older veligers, the velum degenerates and is replaced by a prominent ciliated foot, with its associated byssus gland. The digestive tract becomes increasingly complex, developing a liver, palps, and a long, coiled intestine. A well-formed mantle is present. At metamorphosis, the larva settles and becomes temporarily attached by the byssus.

Nautilus swims near the bottom or crawls by pulling itself along with its tentacles, like the *Octopus*. The great majority of nautilids are extinct, and their shells, showing varying degree of coiling, are abundant as fossils.

Oliver Wendell, Holmes, while amusing over the shell of a chambered *Nautilus*, wrote the following:

Year after year beheld the silent toil
That spread his lustrous coil;
Still as the spiral grew,
He left the past year's dwelling for the new,
Stole with soft step its shining archway through,
Built up its idle door,
Stretched in his last found home, and knew the old no more.
Thanks for the heavenly message brought by these
Child of the wandering sea,

Cast from the lap, forlorn!
From thy dead lips a clear note is born
Then ever Triton blew from wreathed horn!
While on my ear it rings,
Through the deep caves of thought I hear a voice that sings:
Build thee more stately mansions, O my soul
As the swift seasons roll! leave they low vaulted past!
Let each new temple, nobler than the last,
Shut thee from heaven with a dome more vast,
Till thou at length art free,
Leaving thine outgrown shell by life's unresting sea!

** "Sir Oliver Wendell Holmes"

Male and Female Reproductive Investment

Sexual Differences

Most prosobranchs have separate sexes (gonochoristic) with little or no morphological distinction between males and females except for the reproductive organs. Males may mature at a smaller size and not grow as big as females Fig. 9.5, but the difference is usually slight except among sequential hermaphrodites Fig. 9.7. In some species, such as *Olivella biplicata* males are larger than females.

Energetic expenditure on gametes by large male *Littorina littorea* is about two-thirds that of similarly sized females, the difference narrowing towards smaller sizes Fig. 9.5. Data for other species are sparse, but large sexual differences in reproductive effort are theoretically not expected. Each sex is a limiting commodity for the other, and intra-sexual competition will ensure that both males and females exert the maximum reproductive effort compatible with other demands on their resources. Therefore, if males and females are morphologically and ecologically similar, they will exert similar reproductive effort. *Concordantly* with this prediction, testes reach about 40 per cent of somatic weight in large male *Patella migata* and ovaries of large females reach about 37 per cent.

Sex Ratio

Not only reproductive effort but also the frequencies of males and females are theoretically expected to be equal. If one sex is rare, then since every offspring has a mother and a father, the average member of that sex will be parent to more offspring than the average member of the other sex. Any gene increasing the frequency of the rarer sex will

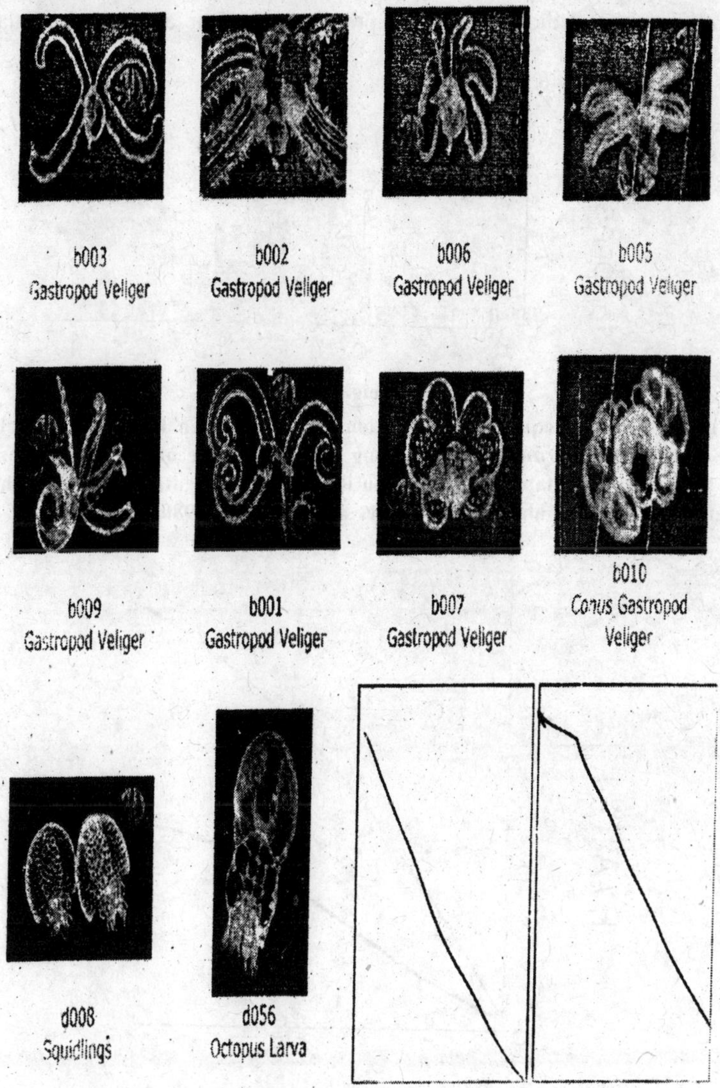

Fig. 9.4 : Larval Molluscs.

therefore be selectively advantageous, but decreasingly so as the frequency rises towards 50 per cent. Sex ratios among prosobranchs confirm quite well to this prediction, but exceptions may occur among sequential hermaphrodites.

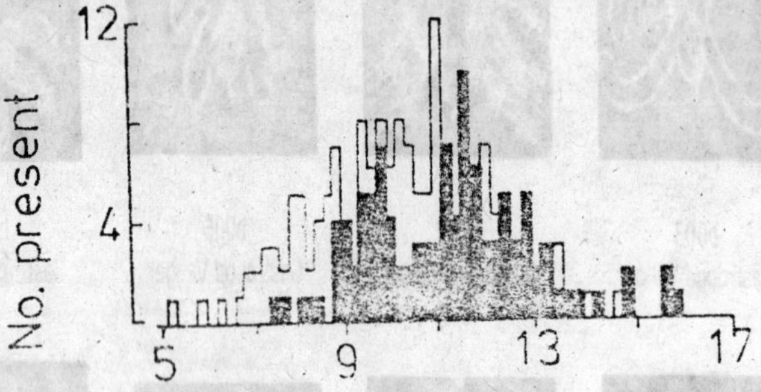

Shell height (mm)

Fig. 9.5 : Size Frequency Distributions of Male (Open Bars) and Female (Solid Bars) *Littorina saxatilis* among a Total Sample of 259 Snails from a Saltmarsh Population. The distribution of males is centred slightly to the left of that of females. After Hughes and Roberts (1980a)

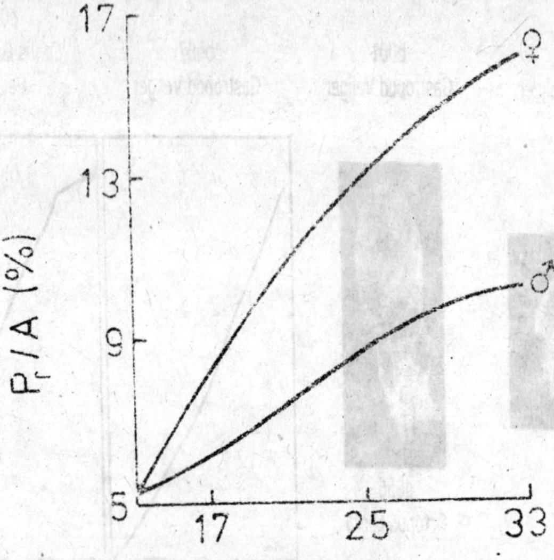

Shell height (mm)

Fig. 9.6 : Reproductive Effort (P.A.). Increases as *Littorina littorea* Grows Larger and Older. As estimated from the drop in mean energy content after spawning, the reproductive effort of males is less than that of similarly sized females. After Grahame (1973).

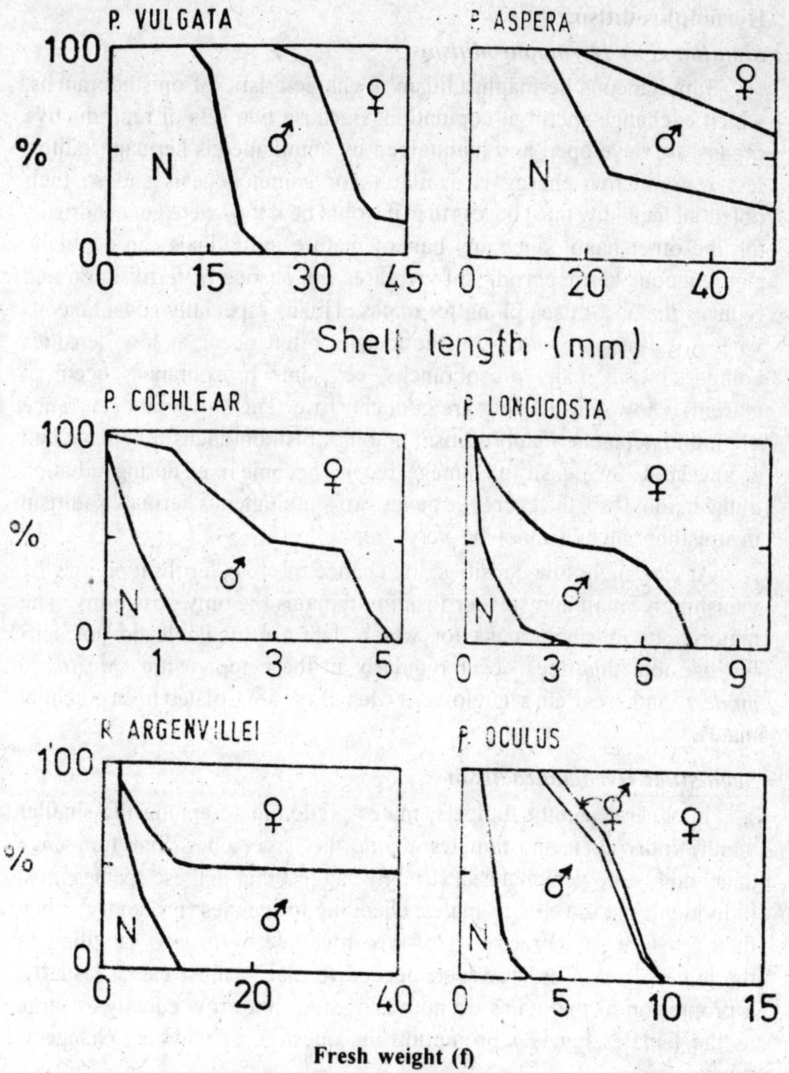

Fig. 9.7 : Protandric Sequential Hermaphroditism is Common Among Species of *Patella*. Discounting individuals with undifferentiated gonads, the ratio of males to females in the population tends to approach unity. The transition of an individual from male to female is usually total and stages with differentiated gonadal tissue of both sexes are very rare, except in *P. oculus*. After Branch (1981).

Hermaphroditism

Simultaneous Hermaphroditism

Simultaneous hermaphroditism is characteristic of opisthobranchs, which exchange sperm at copulation. Because two sets of reproductive organs are developed and maintained by simultaneous hermaphrodites, less reproductive energy is available for gametogenesis and so their potential fecundity must be less than it would be if they were gonochoristic. On the other hand, since any pair of mature individuals can copulate, simultaneous hermaphroditism increases the chance of fertilisation and reduces the cost of searching for mates. This is especially advantageous when meetings are rare. Opisthobranchs often occur at low densities compared with many prosobranchs, yet some prosobranchs occur at extremely low densities but are gonochoristic. The universal occurrence of simultaneous hermaphroditism among opisthobranchs may have had an ancestral, low-density advantage, but has become fixed during radiation of the taxon. If so, the energetic penalty of simultaneous hermaphroditism in opisthobranchs cannot be very great.

At extremely low densities, the chance of cross-fertilisation may be vanishingly small and self-fertilisation remains the only possibility. The majority of opisthobranchs for which data are available do not self-fertilise, but this does occur regularly in the ectoparasitic *Odostomia modesta* and in certain sacoglossans when these are isolated from potential mates.

Sequential Hermaphroditism

In several patellid limpets, males predominate among the smaller mature individuals and females among the larger ones. Size frequency data and histological material strongly suggest that in these species most individuals mature first as males, changing to females (protandry) when they grow larger. Direct tests of this inference by *in vivo* sampling of the gonadal tissue is, however, needed to clarify most cases. Usually, a proportion of the males do not change sex but grow equally as large as the females Fig. 9.7 prompting the question of why sex change is advantageous.

Showed theoretically that sequential hermaphroditism is advantageous when the sexes function best at different sizes. If males function better when small and, or, females function better when large, then protandry is advantageous. Female limpets certainly must exceed a minimum size in order to produce a viable quantity of planktonic larvae but because fertilisation is external, both sexes gain fecundity as they grow and on this account large size would seem advantageous to males also. On the other hand, sperm is cheaper to produce in quantity than eggs, and small individuals perhaps can function effectively as males without detracting too much energy from somatic growth. If males are sufficiently plentiful not to limit fertilisation, reproductive value will be increased at larger sizes by changing from male to female. This is because all eggs are potential offspring and a greater proportion of female than of male reproductive resources is therefore, likely to yield recruits to the next generation. The mixture of gonochoristic males and sequential hermaphrodites presumably represents the counterbalancing advantages of increased male fecundity associated with large size and of becoming female when large.

Protandry is characteristic of most species of *Crepidula*, among which it has clear advantages. *Crepidula fornicata* is a filter-feeder attaching itself to hard objects, such as oyster shells, in muddy bays. Substratum is seldom limiting, because individuals pile up on one another to form stacks Fig. 9.8. Planktonic larvae are attracted to adults and settle among them, developing first as males but becoming female on attaining a certain size. Thus the older, larger individuals in a stack are female,

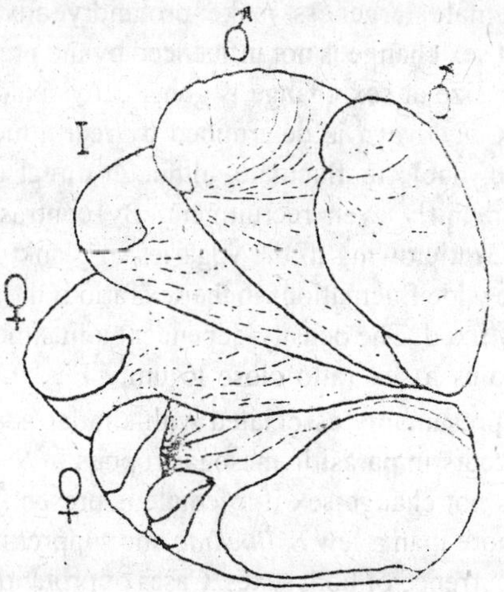

Fig. 9.8 : *Crepidula fornicata* Forms Stacks by the Anraction of Settling Larvae to Adults. Older, larger individuals are female, younger ones are male I-Intermediate stage changing from male to female. After Hoagland (1978).

and the younger, smaller individuals male. The length of the penis increases with distance from the females, allowing all males to reach a female. Moreover, the smaller males are more mobile than the larger ones and can move within reach of most females. Consequently, fecundity in males does not increase with size. Female fecundity on the other hand increases with gonadal volume. Males therefore function best when small and females when large, making protandry advantageous. If larvae fail to encounter adults, they will eventually settle alone, but develop into females that subsequently attract other larvae which will first develop into males and provide the opportunity for fertilisation. There is little risk of inbreeding, because settling planktonic larvae are unlikely to be of the same brood. The size and therefore age at which males become female increases with the density of females Figure 9.9. Apparently there is an underlying physiological propensity to develop as a female but this is suppressed by a pheromone emanating from established females, making adjacent young develop into males.

Crepidula convexa is sympatric with *C. fornicata* but is generally smaller and more mobile. It does not form stacks and males are paired only temporarily with females. Young ones are brooded until they hatch at the crawling stage, but there is no attraction of newly hatched juveniles to the adults, thus promoting outbreeding. The advantages of male mobility and female largeness make protandry advantageous, as in *C. fornicata*, but sex change is not influenced by the presence of females. Since the age or size at sex change is genetically fixed, the sex ratio of populations of *C. convexa* is determined by recruitment and mortality and is, therefore, liable to fluctuate, although direct development will tend to maintain fairly even recruitment. By contrast, recruitment of *C. fornicata* is erratic owing to the vagaries of planktonic development and would cause wide fluctuations in the sex ratio if the age at sex change were genetically fixed. The density-dependent adjustment of sex change, however, maintains a sex ratio close to unity Fig. 9.10.

Protandry, presumably associated with similar advantages to those in *Crepidula*, occurs in parasitic mesogastropods in *Stilifer linckiae*, the small male does not change sex if a female is present. As starfish hosts rarely contain more than a few *S. linckiae*, the suppression of sex change ensures a co-occurrence of both sexes. Cases of protandry have also been recorded in the lanthinidae, the Scalidae, and in the mesogastropod *Lora kurncula.*

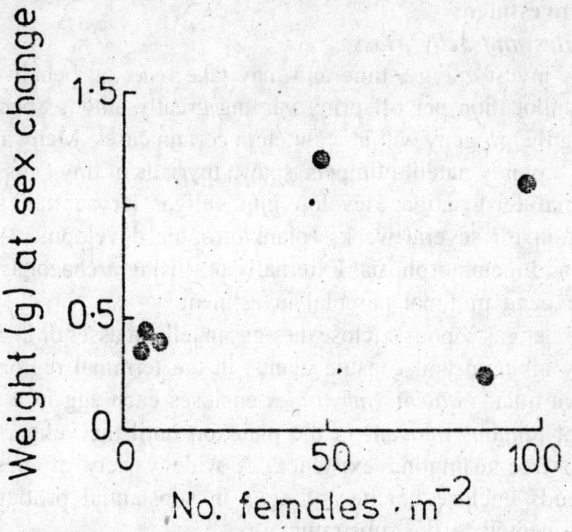

Fig. 9.9 : As they grow. *Crepidula fornicata* Change from Male to Female but Tend to Do so at Larger Body Sizes when Females Are More Numerous. The data are means for different populations. After Hoagland (1978).

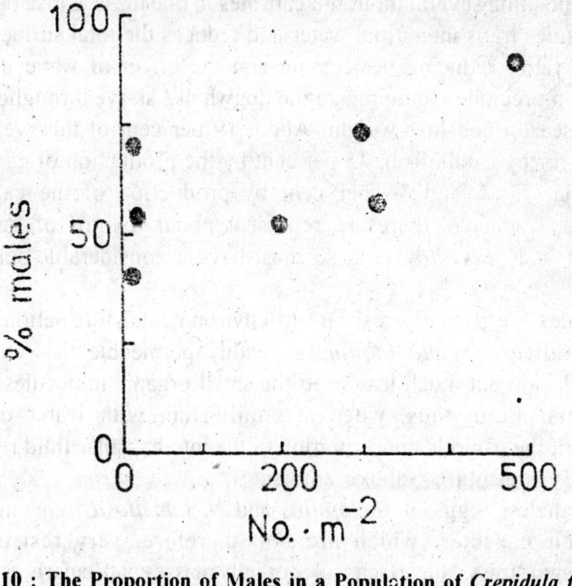

Fig. 9.10 : The Proportion of Males in a Population of *Crepidula fornicata* is Approximately 50% at Moderate Densities, but there is a Tendency of the Sex Ratio to Be Biased Towards Males at High Population Densities. After Hoagland (1978).

Parental Investment

Egg Capsules and Jelly Masses

Parents invest energy, time and may take risks on behalf of their young: the allocation per offspring varying greatly among species and even among the progeny within a clutch in certain cases. Many archaeogastropods, notably patellid limpets, spawn myriads of tiny eggs, which, after external fertilisation, develop into veliger larvae that feed on phytoplankton for several weeks (planktotrophic development) before settlement and metamorphosis. Externally fertilising archaeogastropods therefore expend minimal parental investment.

Most other gastropods enclose the eggs in jelly masses or in capsules secreted by albumin and capsule glands in the terminal region of the reproductive tract. *Littorina neritoides* encloses each egg in a delicate capsule that remains buoyant in the plankton until the veligers match and lead a free-swimming existence. A wide variety of meso- and neogastropods enclose batches of eggs in substantial protienaceous capsules cemented to the substratum.

Dogwhelks aggregate during the winter and communally opposite their egg capsules in large clumps. *Nucella lamellosa* on the Pacific coast of North America, for example, forms breeding groups of up to 200 females depositing several thousand capsules in one mass. Close packing of the capsules traps interstitial water and reduces the total surface area to volume ratio, reducing desiccation and the effect of wave impact. Aggregation precludes foraging, so the dogwhelks starve throughout the spawning season and lose weight. About 19 per cent of this weight is accounted for by catabolism, 45 per cent by the production of eggs and intracapsular fluid, and 36 per cent by production of the capsules themselves. Capsules, therefore represent about a third of parental investment in *V. lamellosa* and so must have a considerable selective advantage.

Capsules are partially resistant to environmental fluctuations. The capsular wall of *Nucella lapillus* is readily permeable to water and sodium chloride but much less so to the small organic molecules in the intracapsular fluid: thus, when at equilibrium with water of 150 mosmlitre^{-1}, the osmotic concentration of the intracapsular fluid remains about 25-30 mosmlitre^{-1} above ambient.

Nevertheless, eggs of *N. lapillus* and *N. lamellosa* frequently die within their capsules; which are not, therefore, very resistant to environmental stress. Nor are capsules or jelly masses particularly resistant to infection by micro-organisms: capsules secreted by *Urosalpinx cinerea* do not always protect the embryos against fungal attack and the jelly masses of cerithiids and nudibranchs sometimes fail to prevent bacterial and protozoan infestation.

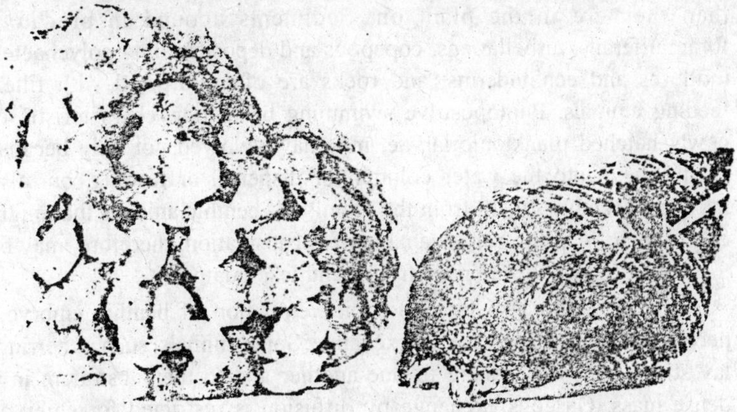

Fig. 9.11 : *Buccinum undatum* Secretes Proteinaceous Capsules, Each Containing a Clutch of Eggs. The capsules are attached to one another in a dome-shaped may that is cemented to the substratum subtidally. Interconnecting spaces around the capsules facilitate ventilation of those beneath the surface of the mass. Deposition of the egg capsules in a cluster may enable the whelk to use small items of hard substratum, such as empty shells, lying on soft sediments or it may confer some protection against predation. Each caosule is about 10 mm long. Several whelks may deposit capsules in the same mass.

Capsules and jelly may offer protection against predation. The egg capsules of *llyanassa obsoleta* spend over a week attached to the substratum during which time over 50 per cent may be eaten accidentally by grazing *Littorina littorea*, but the capsules resist the digestive enzymes, enabling the embryos to survive passage through the gut. Carnivores, however, are more likely than grazers to damage egg capsules; significant proportions of dogwhelk egg capsules are opened by predators and in general capsules appear not to be very effective against predation.

Intracapsular fluid may nourish embryos: intracapsular protein is pinocytosed by kidney epithelial cells of embryonic *Serlesia dira*. But embryos of *N. lapillus* are provisioned with only about 1.1 μl of fluid and they lose weight during development and in *Yenus pennaceus* intracellular fluid represents only 10 per cent of the energy content of the capsule suggesting that nutrition is not generally an important function of encapsulation. What then is the main advantage of this form of parental investment?

Non-motile eggs and embryos, although small, would tend to sink to the sea-bed where risks of predation are probably much greater even than they are in the plankton. Sediments abound in predatory foraminiferans, turbellarians, copepods and depositfeeding polychaetes, molluscs and echinoderms, and rocks are often carpeted with filter-feeding animals. Photopositive swimming behaviour, characteristic of newly hatched planktonic larvae, may have evolved not only because it lifts them into the water column for dispersal or feeding, but also because it removes them from the vicinity of benthic animals that might engulf them. A principal advantage of encapsulation, therefore, may be to protect eggs and embryos from benthic predators.

Encapsulation also facilitates the ventilation of benthic embryos, necessary for adequate gaseous exchange. Some animals, such as herring, lay sticky eggs that adhere to one another and to the substratum in a dense mass. Gaseous exchange by diffusion is restricted for embryos deeply embedded in the mass and this may severely retard their developmental rate or even kill them. However, when embryos are embedded in a matrix of jelly, as with a variety of prosobranchs and most opisthobranchs, they can be spaced out from one another, so increasing their ventilation.

But even when embedded in jelly, central embryos can suffer reduced ventilation if the mass is globular. The bullomorph *Melanochlamys diomedea* lays thick gelatinous egg masses in which the central embryos are less well ventilated, and so develop more slowly than the peripheral ones. This causes asynchronous development among the embryos, but synchrony can be induced experimentally by cutting the egg mass into small pieces (Chaffee and Strathmann, 1984). Another bullomorph, *Haminoea vesicula*, from the same habitat lays a thin gelatinous egg ribbon which avoids problems of ventilation and allows synchronous development of the embryos.

The egg capsules produced by many meso and neogastropods may promote ventilation even further than do jelly masses because the intracapsular fluid can be circulated. If the fluid is well mixed, the number of embryos could increase in proportion to the surface area of the capsule, whereas because of limitations on diffusion, they could increase only in proportion to the radius of spherical jelly masses (Strathmann and Chaffee, 1984). The number of embryos per capsule is correlated with capsular surface area in *Conus pennaceus* (Person and Corpuz, 1982).

Brooding

Some gastropods brood their offspring, protecting them from environmental hazards and predation. Brooding generally occurs more frequently among smaller than among larger species, although the reverse may be true of certain trochaceans (Strathmann and Strathmann, 1982). Theoretical explanations for this trend have invoked two possibilities: first, the incapacity of large adults to accommodate the large numbers of young juvenils from the allometric relationships between fecundity and body size; secondly, the ability of large adults to produce young in sufficient quantities to offset the high mortality rates suffered by independent larvae while gaining advantages of dispersal and, since large adults tend to be long-lived, of achieving high reproductive success in years conducive to larval survival and settlement. These *putative* relationships are, as yet, little understood, and the tendency for brooding to be associated with small adult size may result from several different selective forces in different taxa (Strathmann and Strathmann, 1982).

Clear advantages of brooding, irrespective of adult body size, may be seen among littorinids. *Littorina arcana* secretes jelly around each clutch of eggs, cementing them to rocks; but the jelly gland of the sibling species *L. saxatilis* is modified into a brood chamber in which the eggs are retained until hatched Fig. 9.12a. Whether embryos derive nutrition from the brood chamber is not known, but retention of the eggs reduces the number that can be laid in a season compared with the oviparous species, and is a cost of this type of parental investment. The reduced fecundity is compensated by the ability to rear young in places too harsh for external egg masses, such as occur high on rocky shores and on silty saltmárshes (Hughes and Roberts, 1981).

Pteraeolidia tanthina wraps its body round the egg mass while clinging to the substratum (Rose and Hoegh-Guldberg, 1982), and *Bullia melanostoma* holds the eggs in its foot Fig. 9.12b. Parental investment of this nature may be costly because it prevents foraging or renders the, parent more vulnerable to predators. *Neptunea pribiloffensis*, inhabiting Alaskan rocky shores, deposits large egg masses that are readily consumed by sea urchins if left unguarded. The egg masses, however, are usually laid next to the large sea anemone, *Tealia crassicornis*, which consumes any urchins coming within range. Not only does *N, pribiloffensis* use *T. crassicornis* as a 'baby-sitter' to protect its eggs, but also the anemone may benefit from the association if urchins are attracted to the 'bait' provided by the eggs (Shimek, 1981).

Provision of Yolk and the Evolution of Nurse-eggs

A major form of parental investment among gastropods is the endowment of embryos with yolk. Larger eggs containing more yolk hatch into large juveniles, which generally survive better than smaller ones. However, fewer larger eggs can be produced per unit of reproductive effort and they develop more slowly (Fig. 9.12), partly because yolk retards cleavage.

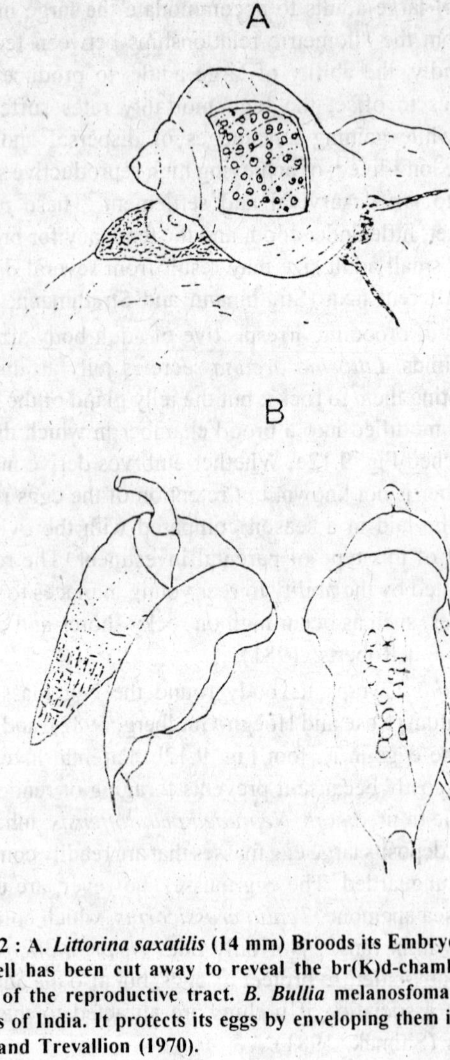

Fig. 9.12 : A. *Littorina saxatilis* (14 mm) Broods its Embryos to the Crawling Stage. The shell has been cut away to reveal the br(K)d-chamber, which is a modified section of the reproductive tract. *B. Bullia* melanosfoma (40 mm) Inhabits Surf-Beaches of India. It protects its eggs by enveloping them in the folds of root. After Ansell and Trevallion (1970).

Gastropods exploit two methods of reducing the risk of mortality associated with prolonged development. First, capsules enclosing larger eggs may be made stronger Fig. 9.12. *Conus vexillum* produces eggs of 140 um diameter, allocating 20 per cent of spawned energy to capsular material, whereas *C. pennaceus* produces eggs of 500 µm diameter, allocating 47 per cent to capsular material, equivalent to a decrease in fecundity of 34 per cent compared with *C. vexillum* (Perron, 1981). Given a fixed amount of available energy (Sibly and Calow, 1982), optimal capsular thickness is likely to be a compromise between added protection, optimal size and number of embryos per capsule. Secondly, prolonged development may be avoided by producing small eggs and provisioning them with extra-embryonic yolk (Spight, 1975).

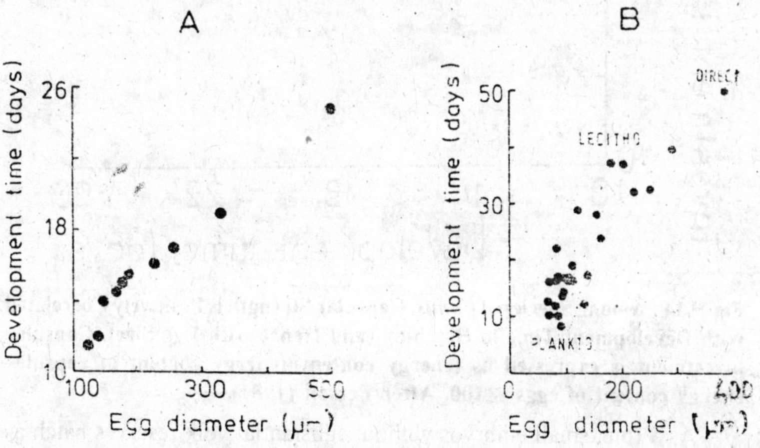

Fig. 9.13 : A. Development Time to Hatching is Positively Correlated with Egg Size among Species of *Conus*. After Perron (1981). B. Development Time to Hatching is Positively Correlated with Egg Size among Nudibranchs, with the Result that on Average those with Direct Development Take the Longest Time to Hatch and those with Planktotrophic Development the Least. After Todd and Doyle (1981).

Deaths are frequent among developing embryos. Mortalities of 20-50 per cent have been recorded among encapsulated muricid embryos and in most cases the abortive individuals are eventually eaten by their siblings. During about the last third of intracapsular development, embryos possess velar lobes with which they are able to enfold and fragment defunct siblings. Consumption of moribund embryos not only helps to keep the capsule free of microbial infection, but also supplies the surviving embryos with extra yolk. Embryos that have consumed yolk develop quickly, commensurate with their small size, but owing to the nutritional subsidy they hatch at a more advanced stage of development.

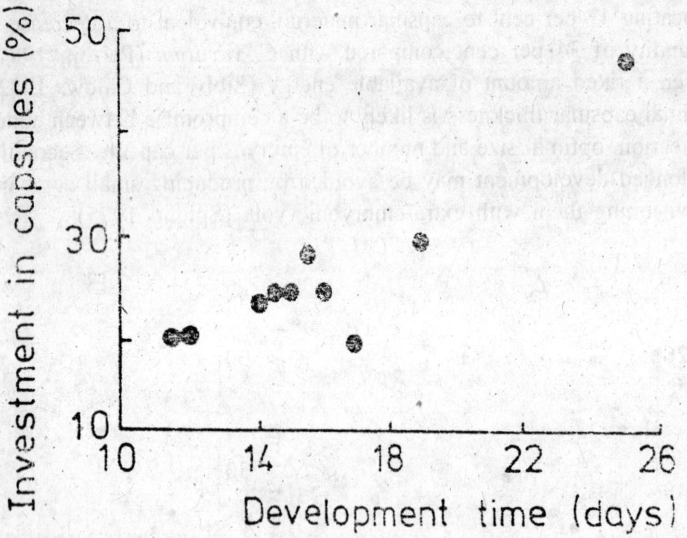

Fig. 9.14 : Among Species of *Conus*, Capsular Strength is Positively Correlated with Development Time to Hatching (and Hence with Egg Size). Capsular investment is expressed as (energy content)/(energy content of capsule– energy content of eggs × 100. After Person (1981).

As a rule, small embryos without substantial yolk reserves hatch as planktotrophic veligers that depend on planktonic food to supply the energy seeded for continued development and survival. Those with intermediate yolk reserves hatch as lecithotrophic veligers that do not require planktonic food but metamorphose within a few days or even a few hours after hatching. Those with large yolk reserves complete metamorphosis within the capsule (direct development) and emerge as crawling young.

Usually each species has a characteristic form of development corresponding to be endowment of yolk to the embryos, but in certain species the endowment seems to involve an element of chance. In certain vermetids and in *Murex imarnatus*, embryos in some capsules develop directly for crawling young, whereas in others of the same brood they emerge as planktonic veligers, depending on whether or not they have eaten abortive embryos. In other cases, the provision of extra-embryonic yolk is more controlled, whereby a fraction of the eggs always abort at some stage in development, acting as 'nurse-eggs' for the survivors.

Nurse-eggs are prevalent among gastropods consistently undergoing direct development, but in some species the proportion of nurse-eggs percapsule varies according to environmental conditions, leading to direct development when the proportion is high and to planktonic veligers when it is low. For example, at certain localities most eggs of *Natica catena* are viable and hatch as veligers, but elsewhere many of them are infertile and are consumed by the viable embryos, which hatch as crawling young. *Elysia cauze* provisions embryos with a yolk ribbon, analogous to nurse-eggs, that ramifies through the egg mass. Development varies from planktotrophy in the spring, lecithotrophy in the summer, to direct development in the autumn, as the amount of yolk ribbon is altered.

Differential parental investment among populations may be genetically determined in some species. *Elysia chlorotica* on New England shores has planktotrophic development in certain populations and direct development in others. When individuals from the two types of population are hybridised, the F_{-1} egg masses contain either small eggs with planktotrophic development or larger ones with direct development, whereas the F_{-2} egg masses have intermediate-sized eggs with planktotrophic development.

Larval Developmental Mode

The relative advantages of planktonic and direct development have long been debated but difficulties of measuring the survivorship and dispersal of larvae have stilled any practical test of the ideas, so logic and circumstantial evidence remain as the only means of appraisal. Dispersal of planktonic larvae could be selectively advantageous when adults have limited powers of dispersal and exploit scattered patches of habitat that vary temporally but randomly in quality. Some of the dispersing larvae will probably colonise newly favourable patches, increasing parental contribution to future generations, while insuring against deterioration of the original site. Larvae, however, drift passively without any means of controlling their direction, and many of them by chance will never encounter a suitable place for settlement. On the other hand, when ready to metamorphose, larvae become photonegative bringing them close to the substratum, which they can explore by touch and olfaction. If the substratum is suitable, the larvae will siere, but if not they will delay settlement for a time, during which a suitable place may be encountered.

This 'delay period' is limited by energy reserves in lecithotrophic larvae but not in planktotrophic ones, among which it shows great interspecific variation. The larval development has a genetically fixed end point and that the potential length of the delay period is governed by the rate at which larvae progress towards it. For example, larval *Ilyanassa obsolete* have lower absorption efficiencies and higher respiration rates than larval *Crepidula fomicata*, causing them to progress more slowly through their developmental programme and so imparting greater delaying capabilities.

Remarkable delaying capabilities are found in certain gastropod larvae. Veligers of *Aplysia juliana* stop growing some 30 days after hatching from the egg mass, having reached the 'competent' stage at which they will respond to the presence of *Ulva*, the preferred food of the adult, by settlement and metamorphosis. In the absence of *Ulva*, the larvae maintain a constant tissue and shell mass; swimming, feeding and remaining competent to metamorphose for over 200 days. An exponentially decreasing survivorship curve Fig. 9.15 throughout the delay period suggest that extrinsic mortality factors account for most larval deaths, out eventually brown pigment accumulates in the larval foot and mantle, indicating senescence, perhaps equivalent to Fechenik's

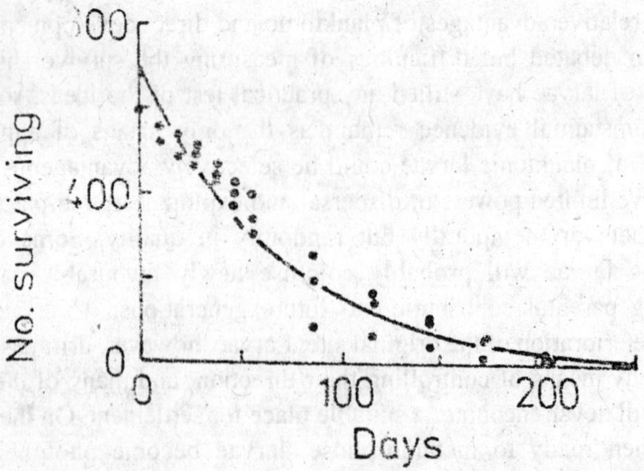

Fig. 9.15 : The Survivorship Curve of *Veliger Larvae* of *Aphysia juliana* Hatched in the Labotatory is Approximately Exponential, Implying a Constant Instantaneous Rate of Molity, in this culture, three veligers were still alive 316 days after hatching. After Keroof (1981).

'developmental end point'. Not surprisingly, *A. julictin* is circumtropical in distribution: currents will carry a drifting larva from Japan to the Hawaiian archipelago in about 77 days, well within the delay period of *A. juliana.*

Veligers able to remain planktonic for a long time (teleplanic) probably traverse the oceans quite commonly. For example, the Caribbean species *Cymatium parthenopeum* produces teleplanic veligers in such quantity that 1.3×10^{12} of them are estimated to cross the North Atlantic each years, having successfully colonised the Azores and occasionally metamorphosed off Ireland, because species with teleplanic larvae have wide geographical ranges, they are probably generalists with wide environmental tolerances. If so, they will be less susceptible to extinction and should have a longer fossil record than species with poorer dispersal capabilities. Fossil shells with protoconchs similar in shape to those of related extant *teleplanic* species would probably themselves have been teleplanic. Although exceptions occur, species with non-planktonic larvae tend to have protoconchs with fewer, wider whorls than those with planktotrophic larvae and of the latter, teleplanic forms tend to have a convoluted rim of the protoconch Fig. 9.16, which accommodates the extended velar lobes. Based on such comparisons, conferred that members of the Bursidae and Cymatiidae with long fossil records were teleplanic.

The maximum delay period of *Aplysia juliana* is at least 6.7 times longer than the developmental time required to reach metamorphosing competence, suggesting that prolonged planktonic life serves only to increase the chance of eventually reaching a suitable place for settlement. This is not to say that long delay periods are advantageous by increasing the mean distance dispersed. Although dispersal is important diminishing returns are likely to accrue to larvae spreading beyond certain limits, which will be greater when patches of habitat are more variable. Beyond these limits, it is the length of the delay, not the distance from the origin, that increases the chance of successful settlement.

Among dispersing larvae, planktotrophy may have two advantages over lecithotrophy. First, because it feeds to sustain itself, the planktotrophic larva can delay settlement for much longer periods than the lecithotrophic larva, which starves after exhausting its nutrient reserves. Secondly, because planktotrophic larvae are not provisioned with nutrients, they are smaller and can be produced in larger quantities. But there are two disadvantages of planktotrophy. First, the small planktotrophic larva hatches at an earlier stage in development and requires a longer period to reach competence, during which time it cannot settle but is exposed to mortality and may be carried away from suitable habitat. Secondly, the planktotrophic larva may find a shortage of food. This risk, however, is reduced when spawning coincides with plankton 'blooms'.

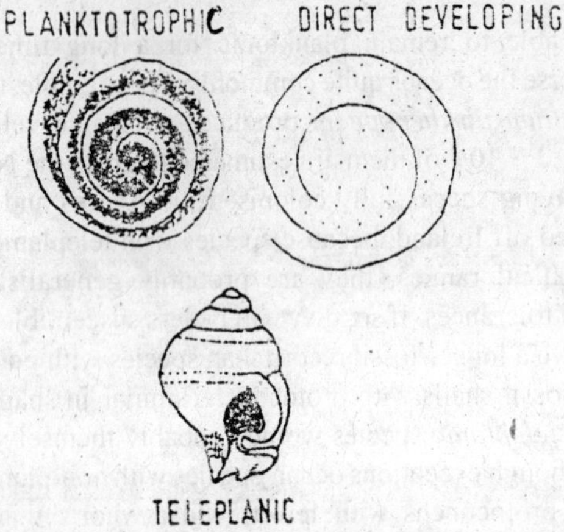

Fig. 9.16 : According to Thorson's 'apex theory', planktotrophic veliger larvae tend to have narrow, many-whorled protoconchs, whereas lecithotrophic or direct-developing veligers of related species have more rounded, wide paucispiral protoconchs. This is illustrated by the protoconchs of *Rissoa guerini* (upper left), which is planktotrophic, and of the closely related *Barleeia rubra* (upper right), which is non-planktotrophic. Both protoconchs are about 1 mm in diameter. After Jablonski and Lutz (1980). Teleplanic veligers, which live for long ceriods in the plankton, often have a folded (sinusigera) lip on the aperture that accommodates the extended velar lobes of the larva. This is illustrated by the protocorch of *Thais haemastoma* (lower centre, 1.4 mm height). After Scheltema (1978).

The suggested that, at least in *Onchidoris bilamellata*, planktotrophic development is advantageous not primarily because it maximises fecundity or increases dispersability, but because it spans the gap between the optimum times for spawning and settlement ('settlement-timing' hypothesis). On local shores, growth and survivorship of *O. bilamellata* interact to give a maximum reproductive value in late February and this is when spawning takes place. Newly metamorphosed *O. bilamellata* are about 0.5 mm long and feed only on barnacle spat, being incapable of handling the larger barnacles eaten by adults. Settlement of *O. bilamellata* larvae, therefore, must coincide with that of *Semibalanus balanoides* in April-May. Planktotrophic development can be prolonged over the

13-week period between optimum times for spawning and settlement, whereas lecithotrophic and direct development would proceed too quickly, causing settlement to fall short of the critical time Fig. 9.18. Because spawning at a certain time of year is associated with a particular range of temperature, altering the egg size is a simple way of adjusting the period from spawning to settlement. This will only work, however, over a seasonal trend of increasing temperatures such as occur in the spring of temperate regions. Larval developmental rate has a Q_{10} of about 2 and laterproduced eggs experiencing higher temperatures will develop quicker, telescoping the progress of staggered clutches so that settlement is synchronised. If spawning occurred when temperatures were falling in the autumn, later clutches would take longer to develop and demographic synchroiiy would be destroyed, with adverse consequences to mating and spawning.

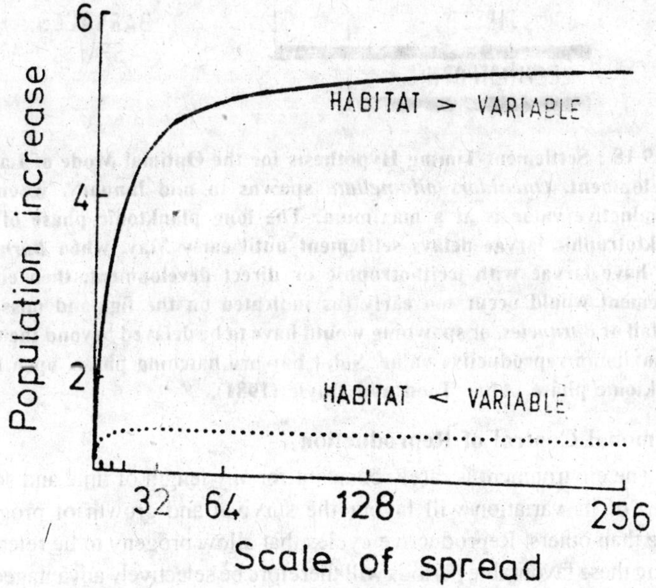

Fig. 9.17 : According to One Mathematical Model, the Advantage of Larval Dispersal Rises Asymptotically (Law of Diminishing Returns) with Increasing Amounts of Spread from the Origin. The advantage is potentially much greater when the habitat is spatially or temporally highly variable. From Palmer and Strathmann (1981).

Physiological constraints, however, may leave no option on developmental mode. Larvae must hatch beyond a minimum size in order to be viable and very small animals may have insufficient resources to produce the large number of eggs necessary to counterbalance the enormous losses in the plankton. Accordingly, tiny gastropods such as *Lacuna parva* and *Littorina neglecta* devote their limited resources to the production of a few well-endowed directly developing eggs, which have a relatively high chance of survival.

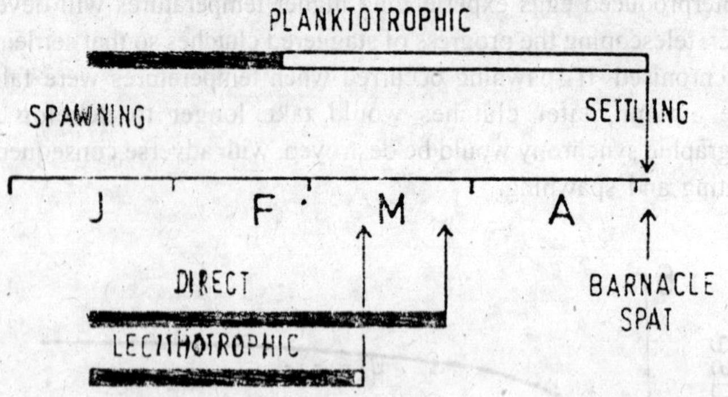

Fig. 9.18 : Settlement-Timing Hypothesis for the Optimal Mode of Larval Development. *Onchidors oilamellata* spawns in mid-January, when its reproductive value is at a maximum. The long planktonic phase of the planktotrophic larvae delays settlement until early May, when *Barnacle* spat have larvae with lecithotrophic or direct development, then either settlement would occur too early (as indicated on the fig) and miss the spatfall of *Barnacles*, or spawning would have to be delayed beyond the time of maximum reproductive value. Solid bar-pre hatching phase, open bar-planktonic phase. After Tood and Doyle (1981).

Hormonal Control of Reproduction

The environment is rarely constant for any length of time and some prizes of its variation will favour the survival and growth of progeny more than others. Reproductive cycles that allow progeny to be released during these favourable periods will therefore be selectively advantageous. Before eggs or young are released into the environment, a complex, protracted series of events takes place involving the mobilisation of stored nutrients, gametogenesis, gamete maturation, fertilisation and perhaps encapsulation or brooding. Each of these events must be appropriately timed if the gastropod is to reach spawning condition when optimum environmental conditions prevail.

Probably all phases of the reproductive cycle are under hormonal control and increasing evidence suggests that reproductive hormone levels in molluscs are influenced by environmental factors such as photoperiod, temperature and nutritional state. A well-documented example comes from work on *Chitons*, close relatives of the gastropods. A decrease in temperature is required to initiate gametogenesis in the *Chiton katherina tunicata* and an increase in temperature is needed to complete it, whereas spawning is stimulated by high phytoplankton levels coinciding with ambient temperatures above a threshold of 12°C. Similar phenomena may be expected among gastropods. Gametogenesis proceeds at temperatures below those necessary for—spawning in *Nassarius* spp., in several Australian prosobranchs and in some opisthobranchs.

It is not yet known to what extent the endocrine system is stimulated directly by environmental factors or indirectly via the nervous system, but there is clear evidence of some interaction between the two systems. The nervous and the endocrine control of molluscan reproduction have been studied most intensively in *Octopus*, in the freshwater snail *Lymnaea stagnalis* and in *Aplysia*. In these normally *semelparous* animals, growth and reproduction appear to be antagonistic processes. Reproductive development of adult *Octopus* is controlled by endocrine activity of the optic gland, itself regulated by the nervous system in response to environmental factors. Activity of the optic gland in young, growing individuals is suppressed by neuronal control from the frontal lobe. Severing the nerves from the frontal lobe allows precocious activation of the optic gland, with consequent sexual development and reduced growth. An analogous neuroendocrinal mechanism controlling growth and reproduction in *Lymnaea stagnalis* involves endocrinal activity of 'light green cells' (so called because of their staining properties in histological preparations) situated within the cerebral ganglia, together with activity of 'lateral lobes' on the cerebral ganglia. The 'light green cells' secrete a growth-promoting hormone, while activity of the lateral lobes modulates this neurosecretion, with the effect that more resources are available for reproduction.

The generality of these reproductive responses to environmental conditions, mediated by the neuroendocrine system, remains to be explored, but differences must be expected in relating to the various patterns of life history. For example, there are apparently large differences in the effect of starvation on reproduction. Starved *iteroparous Chitons* continue to spawn, but at a reduced rate, utilising resources stored before the onset of reproductive activity. In contrast, *Lymnaea stagnalis* is normally a semelparous species which spawns continually after reaching sexual maturity, converting all non-respired assimulated energy into eggs. It does not, however, store any nutrients and, therefore, is incapable of sustaining egg production when starved. There is also considerable variation, among iteroparous species, in the fate of the gonad between successive seasons. On Australian shores *Nodilittorina pyramidalis* completely resorbs unshed oocytes after spawning and the gonad becomes quiescent, whereas *Subninella undulata* stores unshed oocytes until the next spawning season and the gonad continues vitellogenesis.

In addition to controlling the reproductive cycle, hormones also control the expression of gender. As sex chromosomes are apparently absent in molluscs, each individual has the potential to develop into a male or a female. Organ culture experiments have shown that abnormal hormone levels induce penis formation in female *Ocenebra erinacea*, and the development of a vestigial penis by some females is a natural phenomenon in *Urosalpinx cinerea* and *Ilyanassa obsolena*.

Such a labile mechanism for determining gender is conducive to the evolution of hermaphroditism, which evidently has occurred independently in numerous molluscan lineages. In view of this independent evolution, it is not surprising that the mechanism of sex determination varies among taxa. A male-differentiating hormone is produced by neurosecretory activity of the cerebral ganglia in *Calyptraea* spp. but not in *Patella vulgara*. The latter species, on the other hand, produces a hormone in its tentacles which inhibits spermatogenesis but which has not been found in *Calyptraea spp*. Full details of the hormonal basis of sex determination in molluscs have yet to be elucidated.

The neurohormonal control of growth, gender and reproduction offers a potentially useful system for experimentally testing theoretical predictions about life histories, sex ratio, sexual investments and reproductive effort. If the variables considered by these theories, such as the age at first sexual maturity, age of sex reversal, or the number of reproductive episodes, could be controlled under standard laboratory conditions. Often much more direct and, therefore, more convincing tests of the predictions could be made than has been possible hitherto. At the moment there are considerable technical difficulties in culturing many species in large enough quantities, in rearing large cohorts of developmentally homogeneous subjects and in performing operations on the neuroendocrine organs. The benefits to be gained from such an approach are so great, however, that progress may be expected in the future.

Indeed considerable progress has recently been made in understanding the biosynthesis, transport and release of molluscan neurohormones. *Aplysia californica* has proved to be particularly useful for this purpose because on the visceral ganglion are two peripheral clusters of large, protruding nerve cells (bag cells) that are surgically accessible and secrete a reproductive hormone (Egg-laying hormone, ELH) whose effects are easily recognised. Each cluster contains about 400 bag cells and these secrete at least three peptides including ELH. This peptide has 36 amino acids of known sequence and a molecular weight of 4400. It acts like a neurotransmitter on nerve cells and like a hormone on other organ systems. During the reproductive period about half the protein secreted by the bag cells is ELH. The neurological influences of ELH include excitation and inhibition of specific neurons in the visceral and buccal ganglia. This results in the entrainment of a stereotyped pattern of reproductive behaviour, initiated by the inhibition of locomotion and feeding, followed by head-waving and spawning. The hormonal influences of ELH include stimulation of the smooth muscles of follicles in the ovotestis to initiate egg laying.

In addition to its use as a model for neuroendocrinological studies, the bag-cell complex of *Aplysia* has also provided a good experimental system for examining, by techniques of genetic engineering, the structure, expression and modulation of genes that code for a peptide (ELH) of known behavioural function. To do this, a 'library' of the haploid genome was first constructed from the DNA of sperm from a single *Aplysia*. The DNA was digested with an enzyme to break it into short fragments. These fragments were made to attach to a bacteriophage which was then allowed to infect a culture of the bacterium *Escherichia coli*, yielding over a million clones that together would contain and replicate virtually all the fragment types. Messenger RNA. extracted from the bag cells, the visceral ganglion and the digestive gland, was used with the enzyme 'reverse transcriptase' to synthesise DNA identical to the chromosomal DNA from the three sources. Attempts were made to hybridise this radioactively labelled synthetic 'clonal DNA' with the DNA fragments carried by the different clones of *E. coli*. Of 600000 clones screened in this way, two hybridised with bag-cell cDNA and showed no hybridisation with cDNA from the visceral ganglion or digestive gland. The DNA of one of these two clones was mapped using 'restriction endonucleases' to identify the smallest fragment containing homologous material to the bag-cell cDNA. The corresponding fragment of DNA was found by partial-sequence analysis to be comprised of 108 contiguous nucleotides, which encode the 36 amino acids of ELH. This DNA fragment is now being used to analyse the organisation and transcription of ELH genes and for characterising the protein products.

Life Histories

Parental care of development, growth and reproduction constitute life history, and the various conceptual treatments of reproductive effort, semelparity versus iteroparity, sex ratio, hermaphroditism, parental investment and larval development considered in the previous sections are all within the domain of life-history theory. Evolutionary ecologists have tried to produce a general theory of life history that will predict the natural permutations and combinations of these phenomena.

Perhaps the most widely used idea is that of 'r' and 'K-selection' originally proposed by MacArthur and Wilson (1967). The essence of MacArthur and Wilson's reasoning is as follows. If organisms live in situations where their populations are repeatedly reduced to low densities, for example by fluctuating environmental conditions, then genotypes conferring higher rates of population increase during intermittent favourable periods will be selectively advantageous. Rapid population expansion enables organisms to exploit temporarily abundant resources or temporarily benign environmental conditions at the expense of less prolific ones. High potential rate of population increase is associated with high reproductive rate and short generation time, both associated with a particular suite of biological characteristics. High reproductive rate requires high fecundity, and short generation time requires rapid development and early sexual maturity. Since resources available for reproduction are limited high fecundity is associated with high reproductive effort and small progeny. Since weight-specific metabolic rate is inversely proportional to body mass rapid development is associated with small body size. High fecundity coupled with small adult size will depress survivorship and there will, therefore, be a tendency towards semiparity. The opportunistic use of resources will be enhanced by an ability to exploit a wide variety of resources, leading to generalist 'niches'.

Because these properties are advantageous under conditions of unlimited population expansion, MacArthur and Wilson denoted them as characteristic products of 'r-selection' Table 9.1 referring to the symbol for unlimited instantaneous per capita rate of population increase in the 'logistic' growth equation, 'r-selection' may be self-reinforcing, since small size renders organisms more vulnerable to environmental deterioration, making their populations less stable and increasing the advantage of high portential rate of increase.

Table 9.1 : Attributes Theoretically Predicted to Arise from Natural Selection when populations Are Repeatedly Reduced to Low Densities, with the Result that Rapid copulation Increase is Advantages (=selection) and When they Remain Close to on Equilibrium Density with the Result that Competitive Ability is Advantageous (K-selection)

r-selection	**K-selection**
Small body size	Large body size
Early sexual maturity	Delayed sexual maturity
Many, small offspring	Few, well-developed offspring
Brief development	Prolonged development
High reproductive rate	Low reproductive rate
Semelparity	Iteroparity
Short generation time	Long generation time
Opportunistic use of resources	Specialised use of resources
(fluctuating population density)	(steady population density)

Peaks in spawning may occur. These may be due in the first place to a lunar periodicity, The *Chiton, Chaetopleura apiculata, Chlamys opercularis*, and *Ostrea edulis*, lay more eggs between full and new moon than between new and full, although it is uncertain whether it is then the critical temperature for spawning will be reached at different times at different depths *Nassarins reticulatits, Akera bullata*, different age groups may spawn at different times *(Mytilus edulis)*; or each individual may spawn for short periods over the whole breeding season and, if this is arthmical, it may result in a continuous production of young over a long period, however, found *Crassostrea virginica* at different depths and therefore at different temperatures, all spawning simultaneously tight or pressure of water that is the critical factor. In other species a spawning migration may have to occur, as in *Littorina neritoides* in land molluscs there may be a cessation of breeding at the height of the hot or dry seasons so that there appear to be two separate breeding times, one vernal, the other autumnal (*Limax flavus, L. maximus, Lebmannia marginata*).

Other factors play some role in the discharge of the gametes, although they have not been studied at all carefully in molluscs showed that rough weather interfered with spawning in *Chitons*, whereas *Mytilus californianus*, *Patella vulgata* and some other molluscs appear to be stimulated by wave action or mechanical shock.

Effect of Hormones and Neurosecretion

It is likely that hormones and neurosecretion play a more important part in the control of reproductive activity in molluscs than is at present aspect peered. The earliest suggestion of hormonal control was that believed that the production of such secondary sexual characters as the hectocotylus of *Octopus* was under endocrine control. This was denied by Callan (1940) on the grounds that after castration and later amputation of the arm a normal hectocotylus is regenerated. Wells and Wells (1959) have shown that sexual maturity in *Octopus valtags* is brought about by the secretion of a hormone by the optic gland. This is inhibited by a nervous control originating in the subpedunculate and dorsal basal area of the supraesophageal part of the brain. Since the result can also be achieved by section of the optic nerve it seems that this mechanism relates internal function to external conditions in the same way as the eye-hypophysis system of vertebrates. Since all cephalopods except *Nautilus* possess optic glands it may be that this method of control is general throughout the class.

In gastropods one undoubted and some probable examples of sex hormones are known. In *Littorina littorea* males lose the penis after the end of the breeding season and regrow it before the beginning of the next. Similar but 'esser changes affect the ovipositor in females, and the same happens to a small extent in *L. saxatilis*. Other changes also occur in relation to the sex cycle, in that material derived from the regression of the reproductive organs is stored over the winter in connective tissue cells and used later for their reelaboration. The prosobranchs do not lend themselves kindly to many types of experimental work and neither nor, more recently, had any success in attempting to imitate these events by injection of sex hormones. All experimental animals died. Rohlack, however, did find an estrogen (not identical with vertebrate estrogen) in the ovary of *L. littorea*, though she failed to find evidence of androgen production in male winkles have recorded a similar seasonal resorption and redevelopment of the penis in the neogastropod *Nassa obsoleta*, and it seems likely that this phenomenon, as well as the transformation of the genital duct from male to female pattern which occurs in many consecutive hermaphrodites will prove to be under endocrine control.

Retention of the penis in the female phase in Trichotropis may be due to the continuance of some production of sperm in the gonad.

Successfully demonstrated that implantation of pieces of gonad in a number of immature slugs, particularly *Arion subfuscus,* brought about a rapid maturation of the genital glands and ducts, but did not affect penis or genital atrium. The parts affected were mainly those dealing with ova, or sperm received in copulation; this may be correlated with the fact that the implanted gonad was still functional as an ovary but was affected as a testis. The converse experiment, of implanting an immature duct in a mature host, also showed a development of the duct both in size and degree of histological differentiation. Total removal of the gonad in *Limax maximus* was followed 3 months late by a marked regression of the albumen gland and genital ducts, but not of the penis.

Neurosecretory control of spawning has been suggested by Lubet (1957) for *Mytilus edulis* and *Chlamys varia*, on the basis of his finding that the response of these bivalves to other spawning inducers is maximal when there, has been a reduction of the neurosecretory products of the cerebropleural and visceral ganglia, after a maximum coincident with gametogenesis. Removal of the ganglia accelerated spawning. Lubet, therefore, suggested that the neurosecretion inhibits response to the factors which cause spawning.

Effect of Gamones in Fertilization

The existence of endocrine-like substances secreted by gametes and facilitating their union in external fertilization has been known for some time. In echinoderms at least four substances occur: gynogamone I, which activates spermatozoa and is antagonized by androgamone I; gynogamone II (G II) (= Lillie's fertilizin), which agglutinates sperm; and androgamone II, (AII) (antifertilizin) which inhibits the agglutinating effect and dissolves the jelly coat around the egg. The last action may be due to a separate lysin.

In molluscs both GII and AII may occur, but they are absent in certain animals where their presence might be expected. Thus the *Chiton Katharnia tunicata* possesses GII as does the gastropod *Megatbura crenulata*, but the gamone is absent from some other Chitons and from *Haliotis* and *Cumingia.* AII apparently occurs in *Patella vulgata* and *P. coerulea* and in *Megathura cremilata* and *Haliotis cracherodi* at least the sperm of these animals contain a lysin. GII has been recorded from the bivalves *Crassostrea virginica, Ostrea circumpincta Pecten* varius and *Ensis*.

GII appears to be confined to the jelly in which the eggs lie and which seems to be secreted either by follicle cells or by the egg only while it is still in the ovary since its removal stops all further agglutination. The agglutinating substance is precipitated by ammonium sulphate and destroyed by trypsin and therefore seems to be protein although, the nitrogen content (4.5%) is high for a simple protein. In most cases agglutination is reversible, but it is irreversible in *Megathura* and *Katharina.* The agglutinin is heat labile and stable at pH 4 in *Megathura.*

Antifertilizin (AII), which antagonizes GII, is also a protein in *Megatbura.* In *Patella vulgata* and *P. coerulea* it is a basic protein contained in the sperm head. The lysin which destroys the egg membrane was first demonstrated by Tyler (1939) in *Megathura cremilata* and *Haliotis cracherodi.* It is specific in its action a protein and not hyaluronidase. It acts rapidly, dissolving the egg surface in 30 seconds if the jelly has been previously removed, but taking' 4 to 6 times as long if the jelly is present is carried in the acrosome.

In addition to these gamones some molluscs which broadcast their gametes seem to produce substances which activate spawning by other members of the same species: these have been shown to occur in chitons oysters, and mussels. In oysters the spawning agent is heat stable and dialyzable. The females respond only to sperm of the same species, and the activating agent may therefore be another aspect of the gamone. The activating agent inseparable from the agglutinin in *Megathura.* Male oysters, however, are activated by other males, by a variety of eggs or egg waters, and by thyroxin and glutathione. This sensitivity of the males to a heterogeneous collection of stimulants is of value in that it initiates a process of spawning which can then echo and re-echo throughout the population. Nelson and Allison (1940) have shown that the receptors in this case are placed on the gills and that their stimulation induces relaxation of a sphincter on the male duct, resulting in spawning. Experimentally, spermatozoa may be obtained by using sperm or egg water, by electrical stimulation or by heat.

Mention might be made at this point of diantlin, a hormone-like protein discovered by Nelson (1936) in oyster sperm. It has the effect of increasing the ventilation of the mantle cavity by increasing the size of the branchial pores, relaxing the adductors, and accelerating the rate of ciliary beat. Its secretion occurs prior to egg spawning and presumably allows the eggs to pass through the gill passages into the inhalant in chamber more readily. Similar substances occur and act comparably in mussels recorded that the testicular tissue of *Mytilus californianamus* contains a substance causing spawning in females. A similar gonadal stimulant afreets the discharge of sperm in *Tridacna.*

Courtship and Copulation

Gastropods

Copulation is preceded by a period of courtship which may last 2 hours or more in terrestrial pulmonates, whereas it is brief and sometimes hardly recognizable in aquatic gastropods. In some opisthobranchs the period of sexual union is extensive, lasting for several hours and even up to 4-5 days in *Archidoris pseudoargus*, although the actual emission of sperm is of short duration. In many mesogastropods and in hermaphrodites in which exchange of sperm cannot be reciprocal since male and female apertures are widely separate, the partners orientate themselves in the same direction and the male may mount the female, settle on the right side of the body, and even be carried about by her, as in Littorina spp., the markedly sexually dimorphic *Lacuna pallidula*, and *Assiminea grayana*. In some [*Aplysia, Akera, Dolabella, Lymmaea stagnalis* (Bairaud, 1957)] chains of individuals are formed, each acting as male to the animal below. In the majority of hermaphrodites (Pleurobranchidae, Elysiidae, Limapontiidae, nudibranchs, stylommato-phoran pulmonates) copulation is reciprocal. Two animals come to face in opposite directions, approximating the genital apertures on the right sides; this is often preceded by a preliminary recognition when the partners creep round one another. In some opisthobranchs the penis is armed with spines which stimulate the partner and secure a hold, and in the sacoglossans and some eolids the terminal part of the penial duct forms a stylet which in *Acteonia cocksi* and *Limapontia capitata* is driven through the body wall into the bursa. Stimulating organs are more common in pulmonates. The familiar darts of snails such as *Helix, Cepaea, Helicella*, discharged after a period of courtship, rarely penetrate deeply or *cause* harm and are soon removed by muscular movements. In species of *Hehmintboglypta* the dart is not freed, but the partner brings it into action more or less constantly during mating. Some pulmonates enhance precopulatory excitement by bodily contacts by gnawing the partner's body and even the penis. During coition (*coitus*) slugs may have not only the penis everted but also adjacent genitalia. *Agriolimax* has a sarcobelum on the penis which in courtship plays over the surface of the partner; then the atrium is everted, bringing to the surface the duct of the spermatheca, which receives sperm, and also the penis appendage, which links the animals while sperm packets are simultaneously transferred. In *Arion* spp. no true penis occurs and spermatophore transfer is managed by the everted epiphallus with the help of the *ligula*, a special adhesive organ developed on the wall of the atrium or oviduct. In *Limax* the penis is large and complex with one specially thickened fold, the comb, which alone is everted; it may remain so for a period of 90 minutes in *L. tenellus* while the partners revolve clockwise *L. cinereoniger* and *L. maximus* copulate in mid-air suspended from a stout, mucous thread, hanging upside down from a tree or other support with bodies and penis intertwined. At the end of the process they separate, reclimb the mucous rope, and take their own ways. In all these pulmonates sexual biology has become extremely elaborate both anatomically any physiologically.

Because of the rapidity with which the spermatophores explode and because of the nature of the copulatory organ, copulation in cephalopods differs from that of gastropods in a number of ways. The copulatory organ is the hectocotylus, a modification of one or more arms or the spadix, a modification of four tentacles, in *Nautilus*. Only rare genera like *Vampyroteuthis* have no hectocotylus. Normally the hectocotylus receives spermatophores from the funnel and is then thrust into the mantle cavity or buccal region of the female where they are deposited; later it is withdrawn. In Argonauta, *Ocyiboe*, and *Tremoctopus*, however, the hectocotylus (which in these genera carries sperm not enclosed in spermatophores) autotomizes and is left in the female. It contains the distal extremity of the vas deferens dilated into a seminal vesicle, and the whole, normally coiled in a special sac, may, when unrolled, be 10 times as long as the adult male. The hectocotylus may transfer sperm to the neighbourhood of the sperm receptacle on the buccal membrane (*Sepia*), to the mantle cavity (*Loligo*), to a special receptacle near the oviducal opening (*Rossia, Sepiola*), or even into the genital duct itself (*Eledone, Octopus*). To some extent this is variable with the species or with the physiological state of the animal or its eggs. Arms other than the hectocotylus may be used to grip the female during copulation and the process is relatively brief (10 seconds in *Loligo pealei*, 2-5 minutes in *Sepia officinalis*, 10 minutes in *Sepiola atlantica*), although Racovitza (1894) saw a union which lasted for an hour in *Octopus vulgaris*. Speed is necessary not only because of the rapid explosion of the spermatophores but also because the insertion of the hectocotylus into the mantle cavity may interfere with respiration.

Prior to actual union there is often some specialized courtship behaviour, usually involving colour displays. This also helps in sex recognition since the behaviour of the two sexes is different. Thus in *Sepia officinalis* a characteristic black and white pattern of zebra stripes is exhibited by males about to pair, along with an extension of the fourth arm and an opening of an eye toward another squid. If this be a male he displays similarly and this may provoke squabbling and biting, but ultimately one retires. This may happen anywhere in an aquarium, suggesting the absence of a territorial system. If the second animal be female it may display, but it is more likely to swim away, pursued by the male. A pair may then be formed with copulation following. Males treat any animal with resting colour as female and any with a zebra display pattern as male.

Multiple copulations are known, and the process may be repeated up to 4 times in 12 hour.

Aphallic Transfer of Sperm

A number of gastropod species, mainly stylommatophoran pulmonates, show aphallic individuals among the normal euphallic. A small number of prosobranch species among an otherwise euphallic group may be wholly aphallic (Turritellidae, Janthinidae, Cerithiopsidae, Scalidae). This would seem to preclude the possibility of copulation, perhaps even of cross-fertilization; rarely, however, are the consequences so severe. Some prosobranch species achieve an internal cross-fertilization by making use of *spermatozeugmata*, devices elaborated from an apyrene sperm which has attached to it a large number of eupyrene sperm. In *Jantbina* they have been seen to be emitted from the male occur in all parts of the female tract and seem to be a device for ferrying the eupyrene sperm to the female. They are limited to species which are aphallic and which have a gregarious habit. In other species (*e.g., Turritella communis*) aphallism necessitates the sperm being oroadcast, but they seem to be collected by the female out of the same current which bears particulate food into her mantle cavity (since the animals are ciliary microphagous feeders) and stored in sperm receptacles. Fretter has suggested that aphallism in prosobranchs is often associated with a narrow, tightly coiled shell which produces a narrow mantle avacity the respiratory activity of which would be impeded if it had also to house a penis.

In pulmonates partial or complete atrophy of male copularory apparatus occurs in a surprisingly large number of species mostly in the stylommatophoran *Vertiginacea* and *Zonitacea*. These are slow, mainly small animals, not gregarious, and therefore with a reduced chance of copulation. It is also exhibited by a small number of basommatophorans.

So long as self-fertilization can occur, atrophy, patial of complete, of the male organs does not embarrass the species and many permit the establishment of local races showing higher ecological adaptation. Even if self-fertilization is not possible aphallic animals can always act as females, euphallic animals behaving as males. This supposes a rather different copulatory behaviour in those species from that of other stylommatophorans in which copulation may be abandoned if both animals do not act together as male and female; but if it is possible, it is one way in which a trend toward gonochorism could be initiated. Riedel (1955) stated that the absence of copulatory organs and male duct may not be an absolute barrier to copulation because sperm may be passed down the vagina which can evert to act as an interomittene organ allowing some transfer of sperm. The genera in which aphallism has been recorded are: *Physa, Lymnaea, Bulinns, Agriplimax, Vertigo, Truncatellina, Columella, Cbondrina, Acantbimula, Valloina, Retinella, Zonitoides* (Boettger, 1944; Riedel, 1955; Quick, 1960).

Oviposition

Spawn and Capsules

Methods of oviposition vary considerably within the phylum. Primitively the eggs are discharged into the surrounding water as in Amphineura, Archaeogastropoda, Scaphopoda, and most Bivalvia.

Bivalve eggs have only the protection of the vitelline membrane except in the few species in which they are also enclosed in egg sacs (*Nucula delphindonta Modiolaria discors* var. *laevigata, M. nigra Loripes lactens Turtonia minuta*. The secretion for these egg sacs comes from the hypobranchial gland in N. *delphinodonta*, in which the capsules are attached to the posterior end of the valves. In *T. minuta* special, seasonally developed cells in the mantle edge secrete the egg capsules which are attached to byssas threads.

Some archaeogastropods which agglutinate the eggs of one spawning manipulate them with the foot for attachment to the substratum. The secretion for agglutination comes from a variety of sources. In *Diodora apertura* it comes from the egg itself, swelling to a considerable thickness when shed. In Acviaea tessulata the secretion is from the pedal sole which plasters the eggs into a layer one cell thick as in *Diodora*. In *Margarites belicinus* and *Calliostoma zizyphimum* it comes from the enlarged urinogenital papilla and is more copious, protecting the young until they are small snails. Higher gastropods and cephalopods produce egg capsules even when the eggs are retained in a brood pouch, although the wall surrounding the albumen with which each egg is supplied is then very thin. The secretions for making all capsules come from the oviduct and, in cephalopods, also from the nidamental glands which are pallial in origin. The capsules are molded as they pass through the cephalopod funnel, and are then passed to the arms which fix them to a suitable substratum. In gastropods the foot manipulates the spawn, molding it to shape so accurately that for each species every capsule is a replica of the previous one. Some *Littorina* spp. (*littorea, neritoides*) have pelagic capsules extruded from an ovipositor situated near the genital aperture in a position comparable to that of the penis. Here the capsule receives its final form and its outer layers harden in contact with sea water. In other genera there may be a temporary groove on the right side of the foot which conducts the spawn mass from the oviduct to the sole. In the terrestrial *Assiminea grayana* and this groove is permanent and deposits the capsules on the mud as the animal creeps.

The greatest variety of spawn occurs in aquatic gastropods. In opisthobranchs the thin-walled capsules, small and numerous, are embedded in gelatinous secretions which offer sufficient protection for a rather brief embryonic period. In some tectibranchs the spawn is attached by the foot, which grips it as it protrudes from the oviduct. In *Philine* and *Scapbander* spiral strings of eggs in the spawn jelly reflect the circuitous route they have taken through the female duct. In *Bullaria gouldiana*, *Tethys*, and *Aplysia* a tangled egg cord is produced, taking 1-2 hours to leave the oviduct. *Tethys* grips it in a fold of the upper lip, where it is coated with mucus; as the head moves to and fro, the secretion sticks the coils to one another and to the substratum. In the tectibranch *Nauanax inerrnis* the egg string travels forward from the genital aperture between the upturned parapodium and the body, round the anterior part of which it is coiled by rotation of the head; each loop sticks to the preceding to form a skin of threads from which the mollusca later withdraws. Many eolids and dorids gyrate slowly as the spawn leaves the body so that it is anchored in a tight spiral; eolids may attach it to hydroids, seaweed, or bryozoans. In dorids a flat ribbon is formed because the mass is gripped between mantle and foot and the basal edge is plastered to the substratum. In Archidorh pseudoargtis a ribbon 38 × 2 cm contains 5×10^4 eggs. Differences in the body colour of nudibranchs are also present in their spawn. Many sublittoral species migrate onshore to spawn.

Prosobranch capsules have resistant walls and each contains relatively few eggs; with few exceptions the capsules are attached. They may be lens-shaped, globular, triangular in outline, or vase-shaped, sometimes linked to one another in a chain (*Clathrns*), or bunch (*Crepidula*), or piled on one another to form a ball (*Bucciumum*). The wall is often divided by a suture into two halves and sometimes there is also an area plugged with different material. These characteristics derive from features of the glandular oviduct.

The oviduct presumably originated as an open pallial groove with walls thickened, especially laterally, ' y gland cells. In these features it resembles the prostate of *Littorina littorea*, but is more complicated in all females. Production of capsules requires that fertilization occur before the wall is elaborated, so that sperm pouches are associated with the duct. These may be of two types (1) the bursa (copulatrix) into which the penis discharges seminal fluid, and (2) the receptaculum (seminis) where sperm are stored till the time of fertilization. A variety of anatomical relationships may exist between the duct and these pouches. The duct also produces materials to feed and protect the embryos, in areas comprising the albumen, shell, and capsule glands. In some mesogastropods (*Bittium reticulatum, Ceritbiopsis tubercularis, Turritella communis, Clathrus clathrus*) these elaborations occur but the duct is an open one resembling the primitive condition; in the male there is no penis, and sperm are carried in the pallial water current. This condition is probably secondary. Evidence of the closure of the pallial genital duct may be seen in the prostate of *Ocenebra* and *Nucella*. It is incomplete posteriorly, however, leaving an outlet to the mantle cavity in addition to the normal genital pore, perhaps a safety valve for the escape of semen. Actual closure may be witnessed in the protandrous hermaphrodites *Calyptraea, Capulus*, and *Crepidula*, where the open groove of the male becomes the closed oviduct of the female. The ventral wall of the pallial oviduct usually fails to develop glands and so provides an easy pathway for the penis in copulation (*rissoids, calyptraeids, Lamellana*) or for sperm (*Littorina, Nucella, Buccimum*).

Sperm received in copulation are stored in the receptaculum or, if that is absent, in a region of oviduct proximal to the glands (*Cremnoconchus, Pomatias, Lacuna pillidula*). They lie closely packed, their heads embedded in the cytoplasm of the epithelial cells and perhaps deriving nourishment from them. They remain healthy for several weeks; in *Viviparus*, found viable sperm in the receptaculum after 5 months and, in the albumen gland where they are less crowded, after 11 months.

Sperm remain functional in the receptaculum of *Crepidula* for more than a year. In *Littorina littorea.* There are after one copulation. In *Cepaea hortensis*, full oviposition 2 years after copulation and Taylor (1900) mentioned a specimen of *Helix asperrsa* which laid fertile eggs after 4 years in isolation (these are stylommatophoran species which are believed to be self-sterile).

In some prosobranchs special ingesting cells near the receptaculum deal with effete or superfluous spermatozoa (*Acicula, Trivia, Ceritbiopsis, Nucella, Ocenebra, Buccimum, Mongolia*). Excess sperm are also used as nourishment by males: ingestion by the epithelium of the seminal vesicle occurs in the breeding season (*Littorina*) and during oviposition (*Oxycbilus cellarius:*). Other unwanted material may be dealt with in a variety of ways. Yolk from an egg not included in a capsule is taken up by the sperm-ingesting gland (*Ocenebra, Buccinum*). Secretion not used in producing capsules fills a large pouch preceding the brood pouch in Potamopyrgus jenkinsi and corresponding to the bursa of related forms, while in *Rissoella diaphana* a muscular sac at the upper end of the pallial duct collects unused secretions and sperm for discharge to the mantle cavity by a special duct. In many gastropods the bursa is important in this respect. It has a double function, for it harbours the sperm till they transfer to the receptaculum and also retains the secretions in which they were received. In tectibranchs and pulmonates, where the bursa lies freely in the hemocoel with a long duct, the function: may be separate, sperm being deposited in the duct spermatophore evagination, spermatozoid storage and release are still poorly known. The main spawning seasons covers spring and summer, but winter spawning has also been observed. In general terms, the Mediterranean presents much milder winter conditions than the Atlantic coasts of Europe, and the winter phase of the reproductive cycle may be less pronounced, therefore, in the Mediterranean where the deeper water layers maintain a year round temperature of 12-13°C. This is several degrees above the temperature at corresponding depth along the Atlantic coast. In the English Channel inshore water temperatures higher than 12-13°C are limited to about 6 months of the year, whereas corresponding temperatures (with a much higher summer maximum) are recorded over 9 months in the Western Mediterranean. These differences probably account for most of the variation observed between the *Sepia* populations of these areas. On the whole however, the reproductive cycle appears surprisingly similar inspite of aerial differences.

Mortality

The normal life–span of *Sepia officinalis* appears to vary from 18 months to 2 years; some male individuals may attain a greater age. In order to assess the natural rates of mortality, one needs reliable 'cross sections' of the population. Present sampling methods appear inappropriate for this purpose, especially with regard to juvenile stages. Another entirely open question is how much of the total individual fecundity of a mature females is actually utilized under natural conditions. In order to know more about this, one has to define the various conditions lumped together under the label 'spent females'. In a large-sized species like *Sepia officinalis*, the possibility of intermittent spawning with corresponding prolongation of egg maturation must be considered.

Ecology

The seasonal migrations between shallow and deeper waters bring Sepia into contact with various types of soft bottoms. During reproduction in shore water and immediately after hatching, the animals may also live in close contact with rocky bottoms. The ability of very small juveniles to attach themselves to a hard substrate may be very important then because it allows them to withstand strong water movement without being carried away. At the same time, these animals are able to bury themselves in soft bottoms. The method of blowing up sand by water jets from the runnel requires a contraction from the fins in order to keep the animal on the spot. The vigorous undulating movement of the fins that carries the sand already whirled up onto the back of the animal has often been misinterpreted as the actual burrowing process.

Spawning

In the primitive condition the ova and the spermatozoa were emitted in the exhalant water current and fertilization occurred at random in the sea water. This is still true for the Monoplacophora the Scaphopoda, the majority of the Polyplacophora and the Bivalvia, and some of the Aplacophora. It is also true for some of the most primitive prosobranch gastropods. It need not be supposed, however, that such processes of spawning and fertilization are entirely fortuitous, for the condition of the individual and the maturation of the gonads are dependent on the richness of the food supplies which in turn are dependent on the climate. In temperate, boreal, and arctic seas there are marked seasonal changes in temperature and illumination, and the increase in food supplies and in general metabolism in spring and early summer will inevitably impose a limited period of peak breeding activity. In this process there may be two contrasted features; firstly, a period of increasing metabolic activity with growth and maturation of the gonads; secondly, there may be some trigger mechanism, which is sometimes a threshold temperature, which initiates the act of spawning. Nelson noted the following spawning temperatures for various species of bivalves:

10-12°C	*Mytilus edulis, Mya arearia*
15-16°C	*Ostrea edulis, O. lurida, Pecten irradians, Teredo navalis.*
20°C	*Gryphaea (O.) virginica.*
24-25°C	*Venus mercenaria, Mytilus recurvus.*

When the ambient temperature rises to the stated level, spawning is initiated and with the high concentration of garnets in the sea water a maximal level of fertilization is presumably achieved. Hunter (1949), showed that Hiatella spp. Breed a temperatures below 12°C

Special Provision for the Fertilised Ova

The previous section on "spawning" was confined to consideration of those molluscs in which the ova and the spermatozoa were discharged in clouds into the sea water. In a more advanced condition the spermatozoa are received in the mantle cavity of the female, where the eggs are fertilized before being liberated into the sea, or being laid on the surface of the substratum. Characteristically such eggs receive some special provision from the mother in the additional food material, or some from of protective covering. Fertilization of the ova within the mantle cavity is known to occur in some of the Polyplacophora and in a variety of Bivalvia, the sperm being drawn into the mantle cavity in the inhalant water stream. A few specialized bivalves deposit eggs on the substratum instead of emitting them into the plankton. In *Turtonia minuta*, one of the smallest British bivalves, a few large yolky eggs are produced. These, having been fertilized, are laid in a gelatinous capsule with a tough outer skin which is attached to the byssus and thereby to the substratum. The capsule is probably secreted by the glandular margin of the mantle lobes. Again quire a penis situated on the right side of the body at aconsiderable distance from the original genital aperature. In forms with only one ctenidium the respiratory current passes through the mantle cavity from left to right. The spermatozoa are therefore passed out on the right-hand side. It seems possible that an early stage in the evolution of prosobranchs individuals may have congregated in small clusters prior to spawning, which would increase the density of spermatozoa received in the inhalant current,. If individuals next acquired the capacity to recognize a member of the opposite sex that the exhalant sperm-laden water current was

Reproductive Processes in the Prosobranchia

	Archaeogastropoda	Mesogastropoda	Neogastropoda
Eggs set tree singly into the plankton			
(a) Hatching as trochophores	Halitosis tiiberculata Patella spp. Patina pellucida Gibbula cineraria G.umbilicalis Monodonta lineate		
(b) Hatching as veligers		Littorrina neritoides I. littorea	
Eggs laid in gelatinous layers attached to the substratum.			
(a) Hatching as trochophores	Patelloida tessulata		
(b) Hatching as veligers		Lacuna vincta Bittium reticulatum	
(c) Hatching at crawling stage	Diodora aperture Calliostoma zizphinum Gibbula tumida Cantharidus spp.	Lacuna pallidula Littorina litoralis	
Eggs laid in capsules, all eggs hatch as veligers.		Turritella comminits	Nassarius reticulates
(a) Capsules attached to substratum or shells of other specimens		Apporrhais pes-pelecani Hydrobia ulvae Rissoa spp.	Mangelia nebula Philbertia gracilis
(b) Egg "collars" lie free upon the surface of the sand		Natica poliana	

(c) Egg capsules places in a hole cut in the substratum of food material		Cerithipsis tubercularis Trivia monacha Lamellaria perspicua	
(d) Capsules are guaraded by being covered by the shell of the parent		Capulus ungaricus Crepidula fornicate C.nivea (=nummaria)	
	Theodoxus fluviatilis		
Many eggs laid in capsules; only few hatch, remainder begin eatern		Natica catena	Nucella lapillus Buccinum undatum Coins islandicus Neptunea despecta
(a) Capsules laid on surface of substratum: hatch as veligers			
(b) Capsules laid on surface of substratum; hatch at crawling stage			
Few eggs laid in capsules, all hatch at crawling stage.			
(a) Capsules are abandoned		Natica clausa Onoba striata Cingula fulgida C. semicostata C.cingillus Risssoella diaphana	Uroslpinx cinerea Ocenebra erinacea Conus pennaceus
(b) Capsules are guarded by being covered by the shell of the paren		Calyptraea spp. Crepidula williamsi.	
Viviparous forms.		Littorina angulifera	
(a) Young emerge as veligers	Acmaea rubella		
(b) Young emerge at crawling stage			
1. Marine		L. saxatilis Planaxis sulcatus	
2. Fresh water		Hydrobia ventrosa Viviparus viviparous Paludestrina jenkinsi	

directed at the mouth of the mantle cavity of the female. The development of aciliated pathway from the opening of the vas deferens, crossing the floor of the mantle cavity, would serve to concentrate the outgoing stream of spermatozoa. Delivery of spermatozoa into the mantle cavity of the female would be rendered still more if the cilicted pathway were extended on to a projection from the body of the male. It is natural, therefore that a penis bearing a lateral ciliated seminal groove should develop on the right side of the male. Such a condition occurs in *Littorina Calyptraea chinensis, Crepidula unguiformis, capulus ungaricus*, and *Cypraea* spp., in which the vas deference still opens at the posterior of the mantle cavity.

At the same time, the development of a ciliated pathway on the floor of the mantle cavity of the female, on which the cilia beat inwards towards the genital aperture, would aid in carrying the spermatozoa to the point where the ova are to be fertilised. The embedding of this ciliated pathway in a deep groove would isolate the incoming spermatozoa from the exhalant current of the female, and the fusion of the epithelia superficial to this grove would complete the process of formation of the pallial oviduct. The palliate oviduct serves not only to guard the inward journey of the spermatozoa but also to nourish and to encapsulate the freshly fertilised ova. With the development of a closed pallial oviduct in the female, it is natural that the evolving penis of the male should become more highly specialized as an intraomittent organ, with a sub-central closed vas deferens instead of a superficial ciliated serminal groove. This penis would deliver spermatozoa either into a bursa copulatrix, or directly into a receptaculum seminis sited close to the point at which the ova are to befertilised.

WHEN NUDIBRANCHS HATCH, ARE THEY JUST SMALLER VERSIONS?

Molluscs (clams, snails, slugs, squid etc.) basically develop from an egg to a free-swimmimg shelled larval stage, which we call a veliger larva. As with all living things, there are some variations to that plan and Nudibranchs (and other opisthobranch sea slugs) show quite a bit of variation in how they develop from an egg to a crawling slug. There are three main development types and one can make a bit of a guess at what type a particular species will have by looking at the size-of the egg.

As a general rule each species has its own 'typical' egg diameter-very small eggs almost certainly leading to *planktotrophic veliger larvae*

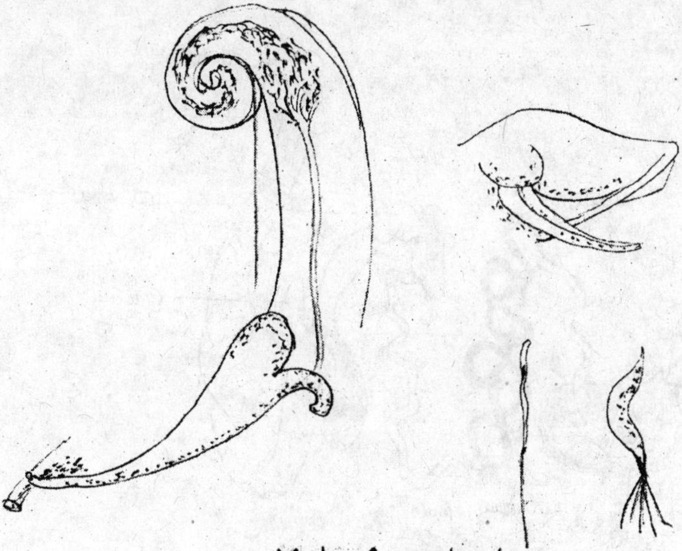

Male Reproductive organs

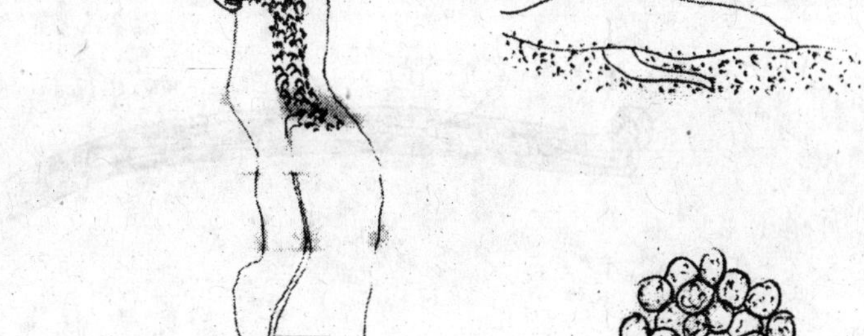

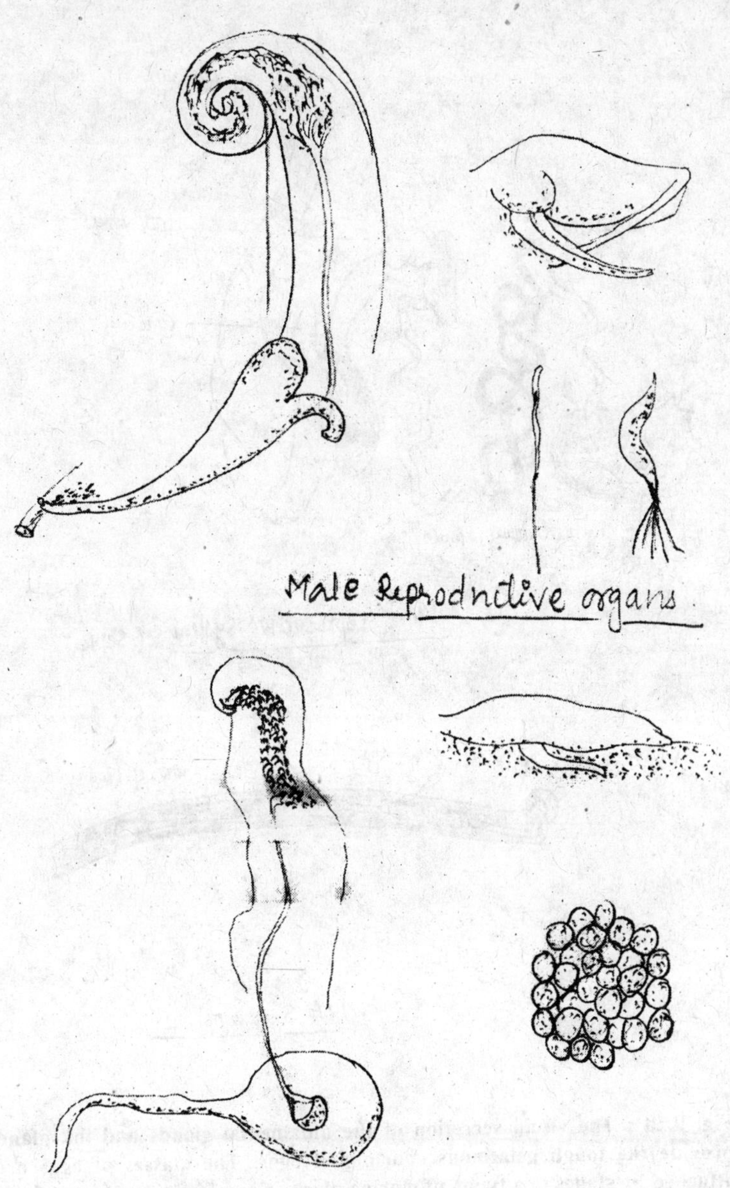

Fig. 9.19 : Female reproductive organs.

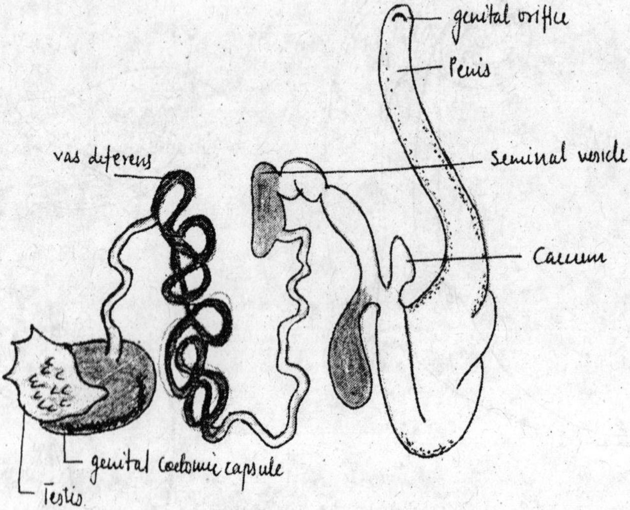

Male reproductive system (Sepia)

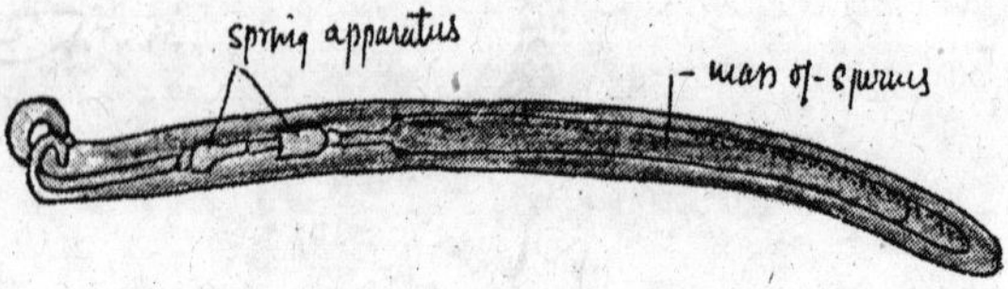

(Sepia) A Spermatophore

Fig. 9.20 : The viscid secretion of the nidamental glands and ink gland provide the tough gelatinous courling of eggs. The masses of eggs are attached by stalles to a twing of marine plant or any foreign body and form the characteristic bunds of 'sea-groups'. The eggs are larger and the developing embryos feed on large amount of stored food yolk. The young are that hatch out are like the adults. No. larvae.

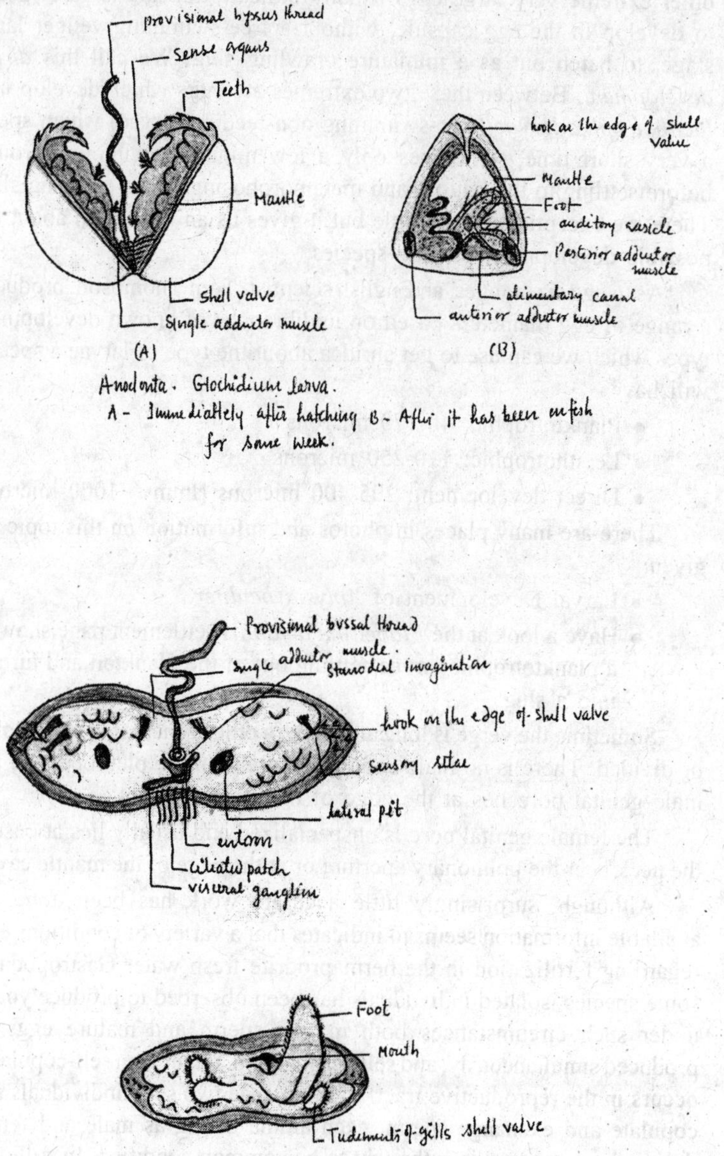

Fig. 9.21 : Anodonta Stages in the metamorphasis of the glochidium.

which spend a long time feeding and growing in the plankton. At the other extreme very large eggs usually indicate that the larva is going to develop in the egg capsule, without a free-swimming veliger larva stage, to hatch out as a miniature crawling slug. We call this *direct development*. Between these two extremes are eggs which develop into *lecithotrophic larvae* (free-swimming non-feeding larvae) which spend a very short time, sometimes only a few minutes, swimming around before settling to the bottom and metamorphosing into a crawling slug. There are exceptions to this rule but it gives us an indication about the possible development for any species.

As to egg size ranges, an English scientist, Tom Thompson, produced a range of egg diameters based on nudibranchs of known development types which we can use to get an idea about the type of larvae a species will have.

- Planktotrophic: 40-170 microns
- Lecithotrophic: 110-250 microns
- Direct development: 205-400 microns [lmm = 1000 microns]

There are many places in photos and information on this topic are given:

- Larval Development of *Aplysia oculifera*
- Have a look at the *Flabellina amabilis* settlement page showing a planktotrophic larvae settling out of the plankton and turning into a slug.

Sometime the verge is long and thin, in other genera it is blunt, lobed or divided. There is no male copulatory organ in the pleuroceridae and male genital pore lies at the edge of the mantle cavity.

The female genital pore is unspecialized and usually lies at base of the neck near the pulmonary aperture or at the edge of the mantle cavity.

Although, surprisingly little accurate work has been done, the available information seems to indicates that a variety of conditions exist regarding fertilization in the hermaprodrite fresh water Gastropoda. In some species isolated indivuduals has been obserbed to produce young, under such circumstances both mature sperm and mature eggs are produced simultaneously, and self-fertilization, rather than self-copulation occurs in the reproductive tract. In other case two such individuals may copulate and exchange sperm, each anima acting as male and female during the process, this is thought to be common condition. In still other instantaneous the ovatestes may produce eggs at one time and sperm at another and therefore, and indivisual may at as either male or female during computation, but not both. Oviposition usually occurs in the spring, although it may continue into summer and early fall. Some species produce few eggs, other hundreds at a time. They are almost invarible deposited in a gelatinous mass on some substrate.

The early development stage occur within the egg man, and by the time the young snail leaves, it has one to two whorls. The viviparidae are ovoviviparus. The individuals being small rather welll developed at birth.

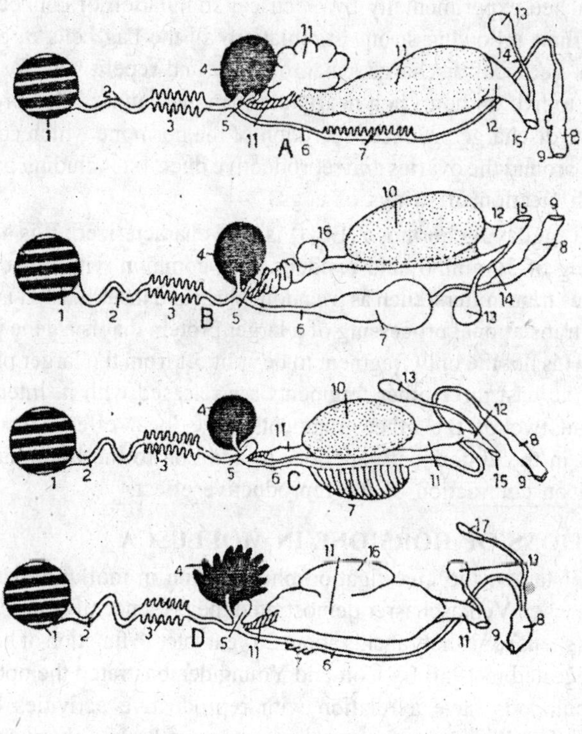

Fig. 9.22 : Reproductive system of male and female individuals of *Bithinia tentaculata* L. F, flagellum, or ovary: T, testis, U, uterus, see legends of foregoing figure for other structure.

HORMONAL SYSTEM

Many molluscs have neuron that have the historical appearance of neurosecretory cells and that changes their apparent secretary activity with conditions such as reproductive state. In a few of these cases, there is experimental evidence for neurosecretory control of reproduction water balance or heart function. We considered as an example the endocrine control of reproduction of *Aplysia.*

The sea hare *Aplysia* is an opisthobranch gastropod. It is a simultaneous hermaphrodite, each in dividable being simultaneously male and female. After copulation, an *Aplysia* lays a long string of more than a million eggs. This egg laying is triggered by a neurohormone secreted by distinct clusters of bag cells in the parietovisceral ganglion copulation presumably leads to activation of bag cells, an effect that can be mimicked experimentally by electrical stimulation of connectives to the ganglion following strong stimulation, all the bag cells of a cluster undergo electrical discharge synchronously and repetitively for several minutes to half an hour, in a nearly all or more manner. This prolonged repetitive discharge produces secretion of the hormone which contracts-muscles around the ovaries and reproductive duct. 1st including ovulation and then ejection of strings of eggs.

The egg laying hormone (ELH) is well characterized. It is a peptide consisting of 36 amino acid residues. It is common with other peptide hormones transmitters, such as β-endonephrin and insulin, ELH is formed by post translational processing of a larger protein that is a gene product. The ELH is not the only fragment to be split off from the larger precursor protein, at least three other fragments are released with it. Interestingly ELH and two of the other fragments have local effects on specific neurons in the abdominal ganglion. In addition to the hormonal effect of ELH on contraction of the reproductive effect.

FUNCTIONS OF HORMONE IN MOLLUSCA

The start of the investigation phenomenon in mollusca dates back as 1935 when Ventaschasrer demostrated the presence of neurosecretory cells in (shells) opisthobranchis about 20 year later Ist function of hormones was suggested in 1956 By Cott and Young demonstrated the optic gland of cephalopods have a relation with reproductive activities of these animals. The Ist report of endocrine control of body volume and ionic metabolism date from 1960.

The concern effects of neurosecretory cells in the pleural ganglion in *L. stagnalis* on body volume of the snail.

Only very recently the role of neurohormones in control of body growth and metabolism has been demonstrated. The activities of animal are co-ordinated and integrated by nervous and endocrine system. Hormones are specific chemical substances produced by particular cells.

These cells secrete the hormonc and pour into the blood or hormone where the hormonal concentration are extremely low. Only a limited number of organs reacts to the presence of there chemical messenger by a change in their activities. These organs are called target organs of the hormones.

Control of activity of the target organs is achieved by feed back mechanism. Process as such as growth reproduction metabolism and regulation of volume and ionic composition of foot blood are under hormone control. There activities have to be regulated with respect to renal change in environment temperature humidity and food availability.

There are some common molluscs in which these functioning have been reported:

- Bivalvia-clams, oysters, mussels
- Cephalopoda—*Octopus*, squid
- Gastropoda-snails, slugs
  - Prosobranchs—*Crepidula*
  - Opisthobranchs—Sea Hare *Aplysia*
  - Pulmonates—*Snails*
    - Stylommatophora-terrestrial-land snails—*Helix*
    - Basommatophora-aquatic snails—*Lymnea*

Nervous system-generally composed of fused ganglia linked by connectives.

Hormonal Control of Reproduction

Prosobranchs

- Endocrine organs (Salcuddin paper)
- Juxtaganglionar organ (JO)
  - Composed of cells scattered in the connective tissue sheath surrounding the cerebral ganglion (*e.g.*)
  - Cells contain large membrane bound neurosecretory granules (peptide containing) that are released during egg-laying
  - Therefore, assumed to have a gonadotrophic role, but not conclusively demonstrated.

Sex Reversal

Marine snail *Crepidula* shows protandric sex reversal

- Protandric—Male maturity phase → hermaphroditic → female maturity phase

- Live in colonies—chain of up to 15 individuals
  - Females on bottom (bigger), males on top and hermaphroditic individuals in middle
  - Members keep contact with tentacles and pallial border of the mantle
  - Secrete factors that control sex change and growth-require contact of individuals - not water soluble.

Masculinizing Factor

- Females release a masculinizing factor from their pallial border
  - Sensed by the tentacles

    Signal goes to cerebral ganglion
    - Neural and hormonal factors produced by cerebral ganglion initiate male pedal morphogenetic activity
    - Pedal ganglion releases a neurohormone that accumulates in the hemal lacunae near the right tentacles (near the penis)

      = penis morphogen
    - Prevents males on top from changing to hermaphrodites and then females

Feminizing Factor

- Males release a feminizing factor from their tentacles
  - Sensed by the pallial border

    Signal goes to pleural ganglion
    - Pleural ganglion releases a neurohormone that results in differentiation of penis
    - Also releases a morphogenetic factor that results in vagina formation
    - Neural signal goes to cerebral ganglion that results in production of a neurohormone that inactivates locomotion (acts on pedal ganglion)

Growth Factor

Males release a factor that triggers somatic growth of lower individuals

Opisthobranchs

All marine hermaphrodites including the sea hare *Aplysia* also have juxtaganglionic organ but function is unknown.

Egg laying hormone ELH (From Scientific American article)

- One of the first invertebrate hormones to be completely characterized
- *Aplysia* egg-laying behaviour - stereotyped series of behaviours associated with ovulation, packaging of the eggs into egg mass and egg-laying
    - Long strings of eggs are produced over a period of several hours and deposited on the substrate.

      Aplysia shows stereotyped head waving, undulation and weaving of the body, head tampts on the substrate associated with egg deposition.
- Bag cells (BC) located in 2 clusters above the abdominal ganglion surrounding the pleurovisceral connective
    - 250-400 cells-large 75 μm diameter multipolar cells contain neurosecretory granules
    - extensive branching - processes end in connective tissues near vascular spaces
    - Extracts of bag cells will elicit egg-laying behaviours when injected into *Aplysia*

Electrophysiology

All bag cells in each cluster are electrically coupled and the two clusters are also connected-this insures that when they fire, all cells do so in union, simultaneously.

- Therefore ⇒ mass release of hormone
- Electrical recording
    - *In vivo* bag cells are normally silent prior to spontaneous egg-laying, then they undergo a period of electrical activity-discharge that is associated with the egg-laying
    - *In vitro* can duplicate this electrical pattern by stimulating the connective or bag cells
        - 60-90 sec of intense high frequency spiking
        - followed by ~20 min of less frequent spikes
        - followed by a refractory period of several hours where the bag cells will not fire - no response to stimulation

Egg-laying Behaviour

- Bag cell hormone release occurs immediately upon the firing of the bag cells

- Ovulation occurs within a few minutes
  - bag cell extracts will induce release of eggs from isolated ovarian tracts *in vitro* within minutes
- Egg-laying initiated –30 minutes after the onset of the electrical discharge and continues over several hours
  - Bag cell hormone = egg laying hormone (ELH) has been isolated (Arch, *et al.,* 1976) and sequenced (Chui, *et al.,* 1979)
    - 36 amino acid peptide m.w. ~4500 Daltons
- 1983 Rothman, *et al.,*-characterized a BCP-9 aa transmitter-like function (autotransmitter)
- 1985 Rothman, *et al.,*-P and γ BCP-pentapeptides also transmitters probably acting on the abdominal ganglion
- Also additional acidic peptide –27 amino acids ~4000 Da sequenced

All five products are released during bag cell discharge and have effects on many different neurons and non-nervous tissues which may be involved in the egg-laying process, *i.e.,*

- L14 A-C in abdominal ganglion inhibited-inking motorneuron
- R15 ""involved in osmoregulation-firing pattern changes
- L10 ""effects heart rate
- B16 and other in buccal ganglion-change feeding behaviour
- ovotestes-causes release of eggs-act on the muscles around follicles
- heart-increases rate and amplitude of contraction

ELH Gene

Gene family of 2-5 members depending on the *Aplysia* species

- ELH precursor prohormone with a hydrophobic signal at one end-the amino terminus
- 25,00-29,00 da peptide produced in the endoplasmic reticulum (ER) and processed in the golgi apparatus and NS granules
- cleaved by a series of peptidases into ELH, α, β and γ BCP and the acidic peptide
- Not all products are in all neurosecretory granules-different neurosecretory granules contain different peptide-different morphological types of granules.

Summary

- Egg-laying-series of behaviours controlled by bag cells
- Coordination involves synthesis of polypeptide precursors containing the various peptides necessary for the full behavioural array. These peptides are proteolytically cleaved from the prohormone and stored in neurosecretory granules.
- NS granules are released during bursts of electrical activity of the bag cells-one of the peptides, a BCP probably acts as an autotransmitter, prolonging the discharge. Bag cells are electrically coupled allowing for the sudden coordinated release of bioactive peptides.

Pulmonates

Lymnaea, Helix, Helisoma

Reproductive Tract

Ovotestes-Ovulation, eggs released, move down hermaphroditic duct where it receives the various secretions from the seminal vesicles, albumen, prostate, muciparous glands and sperm from the bursa copulatrix. Eggs are fertilized and packaged into ootheca for egg-laying.

Brain

Dorsal bodies

Basommatophora-*Lymnaea*

- Non-nervous endocrine cells in two groups near the cerebral commissure (midline)—medial dorsal bodies (MDB)
- Some groups also have lateral dorsal bodies
- Not innervated by CNS

Stylommatophora-*Helix*

- Cells scattered 4-6 groups in connective tissue sheath of the cerebral commissure and subesophageal ganglion
- Innervation from neurosecretory light green cells in the cerebral ganglion-regulate growth also innervated by another cell type containing FMR Famides

Dorsal Body Peptide (DBP)

- DB cells contain electron dense NS granules that are released by exocytosis
- Morphology of cells show features of both peptide secreting and steroid secreting cells

- Therefore, exact nature of DBP is under dispute
  - *Lymnaea*-active DB extract appears to be a polypeptide m.w. 4000-7.000 Da. Activity of golgi apparatus and exocytosis varies with reproductive activity
  - *Helix*-reproductively active DB cells full of lipid droplets, mitochondrial increase-indicate steroid synthesis
    - DB cells synthesize steroid ecdysone *in vitro*
  - *Helisoma*-active DB extract is insensitive to proteolytic cleavage (therefore not polypeptide) and after HPLC separation active fraction contains ecdysone-like material.

Physiological Function

DB control oocyte growth-vitellogenesis, differentiation and growth of accessory sex organs in all pulmonates

- Surgical removal results in retarded growth, reduction in egg-laying and prevents vitellogenesis
- Also affects the activity of the accessory sex organs-especially the albumen gland
  - culture pieces of albumen gland with DB extract, radiolabelled precursors → synthesize albumen
- Role in ovulation
  - DB extract cause mature oocytes to become amoeboid - become mobile
    - EM studies shows rearrangement of actin filaments
    - Motility allows oocytes to move out of the follicle cells and ovarian tract down into hermaphroditic duct where it is moved by cilia and muscle contractions.

Gonads

Basommatophora-remove gonads-no growth, no egg-laying, albumen gland synthesis decrease.

- probably affected via DB

Stylommatophora-more clear-gonads are required for development of accessory sex organs of both males and females.

- Evidence of steroid nature-gonads synthesize steroids *in vitro* especially sertoli cells in *Lymnaea.*
- Also injection of various steroids in vivo influence development of accessory sex organs.

Caudodorsal cells (CDC)

- NS cells produce a CDC hormone (CDCH) which has a very similar function as ELH in *Aplysia.*
- Also stimulate female accessory organs, albumen gland in some freshwater pulmonates *Lymnaea.*

CDC most well known but found in other pulmonates

- Two clusters of cells—Left 20-40 cells. Right 50-100 cells
- 90 μm.
- Axons project to cerebral commissure
  - Some axons form a neurohemal area on the outer connective tissue.
  - Others form a collateral system within the commissure
    - communicate with axons of other neurons in the CNS
- CDC innervated by three synapse-like structures and one true synapse-like structure-2 are cholinergic.
  - This is in keeping with the diversity of factors that will stimulate egg-laying - photoperiod, temperature, water quality, (X content, feeding conditions, etc.

Electrophysiology

CDC are electronically coupled by gap junctions-like bag cells

- Normally silent, in a resting state
- Onset of egg-laying ⇒ electrical discharge
  - high frequency bursting pattern
  - followed by a period of regular firing for ~50 minutes (active state)
  - followed by a 4 hour inhibited state (no firing)
  - and back into the resting state.

Egg-laying

- Ovulation within 5-10 minutes of the burst
- Eggs are fertilized, egg mass formed and eggs are laid after ~2 hours.

Products

Like the bag cells CDC release a number of different peptides required for the complete egg-laying behaviour. CDC hormone alone will not induce whole egg-laying activity.

- single polypeptide precursor-35,000 Da
- cleaved into 3 intermediates
  - CDCH-4500 Da
  - calfluxin-14 Da causes influx of Ca++ to mitochondria of albumen gland accompanied by polysaccharide synthesis
  - α CDCP-probably an autotransmitter-causes maximal release
  - other peptides - at least 3 P CDCP with unknown function
- processing occurs in golgi apparatus and neurosecretory granules-at least two different types of granules.

Lateral Lobes of the Brain

- Contain three types of neurosecretory cells—canopy cells, two droplet cells and the follicle cell.
- Thought to receive photoperiodic cues from the optic nerve
  - coordinate the balance between growth and reproduction by activating the DB and CDC and by inactivating the light green (LG) cells-stimulate growth.

CEPHALOPODS (OCTOPUS, SQUID)

Endocrine organ called the optic gland

- Nervous origin-unlike JO and DB.
- Located on the optic nerve-stellate-presumably produce OG hormone-chemical nature unknown.
- In sexually mature individuals is larger and bright orange.
- Function unknown-can't correlate OG gland activity with reproductive activity.
- OG hormone not sex specific-it regulates yolk production by follicle cells, stimulates growth and synthetic activity of accessory sex organs.
- Under nervous control (inhibitory)—in juveniles severing the nerve tracts causes premature development of gonads.

REFERENCES

Barrington E.J.W. (1987) Cephalopod Taxonmical Studies, Acedmic Press, New York.

Berry S.S. (1912). The Cephalopoda of the Hawaiian Islands, Toronto, Sydeny Pergamon Press.

Boyle P.R. (1983), Cephalopod life Cycle, Academic Press, London, New York, Harcourt Brace Publishers.

Boyle P.R. (1989). "Cephalopod Life cycles". A subsidiary of Hacoyrt Brace lovanavish. Publisher London, New York.

Bullough W.S. (1960). "Practical invertebrates Anatomy" Macmillan and Co. Ltd. New York. St. Martins Press.

Marton Nixon and I. B. Messenger (1977). The Biology of Cephalopodal" The zoological society of London by Academic Press.

Peter V. and Ghramhan H (19992) 144-153.

Purchon R.D. (1968), The Biology of The Mollusca, Pergamon Press London.

Purehon. R.D. "The Biology of the Molluscs"–Pergamon Press Oxfords Newyork, Toronto, Sydny, Paris Frankfust.

Rakshpal Dr. (1971), "Invertebrates structure and function". Prakashan Kendra, New Building and Arnlnabad, Lucknow (India).

Thorson G. (1950) Reproductive and larval ecology of marine bottom invertebrates. Biol. Revs. Cambridge Philson 251-45.

Williamson, D.I. (1951) Studies in Biology of Talitridae (Crustace Amphipoda): Visual Orientation *Talitrus Saltator* J. Marine Bioloassoe. U.K. 30, 91-99.

Young (1962) Biology of Mollusca.

Yonge, C.M. (1960) Mantle cavity, habits and habitats in Blind limpet, *Lepeta Concentrica* Midden droff. Proc. California Academy of Sci 31, 1 03-110.

Zeuthen. E (1947) Body size and metabolic rate in animal kingdom with special regard to microfauna compt. rend. trav. lab. Carlsberg Ser. Chim 26, 17-161.

10

ECONOMIC IMPORTANCE

Molluscs are an economically important phylum.

Mollusca-Classification

Classes

Aplacophora, Scaphopoda, Polyplacophora, Pelecypoda, Monoplacophora, Cephalopoda, Gastropoda

Class	**Sub-Class**	**Order**	**Example**
(1) Aplacophora	–	(a) Neomenioidea	*Neomenia*
		(b) Chaetodermatoidea	*Chaetoderma*
(2) Polypl.acophora	–	(a) Lepidopleurida	*Lepidopleurus*
		(b) Chitonina	*Chiton*
(3) Gastropoda	i. Prosobranchia	(a) Archaeogastropoda	*Haliotis*
		(b) Mesogasrtopoda	*Pila*
		(c) Stenoglossa or Neogastropoda	
	ii. Opisthobranchia	(a) Onchidiacea	*Onchidium*
		(b) Cephalaspidea	*Bulla*
		(c) Parasita	*Thyonicola*
		(d) Anaspidea	*Aplysia*
		(e) Pteropada	*Spiratella*
	iii. Pulmonata	(a) Basommatophora	*Lymnaea*
		(b) Stylommatophora	*Helix*
(4) Pelecypoda		(a) Protobranchiata	*Nucula*
		(b) Filibranchiata	*Mytilus*
		(c) Pseudolamelli-branchiata	*Pinna*
		(d) Eulamellibranchiata	*Unio*
		(e) Septibranchiata	*Poromya*
(5) Cephalopoda	i. Belemmoidaea or Dibramchiata	(a) Decapoda	*Loligo, Sepia*
		(b) Octopoda	*Octopus*
	ii. nautiloidea		*Nautilus*
	iii. Ammonoidea		*Ammonites*

(6) Scaphopoda (Tusk shell)

(7) Monoplacophora (Lamp shells, proboscis worms)

Note: This classification is mluptedfrom Human L. H. (1967) with certain modification from Fiirker & Haswell (1965).

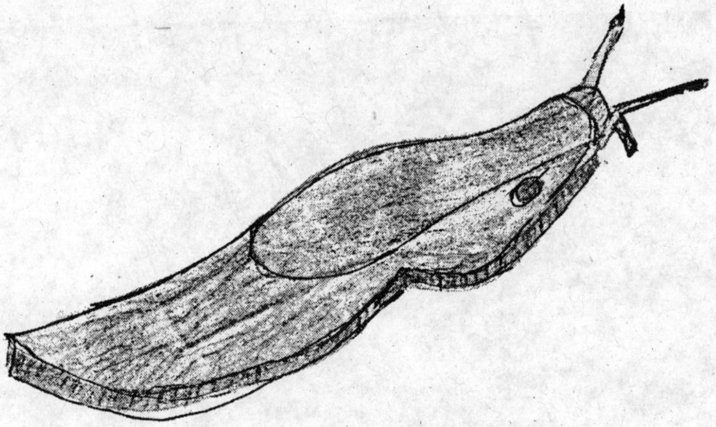

Fig. 10.1 : Slugs are herbivores and can do much damage to garden plants. They have a special long tooth-covered tongue which rasps up hard plant tissues.

Economic Importancc Isopoda

With few exception. Isopods are restricted to small lakes and streams and consequently they are of little importance in the diet of fish. *Sphaeroma terebrans* in perhaps the most important spp. It is borer in salt and brackish water as well as fresh water estuaries of the Gulf-Coast. Causes extensive damage to water waves and piling. *Lirceus brachyurus* some times becomes a pest in commercial beds of water cress. Some isopods serves as intermediate hosts for parasitic nematodes and acanthocephala of birds, fishes and amphibians.

Economic Importance of Pelecypoda

As a result of the utilization of marine shells satisfactory synthetic substitutes, and the gradual depletion of musselbeds, the fresh water pearl button industry has been declining its importance for many years, although the value of the products has held up-well. There are about 40 useful spp and 17 spp of first importance in the manufacture of buttons. There are included in the following genera-Eusconaia, *Megalonaias. Amblena. Quadrula. Actinonaias. Plagioia. Legumia and Lanwsilis.*

The mussels are collected in several ways. The shell of mussels are sold by the tons of the buyers. The beginning of World War II, marked the demise of the pearl button industry and subsequently plastics begin

Fig. 10.2 : This gorgeous sea slug has colourful tassds called cerata on its back. In some sea slug the cerata contain stinging cells, carefully harboured from corals, Hydroidx or similar prey which seem to be functional even though they have passed through the sea slugs gut. In is ân important defence in slugish animal.

to be used almost exclusively for buttons. Mussels are seldom collected primarily for their pearls, the fresh water pearl industry being carried on in conjunction with and in addition to the preparation of shells for the button factories. *Corbicula manilensis*, introduced from Asia, some times is so abundant as to be a nuisance by plugging water ways in pumping plants, irrigation system and the production of sand and gravel. One average pearl, weighing 3.4 grams is reputed to has been sold for $6750. Unfortunately, the finding of really fine pearls is an event of increasing rarity because of the great depletion of our mussel resources.

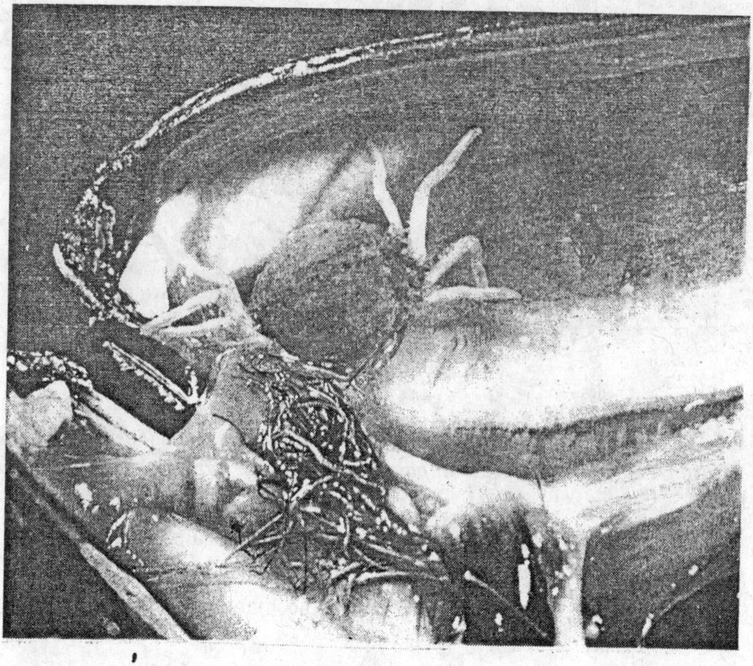

Fig. 10.3 : A common mussel opened to show a pink bodied pea crab living commensally in the mantle cavity. The sea crab does not harm the mussel but uses its hairy claws to thrive food from the mussel's gills.

Economic Importance of Molluscs-(in general)

Molluscs are play role in trophic level or food chain as :

- Important food for various living organisms.

- As host for various parasites.
- Example—Snail acts as host for human blood fluke.
- Some are hervivorous which damage to garden plants.
- Sea—Snails are marine predators feed an mussels and on soft contents.
- They exhibit commensalism.

Fig. 10.4 : Symbiotic association between three sea amemones *Caluactis parasitica* and a Hermit crab. *Eupagurus bernhardus* living in a empty whelk shell.

Example-symbiotic association among three sea anemones *caluactis parasitica* and a hermit crab.

- Peacrab living commensally in the mantle cavity of common mussel.
- Clown fish-*Amphiprion percula*-live in symbiosis with sea anemones.
- As a money earning source.
- European oyster *Ostrea edulis* has been culture since Roman times.
- Oyster are cultured in Japan by various methods.
- The beautiful cowrie *Cypraea histria*, shed of close relatives of this cowrie were used as currency in Africa.

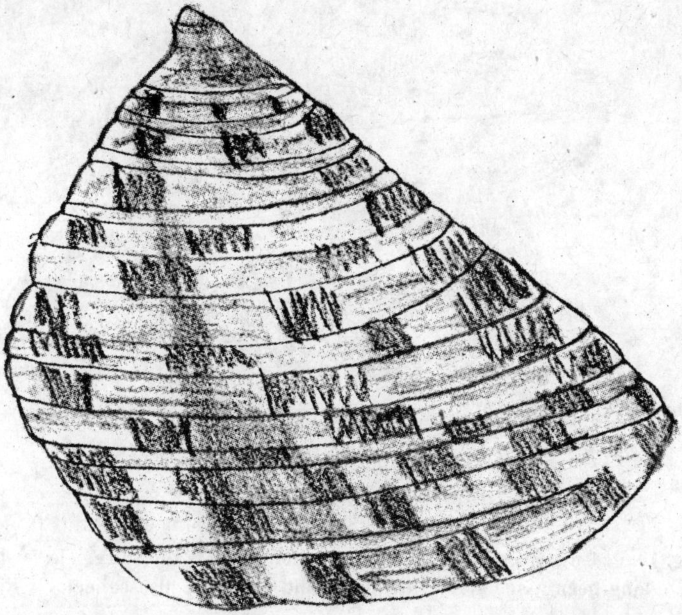

Fig. 10.5 : A painted top shell from the shore owes its pink shell marking to substance called prophyrins.

- Various types of cowries are used commercially for play even.
- These shells have Cony been associated with magic and same are also collectors items.
- *Murex tenuispina* from the Indian ocean is related to the *Murex* snails in the Mediterranean that furnished the framed type purple used as a dye by the Phoenicians and Romans.
- Golden cowries may fetch large prices.
- The shells were frequently imported from the Indian ocean area via England which re-exported them and some 60 tons of money cowries were said to have passed through live paddocks in 1941.

In some families external reproductive structures may be seen. In the amni colidae and valvatidae, for example more or less laterally in the region of the neck is the verge, a projecting male copulatery organ of variable structure and size. In the vivigaridae the right tentacles of the male is much larger than the left and serves as a penis sheath. There are some diagrams of gastropods which showed for beautiful torsion/distorsion thus used in ornamentation.

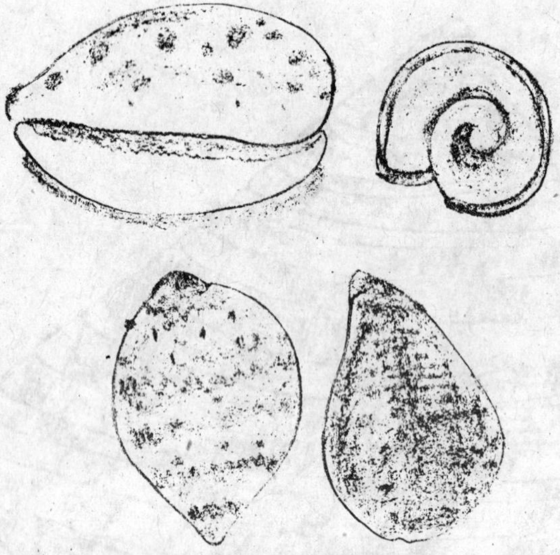

Fig. 10.6 : Cowries, with one seen in section to show the spire. There shells have long been associated with magic and some are also collector's items.

The wentle traps (*Epitonium* spp are superficially resemble to *Turritella* from which the shells are distinguished by closely set ribs or varies ornamenting all whorls.

Cowries (Cypraeidae) are amongst for beauty and variety of colouring of their splendidly polished shells render them conspicuous and valuable. Adults forms are different from others. Young form is elongated with a prominent conical spire, and a long wide mouth aperture bounded by a thin, sharp outer lip. A delicate, periostracur covers the surface. In matured ones mantle flap from each side expands and becomes reflected over the back of shell, edges meeting a little to one side. The mantle and foot are even more vividly coloured than the shell and few objects are more beautiful than large cowry crawling in a coral reef pool the mantle bright with scarlet and yellow and beset with gracefully branched filaments. Cowries live on rocky ground, particularly in and about coal reef. They are often found hiding under boulders at low tide and appear limited on shallow water money cowry. (*Cypraea moneta*) abundant on reef, Eyed cowry (*C. Ocellata*), large Tiger cowry (*C. tigris*) covered with large, bordered spots, Black cowry (*C. mauriliana*) Mole cowry (C. talpa); serpent head cowry (C. caput-serpents), Arabian cowry (*C. arabica*).

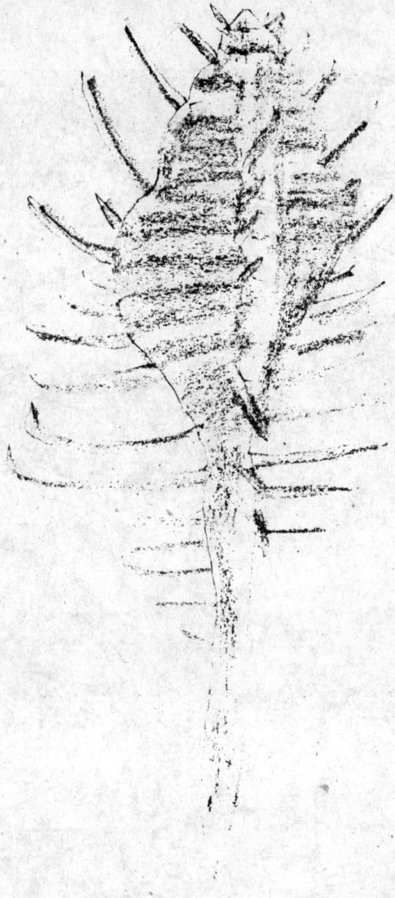

Fig. 10.7 : *Murex temuispina* from the Indian ocean is related to the *Murex* snail in the mediterranean that furnished to framed tyrian purple used as a dye by phoenicians and romons.

Here certain diagrams of molluscan shell are given to show the beauty and diversity of shell torsion and distorsion.

The freshwater mussel is a familiar representative of the Phylum Mollusca. The family Unionidae is widely distributed all over the world and includes nearly all the large freshwater mussels or clams. The family consists of several genera and nearly 1,000 species of which a good number are represented in India. The commonest species in England is the Swan Mussel (*Anodonta cygnea*). The types commonly dissected in India are Unio and *Lamellidens marginalis*. The description that follows

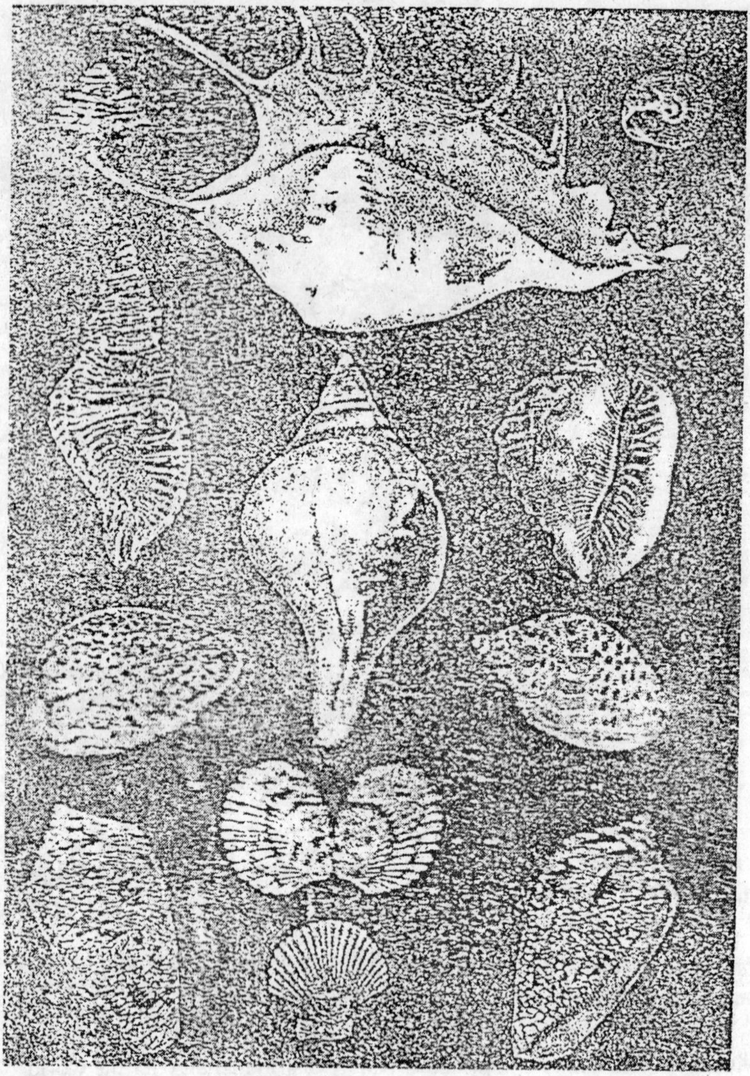

Fig. 10.8 : Some Beautiful Molluscs 1. *Clanculus depictus* (A.Ad.) 2. *Lanibis lambis* (L.) 3. *Cymatium piliare* (L.) 4. *Xancus pyrum* (L.) 5. *Cassis rufa* L. 6. *Cypraea tigris* L. 7. *Babylonia spirata* L. 8. *Cardita bicolor Larn.* 9. *Conus textile* L. 10. *Chlamys senatoria* (Gmelin.)

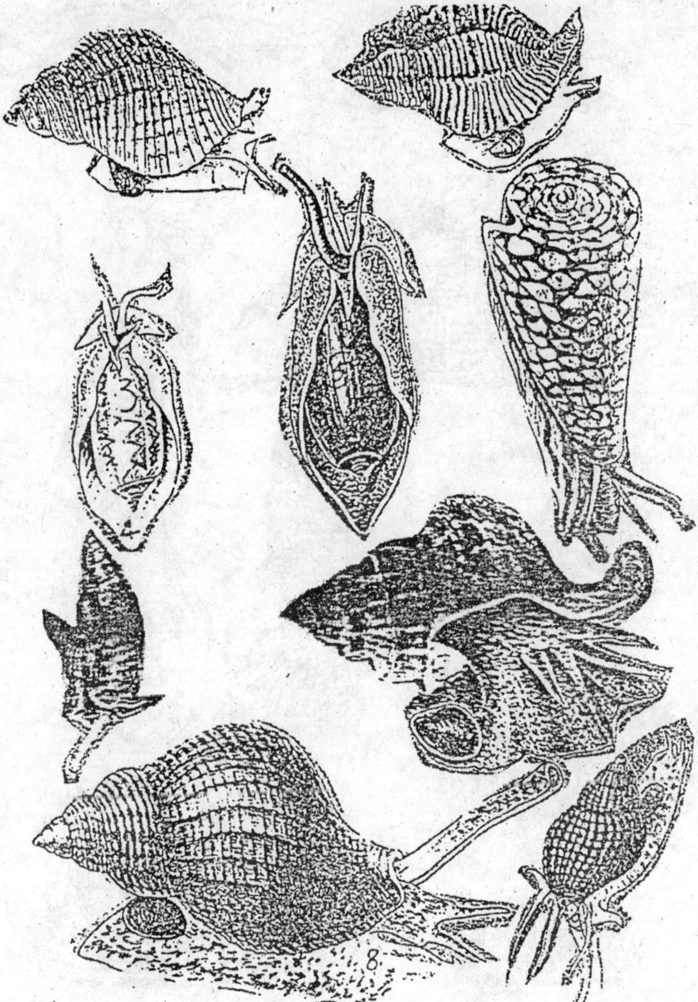

Fig. 10.9 : Marine Prosobranchi 1. Atlanlic dogwink-le *Nucella lapilus*, 2. Ceratostoma *crinaceum* 3. *Conus mormoreus* 4. *Oliva flamulata* 5. Black olive snail *Oliva maura* 6. Myrex dye *Trunculariopsis trunculus* 7. *Mitra cornicula* 8. Waved whelk *Buccinum undatus* 9. *Hinia, reticulate.*

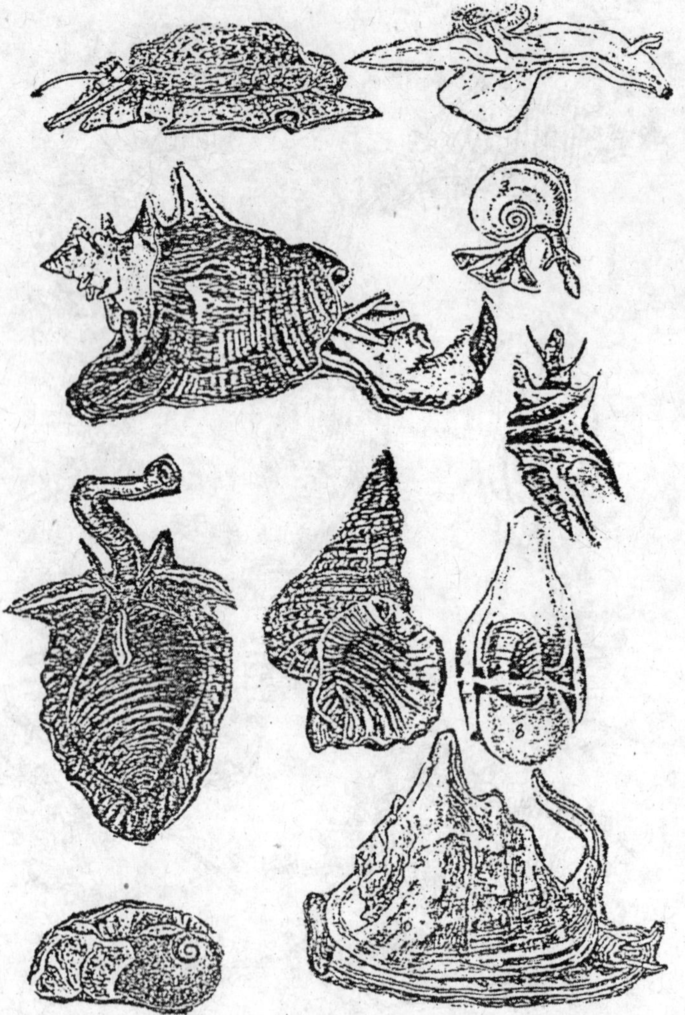

Fig. 10.10 : Marine Taenioglossa 1. Tiger cowries (*Cypraea tigris*) 2. *Carinaria mediferranea* (*pelagic*) 3. *Atlanta per* (*magic*) 4. Pink conch *Stombus gigas* 5. Pelican's foot *Aporrhais pespelecani* 6. Giant tun *Torino galea* 7. Trumpet shell *Charionia tritonis* 8. *Sinum Icachi* 9. *Naticarius uscorum* 10. Large helmet *Cassis cornuta*.

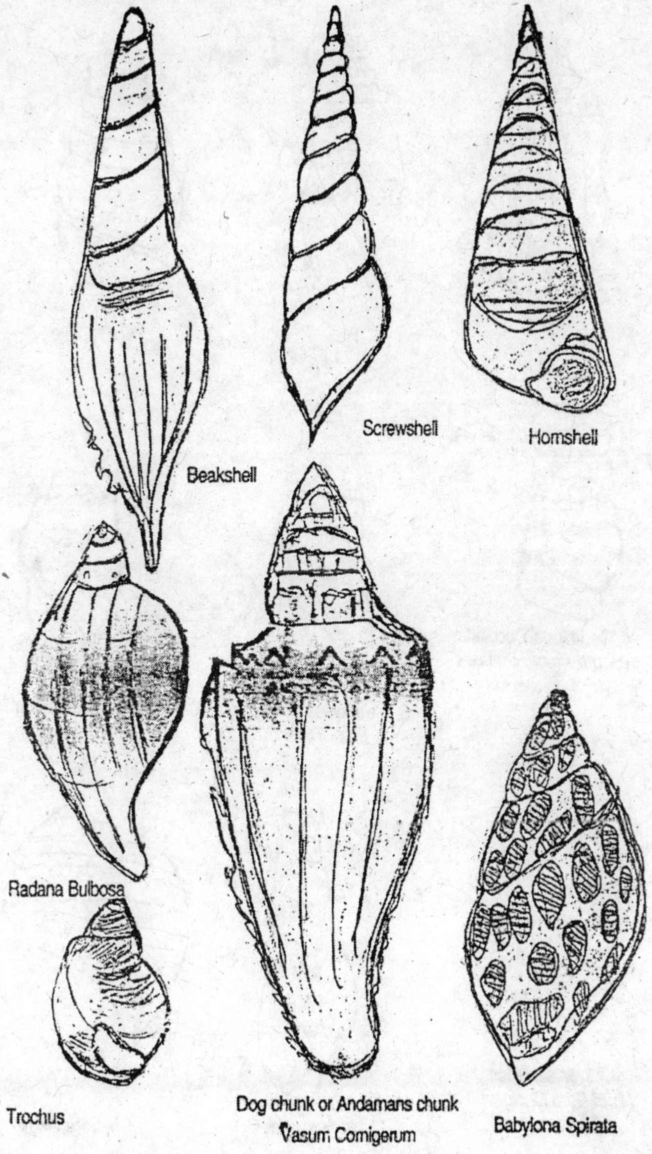

Fig. 10.11 : Few Gastropod Shell's.

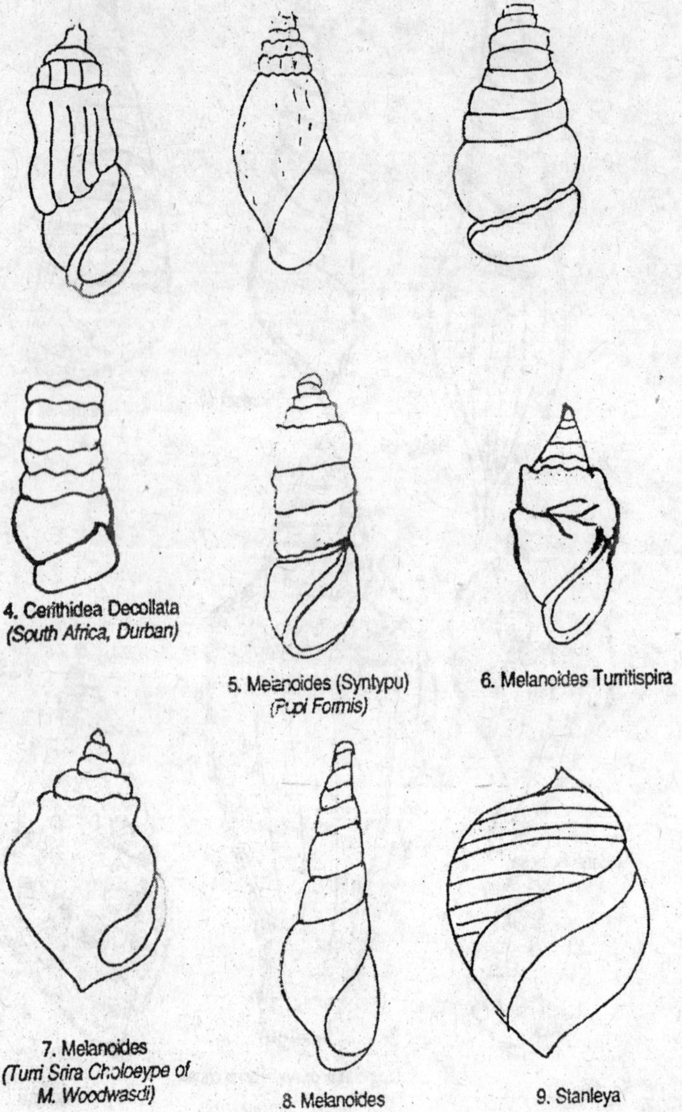
4. Cerithidea Decollata
(South Africa, Durban)
5. Melanoides (Syntypu)
(Pupi Formis)
6. Melanoides Turritispira
7. Melanoides
(Turri Srira Choloeype of
M. Woodwasdi)
8. Melanoides
(***
9. Stanleya

Fig. 10.12

will apply in general to almost any freshwater mussel. The Pelecypoda are molluscs with an internal and external symmetry in which the cephalic region is rudimentary, the mantle is divided into two lobes, right and left, secreting two valves of the shell, the foot is ventral, generally adapted for burrowing, there are two lateral and symmetrical ctendia and there is an elaborate mechanism for feeding by ciliary currents.

The freshwater mussels are found in ponds, lakes, rivers and streams, some in quite and others in flowing waters. They occur more abundantly in water containing lime as this material is necessary for the production of their shell. They live nearly buried in the mud or sand at the bottom, with only the posterior tips of their shell valves exposed. They may also occur wedged in between the rocks and stones, with the values slightly spread, and the two siphons exposed. They burrow slightly slowly crawl by extending their large muscular and plough-like foot between the two valves, or through the gape of the shell, and leave a furrow to marks the path they have followed. They may migrate the shallow places by night and retire to deeper places by day, and may change the habitats with the seasons.

The food consists of microscopic plants (diatoms, etc.), and animals (Protozoa, flukes, etc.) brought in through a current of water caused by the beating of cilia. The water current also brings in oxygen and carries away faeces, carbon dioxide and excretory and genital products.

The siphons open or close in response to light, touch and other stimuli. A slight disturbance causes the withdrawal of the foot and the siphons, and the closing of gape of the shell. Although a clam is usually safe in its tightly coloured shell, it is nevertheless preyed upon by the starfish which has the ability to open the shell and devour the soft body parts.

There are many genera and hundreds of species of there molluscs of which several are found in India. The large lamellibranchs those are available here are found abundantly in the eastern districts of Uttar Pradesh and have now been named Lamellidens marginalis. These mussels live partially buried.

The shell of the lamellibranch consists of two valves, which are united along a straight hinge-line by a tough elastic substance that forms the hinge ligament. The ligament passes transversely from one valve to another. Its elasticity is responsible to open and close the shell. Along the hinge line each valve has teeth like projections that enable the shells to fit with each other. There are the hinge teeth. The hinge is dorsal and the gape is ventral. In Anodonta the connection between the two valves is afforded only by a ligament, but in *Unio* each is produced with strong projections and ridges, the hingeteeth, separated by grooves and sockets, and so arranged that the teeth of one valve fit into the sockets of the other. The same is the case with *Lamellidens*.

The valves are externally marked by a series of concentric lines, lines of growth. They lie parallel to the free edges of the shell. Alternate periods of slow and rapid growth can be marked out by their size. The umbo represents the oldest portion of the values. At this place the shell is the thickest. This is the first part of the shell to be recreated along the edges of which subsequent layers are added. The inner surface of the shell also shows similar markings. On a closer examination it is revealed that the shell consists of three layers:

(a) The periostracum is the outermost layer formed of a chitin-like chemical compound conchiolin. It is dark blue when wet and brown when dry. This thin and translucent layer is partly responsible for some variations in the colour of the shell.

(b) The middle layer lying beneath the periostracum is the primatic layer. It is compared of numerous calcareous prisms embedded in conchiolin.

(c) The last layer of the shell. The nacreous layer or nacre. It is also called mother of pearl and is composed of alternate 10 years of calcium carbonate and conchiolin.

The first two layers are secreted only by the edges of the mantle and hence show concentric markings of discontinuous growth. The inner pearly layer is laid down by the whole surface of the mantle and has smooth lustrous surface. This layer may also secrete a pearl as a protection against foreign body, such as a parasite. When a parasite, say a larval stage of a fluke, enters the mantle it becomes enclosed in a sac formed by the growth scar of the mantle eithelium, which later on secretes thin concentric layers of pearly substance around the body. Except for the concentric lines of growth the shell is smooth and there are no ridges or tubercles which are frequently found in the shells of other bivalve molluscs. The inner surface of the shell presents characteristic markings and structures:

(i) The hinge-teeth that interlock the valves tightly are more clear on the inner side.

(ii) The scars of the adductor muscles are other prominent structures. The scar of the posterior adductor muscle is larger and posterior.

Posterior to the scar of the anterior adductor is the scar of the anterior retractor muscle of the foot. Ventral to this is the of the protractor muscle. Parallel, to the margin between the adductor scars making the attachment of the retractor muscle of the mantle is a fine line, the pallial line indicating the attachment of the mantle.

The mantle consists of two thin lobes attached to one another and to the vinceral mass dorsally. The mantle secretes the two valves of the shell that lie over the two lobes. Normally the mantle is creamy white in colour but, preservation spoils it. Certain parts of the mantle edges are so thickened and rolled as to form a dorsal exhalant aperture and a ventral inhalant aperture. The space inclosed between the two lobes of the mantle is known as the mantle or pallial cavity.' Dorsally the pallial cavity is bound by the visceral mass.

The shells are enclosed agape by the elasticity of the ligament and closed by the contraction of large muscles. There muscles include:

The vascular system in BIVALVES is like GASTROPODS including considerable haemocoelic spaces around the viscera and in the foot.

The heart consists of a ventricle which comes to lie around the rectum during development and into which two auricles discharge. Blood is discharged both anteriorly and posteriorly from the Ventricle but the vessels (anterior and posterior aortae) cannot be homologous with the two aorotae of gastropods. The anterior passes forward and supplies blood to the head, the foot and the visceral mass while the posterior goes backwards and takes blood to the mantle and spihons. Blood is returned from the anterior part of the mantle, the viscera and the foot into a vessel which takes it to the kidney, thence to the gills and back to the heart.

The fresh water clams diocious, *i.e.*, the sexes are separate, but there is no sexual dimorphism. Reproduction involves a parasitic larval stage, called the glochidium in fresh water clams and the veliger in marine forms.

The gonads either testes or ovaries, are a pair of large simple visceral branded structure, lying among the intestinal coils in the visural mass just above the foot. During breeding season, the gonads become greatly enlarged and conspicuous, when it is difficult to distinguish their paired nature. When mature, the testes are whitish and the ovaries reddish and the two sexes can be distinguished though not very distinctly. The epithelium lining the tubules of the gonads gives rise to the spermatozon in male and ova in female. The mature ovum is large, rounded, filled with a finely granular cytoplasm rich in yolk, and containing a vesicular nucleus with vacuole. The gonad of each side has a short duct, the vas deepens in male or the ovidul in female. It leaves the gonad from in upper side and open into the supra-bronchial chamber of the inner gill lamina by a genital aperture, just in front of the renal aperture of the ureter. There are no accessory reproductive organs. Males are more responsive than females to stimulation by rise in temperature and almost always initiates spawning. Presence of eggs stimulate other males.

In male sperms are shed through the genital aperture into the supra-branchial chambers and released from the body with the outgoing water through the exhalant siphon. Some of these sperms will be taken into inhalant siphon of a female purely by chance. The eggs, shed through the genital apertures into supra-branchial chamber, do not leave the body but are carried through the ostia into water tubes of the outer gill laminae, where they are held by nucleus and fertilized by sperm from another clam, drawn in by water currents, thus forming the zygotes.

Development of the fertilized eggs or zygotes proceeds within the outer-gill laminae, which greatly enlarge due to accumulation of a large number of embryos and serve as the breeding pouches or marsupia. The females are easily recognized at this stage by the swollen appearance of their outer grill plates. In anodanta, the eggs are usually fertilized in August and the embryos, which develop from them, remains in the gills of the mother throughout winter, escaping in spring as tiny larvae.

Anodanta will apply almost equally well to it. The later stages in the development, show a complication, due to a temporary parasitism of the larva, "Glochidium".

In larva escapes through the exhalant siphon of female mussel and finds a suitable fish host for further development. The larvae attach to gills or fins of fish for a period of 60-70 days drawing nourishment from host. At the end of parasitic life the juvenile mussel detaches from host and falls to bottom for further growth.

For rearing young mussels ponds should be cleared of weeds and excess bottom mud. To maintain hygiene and enrich status of pond, it should be treated with lime 200-250 kg/ha/yr cowdung (10,000 kg/ha/yr) urea (100 kg/ha/yr) and single superphosphate (100 kg/ha/yr) in instalments. Such doses will help to maintain good algal density in pond. In water having current or ponds having inlet and outlet facility and receiving canal water, metabolites from mussels are washed out. Regulate all water quality parameters specially food, pH and oxygen etc.

and prevent any kind of pollution. The pH of soil and water should both be in range of neutral to slightly alkaline, pH 7.0-7.5 (soil under wet condition) and water 7.0-8.0, dissolved oxygen 5-8 mg/l, total alkalinity should be medium 70-150 mg/l. Calcium is a major ingredient of pearl and oyster shell, so its presence in adequate quality (10-20 mg/l) in water is considered to be essential. Cultured pond should be free from aquatic weeds. The optimum depth for fish-cum-peal culture varies between 2-3 meters.

In shallow pond the water may be very hot during summer thereby causing stress on fishes and pearl oysters while deeper ponds have difficulties in handling and nursing of oysters.

After 60-70 days the larvae get dislodged from fish. In artificial pearl production two methods are used. In first a graft tissue is injected into another mussel. In second method a hard ball like material called bead along with flesh is pushed into its body. The ideal temperature of water should be 15°C-30°C. As the mussels become weak during breeding season this period may also be avoided for pearl culture. The chief ingredient for pearl culture is a healthy above two years old mussel. If a foreign object like an insect or a grain of sand enters the body of mussel some sort of secretion occurs nacre, this deposits layer upon layer leading to the formation of a pearl.

For artificial production of pearl healthy mussel immersed in a pot of water for 24 hours before introduction. Then washed afresh. Choose bigger one which will be donor. A thin and long portion from the side of shell (mantle epithelial layer) is cut out. After cleaning put this portion in glass/slide and cut into small pieces. Add 2-4 drops of water soluble Eosin. Now cover the material for freshness. The other mussel are kept in tub full of water in such a manner that shell valves placed upwardly. This makes valves of the shell to open easily. Sprinkle the methanol (menthol) to easy opening of valves. Then, as a mussel starts to openers its valves a piece of wood is inserted with the help of valve open and mussel is finally placed on mussel holder stand. The pieces of graft tissue may be introduced if mussel is a big sized one. After this operation the mussel should be removed from stand and put into water. The introjected mussels are washed thoroughly and kept in a tub of water for over night. Next day they are again washed before being dropped into pond.

According to second method a small hand round object made from conch shell to inserted along with the piece of graft tissue. At present small beads from conch shell are used. Alongwith the graft tissue this type of beads may be introduced in the mantle epithelial layer of mussel. Dorsal side of the graft should touch the bead. The graft tissue makes a sac within a short period. After that nacre starts formation over the bead to form pearl, this method lakes 18-24 months but in second method only one year is sufficient (Anjana Sen Gutpa 1994).

According to Sen Gupta (1994) there are three basic steps in freshwater mussel.

1. Preoperative management of mussels
2. Implantation management.
3. Post operative management of mussel.

1. Fre-operative Management of Mussels

(i) Selection of mussels

Healthy Donor Age 2-4 years, size less than 14 cm but not less than 9 cm.

Healthy mother age-3-5 years, size 9.5-12 cm.

(ii) Identification of healthy mussel.

(a) Mussel when subjected to shock is able to close quickly.
(b) Shell wall ejecting water in a straight line to considerable distance.
(c) Adductor muscles must be strong.
(d) Strong foot free from any injury.
(e) Fully developed and devoid of injury in pallium and gills.
(f) No empty sound come out from mussel by gentle stroke or finger.

(iii) Collection and transportation

(i) Collected in early spring (Feb.-March) at 20° and transported in 20°C-25°C.
(ii) Mussels are safely transported in semi-dry condition in cane/bamboo/plastic containers polythene with hydrilla.

2. Implantation Management

(i) *Season* : Early spring is the best time. Feb.-April in congenial temperature.

(ii) *Cut piece implantation* should be with sharp knife and tweezers, mantle epithelium of pallium region (posterior to middle) is cut and carefully cleaned by a sponge and stretched and trirmmed to 2 mm^2 or 3 mm^2. Cutpieces are kept soaked in reosporun or physiological saline.

There should be short duration (instant) in cutting the pieces and implantation.

(iii) Care should be taken to maintain proper space between implantation of cut pieces.

(iv) Opening of valves should not be above 1.5 cm.

(v) The region of implantation is washed with water and 0.5-1 cm. wound is made in conformity with diameter of nucleus. After mercurochrome soaked nucleus is securedly placed, the cut piece is placed in mouth of wound, a drop of Betnesol-N may be given in wound.

3. Post Operative Management

(i) *Rearing Tanks :* Water quality and management of aquatic weeds and predators are very important.

(ii) *Rearing method :* Rearing by suspension.

(iii) Rearing time and management are important for pearl.

(iv) *Harvesting time :* December-January best till early April.

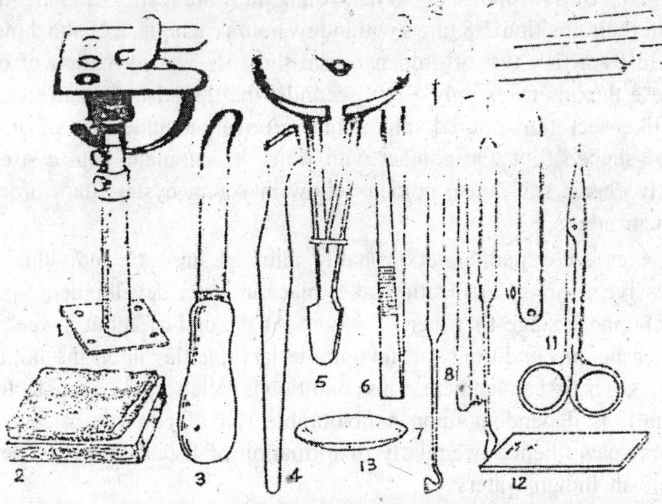

Fig. 10.13 : Mussel holder stand, 2. Wooden piece for tissue cleaning, 3. Knife for mussel cutting, 4. Tissue cutter, 5. Valve opener, 6. Forcep, 7. Bead holder. 8. Bent needle, 9. Bent needle, 10. Mussel opener, 11. Scissors, 12. Glass slide, 13. Watch glass, 14. Wooden peg.

(v) *Washing* : Wash pearl with 3-5% of hydrogen peroxide or diluted HCl.

The large family of Pteridae is of great importance of Indian Ocean (*Pinctada spp*) wingshells (Pterudsproper), Hammer oyster (*Malleus*) and Vulsella. Indian pearl oyster is (*Pinctada vulgaris*) smallest, it seldom exceeds 3.5 inch in height and compares unfavourably in size with huge gold lip *P. maxima*) from mergie and Australian waters. The shell is straight along the hinge line and is produced at each end into a short ear a modification of winglike projections. Ventrally shell is deep and rounded with a series of delicate finger-like at the ends of radial bands made up of older fingers now outgrown, worn and disused. These pearl oysters reach their limit of size in about 31/2 years. If they survive longer the shells become worn, the fingers disappear and overall length and depth actually decrease. The thickness increases until death. This is the best criterion to determine the age. In life history and habits they are akin to mussels. A strong cable of byssal threads attaches them to rocks, stones and shells, they possess the power of casting of their capable at will, they are cable of crawling short distances and this power to shift their foothold sometimes enables them to avoid entombment through an overwash of sand. When young, they are restless and inclined to shift their position 7-8 times within few hours each place of attachment marked by a tiny tuft of golden byssal threads. The formation of one of these threads takes only a few seconds; the thread is formed from a glue-like secretion poured into a fine groove on underside of small tongue-shaped foot. On contact with water it coagulates into a strong slightly elastic substance, golden yellow in young oysters dark bronze green in adults.

Sexes are separate but probably alternate in sane individual in successive seasons, fertilization takes place in water development rapid, a trochophose stage first then a veliger. At the end of about a week or even earlier if conditions are favourable larva settles upon the bottom and makes is first attachment. This condition is called 'spat', a satisfactory 'spatfall' is dependent upon a favourable run of currents at time of oysters spawn hence irregularly of formation of fishable beds of pearl oysters in Indian waters.

Indian pearl oyster is of comparatively rare occurrence; and grains causation is much more common and is to be reckoned important; a more prolific source, however, is that due to irritation caused by fragments of nacre dislogged from the inner lining of the valves by muscular strain, which pass into the oyster soft tissues. These fragments of nacre may wander far; in their course they become invested with a containing sac of nacre-producing cells immigrant from mantle which lay down successive coats of nacre upon the nucleus thus forming a pearl, other pearls mainly of seed pearl kind are formed around nuclei consisting of microscopical concretion which arise within muscles at the place of insertion on shell. These may also wander into oyster's tissues and may even be found in 'nests' of as many as a dozen or more within the visceral mass. Thus pearl formation may be induced by several kinds of irritation.

The solitary handsome stoutly built shell larger black lipped pearl oyster *Pinctada margaritefera* with a shade of green on very dark ground colour decorated with several radial lines of white spots passing outward from umbo. Ears disappear in adult and outline becomes suborbicular. Nacre of this species is thick enough to be of value in button making but its dark smoky colour greatly impairs its market price, it is common in Persian Gulf.

Water depth, bottom characteristics, protection from wave action, tidal flow, height, tubidity, water quality, chemical parameters, predation, fouling, pollution and accessibility are the main factors for site selection. From intertidial region to areas extending upto 5 m depth can be considered for adopting suitable culture method salinity required 22-35 ppt and temperature 21-31°C. Special nursery ponds in intertidal region (can be constructed. Oyster spat out of hatchery are too small, to be grown in field without protection. They should be enclosed in velom screen bags of suitable mesh size and suspended from racks. A string can hold six shell valves containing 80-100 spats and 3-4 strings are enclosed in a bag. Clean the bags with regular intervals. After 40-50 days they are transferred to farm.

They are broadly grouped as bottom (on bottom) culture and off bottom culture practices. The off bottom culture methods are advantageous over the bottom culture in various ways:

1. Relatively rapid growth and good meat yield.
2. Facilities three dimensional utilization of culture area.
3. The biological functions of oyster such as filtration, feeding etc. are carried out independent of tidal flow, and
4. It minimises the silting and predatory problem. Other methods are:
   (i) Raft and string (ren) method.
   (ii) Rack and tray method.

(iii) Stake method
(iv) Raft method.
(v) Long line method.

For farm maintenance one has to control crab, fishes, starfishes, polychaete and gastropods. There are mainly *Scylla Serrata Pagurus, Thais rudolphie*. They kill young oyster gastropod, *Cymatium cingulatum* cause 15% mortality, *Gracilaria* spp alga affects the water flow. *Balanus amphitrite* is a fouler that attached to wooden structure, trays and oysters and compete for food. Farm management includes timely cleaning of oysters and rearing trays, racks, string, rafts. All the devices should be cleaned and poles and other spoiling materials are to be replaced. Harvesting is done manually. Good meat yield can be obtained during March-April and August-September.

The history of Broome Pearling has been one full of high's and low's and after World War II, it was at a particularly low point. Luckily, a demand for cultured pearls developed, no doubt as the economy picked up, and this revival brought about the development of the first pearl culture farm in Australia in 1956.

It was established at *Kuri Bay*, 420 km North of Broome, by a joint venture between Australia and the Japanese. The Kuri Bay venture was formed under the name, *Pearls Pty* Ltd, and consisted of *Male and Co, Broome Pearlers Brown and Dureau Ltd*, and the *Otto Gerdau Co of New York*. The Nippo Pearl Co. of Japan handled distribution and marketing. Tokiuchi Kuribayashi was the principal of this company and was the most powerful and influential man, since Mikimoto's death, in pearl culture. Kuri Bay was therefore named after him.

By 1981, there were five established pearl culture farms in the Broome region—*Kuri Bay, Port Smith, Cygnet Bay,* and two in *Roebuck Bay*.

It is well worth noting, at this point, a little about the history of pearl culture which began in Japan. The first known attempts were in the 13th century, where crude semi-round pearls were the result. Tatsuhei Mise (1880-1924) produced the first round cultured pearl before 1904. Tokichi Nishikawa (1874-1909) became the first person to produce a round cultivated pearl from planned experiments using tiny gold and silver nuclei.

Kokichi Mikimoto (1858-1954) is the man we most associate with Japanese cultured pearls. However his method of depositing nacre from graft mantle tissue has proved too delicate and difficult. The method used by the cultured pearl farmers today is based on T. Mise's basic requirements for the spherical pearl production which is:

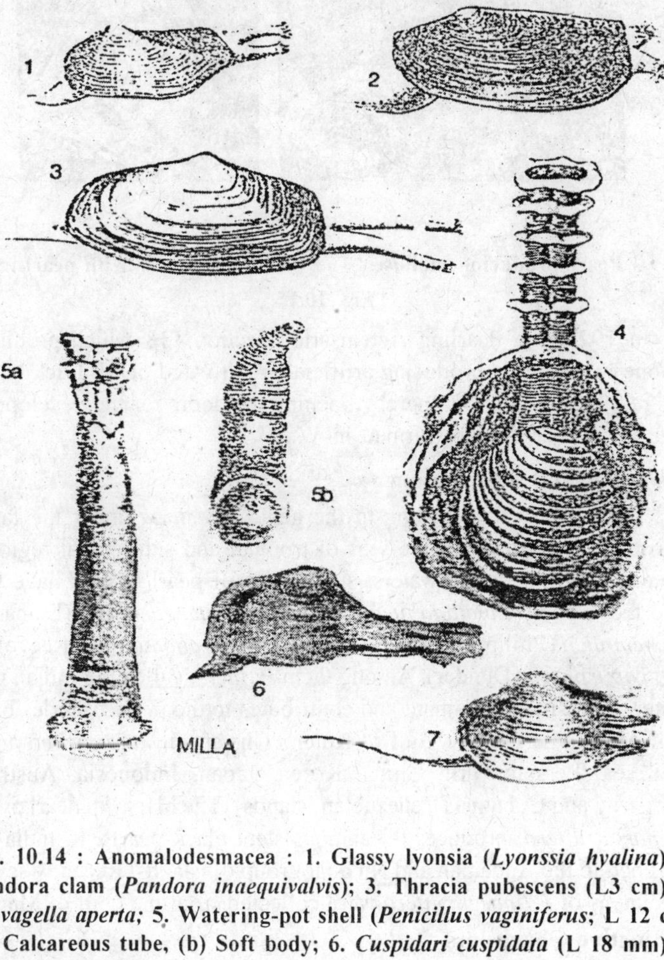

Fig. 10.14 : Anomalodesmacea : 1. Glassy lyonsia (*Lyonssia hyalina*); 2. Pandora clam (*Pandora inaequivalvis*); 3. Thracia pubescens (L3 cm); 4. *Clavagella aperta;* 5. Watering-pot shell (*Penicillus vaginiferus*; L 12 cm), (a) Calcareous tube, (b) Soft body; 6. *Cuspidari cuspidata* (L 18 mm); 7. *Poromya granulata.*

(a) A spherical nucleus.

(b) The introduction of the nucleus into the connective tissue of the oyster.

(c) The nucleus must be inserted along with a piece of mantle epithelium.

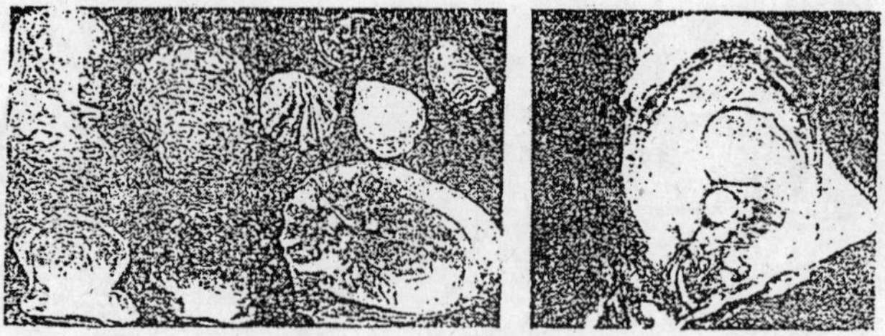

Pearl producing molluscs **Pearl oyster with pearl in situ**

Fig. 10.15

In 1922, the Pearling Act inserted section 113 which prohibited anyone in Australia producing artificially cultivated pearl. Luckily, this was reappealed in 1949, thereby opening the doors for the development of today's pearl culture formed in W.A.

Resources and Distribution

The pearl oysters belong to the genus *Pinctada* under the family Pteriidae. They occur in the seas of tropical and subtropical region of the world. From Indian waters, six species of pearl oysters have been reported, namely *Pinctada fucata* (Gould), *P. margaritifera* (Linnaeus), *P. chemnitzii* (Philippi), *P. sugillata* (Reeve), *P. anomioides* (Reeve) and *P. atropurpurea* (Dunder). Among them *P. fucta*, called the Indian pearl oyster, is the most dominant and contributes to the pearl fisheries of the Gulf of Mannar and the Gulf of Kutch. This species also occurs in the Red Sea, Persian Gulf, China, Korea, Japan, Indonesia, Australia, Margarita and Cubagua (Venezuelan Islands). The black-lip pearl oyster, *P. margarilifera*, produces the famous steel black pearls. In India it is confined to the Andaman and Nicobar group of islands. Recently several specimens of *P. margarilifera* were collected from the Gulf of Mannar.

Distinctive Characters

The genus *Pinctada* oding is characterised by long and straight hinge. The long axis of the shell is not at right angle to the hinge. The left valve is usually deeper than the right. There is a byssal notch on each valve at the base of the anterior lobe. The colouration of the periostracum is variable and often brown.

In *P. fucata* the hinge is almost as broad as the width of the shell, anterior border of the shell does not extend or just extends in front of a perpendicular line drawn from the tip of the anterior ear, posterior ear well developed with moderately developed sinus, byssal notch slit-like, left valve highly convex, outer shell with 6-8 radial bands of reddish brown colour on a pale yellow background, nacreous layer iridescent and of golden, pink or ivory colour and the non-nacreous border on the inner surface of values has brownish or reddish patches.

In *P. margarilifera* the hinge is shorter than the width of the shell, anterior border of shell far in advance of a perpendicular line drawn from the tip of the anterior ear, posterior ear and sinus absent, byssal notch broad, left value moderately convex, outer shell dark brown with greenish tinge and radially distributed white spots, nacreous layer iridescent and of silvery sheen colour and the non-nacreous border is of dark colour; hence the name 'black-lip pearl oyster' for this species.

ECOLOGY, MORPHOLOGY, ANATOMY AND BIOLOGY

Ecology

As is the case with other species of the genus Pinctada, the Indian pearl oyster *P. fucata* is truly marine and does not occur in estuaries of backwaters. The preferred salinity range is 30-36 ppt. It occurs from the intertidal region upto 25 m depth. It prefers clear water and high turbidity is not conducive since it reduces the filtration and food intake.

Morphology

The shell, also called valve, is composed of three layers. The outer layer is formed by a horny substance called conchiolin. The middle layer is the prismatic layer and is composed of minute prisms of inorganic calcium carbonate in the form of calcite crystals separated from each other by thin layers of conchiolin.

The innermost layer is the nacreous layer or the 'mother-of-pearl' layer and is lustrous. It is composed of numerous line lamellae or aragonite crystals. These three layers are formed by the secretion of the epithelial cells of the mantle. The left valve is deeper and more concave than the right one. The shells are joined together on the adductor muscle is noticeable substantially. Thus muscle is attached to both the valves and controls the movements of the valves. Small sears numbering 12 to 15 are present inside the shell for the attachment of pallial muscles.

Ecology Morphology, Anatomy and Biology

The mantle is a fold of skin covering the soft body of the pearl oyster between the valves. It is formed of two lobes on either side and these are fused together dorsally below the hinge line. Seawater enters the mantle cavity through the inhalant aperture and leaves by the exhalant aperture. The mantle can be divided into three zones, namely, the edge or the marginal zone the pallial zone and the central zone. The mantle from the pallial zone is used for the graft tissue preparation.

The foot is a tongue-shaped organ capable of expansion and contraction. It arises in the anterior region of the visceral mass. The byssus gland is lodged at the proximal end of the foot and a bundle of stout laterally compressed bronze green fibre known as byssus originate from this gland.

The gills are paired, filamentous and are attached to the visceral mass in the front; the opposite end is free in the mantle cavity. They actively filter the water, absorb oxygen and retain organic matter useful as food. When the water is highly turbid the gill filaments become choked and filteration is stopped, affecting vital functions.

The heart, consisting of a ventricle and two auricles, is enclosed by the pericardium; it is ventral to the intestine. The auricles, after receiving oxygenated blood from the gills and mantle, pass it to the ventricles. From the ventricle arise two aorta, the posterior aorta supplying blood to the rectum, anus and adductor muscle and the anterior aorta to the rest of the body. The deoxygenated blood collected in the veins reaches the kidneys and gills where it is purified.

Two curved leaf-like structures known as the labial palps are located on the mouth opening. They select the food particles, pass them on to the mouth and reject the unwanted marker as pseudo-faeces. The selected food is passed through the mouth, oesophagus and the stomach where it is smashed and subjected to action by the enzymes, supplied from the crystalline style. The larger particles are directed towards the intestine: other nutrients are dispatched to the digestive diverticula which surround the stomach. The intestine extends through the visceral mass and forms a loop, ascends above the heart and opens into the long rectum, followed by the anus through which the faeces is ejected out.

The excretory system consists of paired nephridia and numerous small pericardial glands projecting from the walls of the auricles.

The nervous system consists of three pairs of ganglia, *viz.* cerebropleural, pedal and visceral, from which the nerves arise. The mantle is supplied by a complex network of nerves, the pallial plexus and is sensitive to slight stimulus. A branchial eye is present on the first left branchial filament and it can detect variation in light intensity.

Biology

The pearl oyster is a filter feeder. The food mostly consists of microalgae and particulate matter. The food brought inside the mantle gets entangled in mucus sheeth produced by the gills. The labial palps direct these entangled food into the mouth. The food sorting, mechanism in the pearl oyster is considered to be not efficient since several organisms which cannot be digested find their way into the stomach and intestine. Alcyonarian and sponge spicules, lamellibranch larvae, gastropods, heteropods and crustacean appendages are some of the items, apparently not of food value, but found in the stomach of pearl oysters.

P. fucata grows fast in the first year. At the Krussadai Island the estimated length at the end of the first year is 4.5mm, second year 55mm, third year 60mm, fourth year 65mm and fifth year 70 mm. The corresponding estimated weights are 10gm., 30gm., 45gm., 60gm. and 70gm. Hatchery produced oysters have been reported to attain model sizes of 47.5mm, 64.5mm. and 75mm. at the end of the first, second and third years, with corresponding weights of 8.3gm. 31.6gm. and 54.4gm. In the Gulf of Kutch the estimated lengths at ages 1 to 6 are 44.05mm, 61.68mm, 76.20mm., 81.62mm., 85.15mm. and 86.65mm. respectively.

The sexes are separate but male and female oysters cannot be distinguished by the shell characters. Change of sex has been reported in pearl oysters. The reproductive system consists of a pair of gonads which spreads superficially over the hepatopancreas and intestine in mature condition. It is pale yellow in colour in males and is of a deeper shade in females. *P. fucate* attains first sexual maturity at about 15.5mm. at the age of 3-4 months. The eggs or sperms are released through the paired gonoducts ending in the genital openings located anterior to the gills. In the Gulf of Mannar P. fucata spawns twice a year during June-August and November-January coinciding with the southwest and northeast monsoons respectively. Between year variations are known to occur in the spawning of *P. fucata*, and also differences under the wild and farming conditions.

LOOKING AFTER OYSTERS

The shells which have been collected and transferred to the pearl farm are placed in baskets or panels which are attached to long lines connected to the floating rafts.

The line is dropped down into the ocean with the oyster securely inside the basket, where they remain until they become operated on for seedlings.

Seedling takes place when the temperatures become stable of which is usually around July. The oysters are very sensitive and can easily die if handled during changing temperatures. They are, therefore not touched, if possible, in May or September. The oyster can produce more than one pearl in its lifetime by taking good care of it. Regular cleaning of the shells to remove seaweed results in better pearls plus makes them easier to handle. The cleaning is done by a machine which uses water jets and brushes to clean off any seaweed or other growth from the outside of the oyster shell. It may also be cleaned using a tool similar to an axe.

It is most important to regularly turn the oyster shells through 180°, so that the nacre of the developing pearl inside the oysters, will be evenly distributed around the nucleus. The oysters need very tender loving cares so as to be productive when harvested.

THE PEARL FARM

Selecting a pearl farm site is extremely important as there are many important factors which must be considered, of which will determine the type of pearls produced, and the oyster survival rate. Some of these environmental factors are:

(a) *The relief of the region* : Natural features, such as mountains and reefs, are needed to protect the farm from winds, currents, storms, etc.

(b) Constant regularity of temperature

(c) Type of sea bed, such as rocky or sandy.

(d) *Presence of gentle currents* : Currents are essential for the survival of the oysters as they bring food and oxygen.

The whole pearl farm system is based on a series of floating wooden rafts. Ten basic units, each 20 ft. times 20 ft., are joined together in two rows of five, to complete one raft. 100 plastic covered wire mesh baskets, each of which house 10 oysters, are suspended from each unit. 8-10 empty metal drums are used for unit flotation and each raft is moved by iron anchors.

A typical work schedule for a pearl farm can be seen as follows:

January : Prepare for wild shell collection, organise dive crews, fishing gear, paperwork, and licence fees.

February : Fishing for 20,000 wild shell begins, linked to tide patterns. (Note: Tides in the area can vary by 10 m. per day)

March : Collected shell is 'dumped' on the seabed or site leased by the company and allowed to rest. Maintenance of dumped shells, turning and cleaning them. X-ray shells seeded check if implanted nuclei have been rejected. Oysters which reject nuclei are usually re-seeded.

April : Water temperature begins to drop as winter approaches, rest period for the shells.

May : Ongoing farm work, turning and cleaning previous two years' seeded oysters kept suspended in wire panels in the water column.

June : Prepare for operating on oysters to implant nuclei. (Note: Some technicians may come from overseas, and some companies have boats fitted as mobile laboratories so seeding can be done on the pearling grounds.) Seeding and harvesting begin.

July : Normal operating time for pearls, seeding new oyster, re-seeding those which have rejected nuclei. Oysters which produce acceptable pearls are also re-seeded.

August : Harvest of previous year's seeded shell continues, then a two-month turning program follows operations. The oysters are turned over to encourage production of round pearls.

September : Turning operated shell.

October : Turning, cleaning, and change of areas.

November : Transportation of operated shell to grow-out

December : Oysters into longline system. Clean gear.

SEEDING

Once oyster are two-three years old and are healthy, they are able to undergo surgical implanting, otherwise known as seeding. This is a very delicate operation and takes many years to perfect, as it is designed to minimise stress on the oyster. It involves three stages:

Preparation of the Graft

A donor oyster is sacrificed in order to obtain a piece of mantle needed so that the host oyster will accept the nucleus. The mantle is located on the outer section of the oyster and produces the nacre forming pearls. Before a graft is taken from the mantle, the oysters are starved for several days. This allows the oysters metabolism to slow down, therefore decreasing the risk of core rejection as well as making it easier to open the oyster.

Attaching the Graft

The oyster is opened with special wedges and pliers, and a scalpel slit is made in the soft tissue near the reproductive organ. A graft of living mantle, four-five millimetres square, of which is obtained from another oyster, is then inserted into the scalpal slit.

Inserting the Core

A core/nucleus, usually a sphere made of Mississippi mussel shell is placed in the scalpal slit and the oyster is then returned back to the water. The inserted core irritates the oyster, provoking it to gradually coat the core with thin layers of mother of pearl nacre. After some time, the oysters are collected, and x-rayed to see whether the implants have been accepted. Oysters which have rejected the implant are returned to the water and are once again operated on the following year.

The seeder must be extremely careful not to harm the tiny pea-crab which lives unharmed within every healthy oyster. It is presumed that the crab assists the oyster by keeping it clean and by sharing the debris which the oyster sucks in.

HARVESTING

After 2-3 years, the oysters are harvested, as it rakes at least this long for only 2 mm. of nacre to have formed around the core. This is a time of high anxiety for the pearl farmers as it will determine whether the seeding was successful. If so, thousands of dollars worth of fine quality pearls will have been produced.

The oysters are split open the and the pearl bag cut by the scalpel, and the pearl is removed. If the oyster has survived the past three years of gestation without any major problems, the harvester will place another core in the same position as the extracted pearl, without the need for another mantle. This occurs in about one-third of oysters. Approximately 35% of the oysters produce good pearls, 20-30% die, 20% reject the nucleus, and 10% are poor quality pearls.

Although these figures do not seem ideal, they are constantly improving. As time progresses, more successful methods and equipment will hopefully be introduced, which will reduce the number of deaths and increase the number of quality pearls produced.

The high mortality rate amongst oysters is thought to be associated with transportation, the difference in water temperature, and the increase in marine bacteria found in the water of the carrier tanks. Luckily, nothing is wasted in the pearling industry, as excellent prices are paid for the oyster meat as well as the *Pinctada maxima* mother of pearl shell, of which is used for buttons and inlay work.

SELLING PEARLS

The town of Broome has many pearling shops and boutiques offering pearls direct from the sea, as well as pearl jewellery, in a wide range of products, unmatched in the world for variety and quality. Retail sales in Broome totaled several million dollars in 1994 of which is expected to pass $10,000,000 a year before the end of the 1990's. This an extremely important contribution to the towns economy, and the development of pearl retail stores and businesses has promoted Broome as a tourist attraction as well as generating jobs.

Large-scale pearling companies, such as Paspaley Pearls, establish annual auctions which are held at various locations in the world, some major centres being Tokyo, Darwin, Brisbane, and Hong' Kong. They are attended by top buyers from all over the world, who spend as long as 4-5 days examining pearls in minute detail, before the actual auction takes place. When bidding commences, sums in excess of US$ 1,000,000 are paid for exquisite pearls.

These auctions bring the world's best known jewellery agents and buyers together, assisting in widening international exposure to the splendour of Australia's South Sea pearls: In

Hydrabad, Bhuvneshwar, Cochin, Rameshwaram, Madras, pearl culture is very common.

The future of the pearl industry lies in breeding oysters so as to avoid overfishing and thus extinction. Mortality rates are too high and therefore, there must be a continuous supply of oysters to ensure the industry survives.

Pearl culture is being carried out only in countries such as Japan, Australia, China and Indonesia. If culture, can be done, successfully to produce internationally qualifying grades of pearl, then it would be a great success. Hence the lack of thriving domestic market can be very well compensated for a promising international market.

The pearls are one of the most sought after materials in jewel industry. Thus it is a very important component in the jewellery export market and also in local market.

Normal Mussel gives 31/2 ounce of raw edible portion 81 calories and 2.2gm. fat.

Pearl has Several Importance

1. Jewellery (Pearl 1)
2. Food (Adductor Muscle-soup)
3. Poultry feed (shell)
4. Flooring chips (shell)
5. Medicinal
6. Button.
7. Ornaments.

Mussel

1. Lime Industry (shell)
2. Cement Industry (shell)
3. Food (Meat)
4. Cancer Cure (Meat)
5. Biological filter (aquaculture)

Edible Oysters

In India six species of oysters, namely, Indian backwater oyster *Crassostrea madrasensis* (Preston), Chinese oyster *C. rivularis* (Gould), west coast oyster C, gryphoides (schlotheim), India rock oyster *Saccostrea cucullata* (Born), Bombay oyster *Saxostrea cucullata* (Awati and Rai) and giant oyster *Hyostissa hyotis* (Linn) are found. The first four have commercial value.

The shape and size of oyster shell is highly variable. If grown individually on a soft bottom, shell tends to be smooth and elongate. If grown individually attached to a firm surface the lower value grows attached to surface rather flat thus losing the normal cupped appearance. Proximity to other oysters when grown in a cluster causes distortion in shape of shell called Xenomorphoism shape determined by contours of substrate and is important in oyster culture.

Various general and species are identified based on shape, size and other characteristics of shell, anatomy, promyal chamber, gill ostia, heart and gut and breeding habit.

C. madrasensis : Shell valves irregular in shape, usually elongate, left valve deep and right one slightly concave, hinge narrow and elongated. *Adductor* muscle kidney shaped and dark purple, grows 22 cm. euryhaline, inhabits backwaters, creeks bays and lagoons, found from intertidal region to 17m depth.

C. gryphoides : Shell valves elongate and thick. Left valve cuplike. Hinge are well developed and has a deep median groove with later elevations. Adductor muscale scar is broad, more or less oblong and striations on scar are absent or obscure. Grow 17cm. it is also euryhaline occurring along north Karnataka, Goa, Maharashtra coasts. In Maharashtra it is exploited regularly from several creeks. It occurs in water upto 7m. depth.

C. rivularis Shell valves large, roughly round, flat, thick and with a shallow shell activity. Left valve is thick and slightly concave and right one is about same size or slightly larger. Adductor muscle scar is oblong and white or smoky white Grow 15cm. It found in Gujarat coast and also in Maharastra together with *C. gryphodies* in Mahim, Ratnagiri and Jaytapur creeks.

Saucostrea cucullata : Shell valves hand, stony, trigonal and pearshaped. Hinge straight, devoid of teeth and numbonal cavity well developed. Margin of both valves have well developed angular folds sculptured with laminae. Small tubercles present along inner margin of right valve and corresponding pits in left valve. Adductor scar is kidney-shaped and white or colourful. Occurs in all along the main land coast and Andaman and Lakshadweep Islands. In shallow water they attached to rocks and boulders and can withstand surf and wave action in sea.

Edible oyster food consists of organic detritus and phyto plankters such as diatoms and nanoplankters. The food parties are entrapped in mucus of gills and are passed in water currents towards the mouth by rapidly beating gill cilia (fine hairs). The labial palps, number four located close to mouth, surf the food before it enters the mouth. The unwanted food particles are rejected as pseudofaeces. Close to entrance to intestine is located the 2-3cm. long crystalline style sac which contains a gelatinous rod (crystalline stole) known mucoprotein, contains digestive enzymes which convert starch into sugars. Apart from usual extracelles digestion in stomach intracellular digestion occurs in digestive gland.

The growth is related to food availability and also to environmental conditions, particularly to salinity and temperature.

Condition index of oysters denotes the quality of meat and it is useful to determine the best period for harvest. It is also helpful to assess the suitability of a locality for culture. High condition indicates greater proportion of meat in whole weight of oyster; those prime condition are tasty when compared to flacid and watery meats of oysters in poor conditions. Soft body of oysters changes according to the different stages of reproductive cycle. In maturation gonad size increases in weight causing increase of weight in soft body.

$$\text{Condition factor} = \frac{\text{weight of dry meat (ovendried at } 90 - 100°C}{\text{volume of shell cavity}}$$

In *C. madrssensis* the condition is considered as high if it is above 140 and poor if it is below 70.

In genes *Crossostrea* sexes are separate but occasionally hermaphrodite nay found. Ovary and testes comprise a series of branching tubules, follicles on both sides of body. After spawning fertilization occurs externally. In nonbreeding season gonad is replaced by connective tissue called leydig tissue which mostly have glycogen. In this stage sex cannot be determined. The sex of oyster may change during breeding season. In *Crossostrea madrasensis* young oysters primarily function as males 60-75% and later change their sex. A single female measuring 80-90 mm. spawns 10-15 million eggs at a time. In Kakinada Bay, it spawns during January-June. In Madras harbour, under marine conditions year round spawning has been reported. In Adayar estuary it spawns during October-December and in March-April. In Muli estuary, major spawning activity is confined to April-June and minor in November.

As per survey conducted by CMFRI at some important centres the annual catch is estimated at less than 2,000t. *C. madrasensis* is dominant species from fisheries point of view. Oyster seed may be collected by clutch in water column at appropriate period. During spawning period the spat collectors are laid on bottom on racks or suspended from racks. If collectors are laid in advance of spatfall, they may be covered with silt or settlement of foulers making them unsuitable for oyster larvae to settle. The larval period in *C. inadrasensis* is 15-20 days. The ideal time is 7-10 days after peak spawning. Strong currents interfere with larval settlement and may result in poor spat collection. Different methods are used for different culture method *e.g.* for ren method oyster shell strings are used.

The techniques for mass production of both attached, and free seed of *C. madrasensis* has been developed by CMFRI at its shellfish Hatchery, Tuticorin which is described here. The creteria for site selection and inputs for establishment of hatchery of pearl oyster and edible oyster are similar with minor differences.

For brood selection oyster conditions and age are to be considered. Mixed stock from various area provides good results. Oysters of 60-90 mm, with 30% of them (males) in 60-75 mm. length are desirable. First of all clean them with wire brush to get rid of plants and animals adhering to shell. A batch of 25 oyster can be placed on a synthetic twince-knit PVC frame in 100-1 FRP tank, and raw (unfiltered) pre-cooled seawater (at 20-22°C) is filled in tank and aerated. About 2-3 1 of mixed algal culture (1.5-2 million cells/ml) precooled at 20-22°C is given as feed twice a day and maintained at 10°C below ambient temperature.

Conditioned oyster are subjected to thermal stimulation by suddenly transferring them to sea with temperature 34-35°C. This sudden change causes spawning. The spawning oyster are immediately separated and placed in spawning trays having filtered seawater at ambient temperature. Remove oyster after spawning. Gametes are mixed and gentle aeration provided. Fertilization takes place immediately and first cleavage with 45 minutes. After four hours morulas stage is reached and swimming morula is siphoned and reared in filtered sea water. At the end of 20 hours the straight hinge or D-shaped larval stage is reached.

Stage	**Size (m) m**	**Hours/Days**
Straight hinge	60-70	20 hours
Early umbo	100	3rd days
Mid umbo	150	7th day
Advance umbo	200-270	12th to loth day
Eyed larva	280-290	13th to 17th day
Pedireliger	330-350	14th to 18th day

As larva grow, their density in rearing tanks is reduced and food ration increased. The schedule for rearing density and food requirements of larvae is given below:

Larval stage	**Density of larvae/ml**	**Algal cell concentration in ml/larval/day**
Straight-hinge	5	3000-4000
Umbo	3	400.0-5000
Advanced umboto eyestage Pediveliger	2	5000-8000

Non toxic, hard, chemically, stable and clean spat collectors are used to produce attached spat oyster shells as spat collectors are widely used for the ren method of culture. Brush well the shell, washed in chlorinated water followed by soaking and repeated washing in seawater, so that the pH of water in rearing tanks is maintained. Shell collectors are spread uniformly on bottom of one tone FRP tanks containing filtered seawater. In addition several rows of shell rens are also suspended in settlement tanks to increase the surface area. The eyed oyster larvae are released in tanks at a concentration of 1 larvae/mi in well aerated seawater. In following few days larvae are set as spat on oyster shells. The concave side of shell usually has more settlement.

For production of clutchless spats (free spat or single spat) pretreated polythene sheet is spread as a lining on bottom and sides of a FRP tank. Oyster shell grit of 0.5 mm size are wasted thoroughly sterilized in 10 ppm chlorine further washed in running filtered seawater and dried. These shell grit are spread uniformly at bottom of tank and larvae are released at setting stage. Rear spat for three weeks after setting in hatchery before they are transferred to field for nursery rearing.

Clam plays an important role in economy of coastal areas, their meat is cheap practised for culture in different countries.

The clam species belonging, to families Arcidae, Veneridae, Corbiculidae, Solemidae, Mesodesmatidae Tellinidae and Donacidae, Arcid clam are called blood clams.

Paphia malabarica, Meretrix costa, M. meretrix Katelysia opima, Villorita cyprinoides, Tridacna crocea, T. maxima, T. squamosa. Hippopus hippopus, Anadra granosa, Tridacnid clams are well known. They filter feeder and their biology is similar to oyster as they are also a bivalve. Their bottom culture is quite popular.

The giant clam is potentially dangerous to man because the bivalve is extremely well camouflaged and have tremendous strength of single adductor muscle (which is considered a delicacy also). Once an object is caught between the two valves it can only be released again by either severing adductor muscle or breaking the shell.

In the early 1930s a crude pearl weighing 7 kg and measuring 23 by 15 by 14 cm was extracted from Tridacid clam in Philippines.

The Venus clams (family-Veneridae) are represented by some noteworthy forms in Europe and Americas. They are quite frequently jump a short distance with the aid of their foot. In the venus clams siphons are much longer and are partially fused, and externally these shells are quite distinct from cockles. Their shells are almost always weakly ribbed and lack spines and they are often decorated with dark brown bands. The watery venus (*Venus verrucosa L-4cm*) is one of the few ribbed species. European venus clam are characterized by distinct lamellar ribs which are arranged in concentrating the light brown coloration of shall values is often complemented by three or four interrupted bands which the concentric rings from inside to outside *Venus gallina* is somewhat smaller, numerous in North Sea. Where it is characteristics species of an entire animal ecosystem on the sandy ground at moderate depths. The edible *Callista chione* is equally abundant in Mediterranean. In contrast to true Venus clams, the Carpet shell (*Venerupis perforans*) is clearly more elongated arid therefore its form is reminiscent of a blunt arcid bivalve.

The American Quahog (*Mercenaria mercenaria*) is another member of Venus clams. It was introduced to northwestern France from eastern coast of North America. The Algonguin tribes which formerly inhabited what is now the eastern USA, used to make beads out of shells which they pierced in middle and strung together into necklaces. These strings of shell beads also served as money which was known by Indian name of *Wampumpeag*. The Indians also strung these shell beads into very intricately designed belts known as wampums which were used as token to bind treaties.

The other candidate species with great economic importance are *Perna uiridis* green mussel and *P. indica* brown mussel. The farming is done is coastal water beyond the surface zone at 10 to 15 meters depth in non-polluted area. The spawn of green mussels are reported in July-September. In July then juvenile (spat) form over the rocks and intertidial zones with density of 6550 per m^2. In one hour 18-20 kg spat can be collected. Spat collectors like coir ropes, any nylon string, roof tiles may also be used in west coast of India during heavy monsoon in shallow water 10-15 m depth longline culture of mussel is practised. Longline unit consist of 60 m long horizontal HPD rope of 20-40 mm thickness, anchored at both ends with 150 kg concrete blocks and a series of 100 litre capacity barrels as floats fixed at 3m intervals. Vertial lines of 6m length seeded with mussel spats are hung at a distance of 75cm between two floats in main line. A longline unit of 60 × 60m can accommodate 12 horizontal ropes and 920 to 1,000 vertical ropes. The distance between two horizontal lines in 5 m. At every 20m horizontal lines are connected using additional horizontal lines pole culture and stake culture are done in estuaries at a depth of 1.5-3m. The spats of 15-20 mm. are wrapped around the poles or stake with cotton mosquito nettings. Spats get attached to poles in 72-96 hours and it is enough time for cotton net disintegration. Do periodic thining also. In long line culture two crops can be obtained.

Table Identification of *Paviridis* and *P. indica* taken within a year. The growth in open sea is very fast. The weight increases 6.0-7.4 gm in one month and production reaches upto 10-13 kg. marketable size from one meter of rope. From a longline unit of 360m^2 a total production 54,720 kg shell on mussels can be obtained of which 40 per cent will be the meat (CMFRI Report TTS-I).

Bottom Culture, Stake Culture, Rack Bag Culture Raft Culture.

Character	*P. viridis*	*P. indica*
1. External colour	Green	Dark brown
2. Mantle margin colour	Yellowish green	Brown
3. Ventral shell margin	Higher concave	Almost straight
4. Middle dorsal margin	Accurate	A distinct dorsal angle or lump present
5. Anterior end of shell	Pointed, break down turn	Pointed and straight
6. Number and size of hinge teeth	Two small teeth one left valve and one on the right	One large tooth on left value and corresponding depression one right value.

The mussels also have organ system like other bivalves. Its foot is like clams and small with little mobility. Mussel threads can be secreted more than one time. As mussels are filter feeder, phytoplankters are good for them. Sexes are separate and gonads located in body proliferate into mantle. They have plankto trophic larvae. Male gonads are creamy white in colour while in female it is pink or reddish.

At 4°C above the ambient gave results in inducing spawing, egg are brick red and measured 40-50 mm. morula size 58-60 (μm) in 4 hours, trochophore 65-70 (μm) in 7 hours. Early straight hinge 70-76 (μm) in 17-20 hours. Early umbo 120-140 (μm) 7 days, Eyed larvae 208-260 (μm) in 13-14 days. Prediveliger 280-320 (μm) 16 days onwards. Heavy spat settlement from 20 days onward.

Other molluscs widely like cephalopods used for food purposes in fresh as well as processed form are cephalopods like octopus, sepia, loligo squids etc. Here only squid processing is described with cephalopods distribution (CMFRI report and MPEDA report).

IMPORT STANDARDS OF JAPAN

Form and shape : The form, trimmed or untrimmed, should be good, having no cuts, splits or any other wounds and having had the

CEPHALOPODS: RESOURCES AND UTILISATION

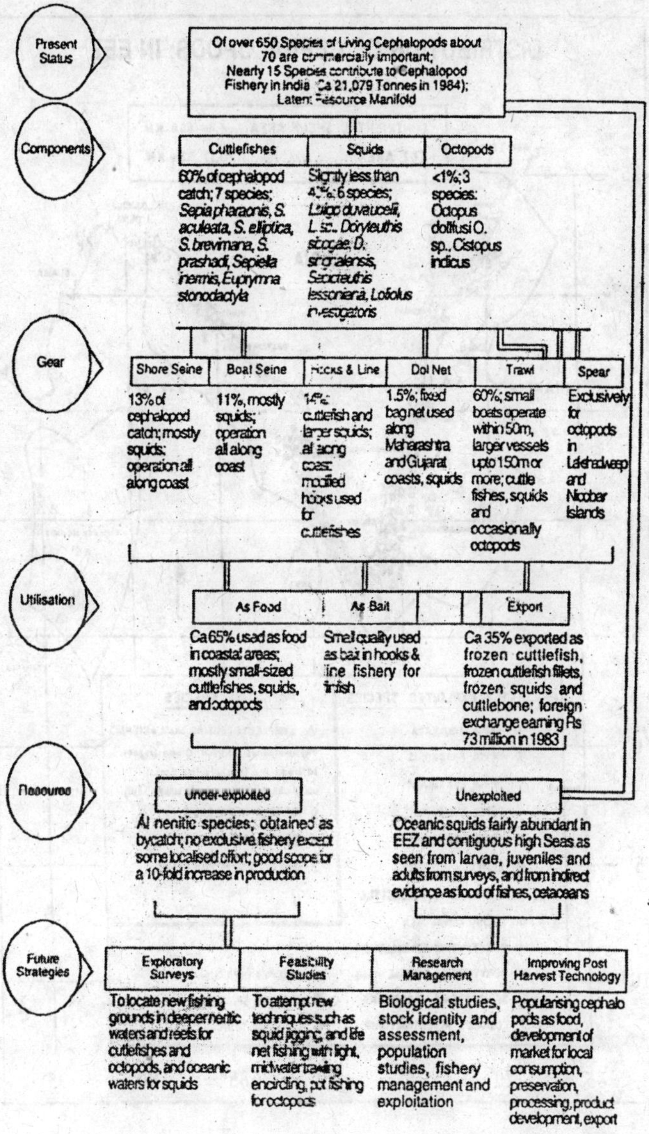

Present Status
Of over 650 Species of Living Cephalopods about 70 are commercially important; Nearly 15 Species contribute to Cephalopod Fishery in India (Ca 21,079 Tonnes in 1984); Latent Resource Manifold
Components
Cuttlefishes
Squids
Octopods
60% of cephalopod catch; 7 species; Sepia pharaonis, S. aculeata, S. elliptica, S. brevimana, S. prashadi, Sepiella inermis, Euprymna stonodactyla
Slightly less than 40%; 6 species; Loligo duvaucelii, L. sp., Doryteuthis sibogae, D. singhalensis, Sepioteuthis lessoniana, Loliolus investigatoris
<1%; 3 species: Octopus dollfusi O. sp., Cistopus indicus
Gear
Shore Seine
Boat Seine
Hooks & Line
Dol Net
Trawl
Spear
13% of cephalopod catch; mostly squids; operation all along coast
11%, mostly squids; operation all along coast
14%; cuttlefish and larger squids; all along coast; modified hooks used for cuttlefishes
1.5%; fixed bagnet used along Maharashtra and Gujarat coasts, squids
60%; small boats operate within 50m, larger vessels upto 150m or more; cuttle fishes, squids and occasionally octopods
Exclusively for octopods in Lakshadweep and Nicobar Islands
Utilisation
As Food
As Bait
Export
Ca 65% used as food in coastal areas; mostly small-sized cuttlefishes, squids, and octopods
Small quality used as bait in hooks & line fishery for finfish
Ca 35% exported as frozen cuttlefish, frozen cuttlefish fillets, frozen squids and cuttlebone; foreign exchange earning Rs 73 million in 1983
Resource
Under-exploited
Unexploited
All neritic species; obtained as bycatch; no exclusive fishery except some localised effort; good scope for a 10-fold increase in production
Oceanic squids fairly abundant in EEZ and contiguous high Seas as seen from larvae, juveniles and adults from surveys, and from indirect evidence as food of fishes, cetaceans
Future Strategies
Exploratory Surveys
Feasibility Studies
Research Management
Improving Post Harvest Technology
To locate new fishing grounds in deeper neritic waters and reefs for cuttlefishes and octopods, and oceanic waters for squids
To attempt new techniques such as squid jigging, and life net fishing with light, midwater trawling encircling, pot fishing for octopods
Biological studies, stock identity and assessment, population studies, fishery management and exploitation
Popularising cephalopods as food, development of market for local consumption, preservation, processing, product development, export

Fig. 10.16

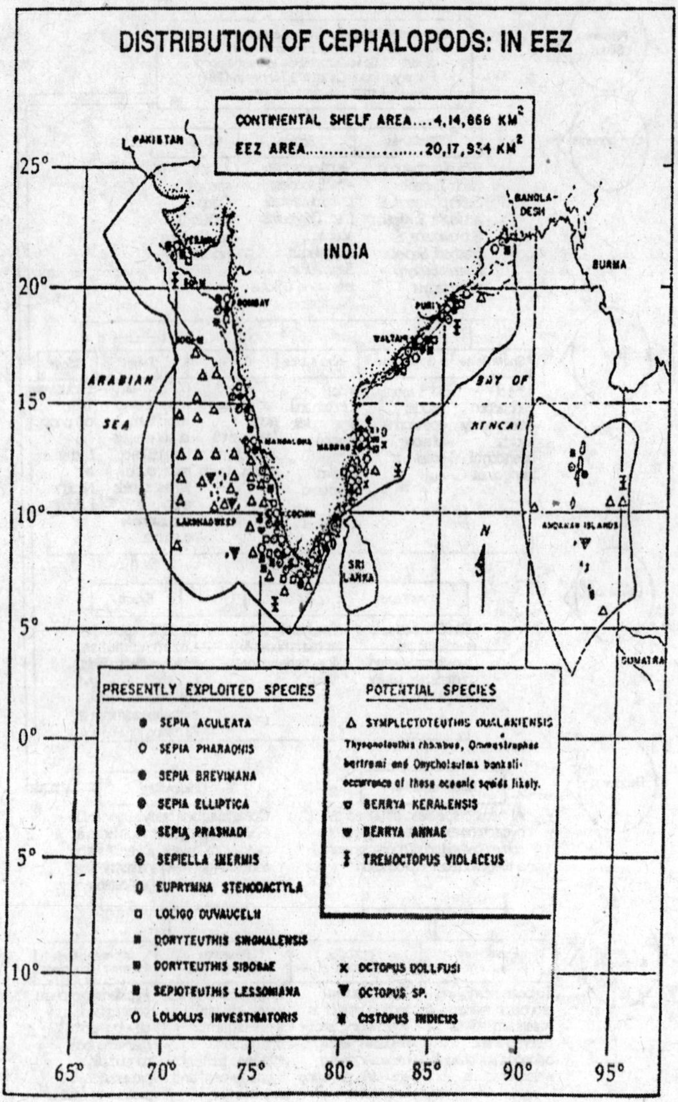
DISTRIBUTION OF CEPHALOPODS: IN EEZ
CONTINENTAL SHELF AREA....4,14,866 KM²
EEZ AREA........................20,17,934 KM²
PAKISTAN
INDIA
BANGLADESH
BURMA
BOMBAY
ARABIAN
SEA
BAY OF
MANGALORE
MADRAS
COCHIN
LAKSHADWEEP
SRI LANKA
ANDAMAN ISLANDS
SUMATRA
N
25°
20°
15°
10°
5°
0°
5°
10°
65°
70°
75°
80°
85°
90°
95°
PRESENTLY EXPLOITED SPECIES
SEPIA ACULEATA
SEPIA PHARAONIS
SEPIA BREVIMANA
SEPIA ELLIPTICA
SEPIA PRASHADI
SEPIELLA INERMIS
EUPRYMNA STENODACTYLA
LOLIGO DUVAUCELII
DORYTEUTHIS SINGHALENSIS
DORYTEUTHIS SIBOGAE
SEPIOTEUTHIS LESSONIANA
LOLIOLUS INVESTIGATORIS
POTENTIAL SPECIES
SYMPLECTOTEUTHIS OUALANIENSIS
BERRYA KERALENSIS
BERRYA ANNAE
TREMOCTOPUS VIOLACEUS
OCTOPUS DOLLFUSI
OCTOPUS SP.
CISTOPUS INDICUS

Fig. 10.17

Processing Flow Chart

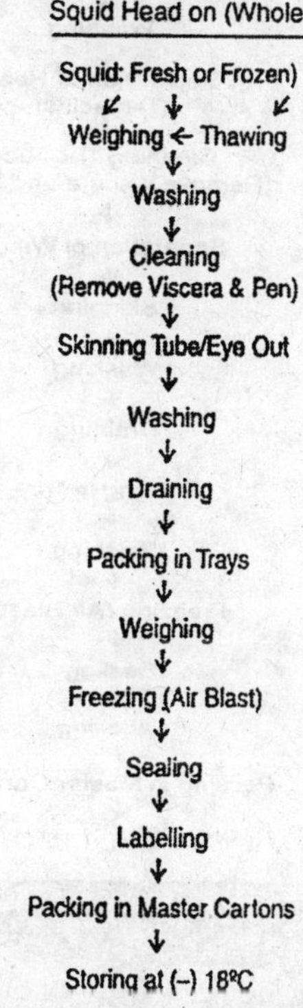

Squid Head on (Whole)

Squid: Fresh or Frozen)

↙ ↓ ↙

Weighing ← Thawing

↓

Washing

↓

Cleaning
(Remove Viscera & Pen)

↓

Skinning Tube/Eye Out

↓

Washing

↓

Draining

↓

Packing in Trays

↓

Weighing

↓

Freezing (Air Blast)

↓

Sealing

↓

Labelling

↓

Packing in Master Cartons

↓

Storing at (–) 18°C

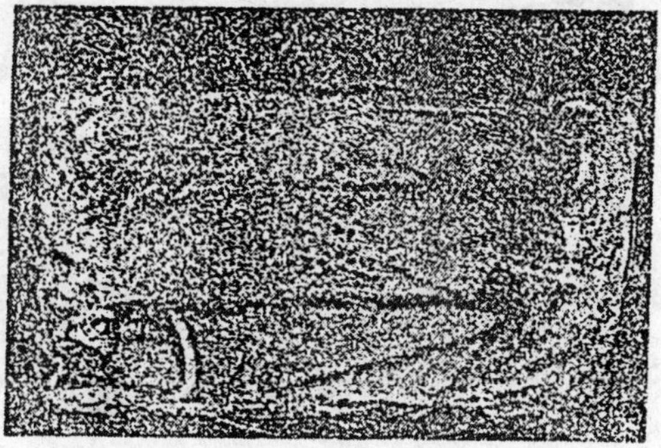

Fig. 10.18

Squid Tubes

Squid: Fresh or Frozen)

↙ ↓ ↙

Weighing ← Thawing

↓

Washing

↓

Separation of Head (Tentacles)

↓

Cleaning The Tube (Remove Viscera and bone)

↓

Separation of Wings

↓

Skinning

↓

Washing

↓

Draining

↓

Packing in Tray

↓

Weighing

↓

Freezing (Air Blast)

↓

Sealing

↓

Labelling

↓

Packing in Master Cartons

↓

Storing (–) 18°C

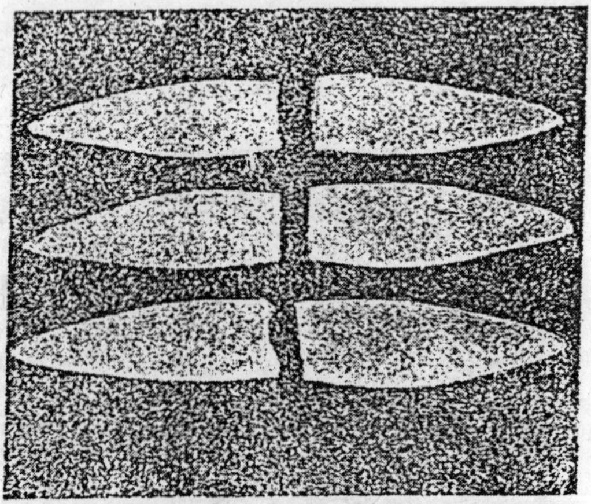

Fig. 10.19

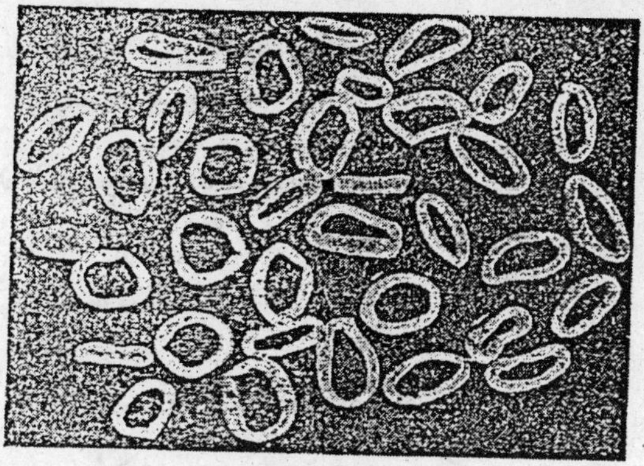

Squid: Fresh or Frozen

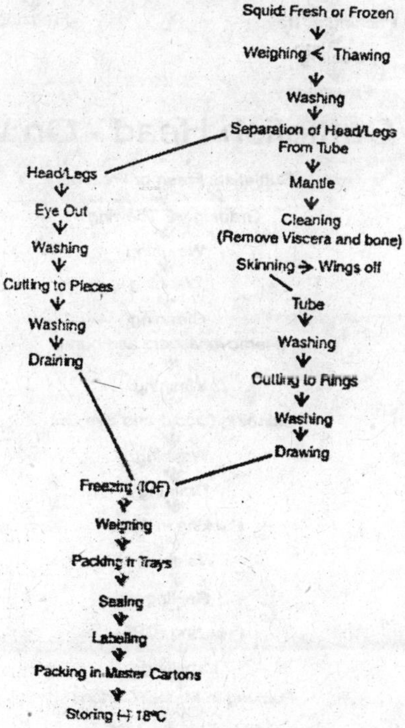

Fig. 10.20

Stuffed Squid

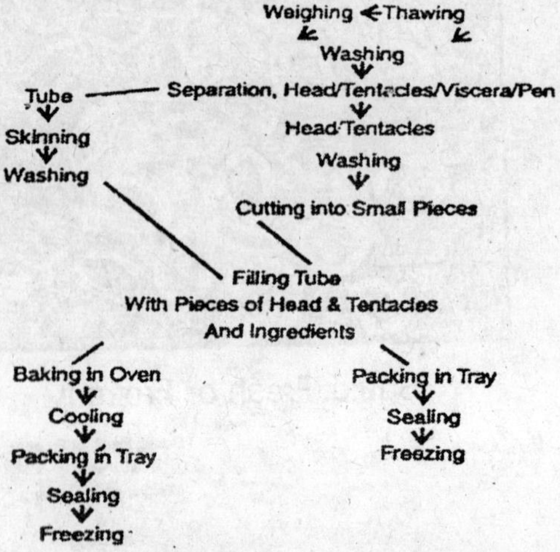

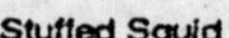

Cuttlefish Head - On Whole

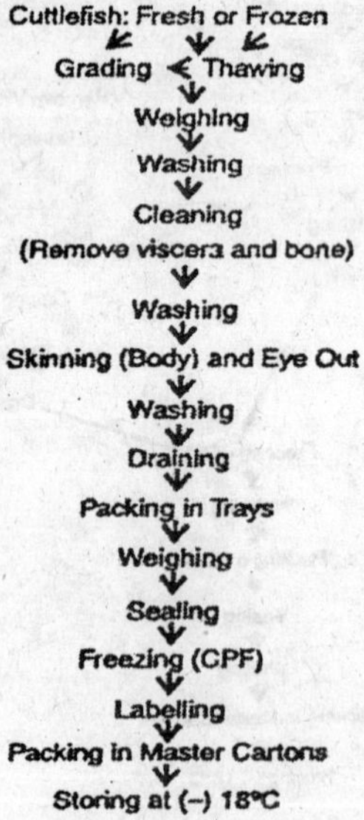

Fig. 10.21

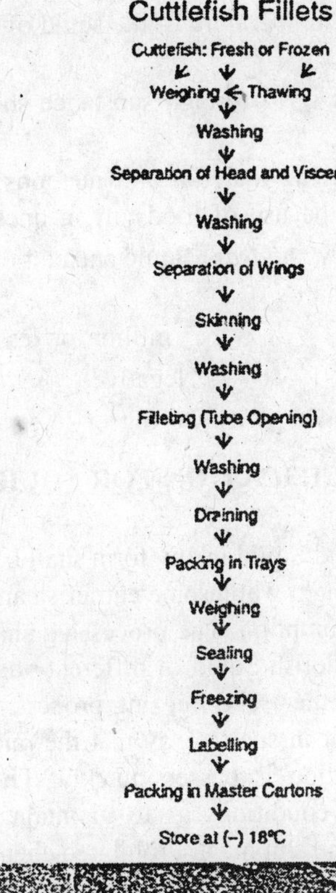

Cuttlefish Fillets

Cuttlefish: Fresh or Frozen
↓
Weighing ← Thawing
↓
Washing
↓
Separation of Head and Viscera
↓
Washing
↓
Separation of Wings
↓
Skinning
↓
Washing
↓
Filleting (Tube Opening)
↓
Washing
↓
Draining
↓
Packing in Trays
↓
Weighing
↓
Sealing
↓
Freezing
↓
Labelling
↓
Packing in Master Cartons
↓
Store at (–) 18°C

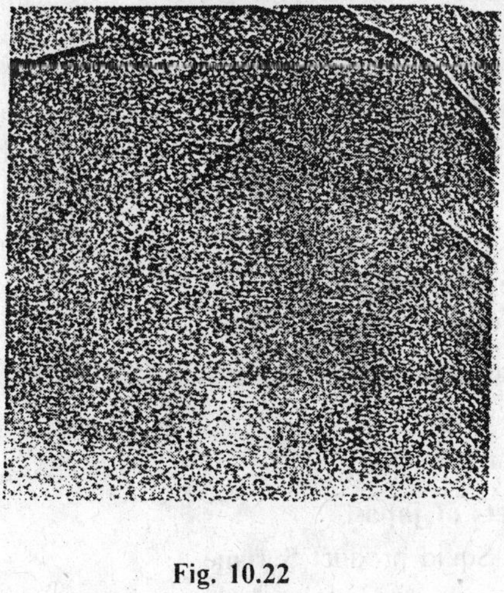

Fig. 10.22

head and tentacled section, the cuttlebone, the fin and the outer skin removed.

Temperature : The temperature of the squid of cuttlefish should be –18°C.

Foreign substances : No foreign substance should be attached to or mixed with the fish.

Packaging : Packaging material and methods should sufficiently satisfy the quality and the use of foodstuff in question.

Sanitary inspection standards : Squid and cuttlefish should meet the following sanitary standards.

No. of bacteria	: 5 million or less per gram
E. Coli	: Negative
Volatile basic nitrogen	: 26 mg or less per 100 grains

EIA'S EXPORT SPECIFICATION FOR SQUID & CUTTLE-FISH

Frozen Cuttlefish and Squid in any form shall be prepared by quick freezing from freshly caught wholesome cuttlefish and squid of a quality suitable for human consumption. The processing shall be carried out in a hygienic manner. Cuttlefish/Squid of different varieties and products shall not be packed together. The freezing process shall be carried out in appropriate equipment in such a way that the range of temperatures of maximum crystallisation is passed quickly. The product shall be maintained under such conditions as to maintain the quality during storage, transportation and shipment. Prohibited chemical additives shall not be used at any stage of processing. Frozen cuttlefish and squid shall have the characteristic colour and odour. They shall not have any discolouration, objectionable odour and flavour. The material on thawing shall be clean, having an attractive appearance and shall be reasonably free from any defects and objectionable foreign matter.

Squid Products of Spain

(i) whole frozen squids

(ii) Block frozen tubes

(iii) IQF squids

(iv) Squid rings

(v) Frozen tentacles

Squid Products of japan

(i) Dried Squid product Surume

(ii) Seasoned Squid products Chinmi-ika

(iii) Fermented Squid Products ika-shiokara

(iv) Smoked Squid products ika kunsei

(v) Pickle squid meat sui-ika

Squids tops all other marine products in terms of ratio of edible to non-edible portions. The edible portion is 80-85 per cent of the weight. Nothing is wasted in processing such as viscera, liver, skin, pen, beak and eyes can be converted into squid meal. Its bone has demand in world market as its use in animal feeds, handicrafts, glass industry etc.

	Bacteriological standards	Cuttlefish/squid	Cooked and boiled
1.	Total plate cunt/gm.	Max. 5,00,000 Max. 2,00,000 (for products) meant for raw consumption)	Max. 1,00,000
2.	E. coli count/gm	Max. 20	Nil
3.	Coagnlase positive staphylococcus/gm.	Max. 100	Max. 100
4.	Salmonella & S. Arizona:	Negative	Negative

Beside this in class cephalopoda a beautiful silvery metallic coiled shelled animal *Nautieus* has great decorative and aesthelic importance.

For Decoration

- Until recently sailers have produced decorative, screen show work by engraving designs on to sheath and whales teeth.
- The Indian chank shell is a large conch that has become the centre of much superstitious custom and is used as trumped at Hindu marriage services and on other religious occasion.
- The shell of the pearly *Nautilus*, a cephalopod used to be mounted in gold and silver and made in to gorgeous chanks and drinking vessels.
- The false trumpet, a tropical snail is the largest of the snail.
- Its shell is used as a water carrier by natives of North West Australia.

For Other Purpose

- An unusual industry based an molluscs was the weaving of cloth from the byssus thread, or beard, of the fan mussel.

REFERENCES

1. Algarswami, K. and Dharmaraj (1984) Manual on Pearl Culture Techniques CMFRI Publication Cochin.
2. Algarswami K (1991) Production of Cultured Pearls 1-111 ICAR, New Delhi.
3. Anonymous, E (1991) Pearl Oyster farming and Pearl Culture Training Manual No. 8. Regional sea farming Development and Demonstration Project, Bankok.
4. Ballantine, D and Morton, J.E. (1965) Filtering, feeding and digestion in the lamellibranch *Lasaea Rubra*, J. Marine, Biol. Assoc. U.K. 35, 241-274.
5. CMFRI Booklets on Oysters. (1995)
6. Elizabath Goling (1196) Bivalves Molluscs, Fishing News Books, Blackwell Science Ltd., Oxford, 200.
7. Eswaran C.R., K.R. Narayanan and M.S. Michael (1969) Pearl Fisheries of Gulf of Kutch, JBNHS 33(2): 338-344.
8. Hancock, D.A. (1973) Kuri Bay Pearls some of the finest in world Australian Fisheries 32(4); 11-12.
9. James PSBR and Narisimham (1997) Mollusca.
10. Jenkins. FEC (1992), Advance Study of Molluscs.
11. MPEDA House, Panampilly Avenue, P.B. No. 4272, Kochi-682035.
12. Meglitsch P. (1967)) Invertebrate Zoology by Oxford University Press. p. 961, Oxford.
13. Morton E.J. (1983) Molluscs Harper and Brothers. New York. p. 5942
14. Mash. K. (1984) How Invertebrates Live, Academic Press, New York, p. 160.
15. Robert W.P.—Fresh Water Invertebrates of the United States. QL141 p. 45
16. Narasimham. K.A. (1991) Present status of clam Fisheries of India. J. Mar. Biol. Ass. India 33(112) 77-88.
17. Pearl culture in Fresh Water Mussels Fresh Water Fisheries Research Station, Kalyani, W.B. (1994).
18. Santhanam R. Ramanathan, N. and Jegatheesan G. (1990), Coastal aquaculture in India, CBH Publishers, New Delhi 92-107 and 117-124.
19. Squid and Cuttle fish—MPEDA Publications Cochin. (1982)
20. Thompson T.E. (1958) The natural history embryology, larval biology and past larval development of gastropods. Phil. Trans. Roy. Soc. B242 1-58.
21. Ward, F (1985) The Pearl, National Geographical 168 (2); 1993-222.
22. Yonge, C.M. (1949) The Sea Shore, Collins Press London.
23. 1978 592'09'297378-8130 ISBN 0-471-04249—and Printed in United States of America.

11

FEW MOLLUSCS

INTRODUCTION

Venus

The clams, mussels and scallops of this group are often referred to as the Lamellibranchs because of their flat body.

Venus mercenaria is known by a variety of common names, such as quahog, hard-shelled clam and littleneck clam. Its entire range includes the shallow, protected waters of the Atlantic coast from the Gulf of St. Lawrence to Texas, but it is most abundant from Cape Cod southward. By means of its powerful foot, this form burrows about over muddy or sandy flats near the lower portion of the intertidal zone. Although, it lies buried a few inches, it keeps its siphonal end above the surface of the mud in order to maintain a current of sea water through its body. This current is essential for respiration and feeding.

SHELL

The shell of *Venus* is composed of two symmetrical halves called valves. Because of this arrangement, the molluscs possessing two valves are called bivalves, as distinguished from one shells, which are called univalved. At the dorsal margin the valves are joined by a brown, external hinge ligament. On either side of this ligaments there is a swelling in each valves called the umbo. This ends in a point to it. The impression of the posterior retractors is usually merged with that of the corresponding adductor. This gives impression the position of the retractor muscles of the siphons.

Along the ventral edge, there are nereis the tiny teeth. This arrangement is the result of secretion by a fringed edge of the mantle. The circulated margin ensures a firmer closing of the values. Along the dorsal edge notice the larger teeth which are of two kinds anterior to the ligament are the prominent cardinal teeth; below ligaments are the less obvious lateral teeth. Which extent posteriorly for same distance. Notice that the teeth fit tightly and form as efficient interlocking mechanism. In *Venus* the this periostracum. Or outer layer of the shell, is partly worn away by constant friction with the sand. The lanner, most layer, the laminated nacreous or pearly layer, is very thick but does not appear as shinning and lustrous as in many molluscs. The middle, or prismatic, layer is present but not easily visible.

MANTLE

The mantle consists of two lobes of tissue, one of tissue is applied to the each, which is directed anteriorly. Growth has proceeded from the beak as an origin and has resulted in concentric lines of growth. The *lunule* is a heart shaped configuration which lies anterior and ventral to the beak. This is a useful character for field identification (lunule absent in Fresh Water mussel). Before opening the valve determine the correct orientation, right, left, dorsal, ventral, anterior and posteriors. Obtain, if possible, a young *Venus* for comparative shell study.

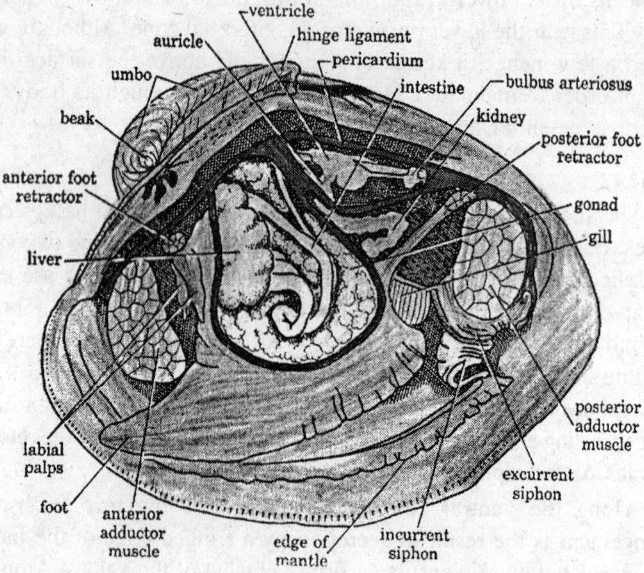

Fig. 11.1 : *Venus mercenaria*: Left valve, mantle and gills removed to show internal anatomy.

Put the two values apart and try to insert your knife between the shell and the mantle which lines it. Detach the mantle carefully, and, at the anteriors and posteriors ends, cut the adductor muscles. Since the contraction of the adductors closes the valves, they will now gape an be readily separated. The gaping is due to the action of ligament. Which opposes the adductor muscles.

On the inside of the shell are various markings, indicating the former attachment of organs, Near each end of a valve is the large round scar of the adductor muscle. Closely associated with these are the two smaller scass of the retractors muscles of the foot. The small impression of the anterior retractor is separated from the larger one of the anterior adductor and lies dorsal and a little posterior inner surface of each valves. These two lobes are continuous dorsally. The free border of each lobe is thickened and contains muscles which attaches along the pallial line. These muscles serve the function of•washing. The posteriors edge of the mantle lobes are also thickened and fused at two points, one above the other, to form a double tube, the siphons. In *Venus* this shiphons are very short but are easily identified by their dark pigmentation. Water enters through the ventral, incurrent siphon, and after circulation through the branchial chambers, leaves by the dorsal, excurrent siphon. A few grains of powdered carmine placed near the siphons of an undissected, living specimen will demonstrate these currents clearly.

VISCERAL MASS AND FOOT

Carefully cut the mantle and reflect its parts. Then ventral portion of the median mass now visible is the foot, which is somewhat hatched-shaped. Identify the small foot muscles, the attachments of which have already been seen on the shell. These muscles may be seen at the extreme anterior posterior margins of the foot, where they extend dorsally toward their attachment on the valves. The dorsal portion of the median mass is called visceral mass.

RESPIRATORY SYSTEM

With the mantle removed, the gills are the most conspicuous organs. They consist of the two pairs of third membranous fields, ridged in appearance. Which lie on each side of the visceral mass. They consists of the two pairs of third membranous fields, ridged in appeared mass. They are attached to the body on each side along a line extending from the wall which separates the two siphons to a point even with the beaks on the exciters. Only the dorsal edge is attached; the ventral edge hinged freely. The outer gills are attached to the mantle lobe near its dorsal region; the inner gills are attached to the visceral mass. The entire space enclose between the right and left lobe of the mantle is called the mantle cavity. Since the gills are attached to the mantle, to each other, and to the visceral mass, their attachments form a continuous horizontal partition separating a small cavity, the branchial chamber is nor visible at this point. It is a median, dorsal tube, extending anteroposteriorly, and connecting with the excurrent siphon.

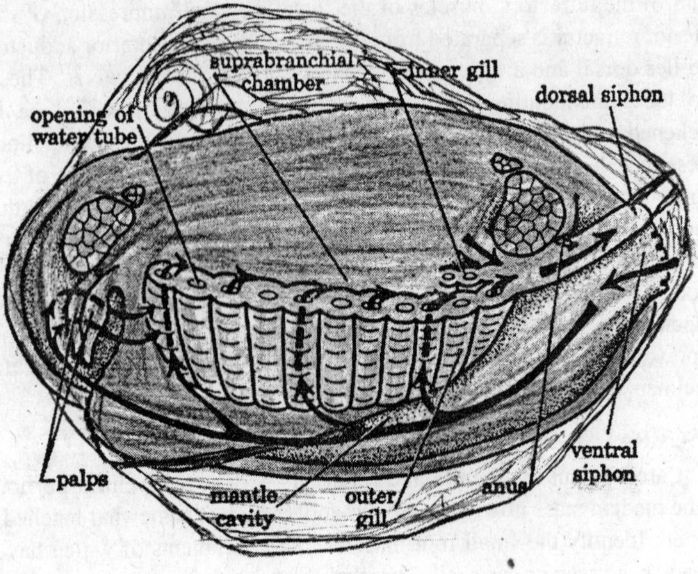

Fig. 11.2 : The circulation of water through the mantle cavity and gills of a *Venus* mussel.

Each gill is a double fold, the inner and outer surface being composed of parallel ridges or filaments. Each surface is referred to as a lamella. Each gill, therefore, consists of two lamellae, on inner and an outer one. The inner and outer lamellae of a single gill are separated by a thin space, which at regular intervals is decided vertically into a series of narrow tubes, the water tubes. These tubes connect dorsally with the suprabranchial chamber. The partitions forming the walls between the tubes are called interlinearly junctions, and they extend dorsoventrally

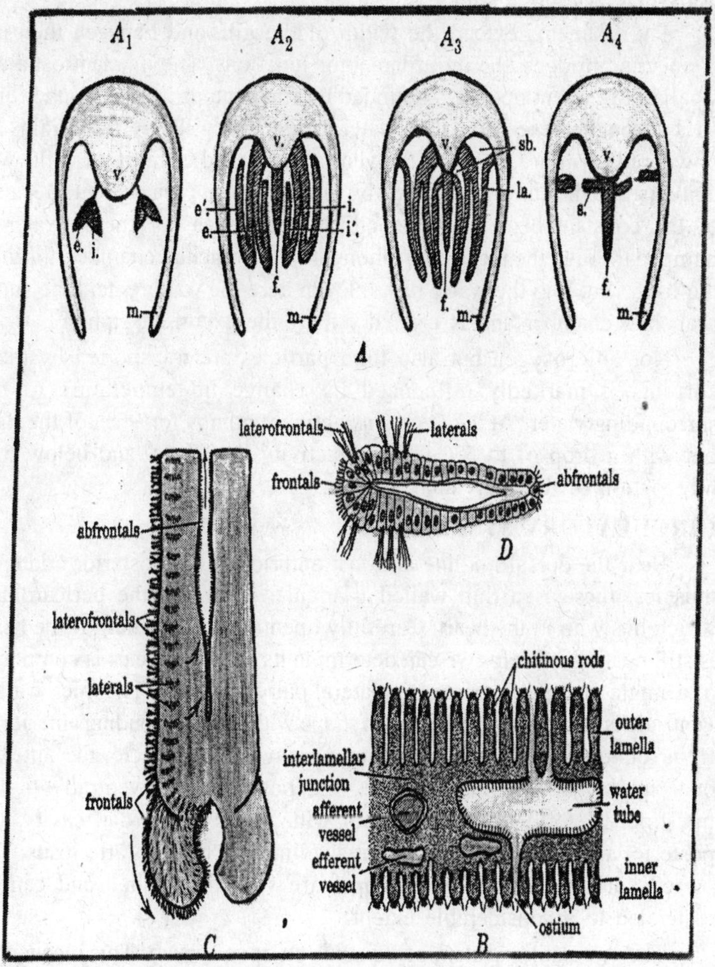

Fig. 11.3 : A diagrams of gill structure of the Pelecypods. Types of gills found in the different orders of Pelecypoda. A_1. *Taxodonta.* A_2. *Anisomyaria.* A_2. *Eulamellibranchiata.* A_4. *Septibranchiata.* v.: visceral mass; m.: mantle; e.: external row of filaments; e'.: external row of filaments turned back; i.: internal row of filaments; i'.: internal row of filaments turned back; s.: septum; f.: foot; sb.: suprabranchial chamber; la.: lamella. (Modified from Lang.) B. Cross section of a mussel gil. (Modifed from Borradaile adn Potts, courtesy of The Macmillan Company.) C. Portion of gill filament, near free edge, to show typieal sets of cilia present and direction of beat. (Modified from Orton.) D. Cross section of one limb of a filament, to show same sets of cilia. (Modified D. from Orton.)

between the lamellae. They contain blood vessels. From one gill, cut a transverse section 1 mm.

The filaments extend the width of the gill, and between then are connecting bridges, the intrafilamentor junctions. The inhalant ostia are small but frequent openings bounded by filamentous and their junctions. It is through these ostia that water enters the gills. A few grains of powdered carmine, placed at the ventral scalloped edge of the gills, will demonstrate the current produced by the prominent cilia present. Because of the constant beating of the cilia, a continues current of water is pumped through the incurrent siphon into the branchial chamber, through the ostia, and into the water tubes. From here the water enters the super branchial chamber and is carried out by the excurrent siphon.

Not only oxygen but also food particles are transported by these currents is markedly influenced by change in temperature of the surrounding water. At 22°C, *Venus* actively pumps for 96% of the day, but with a drop of to 5°C. Ciliary activity decreases, and below 5°C hibernation of the individual, occurs.

CIRCULATORY SYSTEM

Near the dorsal midline and just anterior to the posterior adductor muscles, these is a thin walled triangular chamber, the pericardium, which lies with in the heart. Carefully open the pericardium. If the heart is still beating, perhaps we can determine its rate. The heart is composed of a single median ventricle, and lateral paired auricles. The thick walled ventricle is some what pyramidal in shape with its apex bending anteriorly. It surrounds the intestine. Two aortae leave the ventricle; the anterior one run dorsally to the intestine, the posterior one ventrally to the intestine. Posterior to the heart but with in the pericardial cavity, the posterior aorta forms a conspicuous swelling, the bulbous arteriosus. The paired auricles, triangular in shape, are very thin-walled and can be distended to a considerable extent.

The circulatory system of *Venus* is an open system. This means that arteries communicate with views by way of large spaces called sinuses or lacunae. Blood is pumped by the ventricle of the heart through the two aortae to the viscera and neighbouring sinuses. From here, by a network of veins, blood is returned through the kidneys to the gills and finally brought back to the auricles of the heart.

EXCRETORY SYSTEM

Ventral of the pericardium and between it and the posterior adductor muscle lies a pair of dark glandular organs, the nephridia. Each communicates with the pericardium by a mixture opening which is difficult to find, and also with the branchial cavity by another opening more early seen. Reflect both gills dorsally. This second opening is located on a small papilla which will be describe under the genital system. The nephridia are regionally differentiated in their composition, band consists of a broad U. Shaped tube. The ventral anterior portion communicating with the heart is glandular; the dorsal posterior part is then walled and functions as a bladder.

The coelom is reduced to these small cavities, the pericardial, the gonodal, and the nephridial.

DIGESTIVE SYSTEM

On either side of the anterodorsal edge of the visceral mass lie two triangular flaps, the labial palps. Both the inner and outer flap unite with the corresponding structure of the outer side above the mouth, which

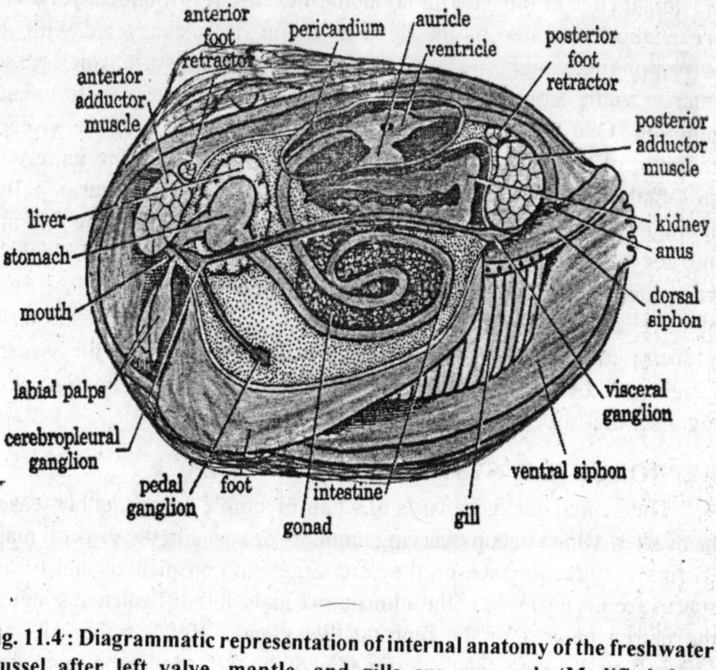

Fig. 11.4 : Diagrammatic representation of internal anatomy of the freshwater mussel after left valve, mantle, and gills are removed. (Modified from Wolcott).

lies slightly posterior to the dorsal border of the anterior adductor muscle. The palps are abundantly ciliated and cause stream of water to pass into the mouth. They, therefore, continue to ingest the particles of food particles already brought in by the currents from the gills.

Venus is a typical filter feeder. This means that food consisting of diatoms, Protozoa and organic detritus, brought or by the respiratory current which is passed anteriorly along the edge of the gills and becomes entangled in strings of mucus. Larger food particles are dropped to the edge of the mantle and discarded. As the mucous strings approach the mouth. Digestion is cheifly intracellular by phagocytes a lining the digestive tract and by wandering leucocytes. Digestive juices from the liver are known to occur, and in addition the crystalline style liberates an enzymes aiding in the breakdown of carbohydrates.

NERVOUS SYSTEM

It consists of three pairs of ganglia and their connectives. The cerebral ganglion will be seen lying posterior to and about 3 mm. From the dorsal edge of the anterior adductor muscles. It is rounded, yellowish organ about the size of the head of a pin. It is connected with the corresponding ganglion of the other side by a commissure which passes anterior to the esophagus. Two conspicuous connectives pass from each ganglion. One of those courses posteriorly to unite with the visceral ganglion of the same side. The other passes into the foot to unite with the pedal ganglion. The cerebral ganglion also furnish the anterior pallial nerve. With a razor make a median sagittal section through the foot and into the visceral mass. This should expose the pedal ganglia which lie embedded in the foot dorsal to the muscular portion. To find the III pair of ganglia, cut the united lamellae of the gills ventral to the posterior adductor muscle. On the anterior surface of this muscle, the visceral ganglia will be seen. They are slightly pear shaped bodies lying so close together that they appear to be single.

REPRODUCTIVE SYSTEM

The genital glands consists of a pair of simple gonads, either teasers or ovaries, which occupy varying amounts of space in the visceral mass. During the breeding season they are large and conspicuous and fill the spaces around the loops of the alimentary canals. It is difficult to distinguish the paired natures of the reproductive glands. Reflects the gills and follow dorsally along the posterior edge of the foot. Under the attachment of the inner gills lies a small projection, the opening of the genital ducts.

A minute opening, the nephridiopre, is also located on the papilla but is difficult to see. It is of course, the external opening for the kidney.

The spawning of *Venus* occurs in the cape cod region from the late June to early August, when the temp of the water has risen to a critical level, about 75°F. Egg and sperm, numbering into the millions, are freely shed into the sea where fertilization takes place. Spawning in these species is not necessarily completed by one discharge, but may occurs several times during the summer. The early embryology of this form is typical for the bivalves. The trochophore and veliger stages are rapidly completed, and during the late summer the young quahog becomes temporarily attached to a substrate by a byssus thread. It soon regains a free existence. For a period of 2 years the young clam lies relatively dormant during the winter but grows rapidly during the warm months. By the third summer, it has matured and is ready to spawn.

Nautilus

Not with standing the phylogenetic controversy, today's living Nautiluses (tetrabranchi) are classified as a separate subclass from all remaining orders since their anatomy shows marked differences from those of the coleoids (dibranchs).

Length varies from 10 to 27cm, the external shell is well developed symmetrical. The shell is planospirally coiled is partitioned into chambers. The siphuncle or tubule extension of the visceral hump extends through the centre of the repute. The modified foot consists of highly two to ninety tentacle with lack suction cups. The funnel consists of two lobes which are not fused. Each transverse row of the radical contains thirteen teeth. There is no ink sac. There are two pairs each of kidneys, gills, osphridia and auricles. The eyes are simple and faceted, without a lens and miracles chamber. The embryology is unknown. There is only one living family : Nautiloids (nautiloids) with a single games by the funnel on the ventral side. Unlike the dibranchs, the nautiloids lack suction cups on the tentacles. Instead, each tentacle is equipped with an *achieve* pad facing the mouth opening with secretes a sticky, glandular substance. Depending on their function, tentacles can be differentiated into chemosensory tentacles and grasping tentacles.

The prominent staked eyes are uncomplicated. A comparison with a simple pinhole corners comes to mind. These eyes lack a lens and an aqueous chamber but posses a well developed retina which receive the light ray that fall through the visual hole. Nevertheless; that compound eye is weak visually. The eyes and the brain are partially located on an it shaped cartilage which supports the bipartite funnel and also serve as muscle attachment. A fingerlike process known as shinopore is situated below each eye. This is the centre for chemosorption. Additional sensory organs are found on each of the four gills. These are plate-like seen as a suture line on the outside. Just as in the primitive gastropod larvae, the shell is coiled towards the head region. This type of coiling is known as exegastric. The *Nautilus* species are also known as exegastric. The entitles species are also known as tetrabranchiata (four-gilled) as opposed to the two-gilled distranchiata-coeloids.

As in all *Nautilus*, the mantle encircles the dorsal surface of the visceral sacs, although in the nautiloids a dorsal flap is also developed which folds back on four lines the shell. The many tentacles, up to ninety are particularly conspicuous. The individual tentacles are short and consists of a thick shaft for a thin, coiled tentacle which can be retracted into the shaft. The shafts of the four dorsal most tentacles are fused into a hood. This beautifully designed shield for the animal is rates conspectus expensive. The other tentacles surround the mouth opening in two incomplete series, the inner having from forty-four to fifty-two, the outer-thirty eight. Some tentacles are modified as copulatory organ. The sense of tentacles is of interrupted organs that also function as olfactory structure. In function they correspond to the osphradia of other molluscs

The pearly *Nautilus* is a nocturnal otherwise bottom sedentary. During the day the animals usually remain hidden, occasionally drilling like the boats near the water's surface it they have been distributed. The shell chambers of the nautiloids contain a fluctuating amount of field in addition to a constant amount of gas. The amount of field can be regulated by the spicule which extends into the last chamber. The amount of fluid influences the gas pressure of thereby its buoyancy potential. The animal can sink or float by regulating the gas/field ratio in the shell chamber. The nautiloids can adjust to the external pressure of the water at various depth. After having observed *Nautilus macromphalus*., the french scientist Rent Catale was led to assume "that it seems, to be the lack of external pressure rather than the proper choice of food which has resulted in the failure of keeping *Nautilus,* six species: Pearly Nautilus (*Nautilus pompilius*: L up to 20cm) 2. Solomon's Nautilus (*Nautilus macromolecule*) 3. New caledonian Nautilus (*Nautilus scrupulosities*). Three other species are found along the Australion Coast.

The uniqueness of the few living nautiloids has learned them a virtual of names in each of several language. Since the animals sometimes profit slowly along near the surface of the water because they have an irridescent naucrouslike inner layer, they are sometimes called "pearly boats". The description 'chambered snails refers to the antidotal's shell structure. A longitudinal section will reveal that the shell consists of reveal successive chambers. These are filled with gas and are connected by a tubelike spleen with the animal proper. The Manila boys is located in the antomatists chamber of the shell. As the animal grow, new chambers are added one at a time to its shell the new growth signs.

these animals in captivity for longer period of time. At depths of from 50 to 650m, pearly *Nautilus* feed on crabs and carrion, they found on the ocean bottom. One can feed this same fare to captive animals without hesitation. When hesitations a prey, the thirteen notched radula probable serve only as a '*shovelling device*.' Often digestion had taken an unusually long time theist fours or more after feeding, prices of food could still be detected in the crop.

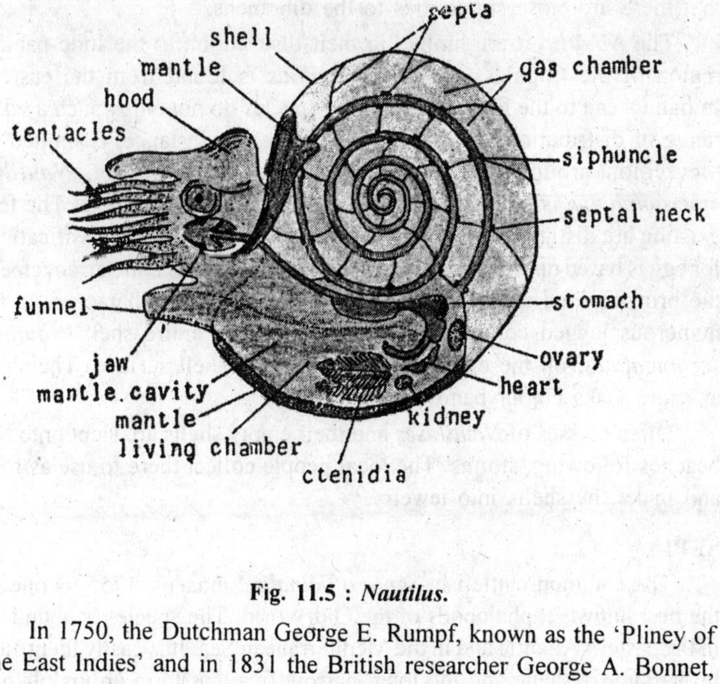

Fig. 11.5 : *Nautilus*.

In 1750, the Dutchman George E. Rumpf, known as the 'Pliney of the East Indies' and in 1831 the British researcher George A. Bonnet, had already researched how these animals swim with jerky jet-propelled motions wherein their tentacles are spread out widely. On the bottom, however, these creature crawl or attach themselves via the tentacles Devil use of the tentacles form crawling still has not been observed in the aquarium, a point emphasized by the British researcher Anna M. Bidder. When the animals are at rest or are swimming rhythmically, all tentacles are with drawn into their shafts. Occasionally these tips may protrude. As soon as one of the external lateral (tasting) loaches touches food, however, all other (gripping) tentacles are extended. There prey is then entangled and pulled underneath the hood. The interior circle of tentacles hold the prey while the jaws tease off piece after piece. The *Nautilus* species do not utilize, the entire mantle cavity as a pressure producing apparatus; only the strongly modified have the funnel, are contracted and relaxed. According to the some observers, the *Nautilus* can move just as quickly as a fish.

By approximately one year of age the *Nautilus* has reached maturity. At this stage its shell consist of twenty three to 27 chambers. Nothing is known about the embryology. On the basis yolky edge one can assume that there are close similarities to the dibranchs.

The *Nautiluses* are limited in their distribution to the Indo-pacific region of the trophics, *Nautilus pompilius* is found from the eastern Indian Ocean to the fine islands. Other species do not enjoy such a wide range of distribution. *Natalus scrobiculatue*, for instance, is limited to the region around the Solomon sea New Guinea. And, *Nautilus macromphalus* is found only near the island of New Cladonia. The few existing are distinguishable only by insignificant features. Classification, for egg is based upon difference in body size and number and arrangement the brownish-red shell bands. The jungle *Nautilus* is characterized by numerous jagged colour strips which cover the entire shell. *Nautilus scrobicalates*, on the other hand, has a rough shell surface. The shell aperture lacks colour bands.

Often masses of *Nautiluses* and their empty shells are slept onto the beaches following, storms. The local people collect there to use as fold and make the shells into jewelry.

SEPIA

The common cuttlefish, *Sepia officinalis* Linnaeus, 1755, is one of the best known cephalopods of the Old World. The species is abundant in the eastern Atlantic and in the Mediterranean Sea. It is easily identified by its slipper shape and the long, narrow fins that form an undulatory margin along each side of the mantle. The largest specimens measure nearly half a metre without the long tentacles. The latter are retracted in special pouches and are ejected only to seize prey. In young individuals the species can be identified from the brown-skin colour. Other features that distinguish these animals from the small-sized, more reddish *Sepia orbignyana* and, *Sepia elegans* include the shape of the chalky cuttlebone and the arrangement of suckers in the tentacle club.

From the North Sea and English Channel southwards, *Sepia officinalis* occurs in coastal waters and on the continental shelf at depths not greater than 150 m. The status of the subspecies *S. officinalis hierredda* Rang, 1837 of the West African coast is uncertain, but the shape of the cuttlebone and other features suggest that it is a separate species rather than a subspecies or geographical race. Further to the South, around Africa, South and East Asia, and Australia, more than one hundred species of Sepia are known. No Sepia occur around North and South America (Voss, 1974) for *Sepia officinalis* generally moves inshore spawning.

The bulk of mantle cavity is occupied by visceral mass consisting of various internal organs. The digestive gland lie interior bordered by retractor muscle of head and funnel. The median rectum opens by central anus at the base of funnel on the either side of the return extends a renal sac opening into mantle cavity by renal aperture

On the left side there is genital aperture, a little posterior two large plume shaped ctenidia, one on each side are present. Two large stellate ganglion lie one of each side on the mantle wall. The round ink sac can be recognized by its metallic cuter. In male, testis lie partially covered with ink sac. In female, the renal sac is hidden from view by a pair of mid mantle gland and a pair of accessory mid-mantle gland.

The genital aperture opens on the left side near the left renal apertures.

In nervous anatomy of animal, different types of ganglions are found.

These ganglions are as followed:

(i) Cerebral ganglion

(ii) Buccal ganglion

(iii) Optic ganglion; From it optic nerve arises which goes to eyes.

(iv) Visceral ganglion; Nerves originating from it goes to visceral organs.

(v) Stellate ganglion; Small branches are supplied to various supplementary nerves of different organ sand fins.

(vi) Gastric ganglion; Nerves supplied to gastric glands and gastric area.

The nervous system of *Sepia* shown a high grade of organization The brain consist of typical ganglionic masses, all concentrated in the head, around the esophagus, behind the buccal mass and protected by cartilaginous 'skull'.

A pair of cerebral or supra-oesophagal ganglia are fused together into a rounded mass lying dorsal to the oesophagus.

Laterally they give off a pair of stout optic nerves which expand into optic ganglia.

A small olfactory ganglia lies on the dorsal side of each optic nerve. Anteriorly, a pair of cylinder, cerebral-buccal connectives connect the cerebral ganglia to a pair of superior buccal ganglia which are situated dorsal to the buccal mass and connected by circumoesophagial connectives to a pair of inferior buccal ganglia lying below the buccal mass A pair of stout circumesophageal connectives connects the cerebral ganglia to rest of brain.

The subesophageal ganglionic mass lying beneath the oesophageal ganglion mass lying oesophagus is partly divided into an anterior lob, the branchial ganglion and a posterior to those, the pedal ganglion.

A pair of pleuro-visceral ganglia is also limited to form a single mass lying in contract with the pedal, behind the oesophagus.

The stellate ganglion by two commissure joins to each mass which joins to the nerves in the fin.

A pair of sympathetic nerves originating from the inferior buccal ganglion, runs posteriorly along oesophagus to join gastric ganglion lying between the stomach and caecum. The gastric ganglion sends nerves to the liver, stomach, caecum and intestine Sepia spawn inshores.

These, living animals have been made in shallow water since remote antiquity. The earliest written observations are those of Aristotle. In modern biological science various aspects of *S. officinalis*, especially its physiology, have been studied over the past 150 years. The earliest detailed account of the embryonic development of *Sepia* was published by Kölliker (1844), His description has been complemented, especially by Nacf (1923, 1928) who proposed a useful staging system. *Sepia* eggs are easily collected in the natural habitat, or they may be obtained by keeping adult animals in seawater tanks until spawning. Embryonic development proceeds normally if the water temperature in the aquarium is kept somewhere between 12 and 22°C. After hatching the young

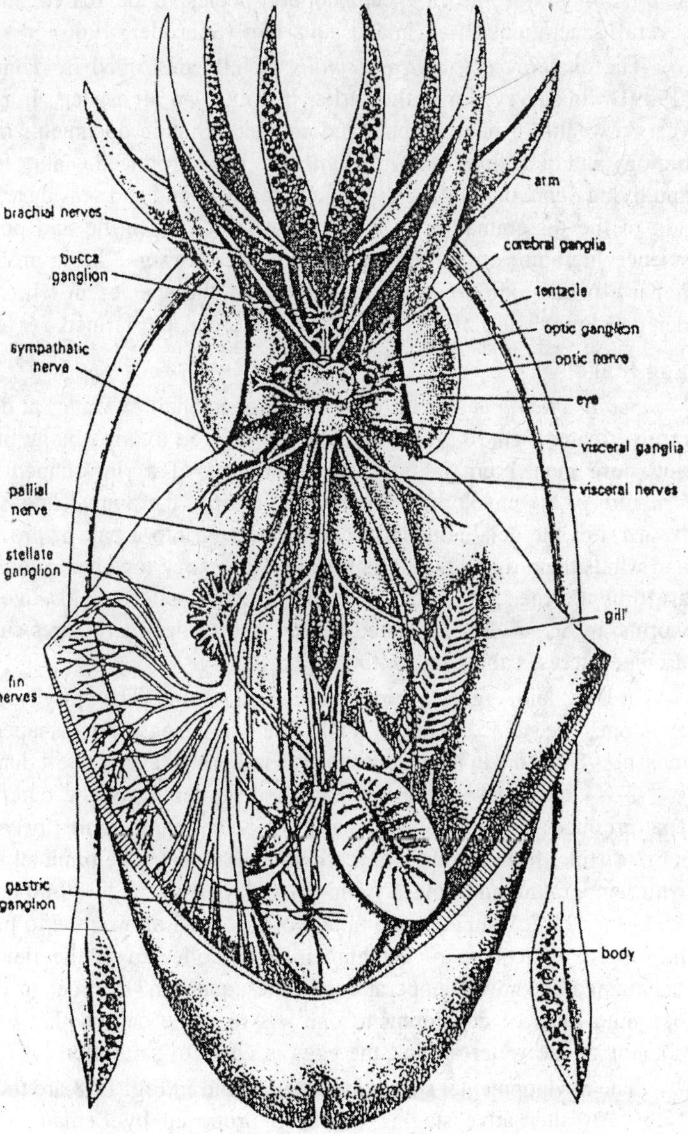

Fig. 11.6 : *Sepia* nervous System.

animals can be reared in the aquarium if live prey is available. Indeed *S. officinalis* was the first cephalopod species to be reared through several generations in an inland aquarium (Schröder, 1966).

The anatomy of *Sepia officinalis* is fully described by Tompsett (1939) who surveyed also the earlier literature on all aspects. In recent years a wealth of photographic and cinematographic documents on the biology and behaviour of *S. officinalis* has been produced (Zahn, 1975), and living *Sepia* on public exhibition in many inland and seaside aquaria add to the dissemination of a particularly rich scientific and popular science literature on the common cuttlefish. Yet some basic problems remain to be studied before the biology of the species in terms of a detailed knowledge of its natural life cycle can be defined.

Egg Stage

Sepia officinalis generally lays its eggs in shallow water, at depths rarely greater than 30 or 40 m. Each egg is fixed to any oblong object, not more than 1 cm in diameter, by means of a ring-shaped basal structure of the envelope. To produce this ring, the animal draws with its arm tips the gelatinous envelope of the egg into a pair of processes and winds them round the support so that they stick together. *Sepia* eggs are thus attached to various sorts of plants, sessile animals like tube worms, to *Sepia* eggs deposited earlier, and to dead structures such as drowned trees, cables or, netting.

Freshly laid eggs are very soft and gelatinous. The spirally coiled envelopes are stained black with ink. The whole egg is flask-shaped and measures 2-5-3 cm in length without the basal ring. The greatest diameter is 1-2-1-4 cm. Subsequent shrinkage and hardening of the outer jelly coat produces a more globular shape, the actual egg chamber becoming more distinct from the apical tip and the basal ring. The minimal varies with temperature and ranges from 40-45 days at 20°C to 80-90 days at 15°C Fig. 11.7. Using a stage-time plot Fig. 11.8 allows one to predict the approximate time of hatching in eggs collected in the field and maintained at known temperatures in the aquarium. In order to obtain optimum rates of development, one has to make certain that oxygen content of the water around the eggs is close to saturation.

The developmental stages indicated as old in Fig. 11.8 are those of Naef. An alternative staging has been proposed by Lemaire (1970) mainly to define cleavage stages in a way similar to the system of Arnold (1965). Stage 9 of Lemaire corresponds to the 'blastula' stage of Naef; the onset of 'gastrulation' marks stage 10 of Lemaire and stage 11 of

Naef. There are only minor shifts away from the 'Naef stage + 9' relation in later development. The hatching stage 30 of Lemaire corresponds to stage xx of Naef.

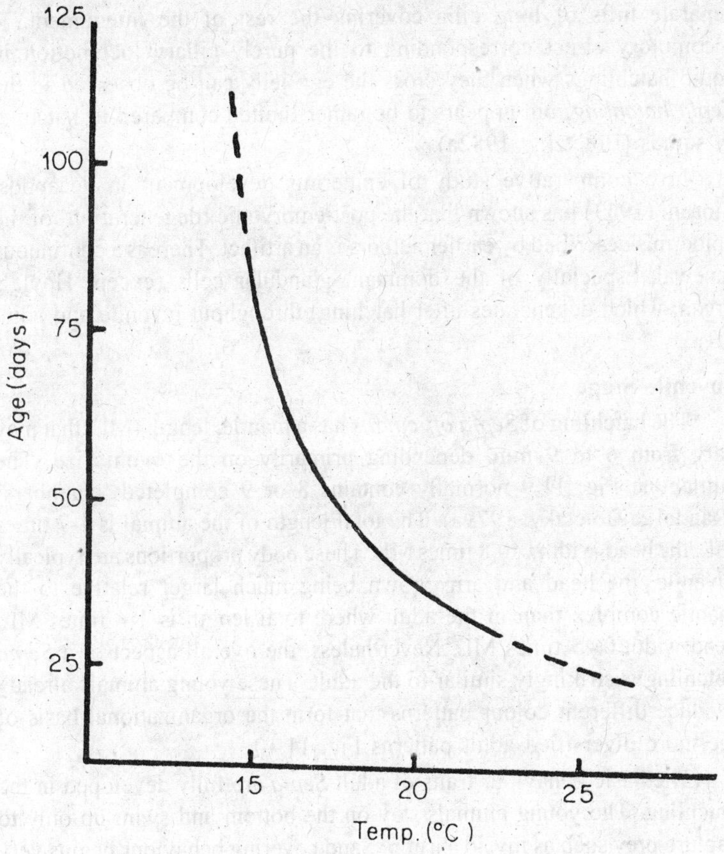

Fig. 11.7 : Length of embryonic development as a function of temperature. (After Richard, 1971; simplified).

The late embryonic stages (xv-xx) following organogenesis, take more than half of the total time of embryonic development, and the final growth from stages xvii to stage xx is as much as 80% of the total (Fioroni, 1964). During these stages of organ growth and differentiation, the integumentary cover of the eye (primary lid) takes on the function of a cornea, the chambered cuttlebone is formed in the shell sac from which a thus in close contact with the chorion, so that after the rupture of the glandular cells (produced by muscular contraction) the envelope in the contact zone is immediately dissolved. The hatching gland is surrounded by ciliary bands similar to those observed in squid embryos. The ciliary beat is directed anteriorly, in these bands as well as in the separate tufts of long cilia covering the rest of the integument. A locomotory effect corresponding to the purely ciliary locomotion in squid hatchlings when they cross the egg jelly can be observed in the *Sepia hatchling*, but appears to be rather limited compared to hatching in squids (Boletzky, 1982a).

In a comparative study of epidermis development in decapods, Fioroni (1963) has shown that the post-embryonic 'degeneration' of the epidermis described by earlier authors is an artifact. There is a continuous renewal, especially of the dominant glandular cells (except Hoyle's organ, which degenerates after hatching) throughout juvenile and adult life.

Juvenile Stage

The hatchling of *Sepia officinalis* has a mantle, length (Ml.) that may vary from 6 to 9 mm, depending primarily on the ovum size. The cuttlebone Fig. 11.9 normally contains 8 or 9 completed 'chambers' (Randel & Boletzky, 1979a). The total length of the animal is 1-7 times ML, the head-width c. 0.8 times ML. These body proportions are typically juvenile, the head and arm crown being much larger relative to the mantle complex than in the adult where total length is 1-4 times ML, head width 0.45 times ML. Nevertheless, the overall aspect of a *Sepia* hatchling is strikingly similar to the adult. These young animals already produce different colour patterns that form the organizational basis of the more diversified adult patterns Fig. 11.10.

The basic behaviour traits of adult *Sepia* are fully developed in the hatchling. The young animals stay on the bottom and swim up only to capture prey such as mysid shrimp. Sand covering behaviour begins very early, and the animals spend daylight hours buried in sandy substrates exactly like the adult. This 'burrowing response', triggered by light, generates a sequence of flushing movements by which the funnel jet whirls up substrate particles which settle on the animal until it is entirely covered. In addition to this ability to bury themselves in soft bottoms, young *Sepia* show a special response to hard substrates to which they can become firmly attached. They use the ventral surface of the mantle and ventral arms, by means of the integumental musculature, as a set of suckers. The body and head with the arms present a perfectly streamlined contour. The ability to become attached to a hard substrate wanes in larger animals although they still show the corresponding streamlined attitude, with rather flattened body, when settled on the bottom. Other behavioural traits develop progressively in young *Sepia*. The initial 'innate' response to prey such as mysids and similar shrimps widens gradually to include a variety of potential prey. It is then secondarily channelled by individual experience (learning) towards the useful prey items. The increasing learning performance is correlated with certain differentiations in the central nervous system. The vertical lobe system, which is known to control learning processes, grows more rapidly than other parts of the brain.

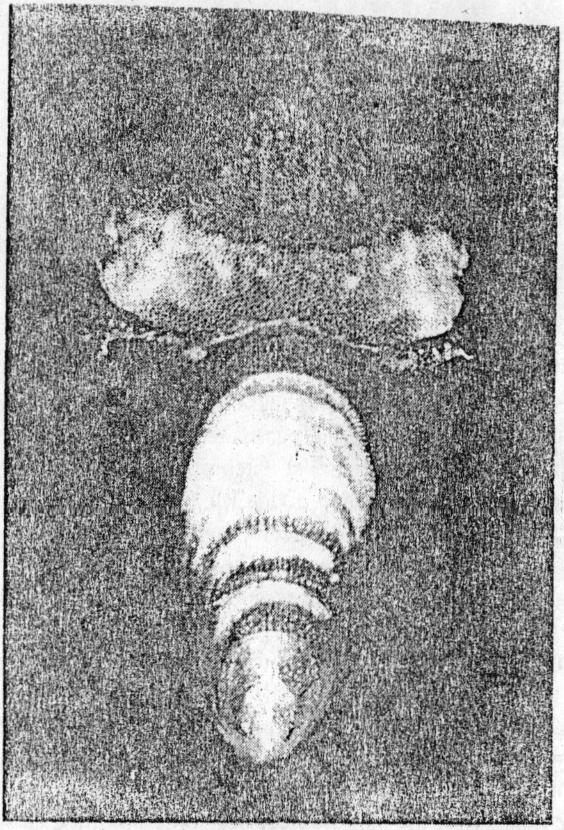

Fig. 11.8 : Newly-hatched *Sepia officinalis*. The animal is anaesthetized and the skin on the dorsal mantle surface has been removed to expose the cuttlebone. The light areas are gas-filled.

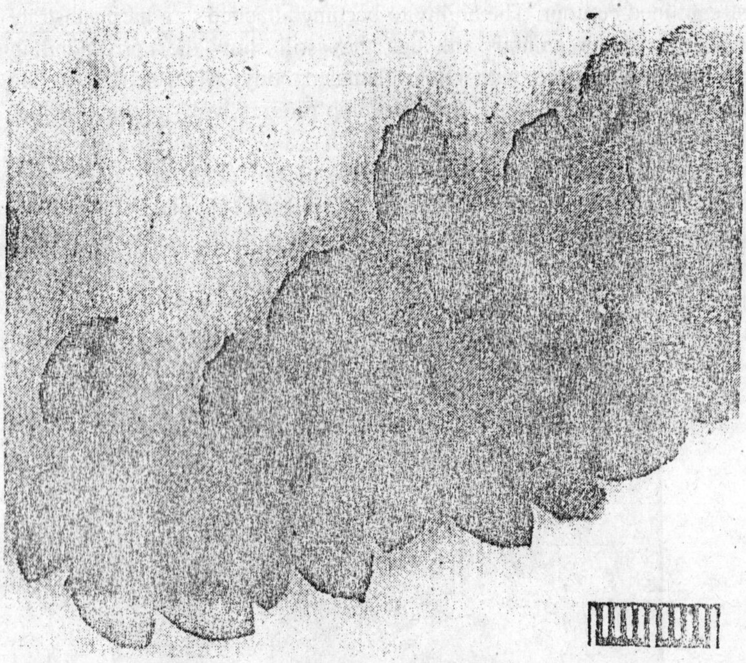

Fig. 11.9 : A batch of newly-hatched *Sepia officinalis* showing different chromatophore patterns. Note also the size difference. (Scale bar 1 cm).

In spite of the gradualness of these developmental processes, it is possible to distinguish underlying phases that are fairly distinct. Richard & Decleir (1969) define a first post-hatching phase 'A' by the presence of yolk in the inner yolk sac and divide it into 3 sub-phases. 'A1' is the end of the actual embryonic development and corresponds to the disappearance of the outer yolk sac. 'A2' covers the beginning of predatory activity and blends into 'A3', representing the final absorption of the yolk reserves. By using electrophoresis of blood proteins, these authors have demonstrated a gradual shift from embryonic conditions, where specific protein bands appear to the adult condition with only one haemocyanin band. Two of the original bands, with the lowest electrophoretic mobility, disappear at the end of A1; at the same time a new band appears. A further band appears at the end of A2 and the definitive band, with the highest electrophoretic mobility, at the end of A3. The appearance of this final protein fraction marks the beginning of phase 'B' during which the remaining intermediate bands disappear one after the other. With only the definitive band remaining, the 'adult' phase 'C' begins at a size of 2 cm ML.

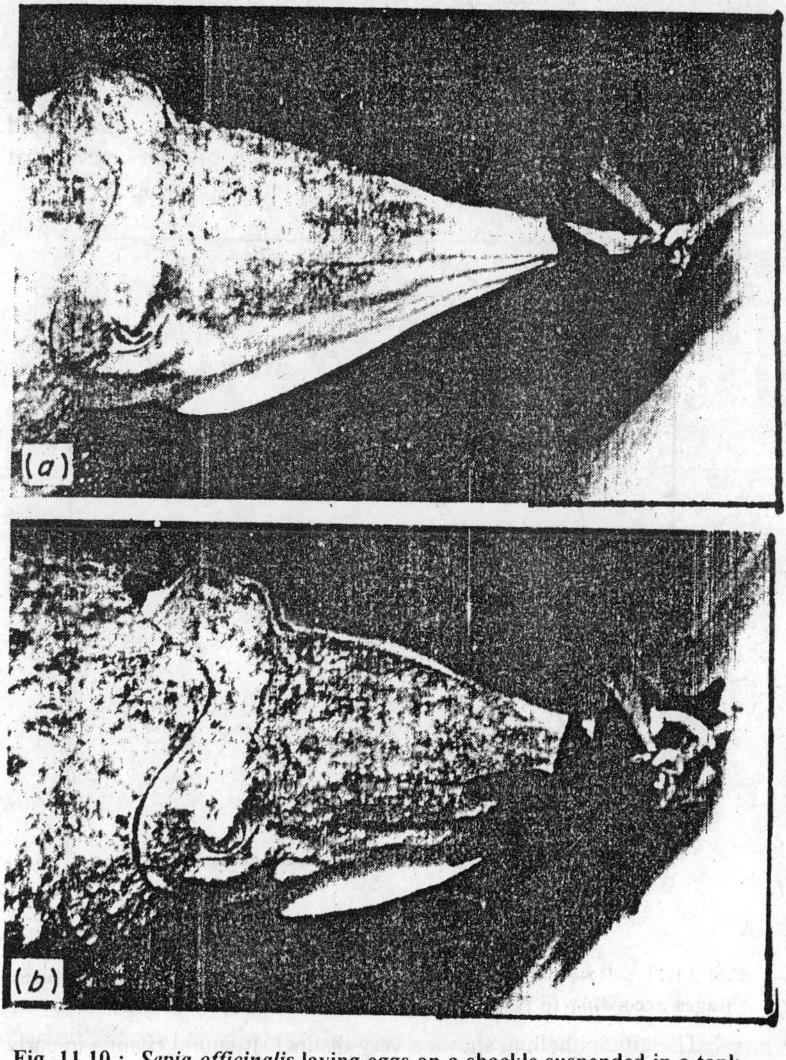

Fig. 11.10 : *Sepia officinalis* laying eggs on a shackle suspended in a tank. (a) the moment of contact between arm tips and egg support; the eyes are turned forward for binocular fixation of the site of egg attachment, (b) a second or two later, the arm tips grab around the support to attach the egg.

The digestive gland is fully functional by the time of hatching. During development a gradual colour change from white, at hatching, through yellow and orange to the typical brown colour of the adult 'liver' reflects further cellular differentiation and the onset of digestive activity. Correlated with these processes are the differentiations observed in the digestive duct appendages ('pancreas'). In the excretory organs, a particular secretion product appears from later post-hatching stages onward in the form of spherical bodies with a very high calcium content that are retained between the infoldings of the renal appendages.

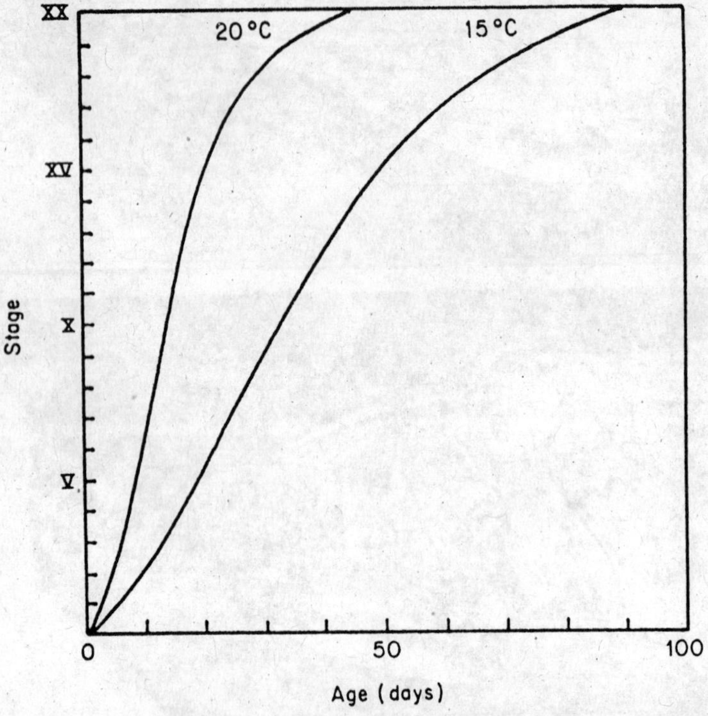

Fig. 11.11 : Rates of embryonic development at different temperatures. (Stages according to Naef, 1923, 1928).

The gill epithelium shows a very distinct structural change in early posthatching development. In the inner part of the secondary branchial folds, the epithelium differentiates into a highly folded layer whose structure closely corresponds to the transport-active epithelia of the excretory organs. Probably the gill takes on an excretory function along with the respiratory function diameter then is about 1 cm for a total length of 2 cm (without the fixating ring). During later embryonic development, the still elastic envelope is dilated by the chorion which expands due to the osmotic pressure of the perivitelline fluid. The envelope reaches a maximum diameter of about 1-5 cm by the time of hatching. Especially at this stage clusters of *Sepia* eggs strikingly resemble dark wine grapes. Evidently this aspect is at the origin of popular names like 'uva di mare' and 'raisins demer' Sometimes eggs are not stained with ink and the envelopes become perfectly transparent when stretched by the expanding chorion. Embryonic development is not influenced by the lack of pigmentation in the envelopes.

The ovum measures from 6 × 5 mm to 9 × 7 mm and is first tightly enclosed by the chorion. The volume of the perivitelline fluid, which is hypertonic to seawater, progressively increases and dilates the chorion. The length of embryonic development pair of lateral pouches differentiates to form an 'articulation' for each fin. Furthermore, the skin of the body surface shows some colouration, and the yellow, red and brown chromatophores become very numerous. Together with iridophores and leucophores, they are integrated in a highly complex system of colour patterning under nervous control that is fully functional by the time of hatching.

The late phase of embryonic development is also characterized-as in other cephalopods-by an accumulation of yolk transferred from the outer into the inner yolk sac. (The latter has a peculiar form, due to the secondary differentiation of a pair of posterior sacs that form during yolk transfer.) Thus the hatchling always has a yolk reserve integrated in the circulatory system of the visceral mass that will continue to release nutritive material after the disappearance of the outer yolk sac. The latter shrinks into a tiny appendage that apparently falls off later. Probably temperature has some part in determining the moment of hatching. Under stress conditions, animals may hatch prematurely and cast off a still sizable outer yolk sac. That this may not happen under normal conditions is due to a tranquillizing factor contained in the perivitelline fluid. It is not yet clear how the threshold is overcome in triggering the specific hatching response.

At late embryonic stages the cuttlebone contains gas; the animals, nevertheless, remain negatively buoyant and generally stay in the upside down position until hatching. The hatching gland (Hoyle's organ) is some species walk on the substrate, enter down. *Streblocerns serricaudatus* walks on the mud with long stiff setae of the second antennae. *Opiryoxiis* walks with the spined tips of the first limbs and the furca of the abdomen, *llyocryptus* crawls in the mud holding its large antennae laterally at right angles to the body and moves them like paddle wheels. Beats of the furca help in propulsion. Other species push themselves through the mud with the furca alone. *Chydorus sphaericus* climbs along filamentous algae by loosely wrapping its valves around the filament, the furca posteriorly forced against the thread and the hooked distal spines of the first legs extending and flexing alternately. *Andnstropus* walks similarly on the trunk and tentacles of *Hydra* without the use of the furca.

Senses : How the chemical nature of the environment is perceived is completely unknown. But *Daphnia* follows increased concentration of O_2, which leads them to areas ideal for its metabolism, with abundant algae and food supply. A strong increase of CO_2 makes Cladocera positively phototactic and they surface.

Temperature differences are perceived by *Daphnia*. Lowering of the temperature causes them to swim up; increased temperature makes them move down.

The naupliar eye is only rarely as large as the compound eye, usually much smaller (except in the Chydoridae). The compound eye itself consists of only a few ommatidia. Among *Daphnia* the three to four cups are much reduced and among predators they are absent.

Daphnia can orient in space with the eyes. It seems to "know" the direction of surface and shore, and can perceive the borders of an area that contrasts with the dark background. Most Cladocera (except some Chydoroidea, but including *Simocephalus* and *Scapholeberis*) swim horizontally with the light above them, a transverse phototaxis indicating that they orient with their compound eyes. *Daphnia*, while maintaining this same orientation of its eyes toward the light, orients its body in any plane. Such freedom of swimming direction is possible because the movable, fused compound eyes of *Daphnia* can be rotated by means of six eye muscles. Even in absolute darkness the top of the eye faces up in all possible swimming positions. The eye rotation probably depends on gravity, and the swimming position on geotaxis, which during the day functions together with phototaxis. The location of the gravity receptors is not known.

The stronger the light above the animal, the more *Daphnia* tips anteriorly, so that in very strong illumination it swims down. In polarized light *Duphnia* swims perpendicular to the plane of light. Many populations, with eyes in intermediate adaptation, move toward the longer wave length, away from blue and violet. Thus they differentiate at least two colours. The biological significance of colour vision is not known.

A general light sensitivity of the integument has been demonstrated, which increases between 700 nm (red) and 400 nm (violet). A clear separation of the light sensory function of integument and eyes has not been possible.

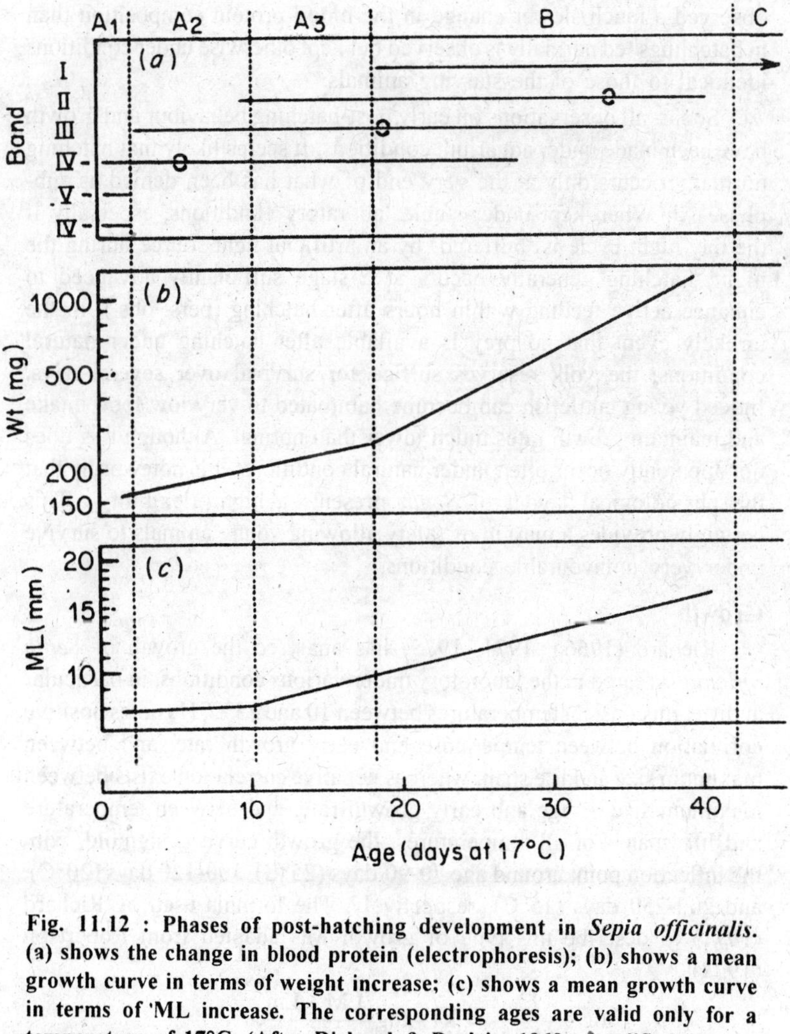

Fig. 11.12 : Phases of post-hatching development in *Sepia officinalis*. (a) shows the change in blood protein (electrophoresis); (b) shows a mean growth curve in terms of weight increase; (c) shows a mean growth curve in terms of ML increase. The corresponding ages are valid only for a temperature of 17°C. (After Richard & Decleir, 1969; simplified).

The length of the early phases of post-hatching development depends on the water temperature and on the feeding conditions clear correlation, independent of the actual age, between electrophoretic protein pattern of the blood and parameters such as weight or length of the animal, quantity of remaining yolk and state of digestive gland development. In under-fed animals, certain differentiations can be greatly delayed independently of water temperature. In unfed, starving hatchlings, observed a much slower change in the blood protein composition than in hatchlings fed normally is observed but kept otherwise under conditions identical to those of the starving animals.

So far, all observations on early post-hatching behaviour and growth have been made under aquarium conditions. It seems likely that hatching normally occurs only at the very end of what has been denied as sub-phase Al. When kept under stable laboratory conditions, especially if the day-night cycle is 'buffered' by an artificial light source during the night, hatching generally occurs at a stage sufficiently advanced to enhance active feeding within hours after hatching (pers. obs.). In the unlikely event that no prey is available after hatching under natural conditions, the yolk reserves suffice for survival over several days. Indeed young cuttlefish can become habituated to very low food intake and maintain growth rates much lower than normal. Although this does not apparently occur often under natural conditions, it is noteworthy that the physiological layout of *Sepia* presents a high 'flexibility'. This certainly provides a margin of safety allowing young animals to survive under very unfavourable conditions.

Growth

Richard (1966a, 1971, 1975) has analysed the growth of *Sepia officinalis* reared in the laboratory under various conditions, in particular at different constant temperatures between 10 and 25°C. He notes positive correlation between temperature and early growth rate, and between maximum size and life span, whereas negative correlation exists between maximum size or age and early growth rate, and between temperature and life span. For all temperatures, the growth curve is sigmoid, with the inflection point around age 70-90 days (25°C), 100-120 days (20°C), and 200-250 days (15°C), respectively. The formula used by Richard (1971) to describe this type of growth was adapted from Robertson (1922):

$$\log \frac{L}{LM - L} = Kt + \log \frac{LM - Lm}{Lm}$$

where

LM = maximum ML in mm;

Lm = ML at hatching;

L = ML at time t;

t = age in days;

K = growth coefficient.

The growth coefficient K varies from 0-0038 at 13°C to 0-0130 at 25°C, according to the equation:

K = 0-0135 log T – 0-0304,

where T = temperature in °C.

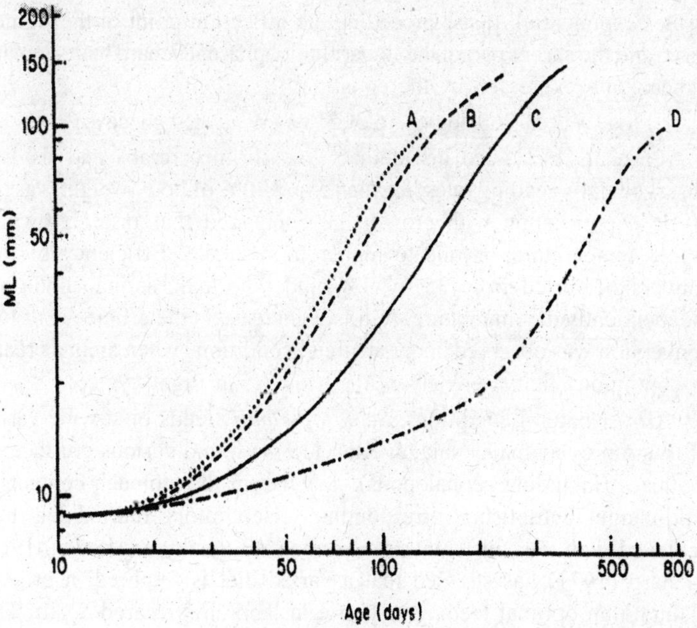

Fig. 11.13 : Mantle length growth curves at three constant temperatures with *ad libitum* feeding (A-C) and at varying temperatures with severe under-feeding during several months after hatching (D). A, 25°C; B, 20°C; C, 15°C. (After Richard, 1971; simplified; (D) after Boletzky, 1979a),

By extrapolation, K can be seen to become zero at 9-5°C, which indicates the lower limit of the normal range of temperature adaptation.

The correlation between length L (ML in mm) and fresh weight W (in g) is expressed by the equation $W = 0.000885\ L^{2.646}$, derived from: $\log W = 2.646 \log L - 3.05284$. Pascual (1978). gives a similar figure:

$$\log W = 2.678 \log L - 3.279.$$

Sepia officinalis reared through three successive generations at temperatures varying from 14 to 26°C. Daily growth rates declined from an initial 5-8% weight increase (2-5-4% length increase ML) in small animals to c, 1% weight increase (0-4% length increase ML) in animals with a mantle length (ML) of about 20 cm, independently of their age. The highest growth rates observed by Pascual are similar to those obtained by Richard (1971) at a constant temperature of 20°C. For the age interval 50 to 150 days (weight increase from 2 to 300 g), i.e. in the later part of the 'exponential' phase preceding the inflection point of the sigmoid curve and the earlier part of the 'logarithmic' phase, Richard has calculated a mean rate expressed by the equation:

$$\log W = 3\text{-}860 \log t - 6.305189 \text{ (W in g, t in days)}$$

Pascual (1978) fed his animals live shrimps, crabs and fish. He observed daily feeding rates between 5 and 20% of body weight, varying in close correlation with growth; he concluded that faster growth at higher temperatures is due to higher food intake. Efficiency of food conversion varied from 35 to 50% under normal rearing conditions, independently of temperature. A drastic increase in the efficiency of food conversion was observed under artificial conditions when animals reared in continuous light were given a 9 h day/15 h night cycle.

Under natural conditions, *Sepia officinalis* feeds on a wide variety of living prey animals. Najai & Ktari (1974) found various crustaceans, molluscs (including cephalopods), and fish in the stomach contents of wild-caught cuttlefish. Correspondingly, laboratory rearing has been achieved with very different prey animals (Boletzky & Hanlon, 1982). Richard (1971) has stressed that a varied diet is required in order to maintain an optimal feeding response in laboratory-reared *Sepia*. Thus availability and quality of food are the essential parameters influencing growth in correlation with temperature.

Under-feeding, no matter whether it is facultative (poor feeding response) or forcible (low food availability), results in slow growth even at high temperatures. One can be certain, however, (hat such under-feeding does not occur normally in healthy animals living in their natural habitat. The cuttlebone with its lamellar structure keeps a reliable record of feeding conditions. In under-fed animals, the rhythm of chamber' formation is slowed to a lesser extent than growth in length, so that more chalky 'lamellae' are formed per unit length ('septal crowding') than in well-fed animals. This condition is expressed in the 'lamellar index' of the cuttlebone which is defined as:

$$LI = \frac{L_s + L_l}{N_s}$$

L_s = length in mm of striated (siphoncular) zone; L_l = length in mm of smooth zone (= last lamella or 'loculus'); N_s = number of striae (= lamellae);

The cuttlebone of wild-caught animals always has an index greater than 1, whereas in rearing the slightest under-feeding already results in an index lower than 1. Under-fed animals maintain a juvenile condition in that L_l remains shorter than L_s (type A), in contrast to well-fed animals in which the relation is reversed with later juvenile growth (type B), before returning to type A conditions with sexual maturation. In demonstrating this normal sequence of growth forms in the cuttlebone, Mangold (1966) has been able to show that the supposed subspecies *Sepia officinalis fillïouxi* is nothing but the final growth stage of *Sepia officinalis*.

General observations suggest that males survive prolonged starvation better than females. Under normal conditions, males grow slightly faster during the adult phase than females, and they attain greater sizes than the latter. In under-fed animals, the sexual dimorphism of the cuttlebone (broader in females) is more marked than in normally fed animals.

Comparison of cuttlebones from wild-caught and laboratory-reared animals clearly shows that the natural growth rates of *Sepia* correspond to the maximum rates obtained under aquarium conditions; these rates vary with temperature. There is one question that remains to be answered, however. In the cuttlebones of *Sepia* from the English Channel and North Sea, one finds a fairly distinct zone of 'septal crowding' resulting in reduced chamber height (0.0-2 mm) lying between earlier chambers reaching a height of 0.5 mm and later ones that are even higher. In large adults, chamber height again decreases. In view of the reduction of chamber height induced by under-feeding, these zones might reflect periods of reduced food intake. Richard (1972) considers this hypothesis, but suggests an alternative explanation related to buoyancy control requirements. It is noteworthy that intermediary zones of narrow chambers (not terminal ones reflecting senescence) are rather indistinct in the cuttlebones of Mediterranean *Sepia*. The total number of chambers allows-estimation of age if the approximate temperature is known.

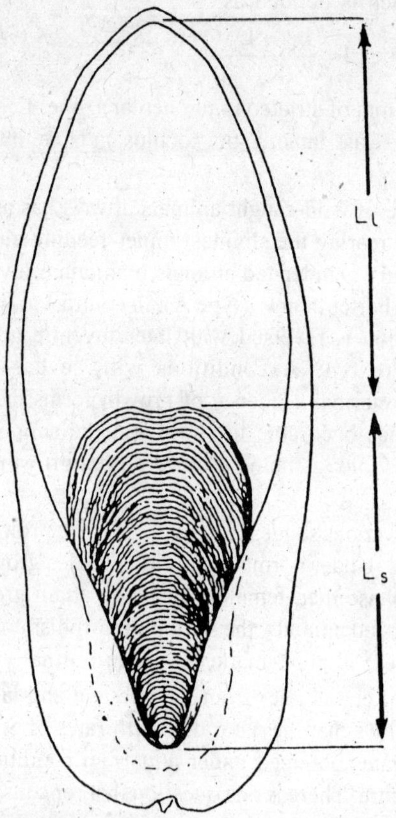

Fig. 11.14 : The cuttlebone of *Sepia officinalis* in ventral view (semi-schematic), showing the striate siphoncular zone (L_s) and the last lamella (L_1). Note that this is type 'B'.

In conclusion, feeding rates are certainly close to the respective (temperature dependent) optimum over most or all of the juvenile and adult growth phases. The highest growth rates of about 0.6-0.9 mm increase in ML per day are observed at temperatures above 20°C. Under natural conditions, they can be expected to occur only in summer in inshore waters. The lowest growth rates under normal feeding conditions so far observed (at temperatures between 10 and 13°C) were a daily increase in ML of 0.2-0.3 mm. Growth analysis of natural populations sampled over several years suggests that actual growth rates vary between the two extremes observed in laboratory-reared animals. In one year, an animal will attain a size of at least 10 cm ML, whereas natural conditions will probably never permit an animal to attain in one year the maximum size of 30 cm ML. This size is only known in male individuals that are probably about two years old. As mentioned before, males grow larger than females by maintaining a high growth rate at the adult stage.

Maturation

The gonads of *Sepia* remain in an undifferentiated state until the end of embryonic development. Histological distinction of ovary and testis is achieved only by the time of hatching. However, the gonoducts in both sexes and the rudiments of the nidamental glands in the females are already distinct before hatching . These organs grow steadily through juvenile life until the gonad weighs about 1 g, which corresponds roughly to 1 % of body weight. At this stage before the onset of sexual maturation, the gonoducts weigh about half as much as the gonad, whereas the nidamental glands weigh twice as much as the ovary. As for the nidamental glands, this weight relationship is reversed during maturation; at the same time, the disparity in weight between the oviduct and the ovary increases about ten-fold. The maximum weight of the ovary in large females is 150 g. The total weight of the reproductive organs may represent up to 16% of body weight in mature females. In the male, the situation is different as spermatozoids are stored in the spermatophores which are transferred to the Needhams sac. The comparatively large weight increase of the male duct is largely due to the storage of spermatophores. However, the total weight of the male reproductive organs never represents more than 5% of the body weight.

Richard (1971) defines six stages of spermatogenesis and oogenesis Table 11.1. This author also shows that a correlation exists between the state of ovarian maturation and the colouration of the accessory nidamental gland. From oogenetic stages 1 to 3, the glands change from colourless via white to beige and yellowish. By stage 4, they have turned orange and then become coral-red (stage 5) and finally pink-red by the time of spawning.

Table 11.1

Phase			Stage
Spermatogonium: mitosis			1
Spermatogenesis	Spermatocyte I. growth, meiosis		2
	Spermatocyte II.		3
Spermiogenesis	Spherical spermatid		4
	Lengthening spermatid		5
	Differentiation of spermatozoid		6
Phase			**Stage**
Oogonium: mitosis			1
Oocyte	Feeble growth of oocyte: prophase of meiosis		2
	Strong growth of oocyte	Previtellogenesis (follicle formation)	3
		Vitellogenesis (secretory activity)	4
		End of maturation (follicle degeneration)	5

Oocyte II: extrusion of polar bodies after fertilization

Cuttlefish may attain sexual maturity at very different sizes. Males can be fully mature, the Needhams sac filled with spermatophores, at a mantle length of only 6-8 cm. Mature males of even smaller sizes have been observed among laboratory-reared animals. On the other hand, male individuals measuring over 10 cm ML may still be immature. A similar situation exists in female *Sepia*; they may become fully mature at sizes varying from 11 to 25 cm ML. Ovary size and egg numbers vary correspondingly. Smaller females have only about 150 mature ovarian eggs, whereas large individuals may have more than 500. The total egg numbers (including immature eggs at various stages from a size of 1 mm upwards) vary from about 500 in small individuals to over 1000 in large ones. The size of mature ovarian eggs also varies with the size of the female. Small individuals lay eggs only 4-5-6 mm long, whereas in large ones the eggs measure 7-9 mm. A similar correlation exists in the males between body size and spermatophore size (5-16 mm).

In order to define the factors responsible for the variations in body size at maturity observed in the natural populations. Reared animals under different controlled conditions. He found a combined effect of temperature and light. Early gonad growth not maturationl is linked to somatic growth and thus is temperature-dependent. Light, and more particularly the spectral range of short wavelengths (blue to bluegreen) has a decelerating effect on gonad *maturation* via the hormonal control system (optic gland), whereas in the mature animals it stimulates mating and spawning.

High temperatures and feeble light intensities or short days would thus result in high rates of growth and gonad maturation. Under natural conditions, however, temperature and light are inevitably acting in an antagonistic manner. Under shallow water summer conditions growth is accelerated by high temperatures, whereas maturation is decelerated by high light intensities and long exposure times within the natural day-night cycle. By contrast, in deeper water under winter conditions, growth is slow due to the low temperatures, whereas gonad maturation is largely unhampered (rather than accelerated) as the decelerating effect of light is at its minimum. The feeble light intensities are nevertheless sufficient for the functioning of the visual system, especially in feeding.

The length of time spent under optimal growth conditions in the early juvenile phase (inshore summer conditions) determines whether or not an animal becomes sexually mature during its first winter. It now remains to define the factors responsible for the seasonal vertical migrations that enhance the antagonistic effects of temperature and light conditions. In *Sepia* populations studied so far there is a general tendency for animals to migrate inshore in spring and summer and move offshore in autumn. However, not all the animals migrate at the same time. Mature males and large mature females are the first to move inshore in spring for reproduction. These females are about 18 months old. Later in spring and early summer, mature females of smaller size appear in the shallows and spawn until late summer. These are only about 14-16 months old Fig. 11.15 they are the offspring of large animals that have spawned early in the year before. Thus. there appears to be a cycle of alternating shorter and longer generations, at least as far as the female individuals are concerned. In the males, the situation is not as clear, because maturation is attained earlier than in females, whereas reproductive activity apparently covers a much longer time. Indeed the largest body sizes (up to 30 cm ML) are observed in males, and although the adult growth rates are higher than in females, the largest males must be at least 2 years of age.

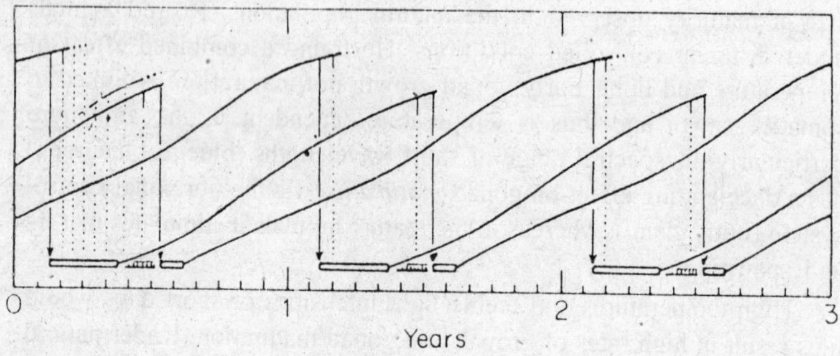

Fig. 11.15 : A schematic presentation of the life cycle in *Sepia officinalis*. Arrows and dotted lines represent spawning, 'white' bars represent embryonic development and, the hatched bars represent the overlap between early and late spawns.

Reproduction

The reproductive behaviour of *Sepia officinalis* has been studied thoroughly by several authors (Richard, 1971; Zahn, 1975). The most spectacular part is the elaborate courtship behaviour of male *Sepia* and the corresponding colour patterns. Both sexes show a typical zebra pattern that is particularly conspicuous in the courting male Fig. 11.16. When approaching a potential partner for copulation, the male extends one ventral arm towards the courted individual. If the latter happens to be another male, it will signal so by stretching out the ventral arm facing the approaching male; if it is a female not willing to copulate, it will swim away, whereas a female willing to copulate will stay. At the start of copulation the male crabs the female from the side and manoeuvres her into a head-to-head position, holding her with dorsal and lateral arms *interwined*. The male then reaches with his left ventral arm, the hectocotylus, into his own mantle cavity to seize spermatophores extruded from the Needhams sac through the penis. The proximal part of the hectocotylus is modified in that normal-sized suckers are replaced by a series of minute suckers scattered over a large area of transverse folds. The spermatophores are transferred to a special pouch under the buccal mass of the female. The details of spermatophore evagination, spermatozoid storage and release are still poorly known.

Spawning generally begins soon after copulation, and the male remains in visual contact with the female, chasing away other males when they come too close. Eggs are laid in regular intervals of 2-3 minutes over several hours. Thus a female may empty her ovary of mature eggs within a few days. In the aquarium, some individuals tend to spawn smaller portions over a longer time, in some cases over several months.

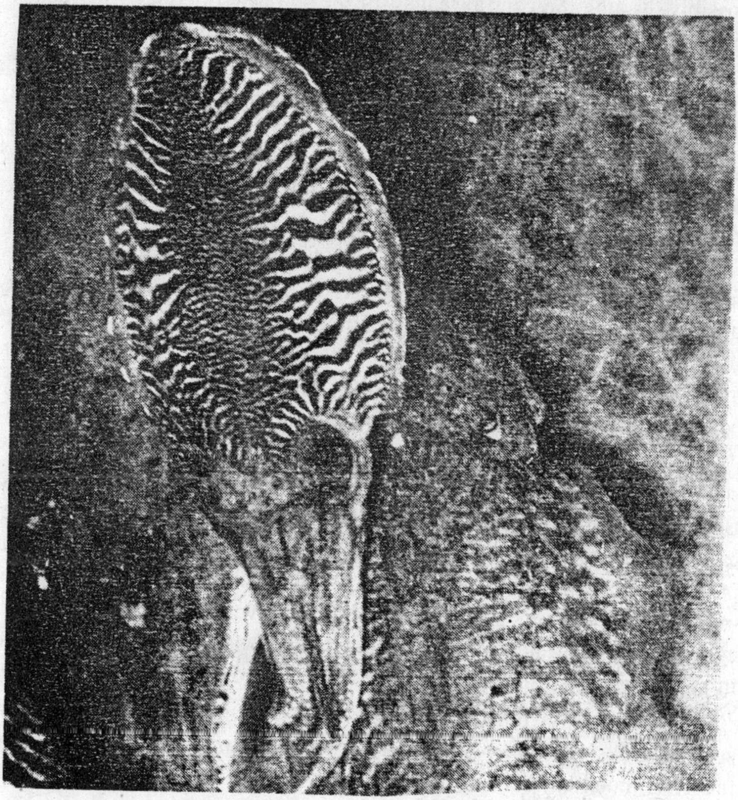

Fig. 11.16 : Male *Sepia officinalis* hovering over a female, and signalling to another male (only partly seen) not to approach

It has so far proved impossible to follow *in vivo* the process of ovum release and embedding in the nidamental jellies. Apparently the ovum, surrounded only by its chorion when breaking from the follicle, receives first a gelatinous envelope in the oviduct from the oviducal gland. After leaving the oviduct, it is embedded in a gelatinous band produced by the nidamental glands. This band is wrapped spirally around the egg, probably with the assistance of the accessory glands which together form a spoon-shaped 'mould'. The nidamental jelly is stained with ink that is released simultaneously from the ink sac. The embedded egg is then driven into the funnel tube, the opening of which is held between the bases of the ventral arms. Spermatozoids are released from the copulatory pouch under the mouth; they penetrate into the very soft envelopes, possibly through the central 'canal' of the spirally rolled jelly that leads to the area of the micropyle. These details remain to be clarified. Finally the egg is attached by the arm tips to a substrate. If no suitable structure for the fixation of the eggs is available in an aquarium, the animal drops the eggs without producing the processes from which the fixation ring is normally formed. This suggests that the final phase of the egg laying process forms a complex of strictly interrelated actions: 'aiming' with binocular fixation of the target structure with simultaneous preparation of the envelope for fixation. Eggs deposited on fine-mesh netting generally spread over larger surfaces than egg masses laid on more distinct support structures. This may be due to an image-disturbing effect of the net structure during the visual fixation of the site 'selected' for the fixation of an egg. A more surprising detail in the egg laying process is that perfectly normal egg cases can be produced and laid in the absence of ova.

Although the average number of mature ovarian eggs in large females is about 500, it appears from aquarium observations that a single female can lay many more eggs if conditions permit the full maturation of most or all the ovarian eggs. Very large egg masses observed in the field are more likely, however, to be the product of several individuals. Indeed eggs are known to act as an optical stimulus to females ready to spawn in selecting the site of egg deposit.

The main spawning season covers spring and summer, but winter spawning has also been observed. In general terms, the Mediterranean presents much milder winter conditions than the Atlantic coasts of Europe, and the winter phase of the reproductive cycle may be less pronounced, therefore, in the Mediterranean where the deeper water layers maintain a year round temperature of 12-13°C. This is several degrees above the temperature at corresponding depths along the Atlantic coast. In the English Channel inshore water temperatures higher than 12-13°C are limited to about 6 months of the year, whereas corresponding temperatures (with a much higher summer maximum) are recorded over 9 months in the Western Mediterranean. These differences probably account for most of the variation observed between the *Sepia* populations of these areas. On the whole, however, the reproductive cycle appears surprisingly similar in spite of area differences.

Mortality

The normal life-span of *Sepia officinalis* appears to vary from 18 months to 2 years; some male individuals may attain a greater age. In order to assess the natural rates of mortality, one needs reliable 'cross sections' of the population. Present sampling methods appear inappropriate for this purpose, especially with regard to juvenile stages. Another entirely open question is how much of the total individual fecundity of a mature female is actually utilized under natural conditions. In order to know more about this, one has to define the various conditions lumped together under the label 'spent females'. In a large-sized species like *Sepia officinalis*, the possibility of intermittent spawning with corresponding prolongation of egg maturation must be considered.

Whereas mass mortality, apparently at the end of the spawning period, occurs on the Atlantic coast, nothing of comparable intensity is known in the Mediterranean. The actual impact of predation by large fish, cephalopods and marine mammals and birds at different stages of the life cycle of *Sepia* is unknown. Yet predation can be assumed to be the major cause of mortality at all stages. Although various parasites are known in juvenile and adult *Sepia officinalis*, most of these (except the *Coccidian*, *aggregata*) do not appear to be very important as mortality factors at pre-reproductive stages.

Ecology

The seasonal migrations between shallow and deeper waters bring *Sepia* into contact with various types of soft bottoms. During reproduction in inshore waters and immediately after hatching, the animals may also live in close contact with rocky bottoms. The ability of very small juveniles to attach themselves to a hard substrate may be very important then because it allows them to withstand strong water movement without being carried away. At the same time, these animals are able to bury themselves in soft bottoms. The method of blowing up sand by water jets from the funnel requires a counteraction from the fins in order to keep the animal on the spot. The vigorous undulating movement of the fins that carries the sand already whirled up onto the back of the animal has often been misinterpreted as the actual burrowing process. Indeed Jaeckel (1958) suggests that fins might primarily serve burrowing!

Live observations of *Sepia* show that fine reversible undulating movements of the fins are essential in slow swimming and manouevring in narrow passages. The fins need not be very broad as they do not provide 'dynamic lift'; the animals are nearly weightless in water due to the buoyant effect of the gas-filled cuttlebone. The entire construction of *Sepia* reflects adaptation to life on the bottom and near the bottom in a very complex environment. It is noteworthy that the methods of concealment are as elaborate as in *Octopus*. In particular, *Sepia officinalis* has a considerable repertoire of temporary skin textures, produced by raising various papillae combined with a mottled chromatophore pattern. The ways of capturing prey also reflect the complex conditions to which *Sepia* is adapted. Whereas very young *Sepia* seize prey always by ejecting the tentacles at an animal which it has fixated with both eyes, larger juveniles and adults tend to seize certain prey animals, especially large crabs, by bouncing upon them, arms spread out, without using the tentacles. A prey animal trying to escape sideways can be seized by only one or two arms, and it appears that under such circumstances *Sepia* quickly switches to monocular fixation.

Insight into the coordination of these complex processes is possible by using physiological methods of stimulation and inhibition of nervous centres.

Life in inshore waters exposes *Sepia officinalis* to hydrologically unstable conditions. The species is known to be relatively tolerant to salinity variations. In the Mediterranean, animals have been observed in coastal lagoons at a salinity of $27°/_{oo}$. The young *S. officinalis* can survive for some time at salinities around $18°/_{oo}$ if slowly acclimatized, but normal embryonic development requires salinities above $25°/_{oo}$. Accordingly no Scpiu occur in the Black Sea or Baltic Sea.

The temperature limits of *Sepia officinalis* are not known exactly. At temperatures below 10°C growth is totally arrested. Temperatures above 25°C are not known to be encountered by the species.

At temperatures around 17°C oxygen consumption in relation to body weight is expressed by the equation:

$$CO_2\ (ml.O_2 \,.\, h^{-1}) = 0.146 \times W^{0/913},\ r = 0.982.$$

W = body weight; the slope of the regression line between O_2 uptake and body weight is 0.913. Very small animals show a higher tolerance to hypoxic water than larger ones.

Sepia officinalis is an important species for the commercial fisheries of many countries; it is caught with simple gear such as lures gigs and with trawl nets. Otter trawls have provided most of the material on which fisheries statistics are based. Biological information on seasonal migration and reproduction in the natural habitat also have been assembled by the use of this type of net. For the Mediterranean, fishery data published by the FAO suggest a distinct decline in the cuttlefish stocks. The recorded annual catches have decreased fairly steadily from nearly 19 000 t in 1968 to about 13 000 t in 1978. Although, these data take all three species of Mediterranean *Sepia* together, it is nevertheless likely that they reflect chiefly a decline in the large-sized species *S. officinalis*, although *S. orbignyana* and *S. elegans* may also be affected.

Conclusion

Under natural conditions, the life cycle of *Sepia officinalis* covers between one and two years, with a fairly wide range of variation according to environmental conditions. Whereas the female cycle is generally (not always)-'adjusted' to seasons by a complex of interdependent intrinsic (hormonal) and extrinsic factors (environmental stimuli), the male cycle is apparently more flexible. Earlier maturity and longer duration of individual reproductive activity of the male individuals probably maintains a high rate of genetic exchange within the population. The physiological aspects of sexual 'dimorphism' in Sepia are closely related to the migratory cycles and their control mechanisms, which are poorly understood. In this context, *S. officinalis* is an ideal candidate for combined field and laboratory experiments. Particularly interesting aspects are opened for field experiments by the presence of an easily 'readable' periodic structure, namely the lamellar cuttlebone. Although the counts of lamellae can only be used for age determination if environmental temperatures are known for all growth phases, they can provide a reliable indication of growth conditions.

The economic importance of the species in the fisheries of various countries generates the danger of overfishing. On the other hand, the uncomfortable outlook on possible stock depletion may provide the means and motivation for more basic research on the biology of the life cycle of the species. *Sepia* has its great importance in processing industry.

Identical problems arise with other cephalopods living in coastal waters. Work currently under way on *Sepia hierredda*, a species in every respect closely related to *S. officinalis*, is providing interesting aspects of fishery biology under different conditions.

Finally, *Sepia officinalis* and other medium- to large-sized *Sepia* species are the only serious candidates among decapod cephalopods for commercial aquaculture if liver prey can be replaced by cheaper food items. In the meantime, *S. officinalis* continues to draw attention as an ideal laboratory animal for various experimental purposes.

Clearly the life cycle of *Sepia* is interesting in two different respects. One is concerned with basic biological study; the other with economic considerations. Both approaches should provide opportunities to refine the present picture of cuttlefish life cycles.

Octopus

An *Octopus* is a cephalopod molluscs belonging to the order, Octopoda. Cephalopods consist of groups of free swimming invertebrates, such as squids, cuttlefish and octopuses, that lack a hard shell.

History and Lifestyle

It is thought that through the process of evolution, the *Octopus* lost its hard shell and has learned to live quite successfully as a soft bodied, tough skinned animal. There are approximately 200 known species of octopuses. Larger species are most often found in cold, northern waters. Most make their homes in shallow coastal water and live inside dens or small caves at the ocean floor level. If no such items are available, octopuses will happily live inside old cavities, pots, jars and other debris.

Octopuses are considered anti-social, and live and travel alone. Octopuses frequently block the entrance to their homes with rocks and debris to keep others at arms length.

Physical Appearance

The *Octopus* does not carry a shell and typically has eight arms or tentacles, a round or pouch-like body and two large, very distinct eyes. Unlike other animals in its class, *Octopus* can see objects.

Octopuses come in all sizes from 2-inches long to 18-feet in length. The most widely known *Octopus* is the Common *Octopus*, which lives in the Mediterranean and the Atlantic oceans. The Common *Octopus* reaches 10-feet in length. The larger. Giant *Octopus*, makes its home in the Pacific, and has a diameter of over 30-feet. Common North American *Octopus* include the Common *Octopus*, Giant *Octopus*, and the American devilfish.

Octopuses are perhaps best known for their long, sucker carrying arms. The typical *Octopus* has 8 arms, with each arm holding two rows of fleshy suckers. Each *Octopus* has a total of 240 suction cups on the underside of each arm. If an arm is bitten off, diseased or otherwise missing from the *Octopus*, a new arm grows back in its place. The arms are joined at the base of the *Octopus* in an area known as the "skirt."

In the center of the skirt, lies the *Octopus* mouth. The mouth of the *Octopus* contains a pair of sharp, horn-styled beaks and the radula, an organ used to drill shells apart and suck away fleshy meat.

Movement

Octopuses move by crawling along the bottom of the ocean with their arms. The *Octopus* can use its suckers to grab on to rocks and pull itself with considerable force and speed.

Feeding Habits

The *Octopus* seizes its prey by means of its sucker-bearing arms. It then pulls the prey into its mouth. A poisonous salivary secretion is emitted immediately, paralyzing the prey and partially digesting it. The *Octopus* then chews, using both its horny jaw and radula.

Common prey of the *Octopus* include crabs and lobsters. Several species of *Octopus* feed on other shellfish, plankton and marine fish.

Defense Mechanisms

The *Octopus* is thought to be the most intelligent of all the invertebrate animals. In an effort to protect itself, the *Octopus* often constructs barricades made of large stones on the ocean floor. The *Octopus* hides behind the barricade or in its crevices.

Much like squid, the *Octopus* is also capable of producing a dark ink-like fluid from their internal ink sacs, which clouds the water and gives a message to approaching danger.

Octopuses are also capable of changing their colour, depending on mood and environment. Rapid waves of colour can sweep over the body of the *Octopus* in seconds, and range from pink to brown. When in danger, the *Octopus* often camouflages itself, by matching its body colour to that of its surroundings.

Breeding

Male and female *Octopus* reproduce sexually. Usually, Tie is equipped with a sexual organ (on a modified arm) that deposits packets of sperm (known as "spermatophores") into the mantle cavity area of the female. Eggs, about 1/8-inch in size, are then encased in capsules and attached by the female to rocks or in holes. The female may deposit as many as 100,000 eggs at a time. It takes 4-8 weeks for eggs to hatch, during which time the female guards the eggs around the clock. Young octopuses hatch without a larval stage and appear as drifting plankton.

Life Span

Although, science has documented the existence of octopuses over the course of time, very little is known about certain areas of the *Octopus*. Lifespan, for example, is not entirely understood. Studies in which *Octopus* were raised from eggs in an aquarium produced varying results on the exact lifespan of the *Octopus*. However, it is believed that the average lifespan of the *Octopus* is just one year. Many believe that the *Octopus* "lives fast and hard," always having the upper hand, because of their shortened life cycle.

Octopuses reach sexual maturity at the age of 5-months. Science has shown that Octopuses possess an advanced brain that can be taught to do various things. Researchers have, among other things, successfully trained Octopuses to remove lids from jars in order to gain access to food.

Octopus meat has been considered a culinary delicacy in many parts of the world prepare and cook *Octopus* as a food source.

Octopus (*Octopus* Vulgaris Cuvier), A New Candidate for Aquaculture

Octopus (*Octopus* vulgaris) is in high demand in Europe respecially Spain with a market cost considered to be on the increase. This organism is distributed world-wide, except in polar and subpolar regions, and it presents a large market. Easily adapted to captivity, its life cycle from the egg stage to commercial adult size is estimated at two to three years. The species' high protein content, able to exceed 70% of its body composition, sets it out as a potential candidate for rearing on an industrial scale.

The Spanish Institute of Oceanography (IEO), located at Cabo Estai, Vigo, Spain, is currently conducting a three year project to analyse the feasibility of *Octopus* rearing in Galicia. In the same manner as these researchers began with turbot some ten years ago, the objectives initially posed involve gaining knowledge of the growth rates for this species in captivity and obtaining spawns under controlled conditions in order to complete the full rearing cycle.

On-growing

On-growing data for this species, obtained over the first two years of the project, were promising. Adult individuals of 1.3 kg had grown 12 kg fresh weight in 10 months, whereas the 600 g individuals reached 6 kg in eight months. The smallest, from 300 to 500 g, need four months to reach 2.5 kg.

Food supply, with a low commercial value to make possible industrial application highly viable, was based on 75% of crustaceans (*Carcinus maenas*, *Polybins henslowii*), 20% of frozen fish (*Micromesistius poulasson*) and 5% of mussels (*Mytilus spp.*). If stabling is conducted with similar sized individuals, no problems with cannibalism are noted when in captivity. Mortality has been low in all experiences. Behaviour has involved defence and camouflaging under artifacts placed in the farming tanks.

Current experiences point towards prior separation of males from females to optimise fattening, thus avoiding the fertilisation and natural death of females as occurs once the paralarvae are born. Taking into account the growth rates for each sex and the mortality observed, it is advisable to fatten the males up to 3 kg and the females up to 2.5 kg.

Using frozen fish, such as sardine (*Sardino, pilchardus*), horse mackerel (*Trachurus trachurus*), blue whiting (*Micromesistius poutassoii*), sea bream (Boops boops), mackerel (*Scomber scombrus*) and molluscs (*Mytilus sp.*), similar fattening results were obtained with growth rates of 0.3 to 0.8 kg/month, and a mortality rate of 5.7%.

The fact that from the 500-700 g individuals and upwards it is possible to attain the commercial size of 2.5-3 kg in a period of 4 months of fattening, it has been established that *Octopus* is currently considered as a priority interest species for aquaculture in Spain, and a good deal of interest is being shown in various Autonomous Communities in terms of developing fattening projects, both in floating structures and in tanks. Specifically in Galicia there are at present several companies engaged in fattening *Octopus* in floating cages. The general exploitation layout in a company where these experiences are underway involves cylindrical or square shaped cages, with individual dens (on the walls or in the centre) with a capacity for 150 *Octopus*. These cages may be individual units with an independent floatation system, or they may involve a common floating platform. The fattening process lasts 4 months, and three fattening cycles may be conducted during the year, attaining the commercial size (2.5-3 kg), so that a company with 25 cages may fatten about 11,000 octopi per year.

Reproduction

Female *Octopus* deposit and attach spawns on the internal ceilings of natural rocky caverns. With this in mind, IEO researchers have designed several artificial devices and systems orientated towards obtaining spawns in captivity conditions.

One of the characteristics of *Octopus* is its high fecundity, each female being able to produce from 200 thousand to 500 thousand eggs. Experiences in 8 m^3 tanks have shown that fecundation occurs spontaneously in captivity, and in a few weeks, once the females have spawn and the embryonic development has been completed, high number of paralarvae are obtained. Recent experiences have shown that the feed for reproducers may be the deciding factor in terms of the strength and survival of the newly born *Octopus*. It is advisable to separate females which have already spawned from the tanks with reproducers, in order to control and conduct individual follow-up

Spawn

Females with spawn stop eating, losing as much as 60% of their initial weight, focusing all their attention on cleaning egg clusters and defending against any external danger. For one or two months, depending on temperature, they consume all their reserves, and at the end of the process, when the last paralarvae have born and clusters are completely empty, they die. This is a natural process which occurs in the same manner in the natural environment.

Paralarvae Rearing

The bottleneck for making it possible to control the entire rearing process of this species lies in the high mortality rates noted during the larval stage. At this stage, newly born *Octopus*, measuring only 3 mm in length, need a live prey to cover their vital requirements. In their natural environment, over the first two months of life, they feed on larvae of prawns, crabs and other crustaceans.

In the experimental trials conducted by the IEO, food items have included different types of live prey such as crustacean zoeas, copepods, euphausiaciae and *Artemia*. The best results were obtained with adult *Artemia* (from 2 to 4 mm total length) reared on phytoplankton. Under these conditions, *Octopus* paralarvae triple their dry weight in a period of 30 days.

Despite all these results, paralarvae mortality is very high (up to 100% after one or two months of the beginning of the paralarvae culture), so that in order to progress in research on full *Octopus* rearing, it is necessary to focus maximum effort on solving the technical drawbacks involved in rearing paralarvae. The final aim would entail improving the optimum rearing conditions and to identify a live prey with proper size and nutritional content to reduce the current high mortality rates.

Conclusion

Substantial interest is evident in the farming sector for rearing *Octopus*, both at national and at international level, particularly in Mediterranean and Latin American countries and Japan. However, taking into account all the above premises considered so far, could *Octopus* be a serious candidate for developing and strengthening its industrial rearing? Research conducted to date at the IEO points to the fact tlihat ongrowing *Octopus* is already industrially viable, but obtaining juveniles to complete the cycle is a matters still to be resolved, and with difficulty, and should be approached jointly by various interdisciplinary research teams. Ideally, this would entail co-ordinating all the research effort developed in order to achieve a rational transfer of this knowledge to the sector.

12
ENERGY METABOLISM

The immediate source of energy in biological process is ATP. Its connection in cell fluctuates.

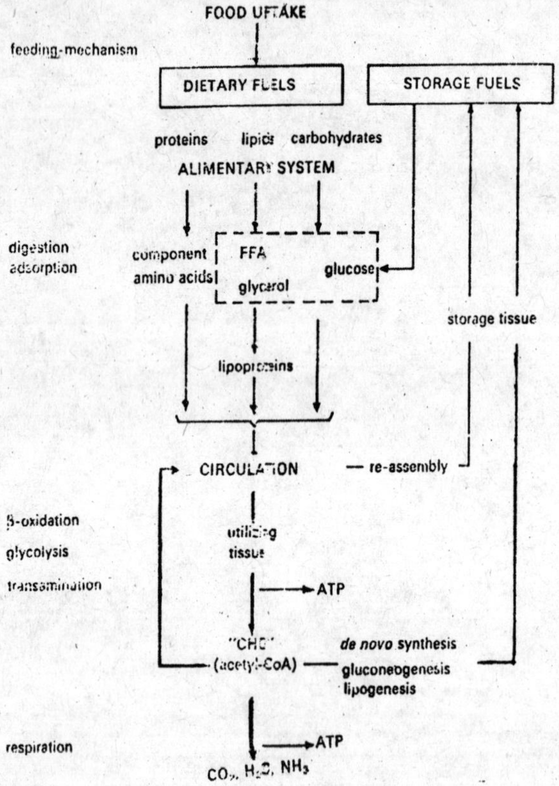

Fig. 12.1 : Interrelations between energy metabolism, nutrition and feeding.

The sum of the adenylates determines the storage potential for chemical energy and the number of energy rich phosphate bounds (two for each mole of ATP, one for each mole of ADP) determines the energy status of the cell. This has been termed by Atkinson (1968) the energy charge and can be expressed by the following equation: ATP + ½ ADP/ ATP + ADP + AMP. Each cell contains adenylate kinase (or in muscle myokinase), which catalyses the reaction: 2 ADP $\rightleftarrows$ AMP + ATP. Therefore, the concentration of the individual adenylates is interdependent and the value for the energy charge represents a fixed ratio between the three adenylates (Fig 12.3a). The role of the energy charge in the control of cellular metabolism is shown in the lower panel of Fig 12.3b. The energy charge is usually maintained between 0.7 and 0.9, but within this range small fluctuations can cause large changes in the velocities of enzyme reactions and this results in a feedback control for ATP production.

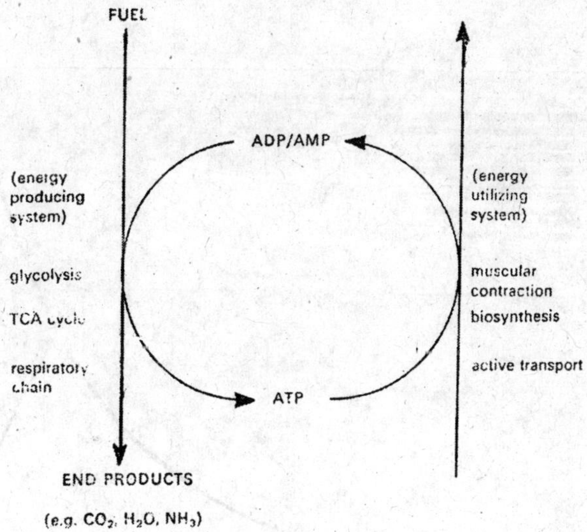

Fig. 12.2 : The role of ATP as an energy transfer compound.

The controi of metabolism by ATP is based on feedback signals to key regulatory enzymes. It is essential that the ATP concentration does not exceed the concentration at which saturation of its site on the enzyme

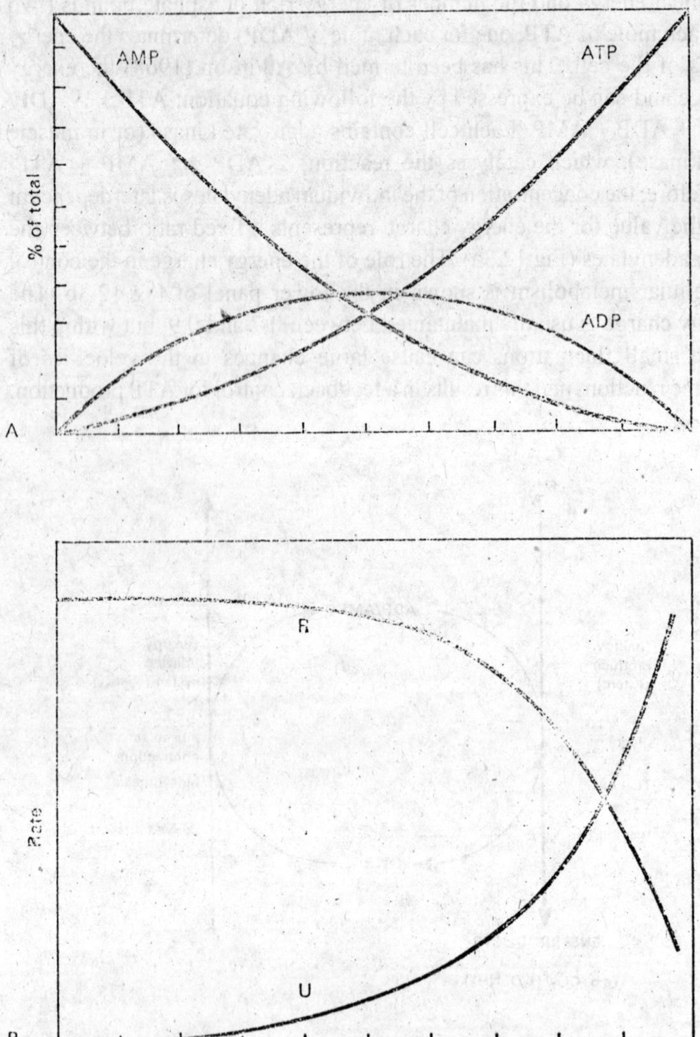

Fig. 12.3 : (A) Relative concentrations of the adenine nucleotides as a function of energy charge; (B) Response of enzymes involved in regulation of ATP-generating (R) and ATP-utilizing (U) pathways (after Atkinson, 1968).

is reached. This explains why the main strategy of energy storage is not based upon ATP accumulation hut instead upon stored fuels. Muscles, which are characterized by large and sudden changes in energy demand, often have additional stores or energy in the form of phosphagen. In the invertebrates this is arginine phosphate, a compound which can easily transfer its phosphoryl group reversibly to ADP. The reaction is catalyzed by arginine kinase: arginine-P + ADP ⇄ arginine + ATP. Table 12.1 shows data for ATP and arginine phosphate levels in some molluscs. Muscles with the highest energy demand, *e.g.,* in *Pecten,* show the highest arginine phosphate levels. High levels of arginine-P in the range of 30-50 mM, have also been found in the mantle of cephalopods. When comparing ATP levels in different tissues of the same species, the muscles show the highest values. This indicates that in muscle tissue ATP may play a (minor) role in energy storage.

In the subsequent discussion we will focus on three main aspects of energy metabolism in molluscs: which fuels are available, which fuels are used and how do the metabolic transformations take place?

The first question can be approached by studying which nutrients are absorbed, which are stored and which fuels (for storage) can be synthesized *de novo*. Most bivalves and a number of gastropods are herbivorous, detritus or plankton feeders and the bulk of their food source consists of carbohydrates. Besides starch and glycogen other polysaccharides, indigestible by vertebrates, can be utilized. This is reflected by the fact that many carbohydrates, some of which are absent in vertebrates, can be found in molluscs. Endogenous cellulase and cellobiase are secreted by herbivorous gastropods. Wood-boring bivalves

Table 12.1 : Concentrations of ATP and arginine phosphate (μ mol/g wet tissue) in muscle tissue of some molluscs and concentration of ATP in various tissues of *Mytilus edulis*.

			ATP	Arg. P
*Helix pomatia**:	foot		0.76	1.44
Pecten maximus:	adductor		6.63	52.16
Mytilus edulis:	adductor		2.02	1.99
*Mytilus edulis***:	adductor	digestive gland	mantle	residue
ATP:	4.76	1.31	2.20	1.84

* After Beis & Newsholme (1975).

** After Wijsman (1976).

also secrete cellulase. Alginase and β-glucuronidase are secreted by the digestive gland of various molluscs. The presence of such a wide range of carbohydrases represents a biochemical adaptation to a diet in which carbohydrates predominate (Barnard, 1973). All cephalopods are predacious carnivores and the bulk of their food consists of proteins. In vertebrates, the pancreas is the most important gland for the production of enzymes for the digestion of proteins. It is interesting that cephalopods have a well developed, clearly distinct pancreas.

When the level of food is below the maintenance ration the animal becomes dependent on its fuel reserves. Lipid is stored in many molluscs. In chitons (Polyplacophora) lipid is often present in relatively large amounts, especially in the gonads. In bivalves the lipid levels are in general relatively low. The same applies to gastropods, but in some species, for instance in the abalones (Haliotidae), lipid is prominent in the ovaries. Cephalopods also show a high level of lipid in most organs; the levels are extremely high in the integument and brain of the squid. Most adult bivalve and gastropod molluscs *e.g.* oysters, mussels, limpets and abalones store large amounts of glycogen, but the glycogen levels in cephalopods are usually low.

In general we may state that all molluscs store—at least—some lipid. In the Polyplacophora and Cephalopoda lipid dominates over carbohydrate, whereas in Gastropods and Bivalvia the opposite is true. A variety of molluscs can incorporate ^{14}C-labelled compounds into fatty acids and this demonstration of lipogenesis may be related to the presence of lipid stores (Voogt, 1972). Gluconeogenesis in molluscs has been poorly studied. In vertebrates the mobilization and storage of fuels is under endocrine control, but no direct evidence for such a control mechanism has been found for molluscs.

Utilization of stored energy reserves has been studied in relation to starvation, gametogenesis, anaerobiosis and muscular activity. In this paper we will restrict our attention to muscular contraction and anaerobic metabolism. Molluscs possess a variety of muscles which differ in their physiological function, their contractile mechanism and their dependence on available oxygen. There are also differences in oxygen dependence between various molluscan species. In general, bivalves, especially in the littoral zone, are most tolerant to anoxia. Several marine and freshwater gastropods tolerate anoxia quite well, whereas all cephalopods are intolerant to lack of oxygen. Scallops are an exception, they are also intolerant to lack of oxygen.

As examples of biochemical adaptation we will consider the relationship between the biochemical characteristics of ATP formation on the one hand and the physiology of the muscles and anoxic tolerance on the other hand in general, sustained acrobic muscle activity is based on fat as the primary energy source, whereas burst activity is based on carbohydrate fermentation. During anoxia the animal relies on carbohydrate reserves. Whatever the type of fuel, degradation can be divided into two main processes: the dehydrogenation and the oxidation of reduced products. Degradation of organic substrates by dehydrogenation reactions results in the production of NADH and in the case of carbohydrates in a low yield of ATP. Complete breakdown of carbohydrates into CO_2 and H_2O involves glycolysis and the tricarboxylic acid cycle (TCA-cycle). Oxydation of NADH by oxygen is mediated by the election transport chain and results in (he production of relatively large amounts of ATP. Up to the level of the TCA-cycle (acetyl- CoA) the dehydrogenation steps also result in the formation of small amounts of ATP, linked to substrate conversion. This ATP production plays a dominant role in anaerobic metabolism. One of the problems during anaerobic metabolism is the maintenance of redox balance. Two principles are followed by molluscs to meet this problem:

1. The electron acceptor is an end product, or an intermediate of the pathway itself, *e.g.*, pyruvate, oxaloacetate and fumarate (Fig. 12.4). This leads to a metabolic *cul de sac* and accumulation of organic end products;
2. The synthesis and storage of high electronegative substances in the preanaerobic condition *e.g.* the lipochrome pigment present in the nervous tissues of gastropods and bivalves.

In molluscs we can distinguish three pathways for the utilization of carbohydrates, which are shown in Figure 12.4. The upper panel of the left diagram represents glycolysis and the lower panel the TCA-cycle. The width of the arrows indicates the relative maximal flow of substrate along the pathway. Hexokinase (HK) and phosphorylase (Ph'ase) determine whether glucogen or glucose is the immediate energy source used in the pathway. Glycogen is generally stored in the tissue itself whereas glucose is supplied by the blood. Phosphofructokinase (PFK) catalyzes non-equilibrium reaction. In general it has the lowest specific activity in the overall pathway and therefore the flow of substrate depends highly on this enzyme. Glyceraldehydephosphate dehydrogenase (GAPDM) catalyzes the oxidative step in the pathway and at this point glycolytic NADH is formed. What ultimately happens to phosphoenolpyruvate (PEP) depends on the relative activities of phosphoenol pyruvate carboxykinase (PEP-CK) and pyruvate kinase (PyK). The PEP-CK converts PEP into oxaloacelate, and this is the initial reaction of the 'succinate-propionate pathway'. PEP-carboxykinase produces GTP or ITP whereas ATP is requires for muscle contraction. Tissues which rely on the 'succinate-propionate pathway' are also rich in nucleoside dipliosphokinase (NDPK). This enzyme brings about a rapid equilibrium of the phosphate groups among the various nucleotides:

$$N_1TP + N_2DP \rightleftarrows N_1DP + N_2TP.$$

Pyruvate kinase catalyzes the conversion of PEP into pyruvate. Pyruvate may enter the TCA-cycle, followed by complete oxidation into CO_2 and H_2O or it may be converted into lactate, N-(1-carboxyethyl) alanine or octopine. Octopine dehydrogenase (ODH) catalyzes the condensation of arginine with pyruvate:

$$\text{pyruvate} + \text{arginine} + \text{NADH} + H^+ \rightarrow \text{octopine} + \text{NAD} + H_2O.$$

Citrate synthase (CS) and isocritrate dehydrogenase (ICDH) are key enzymes of the TCA-cycle.

The activities of the enzymes described so far are known from a great number of molluscan tissues and it is possible to predict which of the three pathways shown in Figure 12.4 is followed in a particular tissue and in the different types of muscles (Zammit & Newsholme, 1976).

In pathway la (Fig. 12.4) glycogen is converted into lactate or octopine and the pathway can be characterised as follows:

1. oxygen independent,
2. a high glycolytic potential, achieved by high enzyme activities and a large glycogen store,
3. a large-percentage change from low to high activity state. This is because signals such as the energy charge can reach all key points of the pathway at the same time since glycolysis is entirely located within the cytosol. There is a tight integration of all steps in the pathway giving control of the overall process,
4. tolerance to the accumulation of end products.

Octopine formation from pyruvate and arginine has been demonstrated in muscles of various groups of molluscs. Both lactate dehydrogenase and octopine dehydrogenase catalyze readily reversible reactions. Prolonged muscle action will lead, in addition to octopine and lactate, to an accumulation of pyruvate which in turn decreases the rate of NADH reoxidation and of glycolysis. In vertebrates lactate can be washed out by the blood, but in molluscs which, in general, have a poorly developed circulatory system, this does not occur so easily. The condensation of arginine and pyruvate into octopine keeps the level of both metabolites low, thus serving to continue glycolysis. Moreover, this has the advantage of switching on the arginine kinase reaction during high energy expenditure (*e.g.* during, escape reaction, catching prey). Due to the rapid depletion of the energy stores and the accumulation of end products this pathway is not suitable to sustain long term anaerobic metabolism. But because of the high ATP output per unit time (Table 12.2) it is perfectly adapted for short bursts of muscular activity.

This pathway for glycogen degradation operates in the following molluscan muscles:—cross striated, translucent or phasic adductor muscle of scallops, used during escape response,—pedal retractor muscles, used by prosobranch molluscs for rapid retraction into the shell,—mantle muscle of benthic and pelagic cephalopods, used during burst activity, in escape reaction or when attacking prey.

Land snails accumulate lactate during experimental anaerobiosis and this may explain the relatively low anoxic tolerance of these animals.

Next we consider the conversion of glycogen into succinate and propionate, pathway 1b (Fig. 12.4). This pathway can be characterized as follows:

1. oxygen independent,
2. a low glycolytic potential reflected by low enzyme activities. Nonetheless, tissues in which this pathway operates show a moderate or even high accumulation of glycogen,
3. there is no substantial change in substrate flow when this pathway comes into operation (no Pasteur effect). Compared to the previous route (1a) this pathway is relatively complex, which prevents a tight metabolic integration of the different steps in the pathway. It is not possible to induce a sudden and many fold increase in the flux through the pathway,
4. low substrate flux results in a slow accumulation of end products,
5. involvement of mitochondria results in a relatively high ATP yield (Table 12.2), when compared to the previous pathway.

Table 12.2 : Summary of the characteristics of three pathways for the utilization of glycogen (for details see Fig. 12.4).

Glycogen →	Lactate or Octopine	Succinate Propionate	$H_2O + CO_2$
ATP yield/glucose-1P	3	3-7.4	39
ATP output/time	high	low	moderate
Fuel consumption	high	low	moderate
Integration	high	low	moderate
Compartmentalization			
within cell	no	yes	yes
within tissue	no	yes	no

These characteristics, especially those under points 4 and 5, explain why the 'succinate-propionate pathway' is well adapted for long term anaerobiosis. A rapid accumulation of noxious organic acid end products is avoided, partly because of the low energy demand. This is true for sessile or sluggish organisms. The pathway operates in littoral bivalves during exposure at low tide. It also operates in littoral gastropods, whereas freshwater snails generally show a mixture of the type 1a and type 1b pathway and this is reflected in both lactate and succinate accumulation.

The third pathway, 1c (Fig. 12.4), results in the total oxidation of glucose into CO_2 and H_2O. In contrast with the other pathways, this route is oxygen dependent and is not involved in anaerobiosis, but it appears to be the principal way to generate ATP to power a long term muscle activity. The pathway results in a high ATP yield per glucose molecule, and it is adapted to sustain muscle action for extended periods of time with a more or less uniform activity. The mantle muscles of swimming cephalopods form a good example; these muscles have a high respiratory potential due to high mitochondrial densities. The animals have a well developed circulatory system, including respiratory pigments, for the supply of oxygen and glucose. Fatty acids may also be used as an energy source.

A variant of this pathway, which can operate under anaerobic conditions, is found in the nervous tissue of gastropods and bivalves.

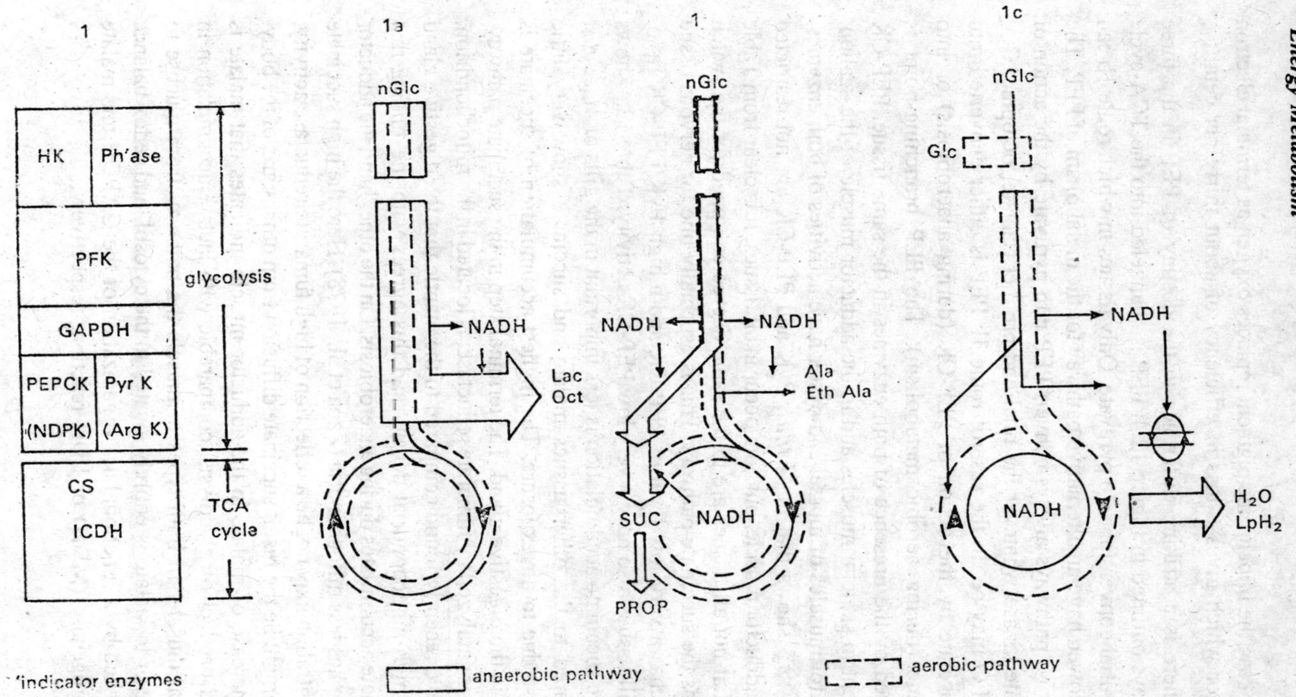

Fig. 12.4 : Three main pathways for the utilization of carbohydrates in molluscs (for explanation see text).

In this case the lipochrome pigment replaces oxygen as terminal electron acceptor which gives a loss in efficiency of about thirty per cent.

There is a striking difference in the destiny of PEP in the three pathways outlined in Figure 12.4. PEP is channeled into the TCA-cycle after initial conversion into pyruvate. Only tissues in which route 1b can be followed have an alternative pathway for the metabolism of PEP. The other two pathways always convert PEP into pyruvate by the action of pyruvate kinase whatever the final end product (lactate, octopine, CO_2 plus H_2O may be. In the case of route 1b PEP is either converted into oxaloacetate by the action of PEP-CK (during anaerobiosis) or into pyruvate (during aerobic metabolism). The PEP branching-point is reflected by the presence of both enzymes in the same tissue. PEP-CK is absent in skeletal muscles and in the adductor muscle of the scallop. Adductor muscles of mussel and oyster have activities of both enzymes. Table 12.3 shows that in *Mytilus*, PyK and PEP-CK are not restricted to the adductor muscle, but both occur in all tissues. It is clear from Table 12.3 that not all tissues are to the same degree equipped for the switch towards the succinate pathway. This is especially line for nervous tissue which has a very low level of PEP-CK and a high PyK/PEP-CK ratio. The gill tissues also have a very low PEP-CK activity. Table 12.4 shows that the succinate accumulation is less important in the gills of *M. edulis*; compared to the other tissues malate and succinate show very little increase due to air exposure. The highest accumulation of succinate is found in the digestive gland. The terminal step in the succinate pathway, which is catalyzed by fumarate reductase, is located in the mitochondrion. We have recently found compared to the mantle that the digestive gland has a high mitochondrial density and this agrees with the finding that succinate accumulates during anaerobiosis. On the other hand the adductor muscle has less mitochondria (Zaba et al., 1978). Here the high succinate concentration may not be a reflection of high fumarate reductase activity, but it might be the result succinate diffusion from other parts of the body. The increase of malate in the adductor muscle indicates that malate is an end product of glycolysis, comparable with lactate accumulation in skeletal muscle. In both types of muscle the pathway from glucose to malate or to lactate is entirely located in the cytosol, and redox balance is reached by a functional 1:1 organization of the GAPDH and malate dehydrogenase (MDH) or LDH reaction, respectively.

Table 12.3 : Maximal activities (mmol/min/mgprotein, pH 7.0 25°C) of pyruvate kinase and PEP-carboxykinase in various tissues of *Mytilus edulis*. Prior to dissection the mussels were kept for 9 weeks in the laboratory In recirculating sea water of 13°C without food supply. 40 animals were dissected and the tissues pooled in two equal portions which were analysed separately. The data show the mean values. In the PEP-carboxykinase assay Zn^{++} was the cofactor.

	PyK	PEP-CK	PyK/P-CK
Nervous tissue	0.412	0.011	37.5
Gill	0.248	0.009	27.5
Intestine	0.514	0.021	24.5
Foot	0.709	0.029	24.4
PBRM	0.514	0.027	19.0
ABRM	0.384	0.040	9.6
Ventricle	1.115	0.128	8.7
Mantle	0.225	0.030	7.5
Hepatopancreas	0.227	0.043	5.2
PAM	0.621	0.174	3.6

PyK, pyruvate kinase; PEP-CK., phosphoenolpyruvate carboxykinase; PBRM, posterior byssus retractor muscle; ABRM, anterior byssus retractor muscle; PAM, posterior adductor muscle.

Table 12.4 : Levels of malate and succinate (μmol/g wet weight) before and after 24 hours of exposure at 13°C in various tissues of *Mytilus edulis L.*

Tissue	Control		Exposed	
	Malate	Succinate	Malate	Succinate
Posterior adductor	0.22	0.12	0.54	3.45
Mantle	0.04	0.19	0.13	2.40
Gill	0.06	0.03	0.09	0.91
Hepatopancreas	0.12	0.19	0.28	5.31
Residue	0.09	0.10	0.20	2.54

Malate may be redistributed by the circulatory system and by diffusion. When entering tissues which are rich in mitochondria, it can be further converted into succinate. Succinate accumulated in various tissues may enter the gills which may be the main site for converting succinate into propionate (Fig. 12.5). During long-term anaerobiosis the gills show a decline in succinate levels and a concomitant increase in the level of propionate. Propionate is the only anaerobic end product to be excreted (Kluytmans et al., 1978).

Figure 12.6 summarizes our present views on the anaerobic pathway in *Mytilus edulis*. Figure 12.7 shows that alanine, and lactate, are initial end products of anaerobic metabolism. Succinate is formed after a short lag (less than two hours) and propionate after a period of about twenty hours. Besides these three end products some glutamate may accumulate. Initially, besides glycogen utilization, there may be some conversion of aspartate into oxaloacetate in the mitochondria (Fig. 12.6). The gradual shift from alanine towards propionate is accompanied by an increase in the ATP output per mole of glucose (Table 12.5). The longer the anoxic stress continues the more complete is the biochemical adaptation. When propionate is the anaerobic end products some metabolic CO_2 is formed. The overall reactions given in Table 12.5 are the sum of a number of steps, some of which occur in the cytoplasm and some in the mitochondrion. In order to show how redox balance is maintained independently in both cell compartments (the inner mitochondrial membrane is impermeable to NAD) the overall reactions leading to propionate will be given more in detail.

Reactions occurring in cytoplasm:

$$3\frac{1}{2}\text{ glc} \rightarrow 7\text{PEP} + 7\text{ATP} + 7\text{NADH}$$

$$7\text{PEP} + 7CO_2 + 7\text{NADH} \rightarrow 7\text{Mal} + 7\text{ATP}$$

+ ________________________________

$$3\frac{1}{2}\text{glc} + 7CO_2 \rightarrow 7\text{Mal} + 14\text{ATP}$$

Reactions occurring in mitochondria:

$$1\text{ Mal} \rightarrow 1\text{ Pyr} + 1\text{ NADH} + CO_2$$

$$1\text{ Pyr} \rightarrow 1\text{ acetyl-CoA} + 1\text{ NADH} + CO_2$$

$$1\text{ Mal} \rightarrow 1\text{ OxA} + 1\text{ NADH}$$

$$1\text{ OxA} + 1\text{ acetyl-CoA} \rightarrow 1\text{ Sue} + 2\text{ NADH} + 2\,CO_2 + 1\text{ ATP}$$

+ ________________________________

$$2\text{ Mal} \rightarrow 1\text{ Sue} + 5\text{ NADH} + 4\,CO_2 + 1\text{ ATP}$$

$$1\text{ Sue} \rightarrow 1\text{ Prop} + 1\text{ ATP} + 1\,CO_2$$

$$5\text{ Mal} \rightarrow 5\text{ Fum}$$

$$5\text{ Fum} + 5\text{ NADH} \rightarrow 5\text{ Sue} + 5\text{ ATP}$$

$$5\text{ Sue} \rightarrow 5\text{ Prop} + 5\text{ ATP} + 5\,CO_2$$

____

$$7\text{ Mal} \rightarrow 6\text{ Prop} + 12\text{ATP} + 10\,CO_2$$

Overall reaction (cytoplasm plus mitochondrial):

$$\text{glc} \rightarrow 0.9\,CO_2 + 1.7\text{ Prop} + 7.4\text{ ATP}$$

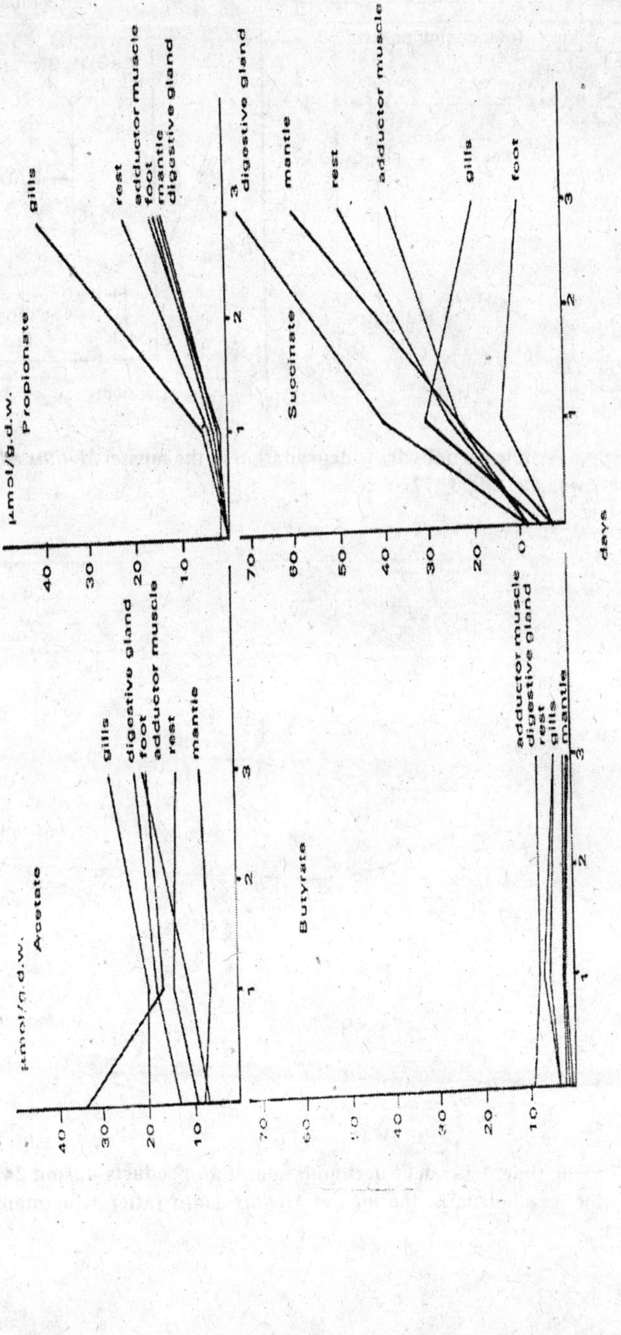

Fig. 12.5 : Accumulation of volatile acids and succinate in the mussel *Mytilus edulis during* 0.1 and 3 days of exposure to water vapour saturated N_2 atmosphere at 12-13°C (after Kluytmans et. al., 1978).

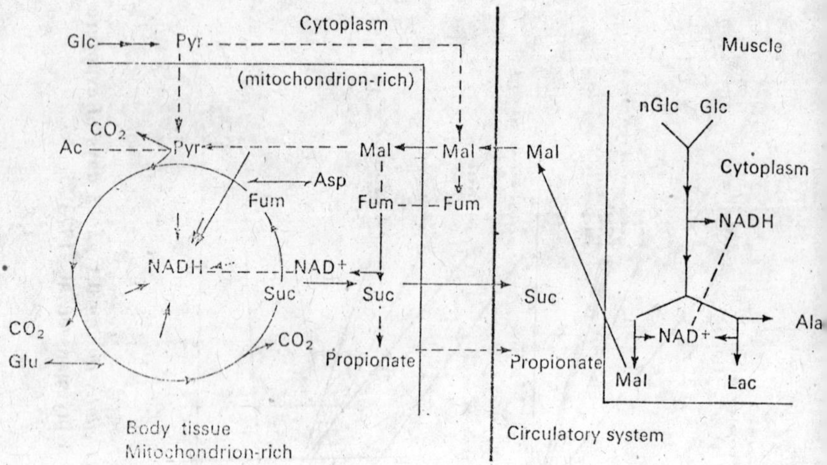

Fig. 12.6 : Anaerobic carbohydrate degradation in the mussel *Mytilus edulis* (after De Zwaan et al., 1977).

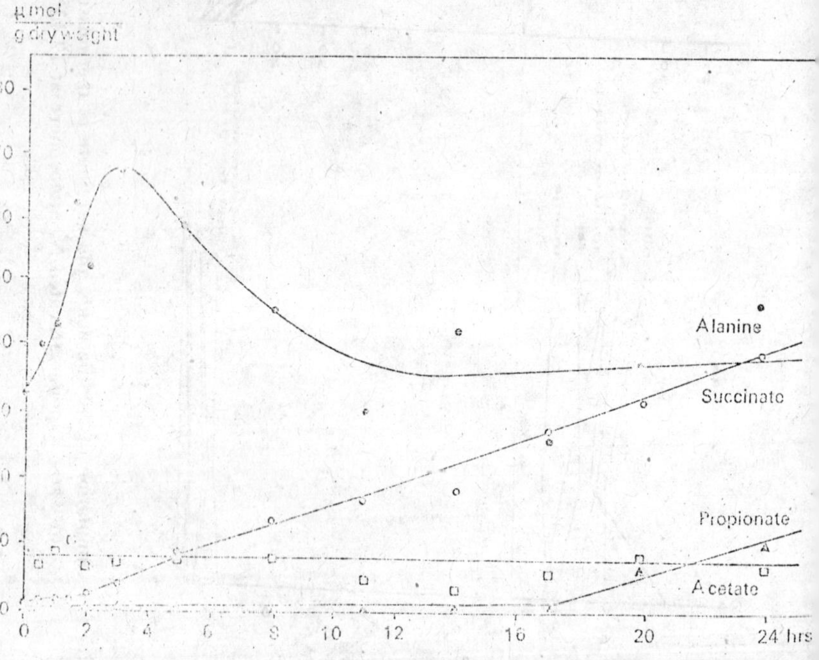

Fig. 12.7 : The time dependent accumulation of an products during 24 hr. of anaerobic metabolism in the mussel *Mytilus edulis* (after Kluytmans et al., 1977).

The equations show that redox balance in the mitochondrion is maintained when of every seven moles of malate, two moles are used to form one mole of citrate which is further metabolised via the TCA-cycle, and five moles are used to form fumarate which acts as electron acceptor. The TCA-cycle thus produces the reducing equivalents (NADH) for the fumarate reductase step. For each complete cycle six moles of succinate will accumulate (five from the succinate pathway and one from the TCA-cycle).

Table 12.5 : The overall reactions for glucose conversion in molluscs. The reactions may occur in this order during successive periods of anaerobiosis. The longer the anaerobic condition the higher the ATP yield.

1.	Glc	→	2 Lac	3 ATP
2.	Glc + $2NH_3$	→	2 Ala	3 ATP
3.	Glc + 0.5 Asp + 0.5 CO_2	→	0.5 Glu + 1.5 Suc +	4.5 ATP
4.	Glc + 0.4 NH_2 + 0.8 CO_2	→	0.4 Glu + 1.2 Suc +	5.2 ATP
5.	Glc + $0.8CO_2$	→	1.7 Suc +	5.7 ATP
6.	Glc	→	0.9 CO_2 + 1.7 Prop +	7.4 ATP

It is now possible to summarise the properties of the pathways given in Figure 12.4, 1a and 1b, more in detail.

	Pathway 1a	*Pathway 1b*
	(The lactate or octopinc pathway)	(The succinate pathway)
(1)	anaerobic metabolism is dependent on glycogen	anaerobic metabolism relies predominantly on glycogen, but amino acids (*e.g.* aspartate) may contribute to the carbon flow.
(2)	pathway operates entirely in one compartment	pathway operates in the cytoplasm plus mitochondrion
	–NADH does not need to cross the inner mitochondrial membrane	–NADH must move across the mitochondrial membrane or redox balance must be maintained independently in the two compartments
	–one redux couple (GAPDH coupled to LDH or ODH)	–various redox couples (e.g. GAPDH/MDH; FR/TCA-cycle)

(3) lactate or octopine is the sole end product	there are multiple end products (alanine, malate, acetate, succinate, propionate, glutamate, CO_2)
(4) only substrate linked phosphorylation of ADP; low ATP yield per mole of glucose 1P	substrate linked and electron-transport coupled phosphorylation (the fumarate reductase electron transfer system); higher ATP yield per mole of glucose-1P; time dependent
(5) no time dependent changes in pathway	changes in pathway (alanine initial end product, propionate accumulation after a lag)
(6) species and tissue specific differences; adapted to 'fast' muscle activity *e.g.* in scallops and cephalopods	species and probably tissue differences (*e.g.* land snails accumulate mainly' lactate, freshwater snails both lactate and succinate, intertidal gastropods and bivalves mainly succinate.
(7) no deviations in the metabolic pathway (a non-allosteric pyruvate kinase isozymes)	deviations in the pathway at the PEP-branching-point and malate branching-point (allosteric pyruvate kinase isozymes)
(8) operates when there is a high energy demand (high ATP output per unit time)-	operates when there is a low energy demand (low ATP output per unit time).

Point 7 needs some further explanation. Tissues in which the succinate pathway is operative, possess both pyruvate kinase and PEP-carboxykinase activities. Both enzymes operate in a competitive way and this resembles the situation in the vertebrate liver during gluconeogenesis. In order to supply the brain with glucose, when liver glycogen is depicted, amino acids originating from muscle proteins are convened in the liver into glucose. During gluconeogenesis there must be a mechanism to prevent any conversion of PEP (produced via PEP-CK) into pyruvate by pyruvate kinase. Liver contains regulatory (allosteric) pyruvate kinase isozymes in which non-reactants control its activity (see Marco & Sols, 1969). The same process occurs in the adductor muscle of oysters and mussels in order to depress pyruvate kinase activity during anaerobiosis. The enzyme of both the Liver and the adductor muscle is sensitive to the same signals, *e.g.* fructose-1, 6-diphosphate, alanine and pH. There is, however, a markedly different response to lowering the pH; in rat liver it results in an increase in activity of PyK whereas in the adductor muscle it strongly inhibits the enzyme activity. The pH drop which occurs when bivalves are exposed to air during low tide, is considered to be the main signal which channels PEP away from the PyK reaction and towards the succinate pathway. L-alanine, which also strongly inhibits PyK, potentiates the pH effect.

Finally, we can summarize the reasons with the succinate pathway can be considered a biochemical adaptation towards life without oxygen:

(1) The relatively high ATP yield and low level of organic end products.

(2) The TCA-cycle still operates during anoxia and allows the continuation of its role in anabolism. This may be of special importance in molluscs which undergo long term anaerobiosis.

(3) Alanine (and glutamate) accumulation may play a role in detoxification of NH_3.

(4) The accumulation of alanine, succinate and propionate causes a smaller fall in pH than an equimolar amount of lactate.

(5) The accumulation of NADH by fumarate is physiologically irreversible and allows a high accumulation of succinate without accumulation of fumarate and, therefore, allows a continued flow of carbon along the pathway.

(6) Succinate (and propionate) can be easily converted into oxaloacetate, which may serve to increase the capacity of the TCA cycle after a return to acrobic conditions.

(7) Propionate (and acetate) is easily excreted.

REFERENCES

Atkinson, D.E., 1968. The energy charge of the adenylate pool as a regulatory parameter. Interaction with feedback modifiers–Biochem. 7:4030-4034.

Barnard, E.A., 1973. Comparative biochemistry and physiology of digestion. In: C.L. Prosser (Ed.).

Beis, I. and E.A. Newsholme, 1975. Comparative Animal Physiology. W.B. Saunders Co., Philadelphia, London, Toronto:133-147.

De Zwaan, A., 1977. Anaerobic energy metabolism in bivalve molluscs–Oceanogr. Mar. Biol. Ann. Rev. 15:102-187.

De Zwaan, A., J.H. Kluytimans and D.I. Zandee, 1976. Facultative anaerobiosis in molluscs–133-168.

Kluytmans, J.H., A.M.T., de Bont, J. Janus and T.C.M. Wijsman, 1977. Time dependent changes and tissue specificities in the accumulation of anaerobic fermentation products in the sea mussel *Mytilus edulis* J–Comp. Biochem. Physiol. 58B:81-87.

M. van Graft, J. Janus and H. Pieters, 1978. Production and excretion of volatile fatty acids in the sea mussel *Mytilus edulis* L–Comp. Physiol. 123:163-167.

Zaba, B.N., A.M.T. de Bont and A. de Zwaan. 1978. Preparation and properties of mitochondria from tissues of the sea mussel *Mytilus edulis* L–Int. J. Biochem. 9:191-197.

Zammit, V.A. and E.A. Newshorme, 1976. The maximum activities of hexokinase, phosphorylase, phosphofructokinase, glycerolphosphate dehydrogenase, lactate dehydrogenase, octopine dehydrogenase, phosphoenolpyruvate carboxykinase, nucleoside diphosphatekinase, glutamate-oxaloacetate transaminase and arginine kinase in relation to carbohydrate utilization in muscles from marine invertebrates–Biochem. 160:447-462.

Zwaan A De (1976) Energy Metabolism in Molluscs in pathways in malacology edited by S. Vander Spoci A.G. Van Broggen and I Lever W. Junk B.V. Publishers the Hagae.

13

TORSION AND DETORSION

TORSION AND DETORSION

The first gastropods must have encountered some difficulty in balancing the tall spire, for the shell peak tended to sag and coil. The first coiled gastropods were not twisted. Their bodies were bilaterally symmetrical and coiling was largely restricted to one plane. The first coiled gastropods belong to the order cochleostraca, which may be translated as "spiral shell", also wholly extinct. These two orders make up the Protogastropoda and include the fossil species that might have been coiled but were never twisted. They are old-fashioned molluscs.

Protogastropoda: These first gastropods did not undergo torsion and had a posterior mantle cavity. Primitively conical, as the visceral mass became taller they developed a spiral shell.

Examples : Palaeacmea, (ii) Scenella (iii) Pelagiella.

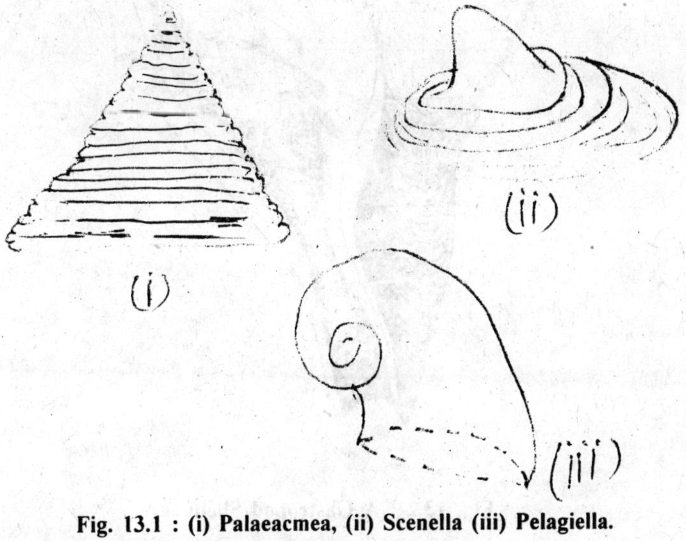

Fig. 13.1 : (i) Palaeacmea, (ii) Scenella (iii) Pelagiella.

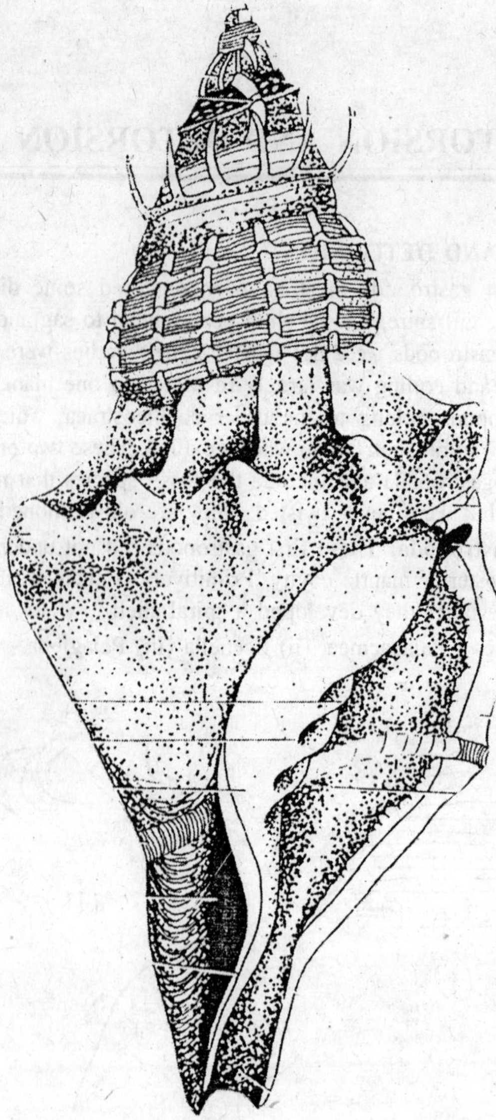

Fig. 13.2 : A Gastropod Shells.

Torsion happens to modern gastropods when they reach the veliger stage. By this time the mantle has appeared, and a broad, ciliated region, the velum, forms the anterior end of the embryo. The body is curved, with both head and foot protruding from the mantle. The visceral hump has already begun to form a spiral. The whole visceral mass revolves 180°. The may take a few minutes, or far longer. In *Haliotis,* torsion occurs in two movements, a fast 90° twist, followed by a much slower second twist at 90°.

The results of torsion are drastic. *Before torsion,* the esophagus opens normally into the stomach, *but after torsion,* it comes into the stomach from behind. *Before torsion,* the mantle cavity, containing the gills and anus, is posterior *but after torsion* they are anterior. *Before torsion,* the visceral nerves and ganglia form a simple loop, *but after torsion* they are twisted into a figure of eight. As the twist is counterclockwise, the left visceral nerve passes dorsal to the right. As the twist occurs between the pleural and parietal ganglia, the pleural ganglia are not affected, but the left parietal is carried upward to be come the supraparietal ganglion, while the right one is carried ventrally to be come the infraparietal ganglion.

There is evidence that there must be, for all of the gastropods that do not undergo torsion have died out, while the twisted ones have become highly successful. It may be that torsion is a mechanical side-effect of forming conical spirals, but this idea is not persuasive, as the first group of gastropods to undergo torsion were predominantly equipped with plane spiral shells.

Torsion-scheme of organization of a gastropod before, during and after torsion. That as the mantle cavity is brought forward, the visceral nerve trunks are twisted and the position of the heart is reversed. The general organization at C is that of the two-gilled prosobranchs belonging to the Archeogastropoda.

Garstang has suggested that the veliger larva gains by torsion, as after torsion it can pull the velum into the shell and fall to the bottom if pursued by enemies. There are serious objections to this idea. As molluscan cilia are generally under nervous central the velum need not be retracted to stop swimming movements. Furthermore, torsion does not occur in other classes of molluscs but their larvae have been able to survive. The mantle cavity is quite plastic, and a far more simple solution would have been to expand the mantle cavity a little to accommodate the velum. If the larva does not profit from torsion,

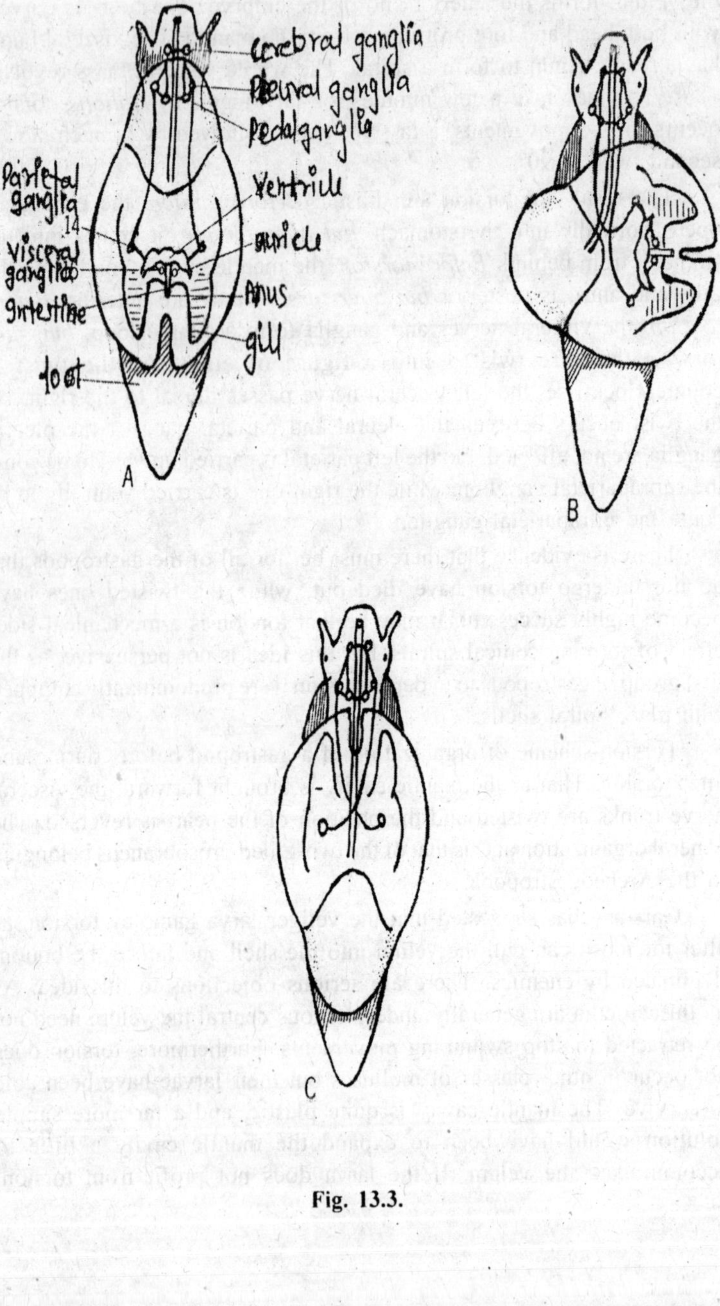
Cerebral ganglia
Pleural ganglia
Pedalganglia
Parietal ganglia
14
Ventricle
visceral ganglia
auricle
Anus
intestine
gill
foot
A
B
C

Fig. 13.3.

presumably the adult must. Torsion brings the mantle cavity to the front of the body. As a result, water currents resulting from forward movement tend to bring water into the mantle cavity, while they work against water entering when the mantle cavity is behind. This would be especially important in freshwater gastropods with the habit of moving against the current in a stream. But there are serious drawbacks to this idea also. Most gastropods are marine, and there is every reason to believe that this was true when the habit of torsion began, so orientation to water currents is probably not an important factor. Most gastropods move so slowly that currents set up by their own movements are not very important factor.

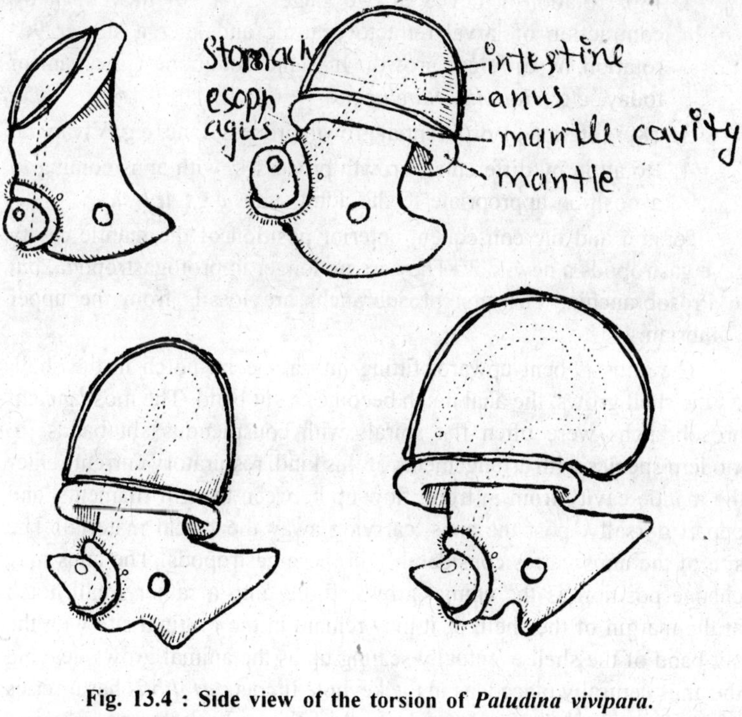

Fig. 13.4 : Side view of the torsion of *Paludina vivipara*.

Besides, many snails have a narrow, tubular siphon for drawing water into the mantle cavity. In these snails, the opening of the mantle cavity is smaller, and any advantage coming from its anterior position is lost. As it stands, there is no really satisfactory explanation for torsion. Perhaps it is no more than a pleiotropic side-effect of some gene constellation that brought other and more important advantages. Be that as it may, there is much to show that torsion created problems.

The most obvious disadvantage of torsion is that the anus is brought directly over the head and threatens to foul the respiratory organs and the head sense organs. It is not surprising to find the first adaptational trend evoked by torsion involved attempts to get the sewer away from the front door. Evidences of this are seen in fossil forms and in some surviving forms that belong to the more primitive groups.

Thomson recognises five ways in which torsion can be brought about:

1. Complete or 180° rotation, achieved by muscle contraction alone, is known only for *Acmaea*.
2. 180° rotation achieved in two stages, first 90° movement by contraction of larval retractor muscle and later a slower 90° rotation by different growth. It is the commonest mechanism today, *e.g. Haliotis, Patella*.
3. 180° rotation by differential growth process done, *e.g.* Vivapara.
4. Rotation by differential growth processes, with anus coming to a position appropriate to the adult state, *e.g. Aplysia*.

Torsion and the consequent anterior position of the mantle cavity gave gastropods a new look. They are no longer in protogastropoda, but in Prosobranchia. The first prosobranchs are fossils from the upper Cambrian.

The anus is bent upward, fitting into a special notch in the shell. As the shell grows, the anal notch becomes a slit band. The most ancient prosobranchs were often flat spirals with conspicuous slit bands. In modern species with arrangements of this kind, respiratory currents enter the mantle cavity from in front, flow up between the gill filaments, and depart dorsally, post the anus, carrying away the faecal material. The site of the anus varies considerably in these gastropods. The anus may change position as the animal grows, fitting into a rather small notch at the margin of the shell, or it may remain in the original site with the slit band of the shell eventually sealing up as the animal grows, leaving the anus centrally placed, as in the keyhole limpets. *Haliotis* has a series of openings in the selenizone. Not all of the archeogastropods have a perforate shell, however. Many modern species have a solid shell, made possible by a new solution to the problem of faces disposal. The gill. auricle, and nephridium on the larval left (adult right) side are reduced and disappear. Why this adoptive tendency should have appeared is not clear, but when completed, a new routing of water through the mantle cavity becomes possible. Water enters on the left, flows over the persisting gill first, and then passes the anus and nephridiopore on its way out of the mantle cavity, sweeping the wastes away neatly enough. It is quite possible that this routing of water through the mantle cavity began before gill reduction, and itself served as a stimulus to reduction of the right gill.

Most gastropods with one gill have also lost the auricle associated with it. They belong to the Mesogastropoda, the largest gastropod order. Mesogastropods appeared first in the lower Ordovician, and may be considered at their peak today. Opisthobranchs and pulmonates are more modern forms that have been derived from the mesogastropoda. As a part of the remarkable adaptive radiation of mesogastropods and the establishment of the opisthobranch and pulmonate stems, several important tendencies have appeared:

1. Many gastropods have further reduced the gill surface. The remaining left ctenidium becomes one-sides, is built into the mantle wall, and eventually may be reduced to a vascular area in the mantle cavity then serves as the respiratory surface, acting as a water or air-breathing lung.
2. Some gastropods undergo detorsion. This reverse movement occurs in development, and while it does not wholly eliminate the results of torsion, it greatly reduces them. Gastropods that have followed this line have no shells or reduced shells, and some have wholly lost the mantle cavity. In this case, the body surface is the site of respiratory exchange, and may bear simple or complex secondary gills.
3. A strong tendency to untwist the nervous system is evident. It grows naturally out of detorsion. But is also achieved by the centralization of the nervous system and the shortening of the visceral nervous.

Subclass Protogastropoda wholly extinct gastropods showing no evidence of torsion, presumably with a posterior mantle cavity, shell conical, plane spirals, or conical spirals.

Order Cynostraca. Protogastropods with conical shells, showing no evidence of spiral coiling.

Examples: Palaeacmea, Scenella.

Order Cochliostraca. Protogastropods with coiled shells.

Example: Pelagielia.

Subclass Prosobranchia. Gastropods which undergo torsion, bringing gills, anus and mantle cavity to the front, primitively with a pair of

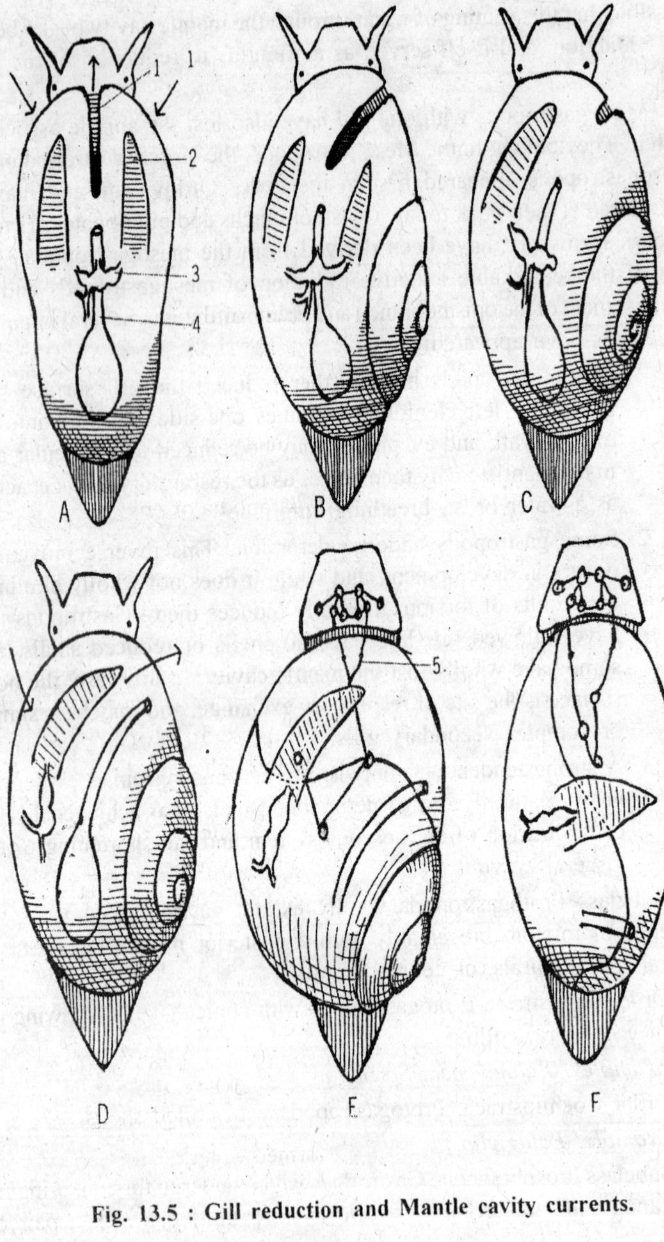

Fig. 13.5 : Gill reduction and Mantle cavity currents.

ctenidia, but usually with a single gill and auricle, primitively with a pair of nephridia, but generally with only the right nephridium retained, nervous system twisted into a figure of eight, which may be partially corrected by some concentration of the nerves and ganglia near the head.

Subclass Opisthobranchia. Marine gastropods with a single auricle and nephridium and with one or no ctenidium, often with surface gills, nervous system untwisted by detorsion and often relatively centralized, with a strong tendency toward shell reduction and a return toward bilateral symmetry.

The shell is a mixed blessing. It provides protection, but at the expense of adding weight. A strong tendency toward the reduction of unnecessary shell weight is seen in many of the gastropods groups. The opisthobranchs show this most clearly. The shell is reduced in association with detorsion, and the amount of detorsion is usually correlated with the mount of shell reduction. They are characterized by shell reduction and loss of spiralling, in some cases, the shell is completely lost. These trends are associated with increased mobility, for many opisthobranchs are pelagic, swimming by means of parapodia or epipodia. The shell is a greater burden on land, and land gastropods usually have very thin shell.

Primitive molluscs have paired digestive glands. Most gastropods have a pair of digestive gland, but it is a mismatched pair as a result of unequal right and left sides associated with coiling. In dextral species, the left digestive gland is larger, while the right is larger in sinistral species. The openings of the digestive glands into the stomach vary greatly Primitively, each gland opens through a separate duct, but the ducts are sometimes united, and in many species the duct openings are subdivided. The digestive glands play an important role in digestion. Even in species where digestion is entirely extracellular, most food absorption occurs in the digestive glands. The digestive glands of carnivores, for example, secrete proteiolytic enzymes that act in the stomach, but most of the food is absorbed in the digestive gland.

After torsion, the intestine emerges from the front end of the stomach. Not much is known of the details of intestinal functions. Some food absorption appears to occur, but the main task is the condensation of the faces. The intestinal wall contains many goblet cells that secrete mucus, helping to compact the faecal materials. Some water absorption aiso occurs, and the faeces are well formed when released. The intestine ends in the anus, which is placed at different points in different types of gastropods. It is far forward and slightly to the right in dextral

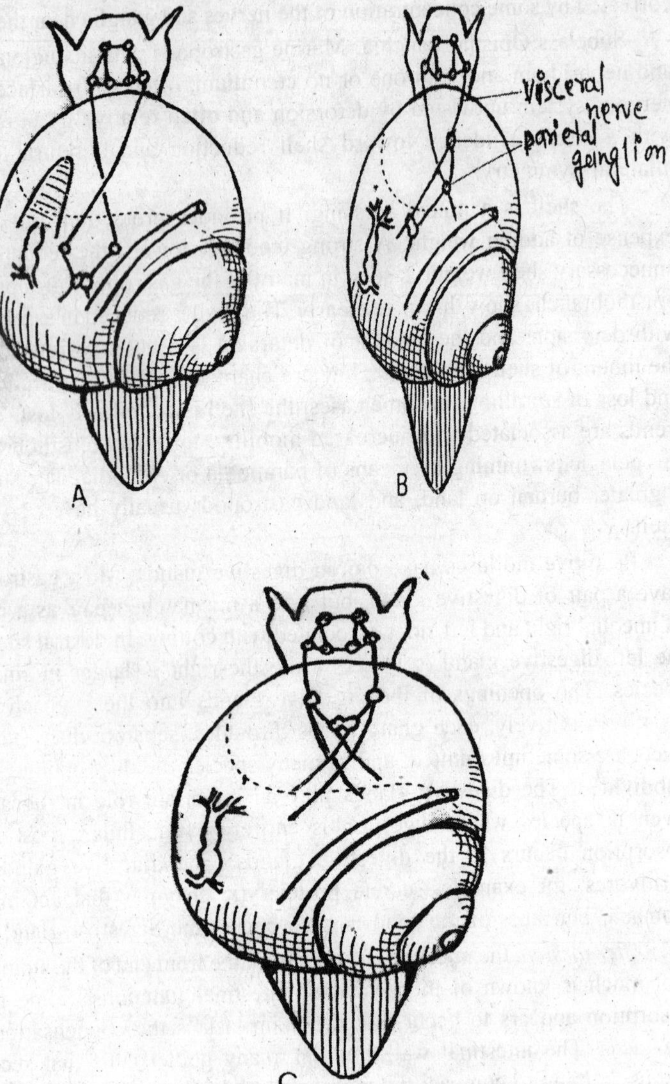

Fig. 13.6 : Modification involved in pulmonate development.

prosobranches. When the slit band disappears and respiratory currents sweep across the mantle cavity, the anus is displaced further to the right. Opisthobranchs arose from prosobranch ancestors and have a single gill. Primitively, the anus is lateral in position, but *detorsion* brings the anus back to a posterior position in some opisthobranchs. A rectal gland, placed some-what in front of the anus, is found in some carnivorous. Its function is uncertain. The intestine of many archeogastropods passes through the pericardial cavity as well as the perivisceral coelom, in much the same manner as in clams. In some of there, the intestine passes through the ventricle of heart.

The primitive respiratory arrangements and the modifications associated with torsion have been described previously. Torsion brings the mantle cavity forward, placing paired ctenidia over the head, with the anus situated between the gills. This arrangement is characteristic of the dibranchiate archeogastropods. In other archeogastropods, the right ctenidium is lost. There monobranchiate archeogastropds, however, retain some traces of the right auricle. The mesogastropods have a similar gill arrangement, but have lost all evidences of the right auricle.

The course of gill reduction is similar as the dibranchiate snails become monobranchiate and as the monobranchiate snails develop a lung. In the dibranchiate haliotus, the right ctenidium is somewhat smaller than the left. The right ctenidium of scissurella is monopectinate, with gill filaments on one side of the axis and adherent to the mantle wall.

As detorsion and shell reduction occur, opisthobranchs tend to lose their mantle cavity, some have a single, reduced ctenidium and a single auricle, demonstrating that they have arisen from prosobranch ancestral stocks. Many, however, have no ctenidium, depending on the body surface or secondary gills derived from the body surface for respiratory exchange.

Primitive gastropods have a pair of U-shaped metanephridia, open to the pericardial cavity through nephrostomes. The right nephridium is also used for the release of gametes. All of the archeogastropods except the neritas retain a pair of nephridia, but in all other gastropods the right nephridium is sharply reduced, remaining only in the form of a gonopericardial canal. This canal connects the gonoduct and pericardial cavity, or has become an integral part of the gonoduct itself. Gastropods are the only molluscs to establish themselves on land successfully, and one of the two groups that adapted to freshwater. Evidently they have some capacity for the regulation of water and salt concentration not shared by other molluscs. As a result, their excretory physiology is of considerable interest. Land snails have considerably more salts in their blood than do fresh-water snails but have poor mechanisms for controlling the concentration. A heavy rain may reduce the salt concentration of the blood by over a half. Water loss resulting from excretion is held to a minimum by excretion of uric acid and by a few adaptations that reduce evaporation at the body surface. Hibernating or estivating snails, for example, secrete a portion over the shell aperture, thus reducing water loss during inactivity.

The arrangement of the primitive circulatory system has been described. Paired ctenidia, located in a posterior mantle cavity, drain into a pair of auricles, and an anteriorly directed aorta leads from the ventricle. After torsion, however, the auricles lie in front of the ventricle, and the aorta extends posteriorly. This is the arrangement found in disbranchiate archeogastropods. As the right ctenidium dwindles and is eventually lost, the right auricle dwindles with it, although the monobranchiate archeogastropods retain some vestiges of a right auricle. In prosobranchs, however, all traces of the right auricle have been lost. The left auricle drains blood from the left ctenidium and the heart retains its position, with the venticle discharging into a posteriorly directed aorta. This arrangement is retained in the pulmonates, although the auricle receives blood from the vascular mantle wall rather than a ctenidium. Among opisthobranchs, detorsion returns the auricle to a posterior position, but they have a single auricle and have evidently arisen from mesogastropods stocks.

Blood returning from the ctenidia secondary gills, or lung is oxygenated, so the heart is an arterial heart, pumping blood to tissues for the distribution of oxygen and pick-up of carbon dioxide. The short aorta gives rise to a cephalic artery conducting blood to the head and foot, and a visceral artery, which empties into lacunal around the kidney. In some of the putmonates, however, blood returns directly to the auricle from the kidney.

A patch of sensory cells, the osphradium, occurs at the base of each gill. The osphradium is often a strand of elevated sensory cells associated with the nerve to the cteridium, but in some cases it is in corporated in the ctenidial axis. Sometimes lateral tracts of elevated cells extend on each side of a central strand, giving the osphradium a bipectinate appearance. Gastropods without ctenidia retain an osphradium if a mantle cavity and secondary gills are present, Aquatic pulmonates have an osphradium, but land pulmonates do not due to the process of torsion and detorsion.

Nervous system. It is easier to understand the form of the gastropod nervous system before torsion.

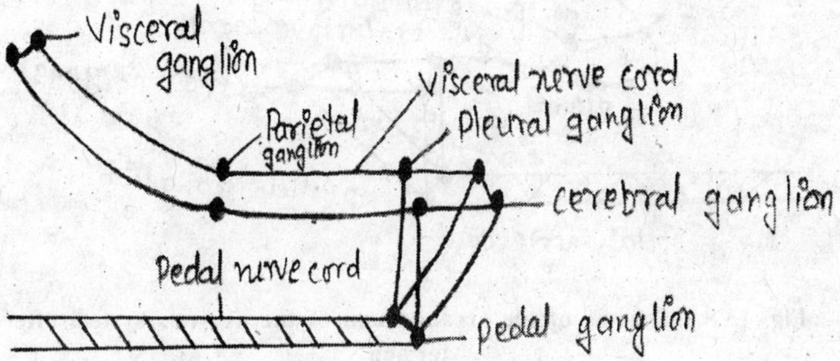

Fig. 13.7 : Scheme of the arrangement of the nervous system before torsion.

Three pairs of ganglia lie near the esophagus. The cerebral ganglia lie above the esophagus, the pedal ganglia lie below it in the anterior midline of the foot, and the pleural ganglia are more laterally placed. Cerebropedal, cerebropleural and pleuropedal connectives link the ganglia, forming a neutral triangle on each side. A pair of small buccal ganglia are attached by buccal connectives to the cerebral ganglia. A pair of pedal nerves arises in the pedal ganglia, and extend back in the foot. They are connected by cross commissures, and are ganglionated at the junction of commissures and nerves. A pair of palliovisceral nerves arise from the pleural ganglia and extend back into the visceral hump. Typically, a parietal ganglion is found on each of the palliovisceral nerves, and each nerve ends in a visceral ganglion. Cross ganglia, the two pedal ganglia, and the two visceral ganglia.

Torsion brings the visceral ganglia forward, twisting the palliovisceral nerves between the pleural and parietal ganglia, but does not affect the rest of the nervous system.

The visceral and parietal ganglia on the larval (pretorsion) right are brought to the left side of the adult. The adult left parietal ganglion is higher than the right and is the supraparietal ganglion, the right parietal is the infraparietal ganglion. All modern gastropods undergo torsion, so it must be considered primitive for them. Undoubtedly the ancient gastropods had a nervous system that was not twisted and they were organized as the larvae are before torsion.

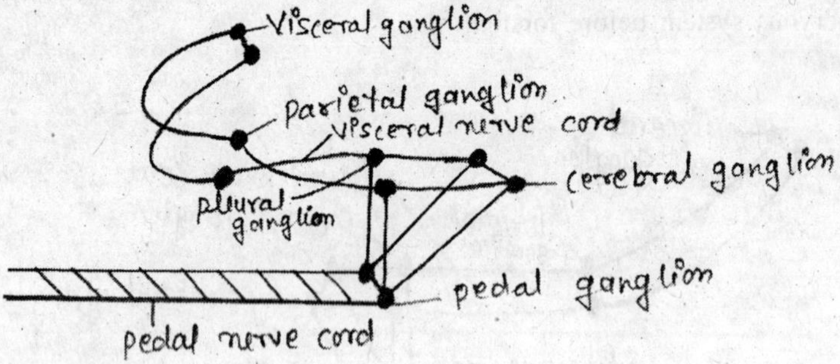

Fig. 13.8 : Scheme of the arrangement of the nervous system after torsion.

A nervous system twisted in this manner is said to be streptoneurous, and is characteristic of the more primitive of the modern gastropods. In many gastropods, however, a more or less complete return to bilateral symmetry of the nervous system is achieved by one of two methods:

1. Detorsion, which untwists the visceral nerve and brings the parietal ganglia back to their original position, and
2. Shortening of the connectives between the pleural and parietal ganglia, which hauls the parietal ganglia back into a symmetrical position. In either case, the nervous system is said to be euthyneurous.

Untwisting of the visceral nerve by detorsion is characteristic of opisthobranchs. The results are seen in *Aplysia*.

Other changes somewhat complicate the picture, however, for the ganglia have tended to be come concentrated at the prominent ganglia have disappeared from the shortended visceral nerves. Pulmonates regain bilateral symmetry by shortening the visceral loop and the connectives. As a result the visceral ganglia unite with the pleurals and both may unite with the cerebral ganglia, forming a complex brain from which peripheral nerves pass to the various parts of the body.

The body is developed by the growing mantle, and the internal organs grow rapidly on the dorsal side. This brings the anus forward, toward the aperture of the larval shell, and gives the digestive tract its characteristic, sharply U-shaped form. It is at about this time that torsion occurs.

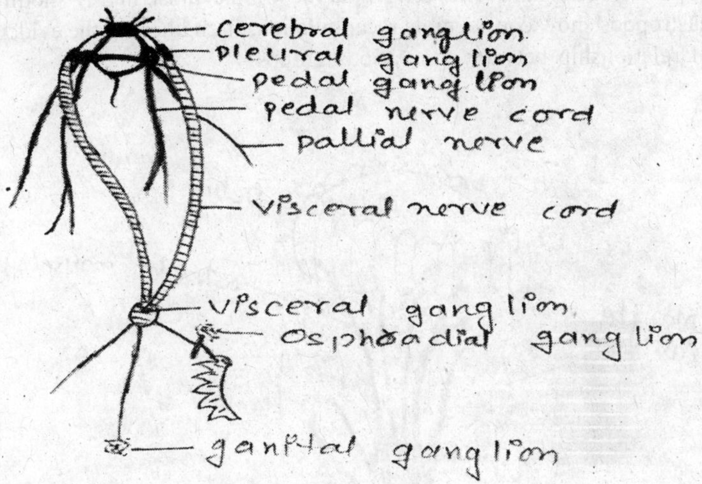

Fig. 13.9 : Untwisting the visceral loop is detorsion as occurs in opisthobranchs. This leavs a long, but untwisted visceral loop.

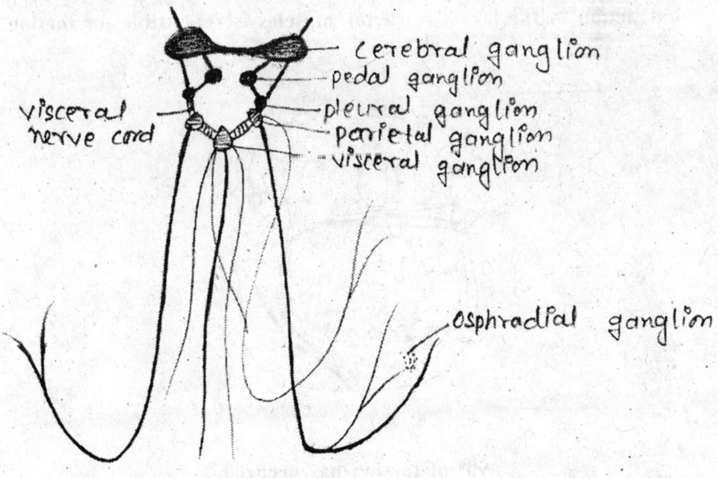

Fig. 13.10 : Straightening the visceral loop is by concentration of the ganglia, bring the visceral ganglia forward.

Most molluscs pass through a veliger larva stage. The most unusual feature of the gastropod veliger is torsion, the details of which need not be described again. The veliger larvae of the most highly modified gastropods show considerable recapitulation, strengthening the evidence of relationship between the various groups.

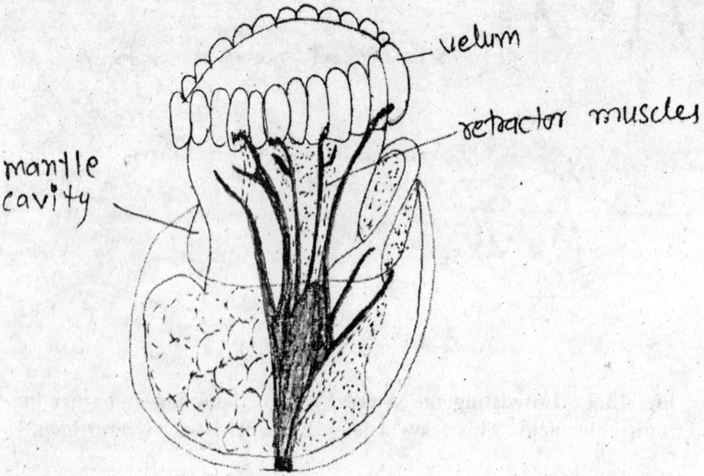

Contraction of the larval retractor muscles is reponsible for torsion.

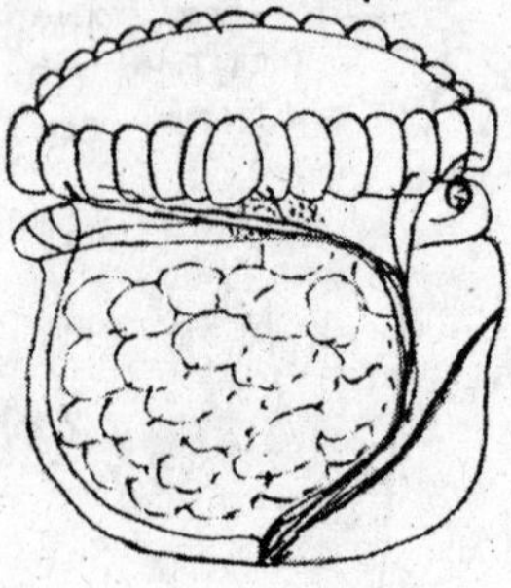

90° of torsion has occured.

. Fig. 13.11.

The young shell beings to form, and mantle folds enclosing a mantle cavity appear. Subsequently six larval refractor muscles develop as seen at figure.

In later years Garstang's theory was strongly supported by Crofts observation that the torsion process in the larval of recent primitive gastropods is indeed due to differences in the development of the two refractor muscles. The (larval) right muscle is formed first and is asymmetrically attached to the head. When is fully differentiated it contracts and the first floor is rotated counterclockwise. The left muscle then develops and this become the permanent shell-retractor muscle, found in all adult gastropods.

Serious criticisms of Garstang's theory can be advanced. Some mirror points are that there in no special reason to suppose that the gastropods arose from snails with a low uncoiled shell, and that it is unknown whether the oldest snails had free-living pelagic larval. Furthermore, Gastrang was unclear with respect to the dangers encountered by the supposed larval. Later authors suggested that the ability to retract the twisted body into the shell and to close it with the operculum was of retract the twisted body into the shell and to close it with the operculum was of adoptive advantage to the larval as a protection from predators, probably by allowing them to survive passage through the gut of these animals. However, no pelagic predators are known to have existed in the relevant period of the Cambrian or Pre-Cambrian.

Besides being rotated the first floor is asymmetrical in several respects, such as the dextral coiling of the shall and many, more or less pronounced, unilateral reductions of the paired organs. In every lower gastropod specimen the torsion-process still occurs during ontogenesis.

The second group of snail from the Cambrian are the Helcionellacea. These are generally referred to the gastropods, which implies that they were supposedly torted. It is later described in details.

Undoubted gastropods from the Cambrian are also known. Most of them belong to the Pleurotomariacea and they had an asymmetric conispiral shell with a sinus, slit or series of holes. All authors agree that these animals were toned because shells of this type can only be held in an endogastric position, with the apex pointing backwards.

Small representatives of all four groups mentioned—the Monoplacophora, the questionable Heleionellacea and Bellerophontacea, and the Gastropoda—already occurred in the Lower Cambrian or even at the end of the Precambrian (Runnegar & Pojeta, 1974; Stanley, 1976). Later on only the gastropods were successful. Why?

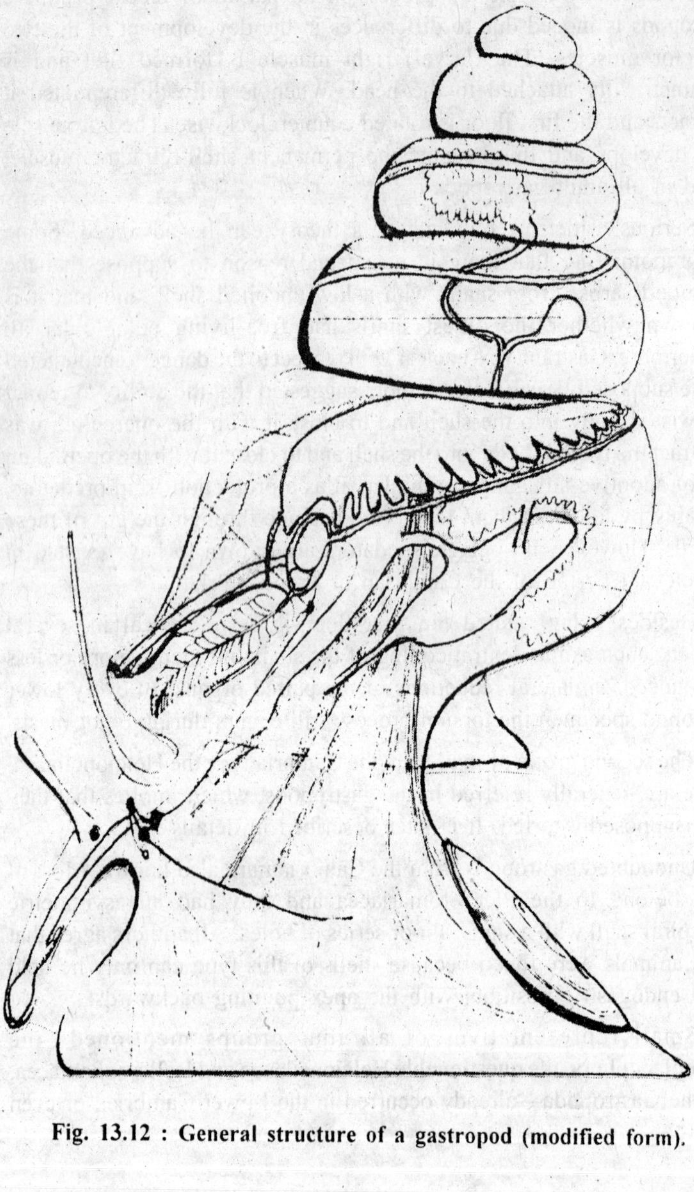

Fig. 13.12 : General structure of a gastropod (modified form).

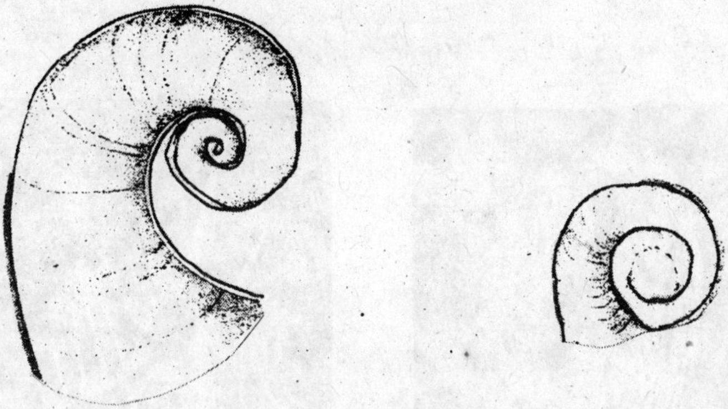

Fig. 13.13.

In recent years some modifications of the old themes have been proposed. Ghiselin (1966) agreed with Naef that torsion arose during the transition from pelagic to benthic life, but he rejected the latter's view that it occurred in adult forms. He thought that torsion originated during the settling of larvae. They would have had trouble in balancing their heavy shells and the exogastric coils of the shell would have interfered with their locomotion. As a consequence the visceral hump and the shell heeled over and rotated toward the rear, thus making locomotion easier. This would have been sufficient advantage to favour these processes selectively. Retention of the rotated position during later stages would have had the same advantages, and a loss of the mechanism for returning the shell to its original, pre-torsional position would explain continued torsion in the adult snails.

This is an ingenious theory, but, when studied in detail, many questions arise: did the early gastropods have pelagic larvae; if settling larvae encounter such severe troubles and if torsion could arise so easily why have there been so many untorted snails in the past; why is the settling of recent torted larvae still such a troublesome affair; why is, as was mentioned before, in many primitive gastropods torsion not completed in the larvae before settling; and the rather vital, inevitable and still unanswered question: what was the advantage of the torted condition for the adult snails so that no back-rotation occurred? Finally, Ghiselin's theory has the same drawback as that of Naef in that it does not account for the direction of torsion. In some larvae the shell would have fallen over to the left, in others to the right. Ghiselin has seen this

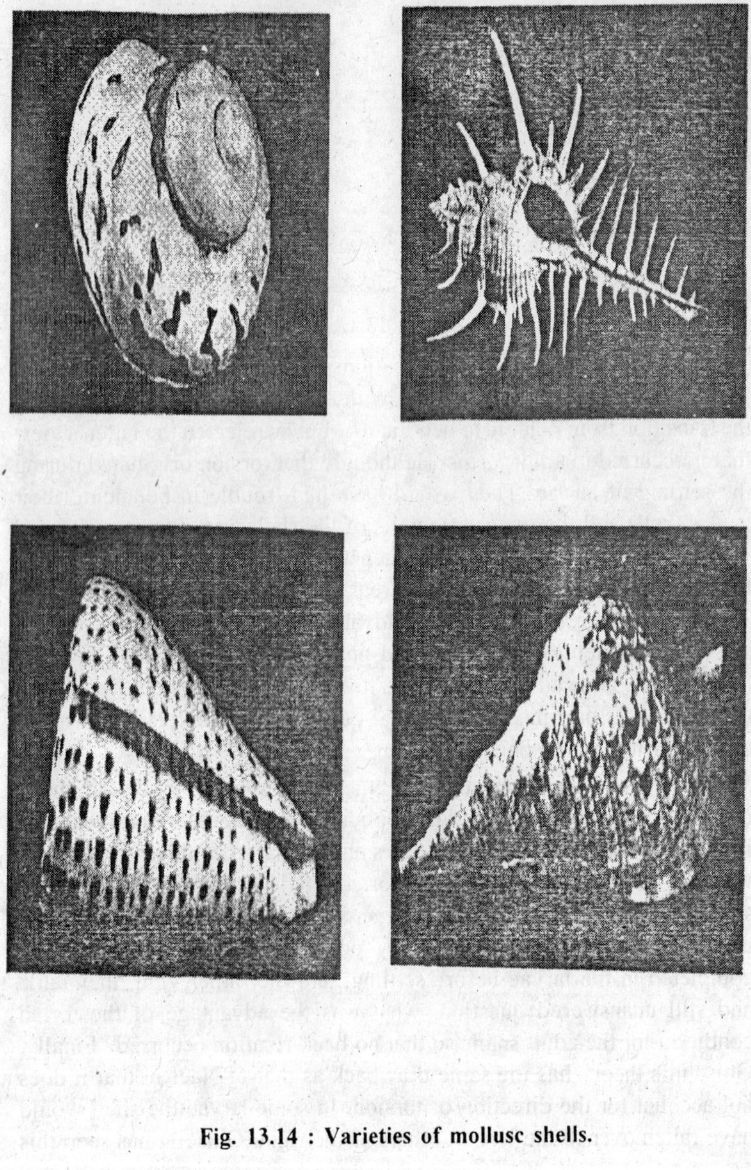

Fig. 13.14 : Varieties of mollusc shells.

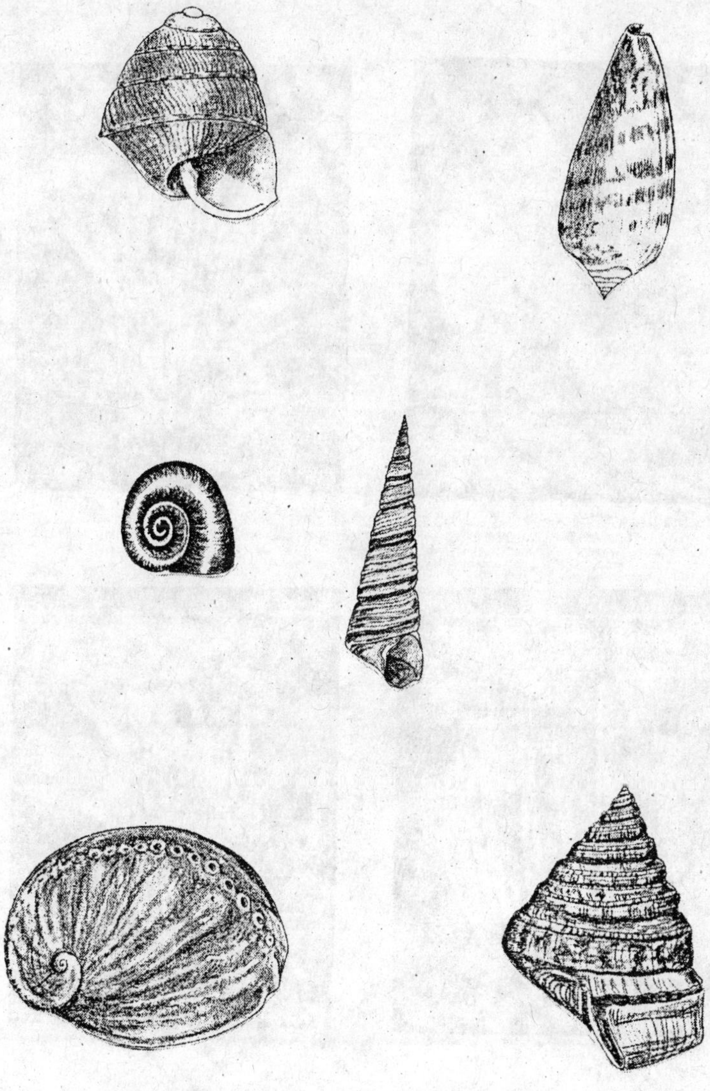

Fig. 13.15 : Some gastropods after complete torsion.

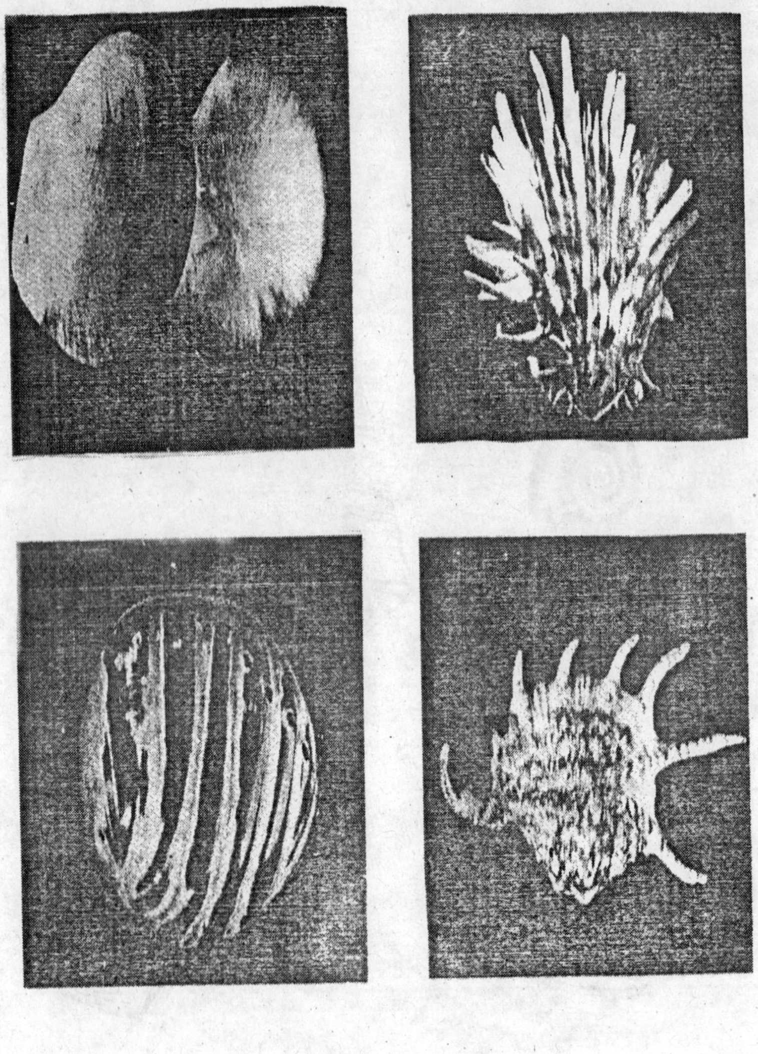

Fig. 13.16 : Varieties of mollusc shells.

difficulty and, therefore, supposed 'a kind of behavioral preadaptation, in which the untorted, settling larva twisted the shell to one side while crawling about.'

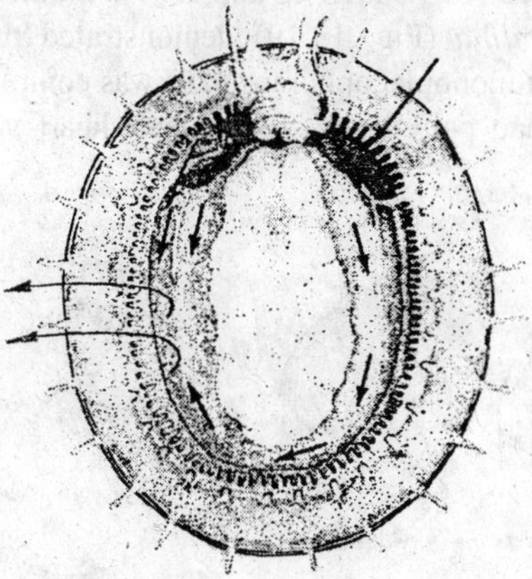

Fig. 13.17 : *Patella vulgata*. Arrows show direction of water currents (modified from Fretter & Graham).

A last theory which will be mentioned here is that of Underwood (1972). He coiled to find the solution by supposing that the first half of torsion may have evolved in the larva where it may have furthered the balance and the relative position of the velum in relation to the bulk of the shell, and the second half as an adaptation to suit unclear needs of the adult. This, in itself interesting, theory introduces few new aspects to the present discussion.

From the survey of the most prominent torsion theories it appears that none of these has satisfactorily answered the questions of the origin and the adaptive function of torsion, either for the larvae or for the adult snails. After so many failures one is tempted to conclude that torsion had no primary function at all, and that perhaps the problem should be tackled from a different angle. Before trying this some palaeontological facts must be discussed.

PALAEONTOLOGICAL FACTS

Many fossils from the group Monoplacophora are known from the entire Cambrian period (Knight & Yochelson, 1960). These animals occurred in shallow epicontinental seas. Scars on the inner surface of the shells show that these snails had many dorso-ventral retractor muscles, or one horseshoe- or ring-like muscle. This indicates that they were definitely untorted. Moreover, the bilaterally symmetrical shells had an anterior apex and tended to be exogastrically curved (Fig. 13.18). Some species were even coiled (Batten et al., 1967). Studies on the recent representative *Neopilina* (Fig. 13.19) demonstrated that the internal anatomy of the adult monoplacophoran snails was completely bilaterally symmetrical: they had paired gonads, kidneys, heart ventricles, etc.

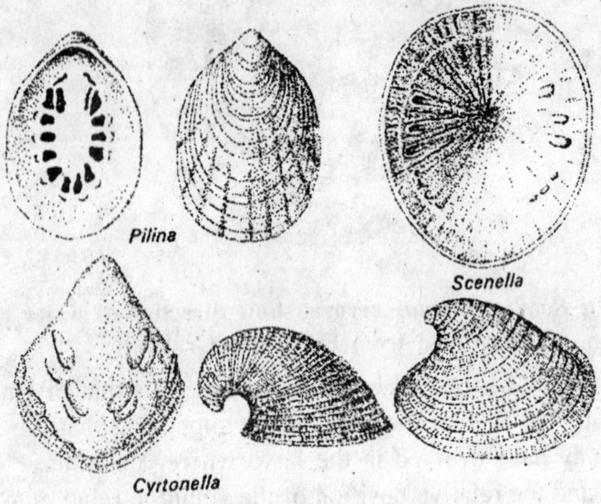

Fig. 13.18 : Fossil Monoplacophora (modified from Knight & Yochelson, 1960).

The second group of snails from the Cambrian are the Helcionellacea (Fig. 13.20). These are generally referred to the gastropods, which implies that they were supposedly torted. In general they had cup-shaped or bilaterally symmetrically coiled shells. The reasons given for the opinion that they were torted are rather vague and weak. The problem is difficult to solve because muscle scars are as yet unknown. Runnegar & Pojeta (1974; see also Runnegar, 1977) recently described a Lower Cambrian helcionellid, *Yochelcionella*, which had a strange tube attached to the concave side of the shell below the apex (Fig. 13.21). This tube probably carried water in or out of the mantle cavity. In a torted snail the tube would point backwards and could not be of use for the anteriorly located mantle cavity, but in untorted animals the tube could be used to channel water into the anterior part of the mantle cavity. From this and other arguments it has been concluded that the Helcionellacea were rnonoplaeophorans, not gastropods.

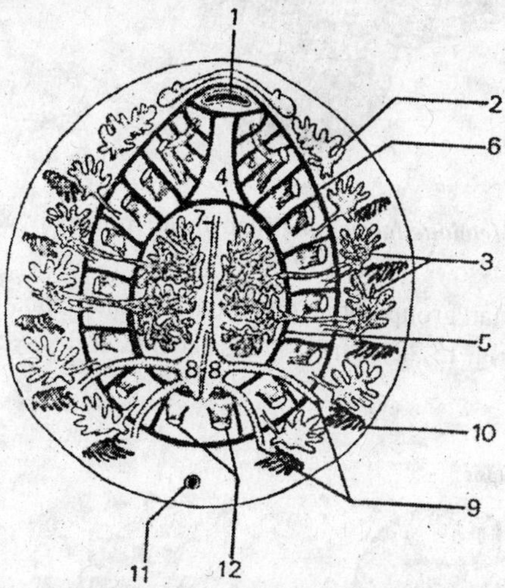

Fig. 13.19 : Anatomy of *Neopilina* (modified from Hyrnan, 1967). 1, mouth 2, nephridia. 3, nephridia serving gonads. 4, ponads. 5. pedal nerve cord. 6, lateral nerve cord. 7, aorta. 8, ventricles. 9, auricles. 10, gills. 11, anus. 12, shell retractor muscles.

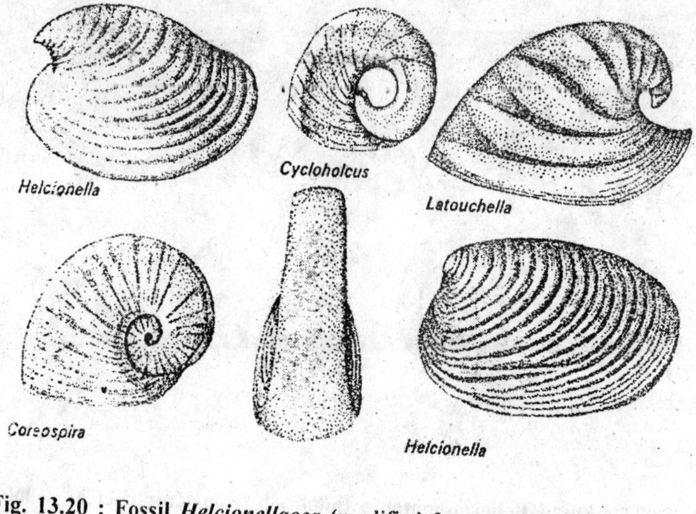

Fig. 13.20 : Fossil *Helcionellacea* (modified from Knight et al., 1960).

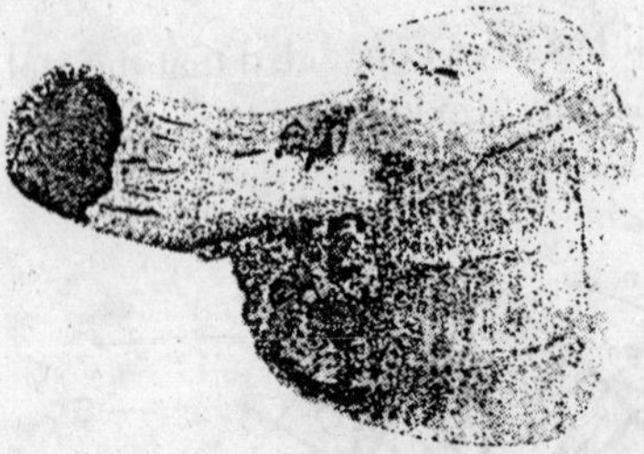

Fig. 13.21 : Fossil *Helcionellacea* (modified from Runnegar & Pojeta et al., 1974).

Another Cambrian group with a disputed systematic position are the Bellerophontacea (Fig. 13.22). These snails had bilaterally symmetrically

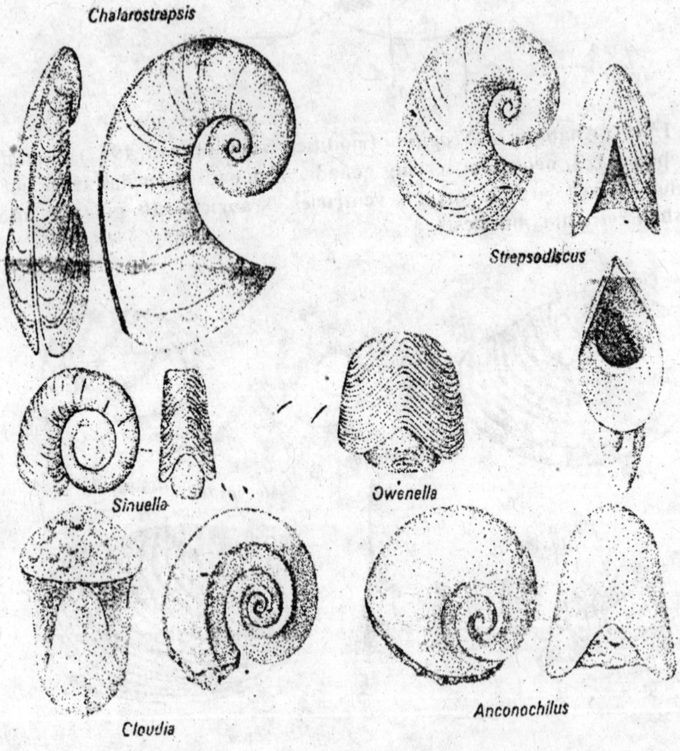

Fig. 13.22 : Fossil Bellerophontacea (modified from Knight et al., 1960).

coiled shells with a median sinus, slit or series of holes. Notwithstanding their completely bilaterally symmetrical shells, it was till recent years generally supposed that Bellerophontaceans were torted (Cox, 1960; Knight et al., 1960). Two main arguments were used. Firstly, like Naef, the palaeoniologists held the view that snails could not live efficiently with exogastric coils overhanging their head.

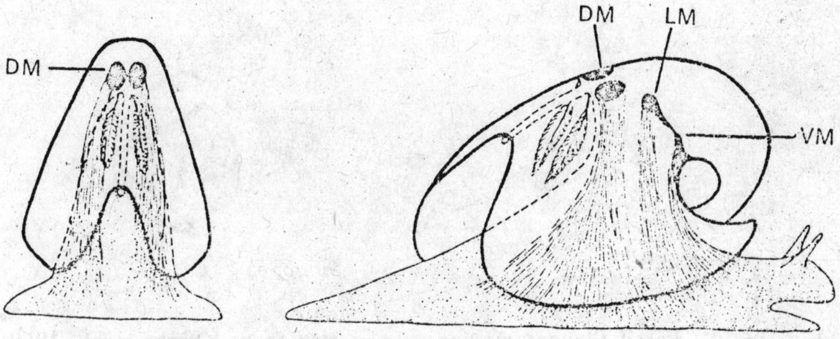

Fig. 13.23 : Reconstruction of *Sinuitopsis acutilira* (modified from Rollins & Batten, 1968). DM, dorsal retractor muscle scar. LM, lateral retractor muscle scar. VM, latero-ventral retractor muscle scar.

Secondly, like Garstang, they thought that the slit must be a post-torsional feature: no untorted mollusc would need a posterior slit. The incorrectness of the latter point is already apparent from the fact that some recent chitons have a posterior slit (Hyman, 1967: 73). However, the whole argument collapsed when fossil bilaterally symmetrical shells were discovered (Fig. 13.23) with several pairs of muscle scars — thus originating from untorted snails—which coiled exogastrically and often had a deep posterior sinus. It is, therefore, quite possible that the whole group of the Bellerophontacea may have been untorted and must be transferred from the Gastropoda to the Monoplacophora (Runnegar & Pojeta. 1974).

Undoubted gastropods from the Cambrian are also known (Fig. 13.24). Most of them belong to the Pleurotomariacea and they had an asymmetric conispiral shell with a sinus, slit or series of holes. All scientists agree that these animals were toned because shells of this type can only be held in an endogastric position, with the apex pointing backwards.

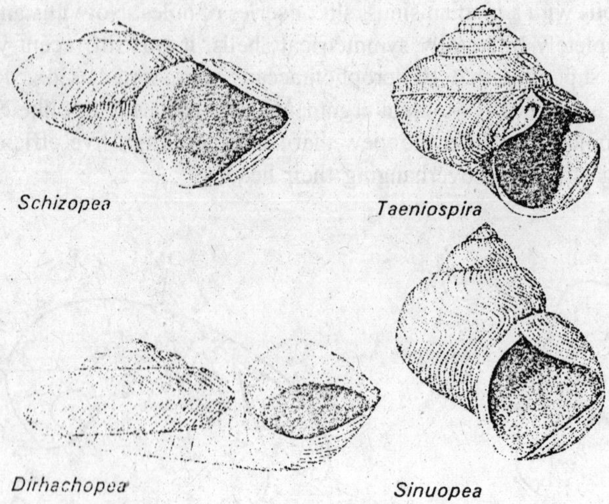

Fig. 13.24 : Fossil Pleurotomariacea (modified from Knight et al., 1960).

AN INTEGRATIVE APPROACH

After analysing the main theories and summarising the palaeontological facts we can now try to find an integrative approach to the problems under discussion. Three introductory points are of importance.

Firstly most recent scientists, in agreement with Naef, are of the opinion that the gastropods arose from snails which had a bilaterally symmetrically coiled shell, therefore, either of the bellerophontacean, the heleionellacean or the monoplacophoran type, and had, furthermore, a reduced number of retractor muscles. Secondly, we should bear in mind that all theories previously discussed primarily attempted to find an immediate adaptive value of torsion. The results were not very satisfactory and it seems that torsion itself might have had no function at all. Thirdly, several theories—those of Lang, Naef and Ghiselin—supposed that as an introduction to torsion the shell of the creeping snail fell over sideways, but they did not explain why the shell fell to the left, so that a counterclockwise rotation was effected. Moreover, these theories met genetic difficulties, because a torsion in only one direction could not become fixed into the blueprints of a bilaterally symmetrical animal if the shell fell at random to the right or the left.

It is, therefore, logical to suppose that some kind of asymmetry preceded torsion. In fact, the theories of Garstang, Solem and Ghiselin imply this: asymmetry of the retractor muscles and of the shell and a 'behavioral preadaptation', respectively, were thought to precede torsion. What kind of asymmetry could this have been? This question is difficult to answer, but it can be expected that many if not all paired organs of the visceral hump were involved. There are reasons to suppose that the asymmetry affected the sequence of development as well as the size of the left and right organs. It has already been mentioned that in *Haliotis* the right retractor muscle precedes the left. Similarly, in the pre-torsional larvae of this snail only the left lobe of the digestive gland is developed and this occupies the whole of the left and most of the dorsal part of the visceral mass, whereas when torsion is completed a larger second lobe originates, which comes to lie on the right side of the stomach and later on curves around to the left of the columellar muscle (Crofts, 1937). On the other hand, in the recent pleurotomariaceans the right kidney is well-developed and the left one is reduced, whereas in all dextrally coiled gastropods the reproductive system is developed only on the right.

Seeing that in recent adult primitive gastropods not only the right parts of the mantle and the shell are larger than the left but also that most of the large organs which occupy the coils are right organs—all together representing, in principle, the pre-torsional left part of the visceral mass–, it can be supposed that the original asymmetry apart from its effects in the embryos, caused a reduction of the right part of the visceral mass of the bilaterally asymmetrical adult snails.

If the asymmetry did not already cause a complete torsion during the embryonic stage, the coiled shell of the creeping snails would have become unbalanced and would have fallen over to the left side, which finally could only have resulted in a counterclockwise torsion. Moreover, the differences in size and growth between the two halves of the visceral mass would lead to a conispiral shell, which in most counterclockwise torted snails is dextral. The relation between the asymmetry of the visceral organs and the directions of torsion and of the coiling of the shell is also indicated by the fact that snails with sinistrally coiled shells are the mirror-images of dextral forms in every respect: torsion occurs clockwise, the reproductive system is only developed at the left side, etc. Raven (1972) showed that the diherence between dextral and sinistral forms is evident as early as the first cleavage of the eggs. Apparently all these asymmetry components are fundamentally interrelated in the basic structure of the gastropods.

The view that torsion merely arose as a consequence of a genetically based asymmetry of the visceral hump and initially might have had no function at all, invites a number of questions. The first is whether the assumption that asymmetry could occur in symmetrical animals without a clear cause or meaning, can reasonably be accepted. The answer is that many animals with externally symmetrical bodies show internal asymmetries. This is also the case in molluscs. For instance, *Nautilus* has paired reproductive tracts, but in both sexes the left tract is reduced. Some recent colcoid cephalopods have paired tracts, but in the others the right one has disappeared. In scaphopods the single gonad always evacuates its products via the right kidney. Bivalves also often show asymmetries in various body parts. That asymmetry could easily lead to torsion in gastropods and not in other molluscs, is due to the fact that they have to balance a visceral hump and a shell upon a flexible column above their head-foot.

The second question refers to the adaptive value of the complex of changes. The problem concerns not so much the initial function of the transformations because in the Late Precambrian and at the beginning of the Cambrian the selective pressure will have been low, since the number of animal species was relatively small and competition negligible and predators were probably lacking (Stanley, 1976). Therefore, the most important question is what the final adaptive function of the more or less complete gastropod structure could have been, compared to that of the original bilaterally symmetrical and untorted snails. Because the discussion of the theories showed that the function of torsion is unclear, the attention must primarily be focussed on the asymmetry of the visceral hump. There is no reason to suppose that the unilateral reduction of the organs could have been very important. Therefore, the form of the shell must be considered. What could be the adaptive value of an asymmetrical conispiral shell? Nearly, all recent archaeogastropods live upon rocky shores and are herbivorous animals, living on algae. They occur mainly in the upper metres of the sea. They and in fact all other snails occurring in this area, have some kind of conispiral or secondarily uncoiled conical shell. Apparently shells of this shape are excellently adapted to strongly moving water from an hydrodynamic point of view. It is evident that bilaterally symmetrically coiled shells, especially if high and large, would be less efficient under these conditions. This means that at all times the gastropod body-structure was better adapted to these conditions than that of high bilaterally symmetrically coiled snails. This already answers the question. There is, however, a reason to suppose that this aspect could have been of special importance in Late Precambrian and Cambrian times. In recent years it has become clear that the primeval atmosphere lacked oxygen (Berkner & Marshall, 1965). After the origin of photosynthesizing organisms the concentration of oxygen gradually increased. At first the concentration was so low that, due to the violent ultra-violet radiation, insufficiently absorbed by oxygen, life could not exist in the upper 10 metres of the sea. Rutten (1971) held the opinion that at that time planktonic life was not possible and that in the oceans living organisms only occurred at a depth of 10-50 metres, in a narrow zone along the shores. It is thought that the upper 10 metres of the sea gradually became inhabitable at the beginning of the Cambrian or at the end of the Precambrian, and that some 200 million years later terrestrial life became possible.

If these new insights are related to the problem of the origin of the gastropods it can be summerised that the earliest snails lived along the coasts at a depth below 10 metres, uninfluenced by the waves above them, so that the form of their shells was relatively unimportant. Gradually, however, the best grazing zone rose higher and higher and, finally, even entered the tidal area in many places. During this important transitional phase of the history of the earth and of the animal kingdom, when empty adaptive zones could be invaded, the gastropods with their asymmetrical conispiral shells had an advantage over the completely bilaterally symmetrical and untorted snails which only could have existed in the littoral and sublittoral rough water zones of rocky cliffed shores as uncoiled conical or as very flat forms. The less well adapted snails stayed behind in the deteriorating lower zones.

In this approach palaeontological facts and current opinions on the environmental conditions at the end of the Precambrian and the beginning of the Cambrian, have been integrated with knowledge of the anatomical tendencies and ecological aspects of recent primitive gastropods.

It must be stressed that these ideas, besides the points already mentioned, do not in themselves contradict all aspects of the older torsion theories: perhaps an endogastric shell is more comfortable to carry than an exogastric shell, or more pleasant for swimming, or easier for settling larvae (Ghiselin, 1966). The problem was, however, how this favourable position could be attained. The opinions of Plate (1896) and Thiele (1901, 1935), who supposed that torsion arose as a result of an enlargement of the left lobe of the digestive gland or of the left gonad, respectively, arc not completely contradicted either, because they correctly saw that asymmetry had to precede torsion. Moreover, Thiele made some excellent remarks on the ecological aspects of the problem.

However, a vital point in which the present approach differs from many of the previous torsion theories is that no speculations on hypothetical ancestral forms have been used.

Although, the ideas which have been discussed provide an answer to the questions of how torsion evolved as a genetically transferable characteristic and why it is in a counterclockwise direction, the present theory does not contain all the answers to the old problem. One must remember that many important questions remain to be considered, for instance: how are the processes discussed related to asymmetries in the early stages of gastropod development, where did operculum come from and how did detorsion of optic the branch snails originates. Torsion in itself had no direct function. Asymmetry of conispiral shell because advantageous for the gastropods in highest grazing zones of rocky shores.

THE ORIGIN OF ASYMMETRY IN GASTROPODS

Asymmetry and Torsion

Among molluscs, the gastropods occupy a unique position on account of the asymmetrical organization of the body. Boutan (1885) was the first to describe that this asymmetry was brought about during development of the embryo by a torsion of the posterior part of the body. The original bilateral symmetry of the embryo is changed secondarily by a rotation of the visceral hump with respect to the head-foot complex, which remains bilaterally symmetrical. The visceral hump, connected by a narrow stalk with the cephalopedal mass, rotates in an anti-clockwise direction in the dextral gastropods, but clockwise in sinistral species. The mantle cavity, originally situated ventrally behind the foot, travels along the right side of the body and becomes situated dorsally, arching over the head. Torsion is preceded by a ventral flexion, by which the gut, which originally is straight, becomes U-shaped. In this way the anal region is displaced ventral and approaches the mouth. During torsion the intestine moves upwards and the anus is displaced dorsally to end in the mantle cavity.

According to Crofts (1955), who studied the torsion process in the archaeogastropods *Haliotis, Patella, Patina* and *Calliostoma*, torsion proceeds in two phases. The first is rather short (3-6 hrs *Haliotis* and *Calliostoma*; 10-15 hrs in *Patella*). During this phase the visceral hump is rotated about 90 degrees (slightly more in *Calliostoma*) with respect to the cephalopedal axis. The mantle cavity, originally situated posteriorly behind the foot, moves to the right side. In the first phase of torsion the larval retractor muscle is supposed to play an important role. It consists of six spindle-shaped cells, all attached to the original right side of the shell near the apex, in *Calliostoma* more anteriorly along the right side. Three of these muscle cells curve dorsally over the primitive gut in a slightly spiral fashion and reach the left side, where two of them become attached to the left side of the foot rudiment. By contraction of this muscle the visceral hump rotates about 90 degrees and the shell becomes half endogastric. The second phase of torsion is a rather slow process, taking about 30 hrs. *Patella, Patina* and *Calliostoma* and about 200 hrs in *Haliotis*. During this period the pallial cavity expands gradually in a dorsal direction above the head, while the shell is increasing in size, especially at the right dorsal side.

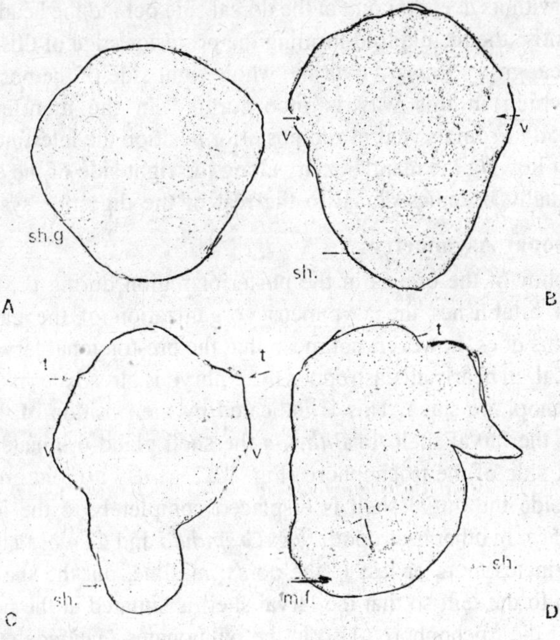

Fig. 13.25. Embryos of the *snail Bithynia tentaculata* stained with silver nitrate, (a) Early trochophore stage, dorsal view. The shell gland (sh.g.) is situated at the dorsal, left side of the posttrochal region, (b) Older trochophore, dorsal view. The shell gland with the larval shell (sh.) is displaced to the left side. (c) Young veliger stage, dorsal view. Shell (sh.) completely at the left side of the posttrochal region, which has impressed appearance, due to a flattening of the right side. (d) Older veliger, viewed from the right side. A mantle cavity (m.c.) formed and extends at the dorsal and right side. f, foot. m.f., mantle fold, t. tentacle, v, veliger.

In most gastropods, the larval retractor muscle does not develop before torsion is completed, and consequently it cannot play a role in the process of torsion. Torsion is brought about by differential growth. In *Paludina* and *Pomatias* the mantle cavity originates from two special grooves on the postero-ventral surface of the visceral mass. The right groove is considerably larger than the left. The mantle cavity extends along the right side and eventually becomes anteriorly situated over the head. In other gastropods *e.g. Theodoxus*, the mantle cavity originates directly at the right side of the posttrochal region. In *Bithynia* the formation of the mantle cavity starts with a flattening of the whole right side, causing the posttrochal region to have a compressed appearance, when viewed from the dorsal side (Fig. 13.25c). A very shallow funnel is formed with its deepest point at the dorsal side behind the head region. Subsequently, the mantle fold forming the posterior edge of this funnel-shaped area, grows forward over the whole right side of the posttrochal region, which in this way is incorporated in the mantle cavity (Fig. 13.25d). From its ventral and posterior position the intestine moves upwards following the mantle cavity along the right side of the embryo, and eventually becomes dorsal to the rest of the digestive system.

Pre-Torsional Asymmetry

The shift of the organs in the posterior region during the process of torsion establishes the asymmetric organization of the gastropod veliger. This does, however, not mean that the pre-torsional larva is still symmetrical. In nearly all gastropods the embryo is already asymmetrical at the trochophore stage. This is indicated by the position of the shell gland and the larval shell. In *Bithynia* the shell gland originates at the dorsal left side of the trochophore (Fig. 13.25a). By a rapid growth of the right side the initial shell is displaced completely to the left side (Fig. 13.25b). In other gastropods, *e.g. Crepidula* and *Ocinebra* the shell gland originates more or less in the dorsal midline, but the shell gland area shifts to the left, so that the larval shell is situated at the posterior left side of the trochophore. Also in the pulmonates *Lymnaea stagnalis* and *Biomphalaria glubrata* a shift of the shell gland takes place. In *Lymnaea* (a dextral species) the shell is displaced to the left side, in *Biomphalaria* (a sinistral species) to the right side. The mantle fold secreting the shell, is better developed on one side, on the right side in *Lymnaea*, on the left side in *Biomphalaria* (Fig. 12.26). According to Raven (1966) this shift of the shell gland is due to an asymmetrical evagination of the shell gland, in dextral species on the left side more rapidly than on the right. As, however, the whole shell gland area is displaced, this shift can only be caused by a faster growth of the right side of the embryo. In this respect the situation in the archaeogastropod *Theodoxus fluviatilis* is of great interest. Here, the embryo is strongly asymmetrical already at an early veliger stage, when the larval shell is still situated in the dorsal midline. The left side of the embryo consists of rather large cells, whereas the right side is composed of small cells (Fig. 13.27). By extension of the right side the shell is displaced to the left, while the large cells move from the left side to a ventral position.

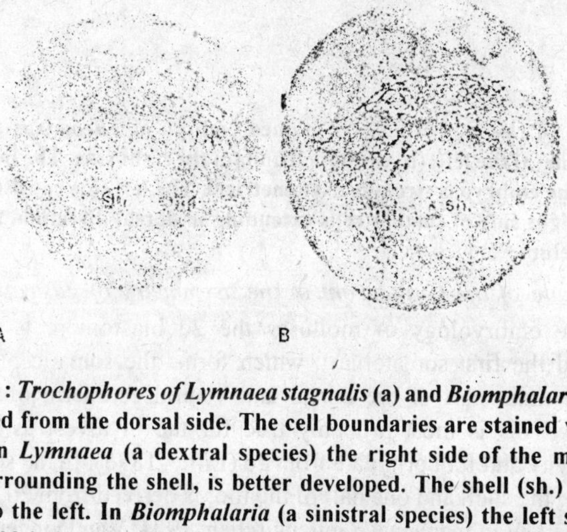

Fig. 13.26 : *Trochophores of Lymnaea stagnalis* (a) and *Biomphalaria glabrata* (b), viewed from the dorsal side. The cell boundaries are stained with silver nitrate. In *Lymnaea* (a dextral species) the right side of the mantle fold (m.f.), surrounding the shell, is better developed. The shell (sh.) is situates slightly to the left. In *Biomphalaria* (a sinistral species) the left side of the mantle fold is larger and the shell is displaced slightly to the right side.

The asymmetry may become apparent even at a much earlier stage, *i.e.* during gastrulation, by a swelling of the right side, which enlarges at the expense of the left side (*Crepidula*, Conklin, 1897; *Trochus*, Robert, 1902; Haliois, Crofts, 1932). In *Trochus* the right margin of the blastopore is larger than the left, mainly due to the size of 4c, which is considerably larger than 4a. In *Crepidula* 5c is formed earlier and lies nearer the dorsal side than 5d, thus indicating a certain degree of torsion in the region of the archenteron.

The above data indicate that in most gastropods the embryo is already asymmetrical long before torsion starts. In dextral species the earliest signs of asymmetry become visible at the right side of the gastrula. It is this side of the posttrochal region of the trochophore that grows faster and extends in dorsal direction. The right side also plays a leading role in the formation of the mantle cavity.

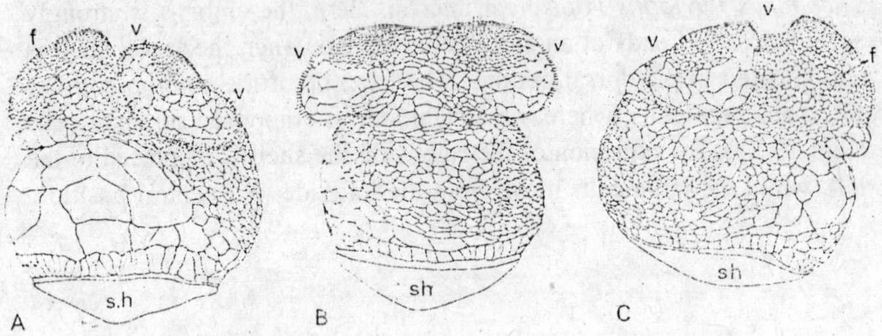

Fig. 13.27 : Drawings of a silver stained embryo of *Theodoxus fluviatilis*, viewed from (a) the left, (b) the dorsal and (c) the right side. The posttrochal region of the embryo is strongly asymmetrical. The left side consists of large cells, the right side of smaller cells extending in dorsal direction, f, foot. sh, shell. v, velum.

The rule of the C-quadrant in the asymmetry of gastropods

In the embryology of molluscs the 2d blastomere is generally considered the first somatoblast, which forms the somatic plate, from which the main part of the dorsal and ventral ectoderm of the body originates. This is most probably true for the bivalves, as the shell, mantle cavity and foot originate from 2d (Lillie, 1895). In the scapliopod *Dentalium* the shell and one half of the foot is derived from 2d, the other part from 3d (Van Dongen & Geilenkirchen, 1974; Van Dongen, 1976a). Experimental data confirm the role of the D-quadrant in the morphogenesis of *Dentalium*. After removal of 2d most of the posttrochal region, except part of the foot, is missing whereas after deletion of the D-blastomere at the 4-cell stage the whole posttrochal region is absent, and foot, shell and mantle cavity are not formed. These data indicate that indeed in bivalves and scaphopods the D-quadrant plays a predominant role in the morphogenesis of the posttrochal region, from which the whole body except the head originates.

It is generally assumed that also in gastropods the D-quadrant plays a similar role in morphogenesis. Recent experiments, however, indicate that in gastropods the contribution of the C-quadrant to the formation of the posttrochal region should not be underestimated. Cather (1967) was the first to show that the shell in gastropods is not only a derivative of the posterior micromere 2d, as was supposed by Conklin (1897) after his cell-lineage study of *Crepidula*. Also the right micromere 2c contributes to the formation of the shell. After removal of 2d in the snail *Ilyanassa* a small shell was formed in 75% of the embryos. After deletion of 2c the shell was reduced to one-half or one-third of the normal size. Only after removal of 2c and 2d a shell was not formed. When the 2d was marked with carbon particles, these particles were found on the posterior region of the trochophore, including the shell. After marking the 2c the carbon particles were observed on the right lateral and ventral side. Later, part of the marked area was incorporated into the mantle

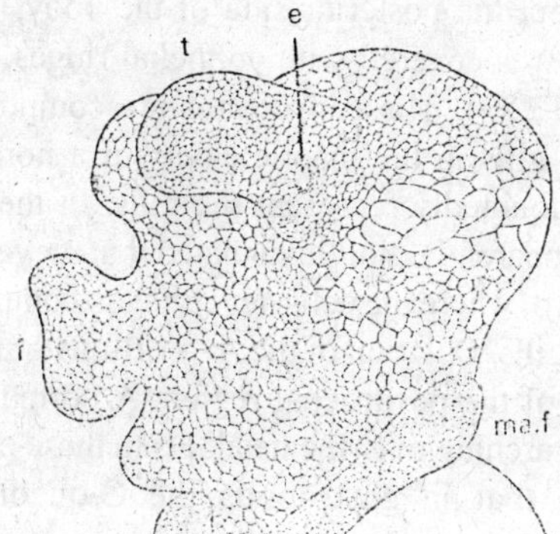

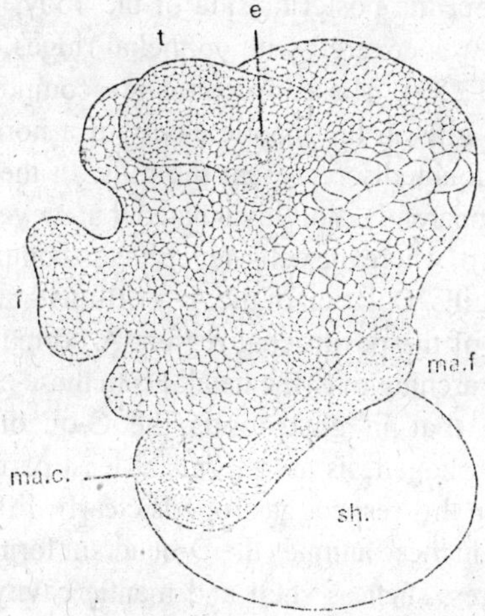

Fig. 13.28 : ABD-embryo of *Bithynia tentaculata*, obtained after removal of the C-quadrant at the 4-cell stage, viewed from the left side. Head with tentacle (t) and eye (e). Foot (f) reduced. The reduced mantle cavity (ma.c.) is situated ventrally behind the foot. Shell (sh) reduced.

edge, while the rest of the area appeared to form part of the mantle cavity. In *Ilyanassa* the C-quadrant is dependent on the D-quadrant for the realization of its potentialities, as is evident from Clement's experiments in which the D-macromere was removed at successive stages of development. In *Bithyynia* the C-quadrant has obtained a more independent position. Studying the development of AB and CD blastomeres, separated at the 2-cell stage, Verdonk and Cather (1973) found that only the CD blastomere formed adult structures. Both AD an BC combinations, separated at the 4-cell stage, however, formed adult structures. This indicates that in *Bithynia* both the C-and the D-quadrant have an organizing capacity. In order to determine the significance of the C-quadrant for the development of *Bithynia* Verdonk and Cather removed this quadrant at the 4-cell stage. The shell appeared to be reduced in size, either cup-shaped or elongate. The mantle cavity, when present, was a shallow groove in the body wall situated at the ventral side behind the foot (Fig. 13.28). Gills, which are the most prominent structures of the mantle cavity, were always missing. Embryos, in which the D-quadrant was removed, developed into curious larvae. They showed a strong ventral flexion, by which the posterior region with a reduced turriform shell becomes situated at the side of the head. A mantle cavity was never formed but the posterior side of the body, bent forward by the ventral flexion, was covered with epithelial ridges, arranged in one or two rows (Fig. 13.29). These structures are composed of the same type of cells as the gills in the mantle cavity of a normal embryo and the ridges must undoubtedly represent the gills. In the–D embryos the mantle cavity is apparently not luvannated but covers the posterior surface of the embryo, where the cells nevertheless complete their normal differentiation into gills. Deletion of the A– or B-quadrant does not affect the morphogenesis of the posttrochal region. A normal shell is formed and a mantle cavity arching over the head. From these experimental data we may conclude that in gastropods the C-quadrant contributes substantially to morphogenesis as it forms at least part of the shell and the mantle cavity. In this respect gastropods clearly differ from bivalves and scaphopods, as in these animals the D-quadrant forms the posttrochal ectodermal structures, such as shell and mantle cavity, all by itself.

As stated above, in bivalves the four quadrants A, B, C and D may be arranged either clockwise or anti-clockwise. This means that the C-quadrant is either to the right or to the left side of the dorsal quadrant. Both configurations, found in nearly equal amounts, give rise to identical embryos, which may be explained by the unique role of the D-quadrant in the development of these animals. In gastropods, however, reversal of cleavage has a profound influence on the organization of the embryo as first stated by Crampton (1894). Dextral snails, in which the shell is coiled according to a right-hand spiral, originate from eggs in which third cleavage is dexotropic; sinistral snails originate from eggs with a lacotropic third cleavage. When third cleavage is dexotropic the quadrants are arranged clockwise; consequently the C-quadrant is to the right, and the A-quadrant to the left of the dorsal D-quadrant. When third cleavage is lacotropic, the arrangement of the quadrants is anti-clockwise and

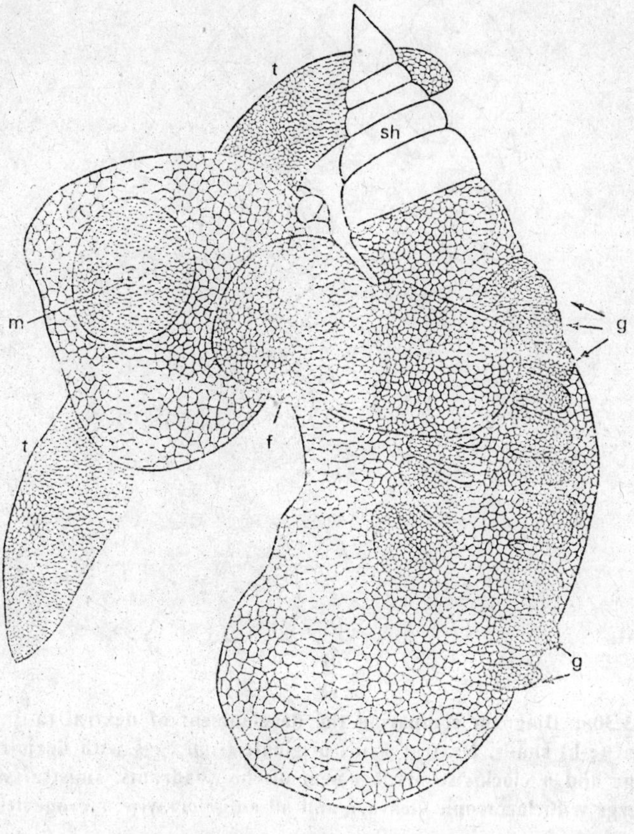

Fig. 13.29 : Typical ABC-embryo of *Bithynia*, obtained after removal of the D-quadrant at the four cell stage, viewed from the ventral side. Head with two tentacles (t) and mouth (m); foot (f) reduced. The posttrochal region is strongly bent, so that the reduced turret-like shell (sh) is situated at the left side of the head. Gills (g) well developed, arranged in two rows.

the C-quadrant is to the left, the A-quadrant to the right of the D-quadrant (Fig. 13.30). Whereas the A-quadrant is of minor importance, the C-quadrant contributes substantially to the morphogenesis of the posttrochal region. It is therefore obvious to assume that it is the position of the C-quadrant, which affects the position of the organs in the posttrochal region of gastropods. Since the work of Crofts (1955) on muscle morphogenesis in *Hdliotis*, *Patella, Patina* and *Calliostoma*, the

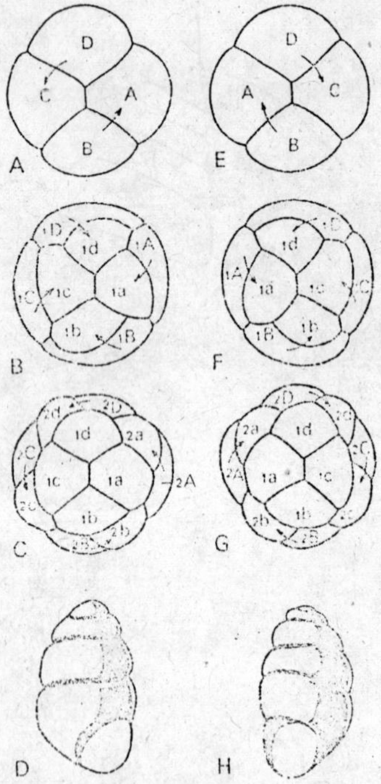

Fig. 13.30 : Diagram comparing the development of dextral (a-d) and sinistral (e-h) snails. Dextral snails originate from eggs with dexiotropic cleavage and a clockwise arrangement of the quadrants, sinistral snails from ergs with lacotropic cleavage and an anti-clockwise arrangement of the quadrants.

role of the larval retractor muscle in torsion has been stressed. According to Crofts the larval refractor muscle originates from cells of the right mesoderm band that arises from 4d. If this is true, then in her theory the origin of the asymmetry must be located in the D-quadrant. Her conclusion that this muscle is derived from the mesentoblas-4d, however, is not based on a detailed cell lineage study. In lobeless embryos *Ilyanassa*, in which a mesentoblast is not formed, large blocks of granular muscle cells are found which resemble the retractor muscle of normal larvae (Atkinson, 1971). After removal of the mesentoblast 4d in *Ilyanassa* normal development may follow and a larval retractor muscle may be present; only the heart and the intestine are absent (Clement, 1960). After removal of the D-quadrant in *Bithynia* muscle cells are formed, radiating from the foot into the body. This indicates that in gastropods as well as in other molluscs an important part of larval musculature originates from larval mesenchyme or ectomesenchyme never originates from D-quadrant. If D-quardent is removed in archeogastropods involvement in torsion may be answered. Origin of asymmetry in gastropods has to be attributed to D-quardant a relation between direction of cleavage and direction of torsion is difficult to explain C-quardant is also involved.

REFERENCES

Ghiselin, M.T. (1966) The adaptive significance of gastropod torsion, Evolution, Malacologia 20, 337-348.

Kerut A.G., (1960) *The Biology of the Mollusca*, Pergamon Press Oxford New York.

Lochman C. (1957) Paleoecology of Cambruim in Monvana and Wyoming-Geo. Soc. Amer. Mamoine.

Meglitsch Paul A., (1967) Invertebrate Zoology, Oxford University Press, London Toronto.

Spoel Vander S., Bruggen Van A.C., Lever J., (1979) *Pathways in Malacology,* Bohn, Schelterna and Holkema, Utrecht Dr. W. Junk B.V., Publishers, The Hague.

Thompson T.E. (1967) Adaptive Significance of gastropod Torsion Malacologia 5, 423-430.

Winchester M.A. and Lovell B.H., (1958) *Zoology,* D. Van Nostrand company, Inc. Princeton, New York, Second Edition.

www.Geologica.Com.

www.Seaworld.Com.